Advances in Industrial Control

For further volumes:
http://www.springer.com/series/1412

Other titles published in this series:

Digital Controller Implementation and Fragility
Robert S.H. Istepanian and James F. Whidborne (Eds.)

Optimisation of Industrial Processes at Supervisory Level
Doris Sáez, Aldo Cipriano and Andrzej W. Ordys

Robust Control of Diesel Ship Propulsion
Nikolaos Xiros

Hydraulic Servo-systems
Mohieddine Mali and Andreas Kroll

Model-based Fault Diagnosis in Dynamic Systems Using Identification Techniques
Silvio Simani, Cesare Fantuzzi and Ron J. Patton

Strategies for Feedback Linearisation
Freddy Garces, Victor M. Becerra, Chandrasekhar Kambhampati and Kevin Warwick

Robust Autonomous Guidance
Alberto Isidori, Lorenzo Marconi and Andrea Serrani

Dynamic Modelling of Gas Turbines
Gennady G. Kulikov and Haydn A. Thompson (Eds.)

Control of Fuel Cell Power Systems
Jay T. Pukrushpan, Anna G. Stefanopoulou and Huei Peng

Fuzzy Logic, Identification and Predictive Control
Jairo Espinosa, Joos Vandewalle and Vincent Wertz

Optimal Real-time Control of Sewer Networks
Magdalene Marinaki and Markos Papageorgiou

Process Modelling for Control
Benoît Codrons

Computational Intelligence in Time Series Forecasting
Ajoy K. Palit and Dobrivoje Popovic

Modelling and Control of Mini-Flying Machines
Pedro Castillo, Rogelio Lozano and Alejandro Dzul

Ship Motion Control
Tristan Perez

Hard Disk Drive Servo Systems (2nd Ed.)
Ben M. Chen, Tong H. Lee, Kemao Peng and Venkatakrishnan Venkataramanan

Measurement, Control, and Communication Using IEEE 1588
John C. Eidson

Piezoelectric Transducers for Vibration Control and Damping
S.O. Reza Moheimani and Andrew J. Fleming

Manufacturing Systems Control Design
Stjepan Bogdan, Frank L. Lewis, Zdenko Kovaccic and José Mireles Jr.

Windup in Control
Peter Hippe

Nonlinear H_2/H_∞ Constrained Feedback Control
Murad Abu-Khalaf, Jie Huang and Frank L. Lewis

Practical Grey-box Process Identification
Torsten Bohlin

Control of Traffic Systems in Buildings
Sandor Markon, Hajime Kita, Hiroshi Kise and Thomas Bartz-Beielstein

Wind Turbine Control Systems
Fernando D. Bianchi, Hernan De Battista and Ricardo J. Mantz

Advanced Fuzzy Logic Technologies in Industrial Applications
Ying Bai, Hanqi Zhuang and Dali Wang (Eds.)

Practical PID Control
Antonio Visioli

(continued after Index)

Tao Liu • Furong Gao

Industrial Process Identification and Control Design

Step-test and Relay-experiment-based Methods

Tao Liu
RWTH Aachen University
Aachen
Germany
liurouter@ieee.org

Furong Gao
Hong Kong University of Science
and Technology
Kowloon
Hong Kong SAR
kefgao@ust.hk

ISSN 1430-9491
ISBN 978-0-85729-976-5   e-ISBN 978-0-85729-977-2
DOI 10.1007/978-0-85729-977-2
Springer Dordrecht Heidelberg London New York

Library of Congress Control Number: 2011941754

AMS Codes: 93B30, 93B52, 93C35, 93D09

© Springer-Verlag London Limited 2012
No part of this work may be reproduced, stored in a retrieval system, or transmitted in any form or by any means, electronic, mechanical, photocopying, microfilming, recording or otherwise, without written permission from the Publisher, with the exception of any material supplied specifically for the purpose of being entered and executed on a computer system, for exclusive use by the purchaser of the work.

Printed on acid-free paper

Springer is part of Springer Science+Business Media (www.springer.com)

To my mother—Zuqin Zhu, and my wife—Ying Zhou

献给我的母亲—朱祖琴，和我的妻子—周颖

Tao Liu (刘涛)

Series Editors' Foreword

The series *Advances in Industrial Control* aims to report and encourage technology transfer in control engineering. The rapid development of control technology has an impact on all areas of the control discipline. New theory, new controllers, actuators, sensors, new industrial processes, computer methods, new applications, new philosophies..., new challenges. Much of this development work resides in industrial reports, feasibility study papers and the reports of advanced collaborative projects. The series offers an opportunity for researchers to present an extended exposition of such new work in all aspects of industrial control for wider and rapid dissemination.

A common engineering approach to complex problems is to look for a set of simple characterising features and then construct an engineering paradigm based on a parsimonious analysis that succinctly and efficiently captures the identified characteristics. In control engineering, a good example is the use of first-order-plus-dead-time (FOPDT) and second-order-plus-dead-time (SOPDT) models to represent the key features of a range of process responses. The elements of this model class have just a few parameters (time constants, second-order model parameters, delay time, zero positions) that are able to depict a wide set of process dynamics.

This particular control engineering approach links model identification to the use of process step responses or to relay experiment data. The simplicity of these two test procedures and their interrelation to the tuning rules for proportional-integral-derivative (PID) controllers has led to an extensive literature that is still developing today, despite the fact that the tuning rules of Ziegler and Nichols were devised over 60 years ago. A modern version of this idea is to widen the class of controllers chosen, to accept that the model representation is not accurate and use robust methods to ensure the fidelity of the control design. This is one of the paths followed in the monograph being introduced here.

A good demonstration of the ingenuity that can be brought to these basic ideas is found in this *Advances in Industrial Control* monograph entitled *Industrial Process*

Identification and Control Design: Using a Step/Relay Test by Tao Liu and Furong Gao. The two questions posed by the authors are:

1. How can we identify models from the FOPDT and SOPDT model class using step response and relay experiment data for stable, integrating and unstable processes?
2. How can we exploit the parsimonious FOPDT/SOPDT model structure in control system designs for: SISO processes, two degree of freedom controllers, cascade control systems, multiloop control, decoupling control and batch process control?

The monograph is divided into two parts that pursue these two questions. Part I (Chaps. 1–6) deals with the identification issues and Part II (Chaps. 7–12) explores the six control design topics of the question above; one topic per chapter. Closing the monograph is a summary chapter (Chap. 13) that looks again at the outcomes of the authors' extensive and comprehensive research and goes on to discuss and list some remaining unresolved issues.

The step test and relay experiment results follow an analytical route that brings rewards in the enhanced clarification of the possible outcomes for the two identification methods when used with different types of processes. The comprehensive set of results presented by the authors look ideal for further use in a possible industrial process identification toolbox.

The sequence of chapters in Part II uses the internal model control (IMC) framework to investigate the six control system design problems. IMC is modified for the various control structures and objectives. In the batch process control chapter, the iterative learning control (ILC) method is used. Inherent within these study topics is the use of the FOPDT/SOPDT class of process models, robust methods to overcome model inaccuracy and PID controllers for implementation where feasible. Each chapter is comprehensive in its coverage of such issues as setpoint tracking, disturbance rejection, robustness and noise rejection, and presents comparative examples to demonstrate performance.

This monograph follows previous *Advances in Industrial Control* monographs for this and the related process identification field, notably the volumes *Identification of Continuous-Time Models from Sampled Data* edited by H. Garnier and L. Wang (ISBN 978-1-84800-160-2, 2008), *Practical Grey-box Process Identification: Theory and Applications* by T. Bohlin (ISBN 978-1-84628-402-1, 2006) and *Autotuning of PID Controllers: Relay Feedback Approach* by C.C. Yu (ISBN 978-3-540-76250-8, 1999 (second edition ISBN 978-1-84628-036-8, 2006)) There have also been *Advances in Industrial Control* volumes on PID and related control aspects including, *Practical PID Control* by A. Visioli (ISBN 978-1-84628-585-1, 2006), *Structure and Synthesis of PID Controllers* by A. Datta, M.-T. Ho and S.P. Bhattacharyya (ISBN 978-1-85233-614-1, 1999) and *Advances in PID Control* by K.K. Tan, Q.-G. Wang and C.C. Hang with T.J. Hägglund (ISBN 978-1-85233-138-2, 1999).

Series Editors' Foreword

Readers from the engineering disciplines of the process industries and academics and postgraduate students from the control field will find this monograph of new results and ideas by Tao Liu and Furong Gao an invaluable companion to these previous volumes.

Industrial Control Centre   M.J. Grimble
Glasgow   M.A. Johnson
Scotland, UK

Contents

Preface		xvii
Acknowledgements		xxi
Abbreviations and Symbols		xxiii

Part I Process Identification

1 Introduction 3
   1.1   The Scope and Objective of Control-Oriented Process
          Identification 3
   1.2   Excitation Signals for Identification Tests 7
   1.3   Open-Loop and Closed-Loop Identification Tests 8
   1.4   Model Fitting Criteria 9
   1.5   Summary 10
   References 10

2 Step Response Identification of Stable Processes 13
   2.1   Implementation of a Step Test 13
   2.2   Model Identification from an Open-Loop Step Test 14
         2.2.1   Frequency Response Estimation 15
         2.2.2   The FOPDT Model or a Higher-Order Model
                with Repeated Poles 18
         2.2.3   The SOPDT Model 20
         2.2.4   A Higher-Order Model with Different Poles 22
         2.2.5   Improving Fitting Accuracy Against Model Mismatch .. 23
         2.2.6   Consistent Estimation Analysis and Model
                Structure Selection 26
         2.2.7   Illustrative Examples 30
   2.3   Robust Identification Under Unsteady Initial Process
          Conditions and Load Disturbance 38
         2.3.1   Implementation of a Step-Like Test 38

		2.3.2	Model Identification	40
		2.3.3	Illustrative Examples	49
	2.4	Piecewise Model Identification Under Inherent Load Disturbance		52
		2.4.1	Partition of Step Response Data	52
		2.4.2	Model Identification	54
		2.4.3	Illustrative Examples	59
		2.4.4	Application to Injection Velocity Control	67
	2.5	Model Identification from a Closed-Loop Step Test		70
		2.5.1	Frequency Response Estimation	71
		2.5.2	Model Identification	74
		2.5.3	Illustrative Examples	76
	2.6	Summary		81
	References			82
3	**Step Response Identification of Integrating and Unstable Processes**			85
	3.1	Practical Implementation Issues		85
	3.2	Open-Loop Step Response Identification of Integrating Processes		88
		3.2.1	Low-Order Model Identification	89
		3.2.2	Illustrative Examples	92
	3.3	Closed-Loop Step Response Identification of Integrating and Unstable Processes		93
		3.3.1	The FOPDT/SOPDT Model for Integrating Processes	94
		3.3.2	The FOPDT Model for Unstable Processes	95
		3.3.3	The SOPDT Model for Unstable Processes	97
		3.3.4	Improving Fitting Accuracy Against Model Mismatch	99
		3.3.5	Illustrative Examples	102
	3.4	Application to the Heating-Up Control of Barrel Temperature in Injection Molding		108
	3.5	Summary		116
	References			117
4	**Relay Feedback Identification of Stable Processes**			119
	4.1	Implementation of a Relay Feedback Test		119
	4.2	Guidelines for Model Structure Selection		122
	4.3	Low-Order Model Fitting Algorithms		125
		4.3.1	The FOPDT Model	126
		4.3.2	The SOPDT Model	134
		4.3.3	Illustrative Examples	159
	4.4	A Generalized Relay Identification Method		170
		4.4.1	Relay Response Expression	172
		4.4.2	Frequency Response Estimation	175
		4.4.3	Model Fitting Algorithms	178
		4.4.4	Illustrative Examples	181
	4.5	Application to Barrel Temperature Maintenance in Injection Molding		188
	4.6	Summary		194
	References			195

Contents   xiii

5	**Relay Feedback Identification of Integrating Processes**	**197**
	5.1 Existence of the Limit Cycle	198
	5.2 The FOPDT Model	204
	5.3 The SOPDT Model	206
	5.4 Illustrative Examples	210
	5.5 Experimental Tests for the Barrel Temperature Maintenance	213
	5.6 Summary	214
	References	216
6	**Relay Feedback Identification of Unstable Processes**	**217**
	6.1 The Limiting Condition for Steady Oscillation	218
	6.2 The FOPDT Model	228
	6.3 The SOPDT Model	231
	6.4 Illustrative Examples	234
	6.5 Summary	239
	References	240

Part II   Control System Design

7	**Control of Single-Input-Single-Output (SISO) Processes**	**243**
	7.1 Control Engineering Specifications	243
	7.1.1 Error Criteria	243
	7.1.2 Time Domain Performance Specifications	244
	7.1.3 Frequency Domain Performance Specifications	245
	7.2 Robust Stability Criteria	248
	7.3 Review of the Internal Model Control (IMC) Design	250
	7.4 Enhanced IMC Design for Load Disturbance Rejection	254
	7.4.1 For FOPDT Stable Processes	255
	7.4.2 For SOPDT Stable Processes	261
	7.5 Proportional-Integral-Derivative (PID) Tuning	269
	7.6 Illustrative Examples	270
	7.7 Summary	276
	References	276
8	**Two-Degrees-of-Freedom (2DOF) Control of SISO Processes**	**279**
	8.1 The Advantage of a 2DOF Control Scheme	279
	8.2 The 2DOF IMC Design for Optimizing the Set-Point Tracking and Load Disturbance Rejection	280
	8.3 A 2DOF Control Scheme for Integrating Processes	281
	8.3.1 Controller Design	283
	8.3.2 Robust Stability Analysis	288
	8.4 A 2DOF Control Scheme for Unstable Processes	290
	8.4.1 Controller Design	291
	8.4.2 Robust Stability Analysis	298
	8.5 Illustrative Examples	299
	8.6 Summary	316
	References	318

9	**Cascade Control System**		321
	9.1 The Advantage of Cascade Control and Implementation Requirements		321
	9.2 Two 2DOF Control Schemes for Open-Loop Stable Cascade Processes		322
		9.2.1 Controller Design	324
		9.2.2 Robust Stability Analysis	327
	9.3 A 3DOF Control Scheme for Open-Loop Unstable Cascade Processes		328
		9.3.1 Controller Design	330
		9.3.2 Robust Stability Analysis	331
	9.4 Illustrative Examples and Real-Time Tests		333
	9.5 Summary		345
	References		347
10	**Multiloop Control of Multivariable Processes**		349
	10.1 Selection of the Input–Output Pairing		349
		10.1.1 Relative Gain Array (RGA)	349
		10.1.2 Singular Value Decomposition (SVD)	351
	10.2 Multiloop Structure Controllability		352
	10.3 Multiloop Control Design		354
		10.3.1 Desired Diagonal Transfer Matrix	355
		10.3.2 Multiloop PI/PID Controller Design	358
	10.4 Robust Stability Analysis		358
	10.5 Illustrative Examples		361
	10.6 Summary		366
	References		366
11	**Decoupling Control of Multivariable Processes**		369
	11.1 Decoupling Control Design for Two-Input-Two-Output (TITO) Processes		369
		11.1.1 Decoupling Control Preconditions	370
		11.1.2 Desired System Transfer Matrix	371
		11.1.3 Decoupling Controller Matrix Design	374
		11.1.4 Robust Stability Analysis	379
		11.1.5 Illustrative Examples	381
	11.2 Decoupling Control Design for Multiple-Input-Multiple-Output (MIMO) Processes		388
		11.2.1 Decoupling Control Preconditions	388
		11.2.2 Desired Closed-Loop Transfer Matrix	390
		11.2.3 Decoupling Controller Matrix Design	393
		11.2.4 Robust Stability Analysis	396
		11.2.5 Illustrative Examples	399
	11.3 A 2DOF Decoupling Control Scheme for MIMO Processes		410
		11.3.1 Desired Set-Point and Closed-Loop Transfer Matrices	412
		11.3.2 Controller Matrix Design	414

Contents xv

		11.3.3	Robust Stability Analysis	417
		11.3.4	Illustrative Examples	420
	11.4	Summary		428
	References			430
12	**Batch Process Control**			433
	12.1	The Implementation Requirements		433
	12.2	An IMC-Based Iterative Learning Control (ILC) Scheme		434
		12.2.1	The IMC-Based ILC Structure	435
		12.2.2	The IMC Design	436
		12.2.3	Robust ILC Design	437
		12.2.4	Implementation Against Measurement Noise	444
	12.3	Illustrative Examples		444
	12.4	Summary		451
	References			451
13	**Concluding Remarks**			453
References				457
Index				467

Preface

In the process industries, model-based control strategies are well known to result in superior system performance in set-point tracking and load disturbance rejection. Accordingly, control-oriented model identification methods have been increasingly explored in recent years. Among various excitation signals used for system identification, the step response test is most widely practised owing to its implementation simplicity and economy. To prevent the process output from drifting too far away from the set-point, closed-loop identification methods from relay feedback tests have been developed on an ad hoc basis in the past two decades. For pioneering works, see Atherton (1982), Tsypkin (1984), Åström and Hägglund (1984) and Luyben (1987). Recent monographs concerned with relay feedback identification can be seen in Wang et al. (2003), Yu (2006) and Sung et al. (2009).

Motivated by the above observation, a series of model identification methods have recently been developed by the authors based on the use of a step response test or relay feedback test. This monograph summarises these results into a systematic identification methodology based on a typical classification of open-loop response characteristics for various industrial processes: stable, integrating and unstable. The low-order model structures of first-order-plus-time-delay (FOPDT) and second-order-plus-time-delay (SOPDT) are mainly studied here, owing to the fact that such models are most widely used for control system design and controller tuning in industrial engineering practice. A few higher-order model identification algorithms are also given to facilitate advanced control design for industrial processes with special requirements. Moreover, identification methods for estimating the process frequency response from a step or relay test are provided, including robust estimation algorithms against measurement noise, in particular for the low-frequency range which is of primary concern for control design and tuning in engineering practice.

In a coherent manner, a series of model-based control methods also developed by the authors are subsequently integrated into this monograph for practical applications – single-input-single-output (SISO) processes, cascade control processes, multiple-input-multiple-output (MIMO) processes and batch processes. These control methods are developed based on the internal model control (IMC) theory (Morari and Zafiriou 1989), robust control theory (Zhou et al. 1996; Skogestad and Postlethwaite 2005) and iterative learning control (ILC) theory (Moore 1993). A common feature of these control methods is that all controllers in these control schemes are analytically derived in the form of design formulae. Each of these formulae, intuitively or essentially, has only a single adjustable parameter that can be monotonically tuned to meet the best trade-off between the control performance and robustness (robust stability), thereby facilitating the control implementation in practical applications.

For ease of reading, the technical development of the proposed identification and control methods are presented in a self-contained manner. Readers are only assumed to have a basic knowledge of linear algebra and complex analysis. Illustrative examples and experimental applications are given for all the proposed methods in an easy-to-follow manner. It is believed that the monograph should be of interest to control engineers and researchers in the process industries, and could also be used for undergraduate and graduate students in control engineering, process system engineering, chemical engineering, mechanical engineering, electrical engineering, biomedical engineering and industrial automation engineering.

The book is divided into two parts – Part I: Process Identification (Chaps. 1–6) and Part II: Control System Design (Chaps. 7–12). Part I provides a basis for applying the control methods presented in Part II. In fact, both parts are self-contained and can be read independently by readers with different demands. A quick preview of the contents is given below:

Chapter 1, the first chapter in Part I, provides an introduction to the scope and objective of process identification, the excitation signals commonly used for open-loop and closed-loop identification tests and the model fitting criteria.

Chapter 2 presents step response identification methods for open-loop stable processes using an open-loop or closed-loop step test. A frequency response estimation method is given. The model structures chosen for identification are FOPDT, SOPDT and a higher-order model with time delay. A robust identification method is proposed for practical applications subject to unsteady initial process conditions and unexpected load disturbance. Moreover, a piecewise model identification method is given for simultaneously identifying the process model and the deterministic (inherent) load disturbance model from a step test.

Chapter 3 presents step response identification methods for integrating and unstable processes using an open-loop or closed-loop step test. Identification algorithms for obtaining the most widely used FOPDT and SOPDT models are detailed, followed by a practical application to the start-up heating control of barrel temperature for an industrial injection moulding machine.

Chapter 4 presents closed-loop identification methods for stable processes using a relay feedback test. The implementation of a relay test of biased or unbiased type is briefly introduced, followed by the guidelines for model structure selection together with a list of different relay response shapes for reference. Analytical relay response expressions are subsequently derived for the most widely used FOPDT and SOPDT models, along with the corresponding model identification algorithms. Furthermore, based on developing a frequency response estimation algorithm, a generalised relay identification method for obtaining a model of any order with time delay is presented, which can be used to identify the process static gain independent of the choice of biased or unbiased relay.

Chapter 5 presents relay feedback identification methods for integrating processes. By deriving analytical relay response expressions for the widely used FOPDT and SOPDT models, the existence of the limit cycle in the use of a relay test is clarified. Based on the developed relay response expressions, the corresponding model identification algorithms are subsequently presented, followed by a practical application to the barrel temperature maintenance for an industrial injection moulding machine.

Chapter 6 presents relay feedback identification methods for unstable processes. A limiting condition to forming steady oscillation under a relay test is revealed by deriving the analytical relay response expressions for a FOPDT model. Identification algorithms for obtaining the widely used FOPDT and SOPDT models are detailed.

Chapter 7, the first chapter in Part II, provides an introduction of control engineering specifications in both the time and frequency domains, along with the closed-loop robust stability criteria used here. Based on a brief review of the IMC design, an enhanced IMC design for improving load disturbance rejection is proposed. The corresponding proportional-integral-derivative (PID) tuning formulae for the use of the unity feedback control structure are given to facilitate practical application.

Chapter 8 presents advanced two-degrees-of-freedom (2DOF) control methods for the separate optimisation of set-point tracking and load disturbance rejection for stable, integrating and unstable processes.

Chapter 9 presents two 2DOF control schemes for open-loop stable cascade processes, and a 3DOF control scheme for open-loop unstable cascade processes.

Chapter 10 provides an introduction of the selection criteria for the input-output pairing of multivariable control, along with the multi-loop structure controllability. An IMC-based multi-loop PID tuning method for the economic operation of such control systems is proposed.

Chapter 11 presents advanced decoupling control methods for multiple-input-multiple-output (MIMO) processes. An IMC-based control scheme is proposed for two-input-two-output (TITO) processes with time delays. An analytical decoupling control design for MIMO processes with time delays is presented in the framework of the unity feedback control structure. Moreover, a 2DOF control scheme for MIMO processes is proposed to improve decoupling regulation performance in both set-point tracking and load disturbance rejection for individual channels.

Chapter 12 provides an introduction to batch process control and the implementation requirements. An IMC-based ILC scheme for realising the perfect tracking of a desired output trajectory in the presence of process time delay and time-varying uncertainties is proposed.

Finally, Chap. 13 summarises the main contributions of this monograph, along with some suggestions and open issues for future research exploration.

Acknowledgements

It is our pleasure to thank the Springer series editors, Professor Michael J. Grimble and Professor Michael A. Johnson, for kindly inviting us to prepare such a monograph in the renowned monograph series of Advances in Industrial Control. The first author is heartily indebted for the Alexander von Humboldt Research Fellowship conferred by the German Government that has allowed him the time to concentrate on writing this monograph. Both authors are also grateful to the Hong Kong Research Grants Council for its financial support under project no. 613107, which has allowed them the opportunities to get together to work on the monograph. For the writing task, the first author would like to express his gratitude to Professor Wolfgang Marquardt for offering an excellent working environment and helpful discussions. The helpful comments and suggestions from Professor Weidong Zhang (the first author's PhD advisor), Professor Hsiao-Ping Huang, Professor Babatunde A. Ogunnaike, Dr. Danying Gu, and Dr. Youqing Wang are very appreciated. Special thanks are delivered to the Springer staffs, Oliver Jackson, Charlotte Cross, and Sunil Padman K.P., for their technical support on the publication. Finally, the copyright permission from the worldwide publishers of Elsevier, IEEE Xplore, ACS, and Wiley for the reproduction of the authors' journal papers is acknowledged.

Tao Liu
Furong Gao

Abbreviations and Symbols

Abbreviations

ARMAX	auto-regressive moving-average with eXogenous inputs
ARX	auto-regressive with eXogenous inputs
CSTR	continuous stirred tank reactor
DOF	degrees-of-freedom
DP	disturbance response peak
FIR	finite impulse response
FOPDT	first-order-plus-dead-time
GM	gain margin
IAE	integral-of-absolute-error
ILC	iterative learning control
IMC	internal model control
ISE	integral-of-squared-error
ITAE	integral-of-time-weighted-absolute-error
ITSE	integral-of-time-weighted-squared-error
IV	instrumental variables
LFT	linear fractional transformation
LHP	left-half-plane
LMI	linear matrix inequality
LS	least-squares
LTI	linear time invariant
MIMO	multiple-input-multiple-output
MP	minimum phase
MPC	model-based predictive control
MSE	mean square error
NMP	non-minimum phase
NSR	noise-to-signal ratio
P	proportional
PI	proportional-integral

PID	proportional-integral-derivative
PM	phase margin
PRBS	pseudo-random binary signal
RGA	relative gain array
RLS	recursive least-squares
SISO	single-input-single-output
SNR	signal-to-noise ratio
SOPDT	second-order-plus-dead-time
SP	Smith predictor
SVD	singular value decomposition
TITO	two-input-two-output
w.p.	with probability
w.r.t.	with respect to

General Symbols

$\Re$	field of real numbers
$\Re_+$	field of nonnegative real numbers
$\Re^n$	real vectors with a dimension of n
$\Re^{m \times n}$	$m \times n$ real matrices
$\mathbb{C}$	field of complex numbers
$\mathbb{C}_-(\bar{\mathbb{C}}_-)$	open (closed) left-half complex plane
$\mathbb{C}_+(\bar{\mathbb{C}}_+)$	open (closed) right-half complex plane
$\mathbb{C}^{m \times n}$	$m \times n$ complex matrices
∞	infinity
j	$\sqrt{-1}$
j$\Re$	the set of imaginary numbers
$L_2(-\infty, \infty)$	time domain Lebesgue space (Hilbert space)
$L_\infty(\mathrm{j}\Re)$	the set of functions bounded on Re(s) = 0 including at ∞ (Banach space)
$H_2(\mathrm{j}\Re)$	subspace of $L_2(\mathrm{j}\Re)$ with functions analytic in Re(s) > 0
$H_2^\perp(\mathrm{j}\Re)$	subspace of $L_2(\mathrm{j}\Re)$ with functions analytic in Re(s) < 0
$H_\infty(\mathrm{j}\Re)$	the set of $L_\infty(\mathrm{j}\Re)$ functions analytic in Re(s) > 0
prefix R	real rational, e.g., RH_∞ and RH_2
$y(t)$	output response in time domain
$y_{\mathrm{sp}}(t)$	desired output trajectory in time domain (or the set-point profile)
$\bar{y}(t)$	model output response in time domain
$\Delta y(t)$	output deviation to the input change
$\hat{y}(t)$	measured output response in time domain
$\zeta(t)$	measurement noise in time domain
$u(t)$	process input (control output) in time domain
$e(t)$	output error in time domain

Abbreviations and Symbols

$d_i(t)$	load disturbance entering into the process from its input side
$d_o(t)$	load disturbance entering into the process from its output side
$Y(s)$	Laplace transform of output response in frequency domain
$\Delta Y(s)$	Laplace transform of $\Delta y(t)$
$\widetilde{Y}(s)$	Laplace transform of $\widetilde{y}(t)$
$\hat{Y}(s)$	Laplace transform of $\hat{y}(t)$
$\xi(s)$	Laplace transform of $\zeta(t)$
$U(s)$	Laplace transform of process input in frequency domain
$E(s)$	Laplace transform of output error in frequency domain
$G(s)$	process transfer function
$\widehat{G}(s)$	model transfer function
$M_T(s)$	maximal peak of the complementary sensitivity function
$M_S(s)$	maximal peak of the sensitivity function
$s = \alpha + j\omega$	Laplace operator, α is the real part and ω is the imaginary part (frequency)
Δ_A	additive uncertainty
Δ_M	multiplicative uncertainty
λ	time scaling factor or a tuning parameter in IMC
e_s	steady-state offset
T_s	sampling period
t_r	the rise time of a step response
t_{set}	the settling time of a step response
t_N	the time corresponding to the N-th sampled data
k_p	process static (or proportional) gain
τ_p	process time constant
θ	process time delay
ω_b (or ω_{BT})	closed-loop system bandwidth
ω_c	cutoff angular frequency
ω_{gc}	gain crossover frequency
ω_{rc}	referential cutoff angular frequency
ω_π	phase crossover frequency
σ_ζ^2	measurement noise variance
I_n	identity matrix of dimension $m \times n$
$\mathbf{0}_{m \times n}$	zero matrix of dimension $m \times n$
$A = [a_{ij}]_{m \times n}$	a $m \times n$ matrix with a_{ij} as the i-th row and j-th column element
$adj(A) = [A^{ij}]_{m \times n}^T$	adjoint matrix of a $m \times n$ matrix (A) with A^{ij} as the complement minor of a_{ij}
$diag\{a_i\}_{n \times n}$	a $n \times n$ diagonal matrix with a_i as the i-th diagonal element
$A > 0$	the matrix A is positive definite
$A \geq 0$	the matrix A is positive semi-definite
(A, B, C, D)	state-space realization of a transfer function
$\square$	end of proof
$\diamond$	end of remark

Operators and Functions

:=	defined as
≈	approximately equal to
≫	far greater than
≪	far smaller than
∠	angle
∃	exist
∀	to any (all)
∈	belong to
⊂	subset
→	tend to
sup	supremum
inf	infimum
min	minimize
max	maximize
$\|a\|$	absolute value (magnitude) of $a \in \mathbb{C}$ or Euclidean norm ($\|a\|_2$) of $a \in \mathfrak{R}^n$
$\dot{y}(t)$	the first derivative of $y(t)$ in time domain
$\ddot{y}(t)$	the second derivative of $y(t)$ in time domain
$L[g(t)]$	Laplace transform of $g(t)$ in time domain
Re(G)	real part of $G \in \mathbb{C}$
Im(G)	imaginary part of $G \in \mathbb{C}$
deg(G)	relative order of a rational transfer function $G \in RH_\infty$ (a order of the numerator over the denominator of G w.r.t. the Laplace operator, s)
$\langle , \rangle$	inner product
⊗	Knonecker product
⊕	direct product (Hadamard product)
$g * f$	convolution of $g(t)$ and $f(t)$ in time domain
$g \perp f$	orthogonality, i.e., $\langle g, f \rangle = 0$
A^T	matrix transpose
A^*	complex conjugate transpose of the matrix A
A^{-1}	inverse of the matrix A
A^+	pseudo inverse of the matrix A
det(A)	determinant of the matrix A
trace(A)	trace of the matrix A
$\lambda(A)$	eigenvalue of the matrix A
$\sigma(A)$	the set of spectrum (singular value) of the matrix A
$\bar{\sigma}(A)$	largest singular value of the matrix A
$\underline{\sigma}(A)$	smallest singular value of the matrix A
$\sigma_i(A)$	i-th singular value of the matrix A
$\|A\|$	spectral norm of matrix A: $\|A\| = \bar{\sigma}(A)$
$\|A\|_2$	2-norm of matrix $A \in L_2$
$\|A\|_\infty$	infinity-norm of $A \in L_\infty$
$F^{(n)}(s)$	the n-th order derivative of $F(s) \in H_\infty$ w.r.t. s

Part I
Process Identification

Chapter 1
Introduction

1.1 The Scope and Objective of Control-Oriented Process Identification

With a wide application of model-based control strategies to pursue superior system performance in set-point tracking and load disturbance rejection in the process industries, control-oriented model identification methods have been increasingly explored in recent years. Process modeling in industrial and chemical engineering has been generally based on the first-principle equations, such as the conservation law related to mass or energy equations, or the equilibrium relationship from thermodynamics, chemical kinetics, equipment geometry and so forth (Seborg et al. 2004). In contrast, control-oriented process identification aims at obtaining a transfer function model that reflects the dynamic response relationship between the manipulated variable(s) and the controlled variable(s) of a process from the viewpoint of system operation. Figure 1.1 shows a typical scenario of industrial process identification. In this scenario the process itself is primarily considered in process modeling, while the augmented process for system operation is completely considered in control-oriented process identification. An excitation signal is added to the process input or the set-point of system operation to cause the process dynamic response for model identification, which is generally not the case in the classical process modeling.

For example, let us consider the industrial stirred-tank blending system shown in Fig. 1.2, where Stream 1 is a mixture of two chemical species, A and B. Through the gate valve, the mass flow rate, denoted by w_1, may be regarded as a constant, but the mass fraction of A, denoted by x_1, may be time-varying with the feedstock and therefore, should be viewed as a source of load disturbance. Stream 2 is composed of pure A, and thus, $x_2 = 1$. The mass flow rate, denoted by w_2, can be manipulated using a control valve. In the exit stream, the mass fraction of A is denoted by x and the mass flow rate by w. The control objective is to blend the two inlet streams to produce an outlet stream that has the desired composition, x_{sp}, which is also called

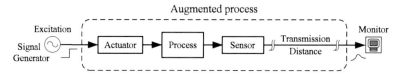

Fig. 1.1 Schematic of control-oriented process identification

the set-point in the framework of a control system. Based on the mass conservation law, a process modeling strategy (Seborg et al. 2004) may give the following state equation to reflect a balanced relationship between the manipulated variable (w_2) and the controlled variable (x),

$$\frac{d(V\rho x)}{dt} = w_1 x_1 + w_2 - wx \tag{1.1}$$

where V and ρ denote the liquid volume of the tank and the liquid density, respectively, according to the assumption of perfect mixing. Correspondingly, an incremental or differential form of the above-stated equation with respect to x and w_2 may be written as

$$V\rho \frac{d(\Delta x)}{dt} = \Delta w_2 - w\Delta x \tag{1.2}$$

with the preliminary assumption that $V\rho$, x_1, and w are fixed for modeling the input–output dynamic relationship. Taking the Laplace transform of both sides of (1.2), it follows that

$$L(\Delta x) = \frac{1}{a_1 s + a_0} L(\Delta w_2) \tag{1.3}$$

where $a_1 = V\rho > 0$ and $a_0 = w > 0$. Obviously, this expression indicates a first-order stable transfer function relating the input increment (Δw_2) to the output increment (Δx).

For a control scheme that uses the composition controller (AC) shown in Fig. 1.2 to achieve the above control objective, the effects from the control valve, the composition transmitter (AT) and the potentially long transmission distance of the control and measurement signals must be envisaged in practice. For instance, a step response test typically turns out the result shown in Fig. 1.3, where an obvious time delay in the output response is observed. In addition, the output response characteristics associated with the control valve or the composition transmitter is often nonlinear in many practical cases. For fitting the experimental data to reflect the real process response for control design, the following model structure is, in fact, more suitable:

$$L(\Delta x) = \frac{k}{\tau s + 1} e^{-\theta s} L(\Delta w_2) \tag{1.4}$$

1.1 The Scope and Objective of Control-Oriented Process Identification

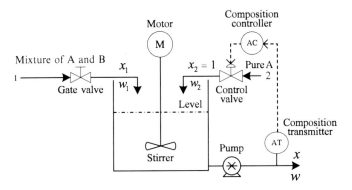

Fig. 1.2 Schematic of a stirred-tank blending system

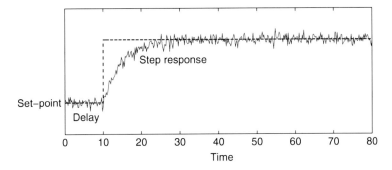

Fig. 1.3 Process response in a step identification test

where θ denotes the output response delay to the input change, k indicates the steady-state gain relating the input to the output, and τ is a time constant reflecting the inertial characteristics of the transient response. Such a model structure, however, can only be conjectured for identification rather than be derived from a process modeling method based on the first-principle methods as aforementioned. Moreover, if the nonlinear dynamics of the control valve and the composition transmitter have to be carefully considered when modeling the transfer function relating Δw_2 to Δx, a higher order or more complex model structure will be derived if a theoretical process modeling is used based on the first-principle equations of physics and chemistry. By comparison, a process identification method based on fitting the real process response data may effectively facilitate determining a suitable low-order model structure for the purpose of control system design.

In fact, there exists no distinct watershed for using process identification or process modeling in practical applications. Both can be alternatively utilized to establish a process model for describing the process dynamic response characteristics. For instance, process modeling may, in principle, determine the manipulated variable(s) and the controlled variable(s) of a process for regulation. Following such a choice, process identification may be used to determine a suitable transfer function model for control system design or online tuning. On the other hand, process identification may be adopted to verify whether the pairing of manipulated and controlled variables chosen from the process modeling is reasonable, or establish a process model with no prior knowledge of the process to facilitate control system design, i.e., "black-box identification" (Söderström and Stoica 1989; Ljung 1999; Pintelon and Schoukens 2001).

For the control objective of set-point tracking and load-disturbance rejection, or for the online regulation of a process in practice, process identification is usually performed around an operating level (or set-point) of the process to establish a linear transfer function model with or without time delay for the convenience of control design. This is because most industrial and chemical processes are essentially nonlinear (Ogunnaike and Ray 1994; Shinskey 1996; Seborg et al. 2004). If necessary, a multiple (or piecewise) model identification strategy may be adopted for a highly nonlinear process by dividing multiple operating regions, and correspondingly, a gain scheduling control strategy may be adopted for system operation.

In practical applications, process uncertainties usually exist from time to time, or from cycle to cycle in batch process operation. From the frequency response view point, such process uncertainties are mainly composed of high frequency components and are in effect similar to random measurement noise (Shinskey 1996; Seborg et al. 2004). Linearized model identification to capture the dynamic response characteristics of such uncertainties is therefore difficult and injudicious. Low-order process models that ignore high-frequency response fitting have mostly been adopted in model-based control methodologies, owing to the fact that model mismatch is allowed to some extent or a prescribed upper bound in the framework of these robust control theories (see Morari and Zafiriou 1989; Ogunnaike and Ray 1994; Chen and Bruce 1995; Shinskey 1996; Zhou et al. 1996; Goodwin et al. 2001; Åström and Hägglund 2005; Johnson and Moradi 2005; Skogestad and Postlethwaite 2005). To accommodate time-varying dynamic response characteristics, online adaptive identification methods in terms of a low-order model structure, such as by using an updating strategy to update the model parameters through online identification tests, have been developed for practical application (Ljung 1999; Pintelon and Schoukens 2001; Wang et al. 2003; Mikleš and Fikar 2007; Sung et al. 2009).

In control system design, industrial and chemical processes are generally classified into three types – stable, integrating, and unstable – based on the observation of the output response to a step change in process operation. Intuitively, the step response of a stable process moves into a new steady state in response to a step change of the input; the step response of an integrating process increases or decreases monotonically, rather than moving into any steady state; the step response of an unstable process, however, is irregular from case to case, but will commonly go beyond the output limit in any case. Note that if the initial process response is in the opposite direction to where it eventually heads toward, such a process is specifically called an inverse response type, which is usually identified in conjunction with any

1.2 Excitation Signals for Identification Tests

of the above three types. Correspondingly, there exists a model type classification in the frequency domain: for a stable process, a transfer function model should have only left-half-plane (LHP) pole(s); for an integrating process, a transfer function model should have at least one pole at the origin ($s = 0$) and may also have LHP pole(s); for an unstable process, a transfer function model should have at least one right-half-plane (RHP) pole, and may also have LHP pole(s) and/or pole(s) at the origin. In addition, an inverse response process corresponds to a model with RHP zero(s). Up to the present, process identification methods have mostly been devoted to obtaining these model types (Söderström and Stoica 1989; Ljung 1999; Pintelon and Schoukens 2001; Zhu 2001; Wang et al. 2003; Åström and Hägglund 2005; Yu 2006: Garnier and Wang 2008; Sung et al. 2009) to meet the needs of advanced process control design.

1.2 Excitation Signals for Identification Tests

A variety of excitation signals have been utilized for model identification, e.g., pulse, step, sinusoid, rectangular wave, and pseudorandom binary sequences (PRBS), in the existing literature (Söderström and Stoica 1989; Ljung 1999; Pintelon and Schoukens 2001; Wang et al. 2003; Åström and Hägglund 2005; Yu 2006; Garnier and Wang 2008; Sung et al. 2009). Each excitation signal has, of course, its own advantage and disadvantage in different system identification scenarios.

Owing to its implementation simplicity and economy, the step response test is most widely practiced for model identification in various process industries. Unlike other tests, no signal generator is needed to perform a step test. The process output response to a step change, however, usually lacks high frequency components, and thus is likely to have a low signal-to-noise ratio (SNR) in the presence of measurement noise. Developing robust step response identification methods is, therefore, among the main contributions of this monograph.

Since the pioneering works of the 1980s, (see Atherton 1982; Tsypkin 1984; Åström and Hägglund 1984; Luyben 1987), the relay feedback test has obviously received increasing attention in the past two decades from both academics and practitioners. Only a cheap relay function module is needed to perform a relay feedback test in a closed-loop structure, which can generate sustained oscillations of the controlled output response for effective identification of its fundamental dynamic response characteristics. Moreover, a relay feedback test will not cause the output response to drift too far away from its set-point, a necessary condition for many practical applications, in particular for highly nonlinear processes with rigorous operating conditions. There are, in general, two types of relay feedback tests: unbiased (symmetrical) and biased (asymmetrical). Using a biased relay test, the process gain can be obtained as the ratio of a periodic integral of the process output to that of the relay output, but gain error may result from unexpected load disturbance. In an unbiased relay test, the influence of load disturbance can be detected intuitively, but the process gain cannot be derived as in the biased case because such a periodic integral is, in fact, equal to zero. To enhance the identification efficacy, research efforts have recently been focused on using a single relay test for model identification (Hang et al. 2002; Atherton 2006). A systematic relay feedback identification methodology that uses a single relay test is, therefore, presented in this monograph for different types of industrial and chemical processes, based on a series of research results explored by the authors in recent years.

1.3 Open-Loop and Closed-Loop Identification Tests

From the viewpoint of interfering with the process operation to perform an identification test, there are typically two types of test, open-loop and closed-loop. An open-loop identification test usually requires stopping the process operation or feedback control to observe the output response by adding an excitation signal to the manipulated input, as shown in Fig. 1.1. An obvious merit is that there exists no correlation between the input and output variables throughout the dynamic response of the test, which may facilitate model identification. For safety and economic reasons, many industrial and chemical processes, such as integrating and unstable processes, are not permitted to run in an open-loop manner. Closed-loop identification methods have, therefore, been developed in the literature (see Ljung 1999; Pintelon and Schoukens 2001; Wang et al. 2003; Sung et al. 2009). A closed-loop identification test is illustrated in Fig. 1.4, where the process actually denotes the augmented process shown in Fig. 1.1 for simplicity. The controller is used for maintaining system stability, which is installed for process operation prior to the identification test. Note that the controller may not be needed in a process with an inherent feedback mechanism. In a closed-loop identification test, there are generally two input ports to which an excitation signal can be added – one is at the process input and the other is at the set-point input.

Based on the use of a step signal, open-loop and closed-loop identification methods will be presented in this monograph for stable, integrating, and unstable processes according to a series of research results recently explored by the authors.

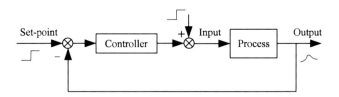

Fig. 1.4 Schematic of a closed-loop identification test

1.4 Model Fitting Criteria

Given the experimental data from an identification test, it is necessary to verify the effectiveness of a model structure adopted for fitting and the accuracy of the corresponding model parameters. A number of fitting objective functions and model validation methods for using different excitation signals for system identification have been presented in the literature (Söderström and Stoica 1989; Åström and Hägglund 1995; Ljung 1999; Pintelon and Schoukens 2001; Wang et al. 2003). Without loss of generality, the following two model fitting criteria are mainly adopted in this monograph for model identification:

1. The time domain fitting criterion

$$err = \frac{1}{N_s} \sum_{k=1}^{N_s} \left[y(kT_s) - \widehat{y}(kT_s) \right]^2 < \varepsilon \qquad (1.5)$$

where $y(kT_s)$ and $\widehat{y}(kT_s)$ denote, respectively, the output responses of the process and the model in an identification test, ε is a user-specified fitting threshold, T_s is the sampling period, and $N_s T_s$ is the time length of the dynamic (transient) response. In a step response test, $N_s T_s$ may be taken as the settling time that is usually defined as the time to move into an error band of 5% with respect to the final steady-state output deviation in response to a step change. In a relay feedback test, $N_s T_s$ may be taken as the limit cycle period, and correspondingly, $y(kT_s)$ and $\widehat{y}(kT_s)$ should be the output responses of the process and the model in the limit cycle.

2. The frequency domain fitting criterion

$$\text{ERR} = \max_{\omega \in [0, \omega_m]} \left\{ \left| \frac{G(j\omega) - \widehat{G}(j\omega)}{G(j\omega)} \right| \right\} < \varepsilon \qquad (1.6)$$

where $G(j\omega) = Y(j\omega)/U(j\omega)$ and $\widehat{G}(j\omega) = \widehat{Y}(j\omega)/U(j\omega)$ denote, respectively, the transfer functions of the process and the model, $Y(j\omega)$ and $\widehat{Y}(j\omega)$ are, respectively, the frequency responses of the process and the model, $U(j\omega)$ is the Laplace transform of the excitation signal added to the inputs of the process and the model, ε is a user-specified fitting threshold, and ω_m denotes a user-specified frequency range for model fitting, e.g., $\omega_m = \omega_c$ is the cutoff angular frequency corresponding to $\angle G(j\omega_c) = -\pi$ which is mostly of concern in the proportional-integral-derivative (PID) controller tuning (Åström and Hägglund 1995, 2005; Yu 2006; Visioli 2006; Sung et al. 2009).

1.5 Summary

The scope and objective of process identification in industrial and chemical processes has been outlined, in contrast with theoretical process modeling based on the first-principle equations of physics and chemistry.

To facilitate model-based control design for system operation, model structures for identification have been presented according to the classification of practical process response types – stable, integrating, and unstable – from the view of control engineering.

The motivation of using the step response test and the relay feedback test for model identification in this monograph has been elucidated, together with the main contributions to be presented for the use of an open-loop or closed-loop identification test in industrial and chemical engineering practices.

For model validation, two fitting criteria, one in the time domain and the other in the frequency domain, that are widely used in practical applications are introduced to evaluate the fitting accuracy and the robustness of identification algorithms presented in this monograph.

References

Åström KJ, Hägglund T (1984) Automatic tuning of simple regulators with specification on phase angle and amplitude margins. Automatica 20:645–651
Åström KJ, Hägglund T (1995) PID controller: theory, design, and tuning, 2nd edn. ISA Society of America, Research Triangle Park
Åström KJ, Hägglund T (2005) Advanced PID control. ISA Society of America, Research Triangle Park
Atherton DP (1982) Oscillations in relay systems. Trans Inst Meas Control (London) 3:171–184
Atherton DP (2006) Relay autotuning: an overview and alternative approach. Ind Eng Chem Res 45:4075–4080
Chen T, Bruce F (1995) Optimal sampled-data control systems. Springer, London
Garnier H, Wang L (eds) (2008) Identification of continuous-time models from sampled data. Springer, London
Goodwin GC, Graebe L, Salgado ME (2001) Control system design. Prentice Hall, Upper Saddle River
Hang CC, Åström KJ, Wang QG (2002) Relay feedback auto-tuning of process controllers—A tutorial review. Journal of Process Control, 12:143–163
Johnson MA, Moradi MH (2005) PID control: new identification and design. Springer, London
Ljung L (1999) System identification: theory for the user, 2nd edn. Prentice Hall, Englewood Cliff
Luyben WL (1987) Derivation of transfer functions for highly nonlinear distillation columns. Ind Eng Chem Res 26:2490–2495
Mikleš J, Fikar M (2007) Process modelling, identification, and control. Springer, Berlin
Morari M, Zafiriou E (1989) Robust process control. Prentice Hall, Englewood Cliff
Ogunnaike BA, Ray WH (1994) Process dynamics, modeling, and control. Oxford University Press, New York
Pintelon R, Schoukens J (2001) System identification: a frequency domain approach. IEEE Press, New York

References

Seborg DE, Edgar TF, Mellichamp DA (2004) Process dynamics and control, 2nd edn. Wiley, Hoboken
Shinskey FG (1996) Process control system, 4th edn. McGraw Hill, New York
Skogestad S, Postlethwaite I (2005) Multivariable feedback control: analysis and design, 2nd edn. Wiley, Chichester
Söderström T, Stoica P (1989) System identification. Prentice Hall, New York
Sung SW, Lee J, Lee I-B (2009) Process identification and PID control. Wiley, Singapore
Tsypkin YZ (1984) Relay control system. Cambridge University Press, Oxford
Visioli A (2006) Practical PID control. Springer, London
Wang QG, Lee TH, Lin C (2003) Relay feedback: analysis, identification and control. Springer, London
Yu CC (2006) Autotuning of PID controllers: a relay feedback approach, 2nd edn. Springer, London
Zhou KM, Doyle JC, Glover K (1996) Robust and optimal control. Prentice Hall, Englewood Cliff
Zhu Y (2001) Multivariable system identification for process control. Elsevier, London

Chapter 2
Step Response Identification of Stable Processes

2.1 Implementation of a Step Test

Generally, an open-loop step test is performed when the process is at zero initial state or in a nonzero steady state, so that an obvious dynamic (or transient) response of the process to a step change of the process input can be observed and measured for model identification. Certainly, a larger magnitude of the step change can facilitate a better observation of the transient response. This, however, is subject to operating constraints of the process in practice. Most existing step response identification methods have been developed based on the above process conditions for a step test, to enumerate a few, Rake (1980), Huang et al. (2001), Wang et al. (2001), Åström and Hägglund (1995, 2005), Ahmed et al. (2007), and Liu and Gao (2010a). Note that a nonzero initial steady state can be normalized as a zero input case by using a steady-state relationship between the process input and output.

In the presence of measurement noise, multiple step tests can be performed to facilitate consistent parameter estimation or model verification according to the statistical averaging principle. To reduce the cost of performing an identification test, robust step identification methods have been developed using a single step test (Bi et al. 1999; Wang et al. 2001; Ahmed et al. 2007).

In practical applications with nonzero initial process conditions, when measurement noise or unexpected load disturbance exists, it is often difficult to tell if the process steady state has been reached for a step test that is suitable for application of the aforementioned identification methods. Moreover, waiting for such a "steady" state to appear for having a step test can be quite time-consuming and troublesome for industrial processes with slow time constants or long time delays. By defining the initial states of the process output and its derivatives as part of the parameters to be identified, Ahmed et al. (2008) developed a robust identification algorithm that can be used under unsteady or unknown initial process conditions. Liu et al. (2007) suggested the use of multiple piecewise step tests for model identification under nonzero initial process conditions or load disturbance with slow dynamics. By comparison, Wang et al. (2008) developed an alternative algorithm to improve identification robustness in the presence of unknown initial process conditions and static disturbance. Liu and Gao (2008) proposed the use of transient response data from adding and subsequently removing a step change of the test, so that an independent least-squares (LS) regression for unbiased parameter estimation can be established to overcome the influence of unknown initial process conditions or unexpected load disturbance.

In a closed-loop step response identification test, the step change is usually added to the set-point rather than the process input because any external signal added to the process input acts like a load disturbance that may be rejected by the closed-loop feedback mechanism. Also, a closed-loop step test is generally performed after the closed-loop system has already moved into a steady state of operation. To facilitate identifying a process model from a closed-loop step test, the closed-loop controller should be prescribed in a simple form like the proportional (P), proportional-integral (PI), or proportional-integral-derivative (PID) type of controllers, so that an analytical or quantitative relationship between the process response and the closed-loop response can be explicitly established (Zheng 1996). Therefore, a closed-loop step test is generally preferred for online tuning with a fixed controller like PID to improve system performance, in contrast with an open-loop step test.

2.2 Model Identification from an Open-Loop Step Test

For the use of an open-loop step test, which is generally performed in terms of zero or nonzero steady initial process conditions, early references such as Luyben (1990), Åström and Hägglund (1995), Shinskey (1996), Rangaiah and Krishnaswamy (1996) and Huang et al. (2001) presented step response identification methods based on fitting several representative points in the transient output response to a step change. Inspired by the graphical area ratio method (Åström and Hägglund 1995), Bi et al. (1999) developed a first-order-plus-dead-time (FOPDT) model-fitting algorithm using numerical integrals to derive the time domain expression of a step response. This approach was further extended to obtain a second-order-plus-dead-time (SOPDT) or a higher-order model (Wang et al. 2001). Applying a linear filter to the step response, Ahmed et al. (2007) proposed an iterative procedure involving LS fitting to determine the optimal time delay model.

A frequency domain step response identification method (Liu and Gao 2010a, b) is presented here for application. By introducing a damping factor to the step response for the realization of Laplace transform, a frequency response estimation algorithm is first presented. Model fitting algorithms are then developed based on the use of the estimated frequency response of the process in a user-specified frequency range for control system design.

For the choice of the model structure, low-order models such as FOPDT and SOPDT are the most widely used in control system design and online tuning in various process industries, since industrial and chemical engineering processes usually contain time delay in response to the set-point change (Shinskey 1996;

2.2 Model Identification from an Open-Loop Step Test

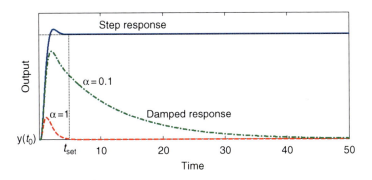

Fig. 2.1 Illustration of choosing α in a step response test

Åström and Hägglund 1995, 2005; Seborg et al. 2004). Correspondingly, identification algorithms for obtaining these low-order models are detailed in the following two subsections. In fact, the true model structure of the process to be identified often cannot be known exactly in practical applications, especially for higher-order processes. Apparently, a higher-order model can give better fitting accuracy on the process dynamic response characteristics. A generalized identification algorithm is, therefore, presented for obtaining an nth order process model ($n \geq 2$) in the subsequent section. Moreover, to improve model fitting accuracy over a user-specified frequency range, e.g., the low frequency range with a phase change from zero to $-\pi$, which is of primary concern in control system design and controller tuning (Shinskey 1996; Goodwin et al. 2001; Wang et al. 2003; Åström and Hägglund 2005; Yu 2006), another generalized identification algorithm is presented accordingly in the final subsection.

2.2.1 Frequency Response Estimation

When a step change is added to the process input in a step identification test as shown in Fig. 2.1, it is well known that the Fourier transform of the output response does not exist because $\Delta y(t) \neq 0$ for $t \to \infty$, where $\Delta y(t) = y(t) - y(t_0)$ and $y(t_0)$ denotes the initial output value in a steady state. However, by substituting $s = \alpha + j\omega$ into the Laplace transform of the step response,

$$\Delta Y(s) = \int_0^\infty \Delta y(t) e^{-st} dt, \qquad (2.1)$$

one can formulate

$$\Delta Y(\alpha + j\omega) = \int_0^\infty \left[\Delta y(t) e^{-\alpha t}\right] e^{-j\omega t} dt. \qquad (2.2)$$

Note that if $\alpha > 0$, there exists $y(t)e^{-\alpha t} = 0$ for $t > t_N$, where t_N can be numerically determined using the condition that $\Delta y(t_N)e^{-\alpha t_N} \to 0$, since $\Delta y(t)$ reaches a steady value after the transient response to a step change.

Therefore, by regarding α as a damping factor of the step response for the Laplace transform, one can compute $\Delta Y(\alpha + j\omega)$ from the N step response data points as

$$\Delta Y(\alpha + j\omega) = \int_0^{t_N} \left[\Delta y(t)e^{-\alpha t}\right] e^{-j\omega t} dt. \tag{2.3}$$

For a step test under an initial steady state of the process, i.e., $y(t) = c$ for $t \le t_0$, as shown in Fig. 2.1, by using a time shift of t_0 (i.e., letting $t_0 = 0$), one can denote the step change to the process input as

$$\Delta u(t) = \begin{cases} 0, & t \le 0; \\ h, & t > 0. \end{cases} \tag{2.4}$$

where h is the magnitude of the step change. Its Laplace transform for $s = \alpha + j\omega$ with $\alpha > 0$ can be analytically derived as

$$\Delta U(\alpha + j\omega) = \int_0^{\infty} h e^{-(\alpha + j\omega)t} dt = \frac{h}{\alpha + j\omega}. \tag{2.5}$$

Hence, a frequency response estimation of the process can be derived using (2.3) and (2.5) as

$$G(\alpha + j\omega) = \frac{\alpha + j\omega}{h} \Delta Y(\alpha + j\omega), \quad \alpha > 0. \tag{2.6}$$

Remark 2.1. Note that it is not applicable to evaluate $G(j\omega)$ in terms of the fast Fourier transform (FFT) and its inverse transform, $G(j\omega_k) = \text{FFT}\{\text{FFT}^{-1}\{G(\alpha + j\omega_k)\}e^{\alpha k T_s}\}$, where T_s is the sampling period for computing the numerical integral in (2.3), as shown in the relay identification monograph (Wang et al. 2003). The reason lies in the fact that the step response is not a periodic signal that can be topologized for such FFT computation. ◇

Note that $G(\alpha + j\omega) \to 0$ when $\alpha \to \infty$. On the contrary, $\alpha \to 0$ will make t_N much larger for computing (2.6). A proper choice of α is therefore required. Considering that all the transient step response data should be used to ensure good estimation of the process frequency response, the following condition is suggested to choose α,

$$|\Delta y(t_{\text{set}})| T_s e^{-\alpha t_{\text{set}}} > \delta \tag{2.7}$$

where $\Delta y(t_{\text{set}}) = y(t_{\text{set}}) - y(0)$ denotes the transient output response to the step change in terms of the settling time (t_{set}), in which $y(0)$ indicates the initial steady

2.2 Model Identification from an Open-Loop Step Test

output value before the step test, T_s is the sampling period for the computation of the numerical integral in (2.3), and δ is a computational threshold which can be practically taken smaller than $|\Delta y(t_{\text{set}})| T_s \times 10^{-6}$.

It follows from (2.7) that

$$\alpha < \frac{1}{t_{\text{set}}} \ln \frac{|\Delta y(t_{\text{set}})| T_s}{\delta}. \tag{2.8}$$

To ensure computational efficiency in frequency response estimation with respect to the complex variable, $s = \alpha + j\omega$, the lower bound of α can be taken with a reference to δ, if there exists no specific limit on the time length of the step test.

Once α is chosen in terms of the above guideline, the time length, t_N, can be determined from a numerical constraint for computing (2.3), i.e.,

$$|\Delta y(t_N)| T_s e^{-\alpha t_N} < \delta \tag{2.9}$$

which can be solved as

$$t_N > \frac{1}{\alpha} \ln \frac{|\Delta y(t_N)| T_s}{\delta} \tag{2.10}$$

Hence, it can be seen from (2.3) and (2.6) that the numerical integral for frequency response estimation depends on the choice of α, rather than the time length of the step response.

Remark 2.2. For a very slow process in practice, any value of α chosen to comply with (2.8) may also satisfy $\alpha < \delta$, which will affect the efficiency in computing (2.6). In such a case, it is suggested that a time-scaled output response be used to perform the Laplace transform, i.e., $L[\Delta y(t/\lambda)] = \lambda \Delta Y[\lambda(\alpha + j\omega)]$, where λ is a time scaling factor. Therefore, the scaled frequency response, $G[\lambda(\alpha + j\omega)]$, can be effectively obtained for model identification. ◇

Note that the following Laplace transform with an initial steady state of the process exists,

$$L\left[\int_0^t \Delta y(\tau) d\tau\right] = \frac{\Delta Y(s)}{s} \tag{2.11}$$

To enhance identification robustness against measurement noise, it is suggested that the frequency response be computed through

$$G(\alpha + j\omega) = \frac{\frac{\Delta Y(\alpha + j\omega)}{\alpha + j\omega}}{\frac{\Delta U(\alpha + j\omega)}{\alpha + j\omega}} = \frac{(\alpha + j\omega)^2}{h} \int_0^{t_N} \left[\int_0^t \Delta y(\tau) d\tau\right] e^{-\alpha t} e^{-j\omega t} dt \tag{2.12}$$

It is seen that, rather than using individual output response points ($\Delta y(t) = y(t) - y(t_0)$) measured from the step test, a time integral for the measured output response points, $\int_0^t \Delta y(\tau)d\tau$, is used to compute the outer-layer integral in the frequency response estimation. This facilitates a reduction of the influence from measurement noise of high frequencies, according to the statistical averaging principle.

2.2.2 The FOPDT Model or a Higher-Order Model with Repeated Poles

An FOPDT model is generally expressed as

$$\hat{G}(s) = \frac{k_p}{\tau_p s + 1} e^{-\theta s} \qquad (2.13)$$

where k_p denotes the process static gain, θ the process time delay, and τ_p the process time constant.

Based on the above process frequency response estimation, a model identification algorithm is presented here for a more general case – a time delay model with single or repeated poles. That is,

$$\hat{G}(s) = \frac{k_p}{(\tau_p s + 1)^m} e^{-\theta s} \qquad (2.14)$$

where m denotes the number of repeated poles that also indicates the process order. It is obvious that $m = 1$ corresponds to an FOPDT model.

To avoid confusion, hereafter denote the nth order derivative for a complex function of $F(s)$ with respect to the Laplace operator, s, as

$$F^{(n)}(s) = \frac{d^n}{ds^n} F(s), \quad n \geq 1. \qquad (2.15)$$

It follows from (2.2) and (2.6) that

$$G^{(1)}(s) = \frac{1}{h} \int_0^\infty (1 - st) \Delta y(t) e^{-st} dt, \qquad (2.16)$$

$$G^{(2)}(s) = \frac{1}{h} \int_0^\infty t(st - 2) \Delta y(t) e^{-st} dt. \qquad (2.17)$$

Hence, by letting $s = \alpha$ and choosing α as well as the guideline for computing (2.3), the numerical integrals in (2.16) and (2.17) can be computed. The corresponding time lengths of t_N can be determined using the numerical constraints,

$$|(1 - \alpha t_N) \Delta y(t_N)| T_s e^{-\alpha t_N} < \delta, \qquad (2.18)$$

2.2 Model Identification from an Open-Loop Step Test

$$|t_N(\alpha t_N - 2)\Delta y(t_N)| T_s e^{-\alpha t_N} < \delta. \tag{2.19}$$

Regarding $s \in \Re_+$ and taking the natural logarithm on both sides of (2.14), one can obtain

$$\ln[\widehat{G}(s)] = \ln(k_p) - m\ln(\tau_p s + 1) - \theta s \tag{2.20}$$

Then, taking the first and second derivatives on both sides of (2.20) with respect to s yields

$$\frac{\widehat{G}^{(1)}(s)}{\widehat{G}(s)} = -\frac{m\tau_p}{\tau_p s + 1} - \theta, \tag{2.21}$$

$$\frac{\widehat{G}^{(2)}(s)\widehat{G}(s) - \left[\widehat{G}^{(1)}(s)\right]^2}{\widehat{G}^2(s)} = \frac{m\tau_p^2}{(\tau_p s + 1)^2}. \tag{2.22}$$

For simplicity, the left-hand side of (2.21) is denoted by $Q_1(s)$ and the left-hand side of (2.22) is denoted by $Q_2(s)$.

By substituting $s = \alpha$, $\widehat{G}(\alpha) = G(\alpha)$, $\widehat{G}^{(1)}(\alpha) = G^{(1)}(\alpha)$, and $\widehat{G}^{(2)}(\alpha) = G^{(2)}(\alpha)$ into (2.22), it can be derived that

$$\tau_p = \begin{cases} \dfrac{-\alpha Q_2(\alpha) + \sqrt{m Q_2(\alpha)}}{\alpha^2 Q_2(\alpha) - m}, & \text{if } \alpha^2 Q_2(\alpha) - m > 0; \\[1em] \dfrac{\alpha Q_2(\alpha) + \sqrt{m Q_2(\alpha)}}{m - \alpha^2 Q_2(\alpha)}, & \text{if } \alpha^2 Q_2(\alpha) - m < 0. \end{cases} \tag{2.23}$$

Consequently, the remaining two model parameters can be derived from (2.14) and (2.21) as

$$\theta = -Q_1(\alpha) - \frac{m\tau_p}{\tau_p \alpha + 1}, \tag{2.24}$$

$$k_p = (\tau_p \alpha + 1)^m G(\alpha) e^{\alpha\theta}. \tag{2.25}$$

Hence, the above algorithm named Algorithm-SS-I for obtaining a time delay model with single or repeated poles can be summarized as follows.

Algorithm-SS-I

(i) Choose α and t_N in terms of (2.8), (2.10), (2.18), and (2.19) to compute $G(\alpha)$, $G^{(1)}(\alpha)$ and $G^{(2)}(\alpha)$ from (2.6) (or (2.12)), (2.16), and (2.17);
(ii) Compute $Q_1(\alpha)$ and $Q_2(\alpha)$ from the left-hand sides of (2.21) and (2.22);
(iii) Compute the process time constant, τ_p, from (2.23);

(iv) Compute the process time delay, θ, from (2.24);
(v) Compute the process static gain, k_p, from (2.25).

By letting $m = 1$, an FOPDT model can be obtained easily from the above algorithm. Note that a referential cutoff angular frequency of the process, ω_{rc}, can be derived from such an FOPDT model by using the critical phase condition,

$$-\theta\omega_{rc} - \arctan(\tau_p\omega_{rc}) = -\pi. \tag{2.26}$$

In practice, ω_{rc} can be used to determine the frequency range in frequency response estimation and model fitting, in particular, in the presence of measurement noise, as will be further interpreted later in the identification algorithms and examples.

In applying Algorithm-SS-I, one should check if $Q_2(\alpha) > 0$ when choosing α, since an FOPDT or a higher-order model with repeated poles may not be suitable for representing those processes with complex response characteristics, as will be illustrated later through Example 2.3 in Sect. 2.2.7.

2.2.3 The SOPDT Model

An SOPDT model with two different poles is generally expressed as

$$\hat{G}(s) = \frac{k_p}{a_2 s^2 + a_1 s + 1} e^{-\theta s} \tag{2.27}$$

where k_p denotes the process static gain, θ the process time delay, and a_1 and a_2 are positive coefficients that reflect the fundamental dynamic response characteristics of the process.

Taking the first and the second derivatives on both sides of (2.27) with respect to s yields

$$Q_1(s) = -\frac{2a_2 s + a_1}{a_2 s^2 + a_1 s + 1} - \theta \tag{2.28}$$

$$Q_2(s) = \frac{2a_2^2 s^2 + 2a_1 a_2 s + a_1^2 - 2a_2}{(a_2 s^2 + a_1 s + 1)^2} \tag{2.29}$$

where $Q_1(s)$ and $Q_2(s)$ are the same as the left-hand sides of (2.21) and (2.22), respectively, and can be computed in terms of the proposed frequency response estimation formulae of (2.6) (or (2.12)), (2.16), and (2.17).

Then, substituting $s = \alpha$ into (2.29) and reorganizing the resulting expression yields

$$Q_2(\alpha) = -\left(\alpha^4 a_2^2 + \alpha^2 a_1^2 + 2\alpha^3 a_1 a_2 + 2\alpha^2 a_2 + 2\alpha a_1\right) Q_2(\alpha)$$
$$+ 2\alpha^2 a_2^2 + 2\alpha a_1 a_2 + a_1^2 - 2a_2. \tag{2.30}$$

2.2 Model Identification from an Open-Loop Step Test

To solve for a_2 and a_1 from (2.30), one can reformulate (2.30) into

$$\psi(\alpha) = \phi(\alpha)^T \gamma \qquad (2.31)$$

where

$$\begin{cases} \psi(\alpha) = Q_2(\alpha), \\ \phi(\alpha) = [-4, \ -2\alpha Q_2(\alpha), 2\alpha^2 - \alpha^4 Q_2(\alpha), 2\alpha - 2\alpha^3 Q_2(\alpha), 1 - \alpha^2 Q_2(\alpha)]^T, \\ \gamma = [a_2, a_1, a_2^2, a_1 a_2, a_1^2 + 2a_2]^T. \end{cases} \qquad (2.32)$$

Therefore, by choosing five different values of α in terms of the guideline given in (2.8) and denoting $\Psi = [\psi(\alpha_1), \psi(\alpha_2), \ldots, \psi(\alpha_5)]^T$ and $\Phi = [\phi(\alpha_1), \phi(\alpha_2), \ldots, \phi(\alpha_5)]^T$, the following LS solution can be obtained

$$\gamma = (\Phi^T \Phi)^{-1} \Phi^T \Psi. \qquad (2.33)$$

It is obvious that all the columns of Φ are linearly independent of each other, so Φ is guaranteed to be nonsingular for computing (2.33), corresponding to a unique solution of γ.

In the sequel, the model parameters can be retrieved from (2.33) as

$$\begin{cases} a_2 = \gamma(1) \\ a_1 = \gamma(2) \end{cases}. \qquad (2.34)$$

Note that there exist three redundant fitting conditions in the parameter estimation of γ, which can undoubtedly be satisfied if the model structure matches the process. In the case of model mismatch, inconsistent parameter estimation may result. To improve the model fitting accuracy, especially when identifying a high-order process, one can use $\gamma(1)$, $\gamma(2)$, $\gamma(3)$, and $\gamma(4)$ together to establish an LS-based fitting solution by taking the natural logarithm of a_1 and a_2, i.e.,

$$\begin{bmatrix} 1 & 0 \\ 0 & 1 \\ 2 & 0 \\ 1 & 1 \end{bmatrix} \begin{bmatrix} \ln a_2 \\ \ln a_1 \end{bmatrix} = \begin{bmatrix} \ln \gamma(1) \\ \ln \gamma(2) \\ \ln \gamma(3) \\ \ln \gamma(4) \end{bmatrix}. \qquad (2.35)$$

Consequently, the remaining model parameters can be derived from (2.28) and (2.27), respectively, as

$$\theta = -Q_1(\alpha) - \frac{2a_2\alpha + a_1}{a_2\alpha^2 + a_1\alpha + 1} \qquad (2.36)$$

$$k_p = (a_2\alpha^2 + a_1\alpha + 1)^2 G(\alpha) e^{\alpha\theta}. \tag{2.37}$$

Hence, the above algorithm named Algorithm-SS-II for obtaining an SOPDT model with two different poles can be summarized as follows.

Algorithm-SS-II

(i) Choose α and t_N in terms of (2.8), (2.10), (2.18), and (2.19) to compute $G(\alpha)$, $G^{(1)}(\alpha)$, and $G^{(2)}(\alpha)$ from (2.6) (or (2.12)), (2.16), and (2.17);
(ii) Compute $Q_1(\alpha)$ and $Q_2(\alpha)$ from the left-hand sides of (2.21) and (2.22);
(iii) Compute a_2 and a_1 from (2.33) and (2.34) (or (2.35));
(iv) Compute the process time delay, θ, from (2.36);
(v) Compute the process static gain, k_p, from (2.37).

2.2.4 A Higher-Order Model with Different Poles

A higher-order model with different poles is generally expressed as

$$\widehat{G}(s) = \frac{b_m s^m + b_{m-1} s^{m-1} + \cdots + b_1 s + b_0}{a_n s^n + a_{n-1} s^{n-1} + \cdots + a_1 s + 1} e^{-\theta s} \tag{2.38}$$

where b_0 denotes the process static gain (k_p), $a_i > 0$ ($i = 1, 2, \ldots, n$) for a stable process, and $n > m$ indicates strict properness of the process transfer function.

Substituting (2.38) into (2.6) yields

$$a_n s^n \Delta Y(s) + a_{n-1} s^{n-1} \Delta Y(s) + \cdots a_1 s \Delta Y(s) + \Delta Y(s)$$
$$= (b_m s^m + b_{m-1} s^{m-1} + \cdots + b_1 s + b_0) \frac{h}{s} e^{-\theta s} \tag{2.39}$$

which can be reformulated into the LS form of

$$\psi(s) = \phi(s)^T \gamma \tag{2.40}$$

where

$$\begin{cases} \psi(s) = \Delta Y(s), \\ \phi(s) = \left[-s^n \Delta Y(s), -s^{n-1} \Delta Y(s), \ldots, -s\Delta Y(s), hs^{m-1} e^{-\theta s}, \ldots, he^{-\theta s}, he^{-\theta s}/s \right]^T, \\ \gamma = [a_n, a_{n-1}, \ldots, a_1, b_m, \ldots, b_1, b_0]^T. \end{cases}$$
$$\tag{2.41}$$

Therefore, by letting $s = \alpha$ and choosing different values of α with a number of M_p, where $M_p = n + m + 1$, and denoting $\Psi = [\psi(\alpha_1), \psi(\alpha_2), \ldots, \psi(\alpha_{M_p})]^T$

2.2 Model Identification from an Open-Loop Step Test

and $\Phi = [\phi(\alpha_1), \phi(\alpha_2), \ldots, \phi(\alpha_{M_p})]^T$, one can solve (2.40) for model parameter estimation with a prior knowledge of the process time delay (θ). If $M_p > n + m + 1$ is chosen, an LS fitting algorithm can be correspondingly established in the form of (2.33).

For implementation, an approximate estimation of the process time delay can be read from the step response test or obtained from an FOPDT model derived using Algorithm-SS-I. This estimation may be taken as an initial value in a one-dimensional search that uses the following fitting condition of the time domain step response,

$$err = \frac{1}{N_s} \sum_{k=1}^{N_s} \left[y(kT_s) - \hat{y}(kT_s) \right]^2 < \varepsilon \quad (2.42)$$

where $y(kT_s)$ and $\hat{y}(kT_s)$ denote the process and model outputs to the step change, respectively, T_s is the sampling period, and $N_s T_s$ is the settling time. Therefore, given a user-specified fitting threshold of ε, a feasible solution of the model parameters can be derived from (2.40) and (2.41) by monotonically varying θ within a possible range for computation. The one-dimensional search step size may be taken as a multiple of T_s for simplicity. The optimal fitting can be determined by deriving such a model that yields the smallest value of *err*.

Hence, the above algorithm named Algorithm-SS-III for obtaining a higher-order model with different poles can be summarized as follows.

Algorithm-SS-III

(i) Choose α_k and t_N in terms of (2.8) and (2.10) to compute $\Delta Y(\alpha_k)$ ($k = 1, 2, \ldots, M_p$) from (2.3);
(ii) Obtain an initial estimate of the process time delay, θ, from the step test, or an FOPDT model derived using Algorithm-SS-I or Algorithm-SS-II;
(iii) Solve the remaining model parameters from (2.40) and (2.41);
(iv) End the algorithm if the fitting condition in (2.42) is satisfied. Otherwise, go back to step (iii) by monotonically varying θ for a one-dimensional search within a possible range as observed from the step test.

2.2.5 Improving Fitting Accuracy Against Model Mismatch

The above identification algorithms, Algorithm-SS-I, Algorithm-SS-II, and Algorithm-SS-III, can ensure identification accuracy if the corresponding model structure matches the process, as will be illustrated by examples later. In case there exists model mismatch when identifying a higher-order process, an accurate fitting may not be guaranteed since the above algorithms establish frequency response fitting only around the zero frequency (i.e., $\omega = 0$). To improve the model fitting accuracy over a user-specified frequency range, e.g., the low frequency range with a phase change from zero to $-\pi$ that is mainly referenced for control system design and controller tuning (Shinskey 1996; Goodwin et al. 2001; Wang et al. 2003; Åström and Hägglund 2005; Yu 2006), another identification algorithm is proposed as the following.

Let $s = \alpha + j\omega_k$ ($k = 1, 2, \ldots, M$), where M is the number of representative frequency response points in the specified frequency range, an objective function for model identification is proposed as

$$J_{\text{opt}} = \sum_{k=0}^{M} \rho_k \left| G(\alpha + j\omega_k) - \widehat{G}(\alpha + j\omega_k) \right|^2 < ER^2 \quad (2.43)$$

where $G(\alpha + j\omega_k)$ and $\widehat{G}(\alpha + j\omega_k)$ denote the frequency response points of the process and the model, respectively, ER is a user-specified threshold for assessing the fitting accuracy, and ρ_k ($k = 1, 2, \ldots, M$) are weighting coefficients for emphasizing the frequency response fitting over the specified frequency range.

To guarantee identification robustness against measurement noise, it is suggested to choose

$$\omega_M = (1.0 \sim 2.0)\omega_{\text{rc}} \quad (2.44)$$

$$\rho_k = \eta^k \Big/ \sum_{k=0}^{M} \eta^k, \quad \eta \in [0.9, 0.99] \quad (2.45)$$

owing to that the step response inherently has a low signal-to-noise ratio (SNR) of the high-frequency part to measurement noise. Note that, if there exists middle- or low-frequency noise, a denoising low-pass filter should be devised based on the process response characteristics and the measurement sensor to exclude or reduce the influence of such noise.

It should be noted that owing to $\sum_{k=0}^{M} \rho_k ER^2 = ER^2$, the objective function in (2.43) can accommodate the widely used frequency response error specification (Pintelon and Schoukens 2001; Wang et al. 2003; Yu 2006),

$$ERR = \max_{\omega \in [0, \omega_c]} \left\{ \left| \left[G(j\omega) - \widehat{G}(j\omega) \right] / G(j\omega) \right| \right\} \quad (2.46)$$

where ω_c is the cutoff angular frequency corresponding to $\angle G(j\omega_c) = -\pi$. In fact, the exact value of ω_c is difficult to find in practice. Using the above ω_M in place of ω_c should suffice for the purpose of computation.

Substituting (2.6) and (2.38) into the left-hand side of (2.43) by letting it equal to zero, and then organizing the resulting expression into the form of (2.40), the following weighted LS solution for parameter estimation can be derived accordingly

$$\gamma = (\bar{\Phi}^T W \bar{\Phi})^{-1} \bar{\Phi}^T W \bar{\Psi} \quad (2.47)$$

2.2 Model Identification from an Open-Loop Step Test

where $\gamma = [a_n, a_{n-1}, \ldots, a_1, b_m, \ldots, b_1, b_0]^T$, $W = \text{diag}\{\rho_1, \ldots, \rho_M, \rho_1, \ldots, \rho_M\}$,

$$\bar{\Psi} = \begin{bmatrix} \text{Re}[\Psi] \\ \text{Im}[\Psi] \end{bmatrix}, \quad \bar{\Phi} = \begin{bmatrix} \text{Re}[\Phi] \\ \text{Im}[\Phi] \end{bmatrix},$$

$\Psi = [\psi(\alpha + j\omega_1), \psi(\alpha + j\omega_2), \ldots, \psi(\alpha + j\omega_M)]^T$, $\Phi = [\phi(\alpha + j\omega_1), \phi(\alpha + j\omega_2), \ldots, \phi(\alpha + j\omega_M)]^T$, $\psi(\alpha + j\omega_k)$ and $\phi(\alpha + j\omega_k)$ are the same as those in (2.41), except for $s = \alpha + j\omega_k$.

It can be easily verified that all the columns of $\bar{\Phi}$ are linearly independent of each other, so $(\bar{\Phi}^T W \bar{\Phi})^{-1}$ is guaranteed to be nonsingular for computation. Accordingly, there exists a unique solution for parameter estimation. For implementation, $M \in [10, 50]$ is suggested for a good trade-off between identification accuracy and computational efficiency.

Also note that a prior knowledge of the process time delay (θ) is needed to derive the remaining model parameters from (2.47). Similar to the above Algorithm-SS-III, a one-dimensional search of θ in terms of the fitting criterion in (2.42) can be implemented for optimal parameter estimation.

Hence, the above algorithm named Algorithm-SS-IV for improving the model fitting accuracy over a specified frequency range can be summarized as follows.

Algorithm-SS-IV

(i) Choose α and t_N in terms of (2.8) and (2.10) to compute $\Delta Y(\alpha + j\omega_k)$ ($k = 1, 2, \ldots, M$) from (2.3), where ω_{rc} for the choice of ω_M can be determined from an initial model obtained using Algorithm-SS-I or Algorithm-SS-II;
(ii) Obtain an initial estimate of the process time delay, θ, from the step test, or a model obtained using Algorithm-SS-I or Algorithm-SS-II;
(iii) Solve the remaining model parameters from (2.47);
(iv) End the algorithm if the fitting condition in (2.42) is satisfied. Otherwise, go back to step (iii) by monotonically varying θ for a one-dimensional search within a possible range as observed from the step test.

Remark 2.3. It can be seen from (2.41) that only a single integral is needed in Algorithm-SS-III or Algorithm-SS-IV for the identification of an nth order model ($n \geq 2$). This single integral can be computed relatively independent of the time length of the step response. Moreover, only M points of the frequency response estimation are needed for model fitting, unlike previous step identification methods based on time domain LS fitting with multiple integrals to a large number of step response data. ◇

It should be noted that minimization of the time domain fitting condition in (2.42) can lead to the minimum of the frequency domain objective function in (2.43), if the model structure adopted matches the process. In the case that there exists model mismatch, the combination of (2.43) and (2.42) for deriving the model parameters can guarantee a good compromise between time domain response fitting and frequency response fitting, but may not realize the global minimization of (2.43) or (2.42).

2.2.6 Consistent Estimation Analysis and Model Structure Selection

For practical application with measurement noise, the capability of consistent estimation needs to be verified for the proposed identification algorithms. Obviously, Algorithm-SS-I and Algorithm-SS-II can ensure convergent parameter estimation, owing to the development of analytical formulae. Whether consistent estimation can be reached or not depends in essence on the unbiasedness of the frequency response estimation algorithm presented in Sect. 2.2.1, which will be analyzed in this subsection. Regarding Algorithm-SS-III and Algorithm-SS-IV, the use of the time domain fitting criterion in (2.42) needs to be clarified for consistent parameter estimation, even if unbiased frequency response estimation is used for model fitting.

To address the unbiasedness of the proposed frequency response estimation algorithm in the presence of measurement noise, the following theorem is given:

Theorem 2.1. *Given Gaussian white measurement noise, $\zeta(t) \sim N(0, \sigma_\zeta^2)$, in the step response tests for identification, unbiased frequency response estimation can be obtained by*

$$Y(\alpha + j\omega) = \lim_{M_s \to \infty} \frac{1}{M_s} \sum_{i=1}^{M_s} \hat{Y}_i(\alpha + j\omega) \tag{2.48}$$

and the estimation error variance is bounded by

$$\lim_{M_s \to \infty} \sigma_e^2 \leq \frac{T_s^2 \sigma_\zeta^2}{1 - e^{-2T_s \alpha}} \tag{2.49}$$

where $Y(\alpha + j\omega) = L[y(t)]$, $\hat{Y}_i(\alpha + j\omega) = L[\hat{y}_i(t)]$, $y(t)$ is the true value of the step response, $\hat{y}_i(t) = y(t) + \zeta_i(t)$ is the measured output response in each test, M_s is the number of the step tests, and T_s is the sampling period.

Proof. Taking the Laplace transform of $\hat{y}_i(t) = y(t) + \zeta_i(t)$ in terms of $s = \alpha + j\omega$ with $\alpha > 0$ yields

$$\hat{Y}_i(\alpha + j\omega) = Y(\alpha + j\omega) + \xi_i(\alpha + j\omega) \tag{2.50}$$

where

$$\hat{Y}_i(\alpha + j\omega) = \int_0^\infty \left[\hat{y}_i(t)e^{-\alpha t}\right]e^{-j\omega t}\,dt = \int_0^{tN_i} \left[\hat{y}_i(t)e^{-\alpha t}\right]e^{-j\omega t}\,dt \tag{2.51}$$

$$\xi_i(\alpha + j\omega) = \int_0^\infty \left[\zeta_i(t)e^{-\alpha t}\right]e^{-j\omega t}\,dt. \tag{2.52}$$

2.2 Model Identification from an Open-Loop Step Test

Without loss of generality, the above Laplace transform of the output response is based on a zero initial process state for the convenience of generating the proof. According to the guideline for choosing t_{N_i} in (2.10), one can see that

$$\int_0^\infty [\xi_i(t)e^{-\alpha t}]e^{-j\omega t}\,dt = \int_0^{t_{N_i}} [\xi_i(t)e^{-\alpha t}]e^{-j\omega t}\,dt. \tag{2.53}$$

For $M_s > 1$, it follows from (2.50) that

$$Y(\alpha + j\omega) = \frac{1}{M_s}\sum_{i=1}^{M_s}\hat{Y}_i(\alpha + j\omega) - \frac{1}{M_s}\sum_{i=1}^{M_s}\xi_i(\alpha + j\omega). \tag{2.54}$$

Let $\hat{t}_N = \max\{t_{N_i}\}, i = 1, 2, \ldots, M_s$, one obtains

$$\sum_{i=1}^{M_s}\xi_i(\alpha + j\omega) = \int_0^{\hat{t}_N}\left[\sum_{i=1}^{M_s}\zeta_i(t)\right]e^{-(\alpha+j\omega)t}\,dt. \tag{2.55}$$

In view of the fact that the Gaussian distribution of measurement noise holds for all the step tests, there exists

$$\lim_{M_s\to\infty}\sum_{i=1}^{M_s}\zeta_i(t) = 0. \tag{2.56}$$

Hence, substituting (2.55) and (2.56) into (2.54) yields (2.48) as shown in Theorem 2.1. Correspondingly, the variance of $\xi_i(\alpha + j\omega) = L[\zeta_i(t)]$, $i = 1, 2, \ldots, M_s$, can be evaluated as

$$\sigma_e^2 = \frac{1}{M_s}\sum_{i=1}^{M_s}\left[\int_0^{\hat{t}_N}\zeta_i(t)e^{-(\alpha+j\omega)t}\,dt\right]\left[\int_0^{\hat{t}_N}\zeta_i(t)e^{-(\alpha+j\omega)t}\,dt\right]^*. \tag{2.57}$$

Let $\hat{N} = \hat{t}_N/T_s$. One can compute (2.57) numerically as

$$\sigma_e^2 = \frac{1}{M_s}\sum_{i=1}^{M_s}\left[\sum_{k=0}^{\hat{N}-1}T_s\zeta_i(kT_s)e^{-kT_s(\alpha+j\omega)}\right]\left[\sum_{k=0}^{\hat{N}-1}T_s\zeta_i(kT_s)e^{-kT_s(\alpha-j\omega)}\right]$$

$$= \frac{T_s^2}{M_s}\sum_{i=1}^{M_s}\left\{\sum_{k=0}^{\hat{N}-1}\zeta_i^2(kT_s)e^{-2kT_s\alpha} + 2\sum_{k=1}^{\hat{N}-1}\sum_{l=0}^{k-1}\zeta_i(kT_s)\zeta_i(lT_s)e^{-kT_s\alpha}\cos[(k-l)\omega T_s]\right\}$$

$$= \frac{T_s^2}{M_s} \left\{ \sum_{k=0}^{\widehat{N}-1} \left[\sum_{i=1}^{M_s} \zeta_i^2(kT_s) \right] e^{-2kT_s\alpha} \right\}$$

$$+ \frac{2T_s^2}{M_s} \sum_{k=1}^{\widehat{N}-1} \sum_{l=0}^{k-1} \left\{ \left[\sum_{i=1}^{M_s} \zeta_i(kT_s)\zeta_i(lT_s) \right] e^{-kT_s\alpha} \cos[(k-l)\omega T_s] \right\}. \tag{2.58}$$

Note that $\zeta_i(kT_s)$ is independent of $\zeta_i(lT_s)$ for $k \neq l$, $i = 1, 2, \ldots, M_s$. That is, they are uncorrelated with each other, corresponding to

$$\lim_{M_s \to \infty} \frac{1}{M_s} \sum_{i=1}^{M_s} \zeta_i(kT_s)\zeta_i(lT_s) = 0, \quad \text{for} \quad k \neq l. \tag{2.59}$$

Besides, there exists

$$\lim_{M_s \to \infty} \frac{1}{M_s} \sum_{i=1}^{M_s} \zeta_i^2(kT_s) = \sigma_\zeta^2. \tag{2.60}$$

Substituting (2.59) and (2.60) into (2.58) and using the mathematical inequality of $e^{-2T_s\alpha} < 1$ to sum up, one obtains (2.49) as shown in Theorem 2.1. This completes the proof. □

It can be seen from (2.49) that σ_e^2 is proportional to T_s^2, and therefore, is far smaller than the measurement noise variance (σ_ζ^2) with $M_s \to \infty$, since $T_s \ll 1$ should be used for the numerical computation of (2.57). If $T_s > 1$ is adopted for a slow process, a time-scaling factor as introduced in Remark 2.2 may be used to rescale T_s to much smaller than unity, so the above conclusion holds as well. For practical applications, it is generally suggested to take $M_s = 5 \sim 20$, for which illustrative results will be shown in examples later.

With the above unbiased frequency response estimation, the following corollary can be drawn accordingly:

Corollary 2.1. *Given Gaussian white measurement noise, $\zeta(t) \sim N(0, \sigma_\zeta^2)$, in the step response tests for identification, based on unbiased frequency response estimation from Theorem 2.1, consistent parameter estimation can be guaranteed by Algorithm-SS-III or Algorithm-SS-IV in terms of using the time domain fitting criterion in (2.42).*

Proof. Denote the measured step response as $\hat{y}(t) = y(t) + \zeta(t)$ in such a step test, where $y(t)$ is the true value of the step response. Assume that $\widetilde{y}(t)$ is the step response of a model identified using Algorithm-SS-III or Algorithm-SS-IV. It follows from (2.42) that

$$err = \frac{1}{N_s} \sum_{k=1}^{N_s} \left[\hat{y}(kT_s) - \widetilde{y}(kT_s) \right]^2. \tag{2.61}$$

2.2 Model Identification from an Open-Loop Step Test

To assess the standard deviation for model fitting, one can take the mathematical expectation of (2.61) as follows:

$$\begin{aligned} E(err) &= E(\hat{y} - \widehat{y})^2 \\ &= E(y + \zeta - \widehat{y})^2 \\ &= E(y - \widehat{y})^2 + 2E[\zeta(y - \widehat{y})] + E(\zeta)^2. \end{aligned} \quad (2.62)$$

Since $\zeta(kT_s)(k = 1, 2, \ldots, N_s)$ is a Gaussian random sequence with zero mean for $N_s \to \infty$, it is certainly uncorrelated with any time sequence such as $y(kT_s) - \widehat{y}(kT_s)(k = 1, 2, \ldots, N_s)$, i.e., $E[\zeta(y - \widehat{y})] = 0$. Meanwhile, there exists $E(\zeta)^2 = \sigma_\zeta^2$ for $N_s \to \infty$. Therefore, one obtains

$$E(err) = E(y - \widehat{y})^2 + \sigma_\zeta^2 \quad (2.63)$$

which indicates that the lower bound, $\min(err)$, can be reached if and only if $E(y - \widehat{y})^2$ is minimized. In other words, only the optimal model can result in this lower bound, regardless of the influence of measurement noise.

Note that given a preliminary value of the process time delay (θ), the remaining model parameters can be uniquely derived from either Algorithm-SS-III or Algorithm-SS-IV based on the above unbiased frequency response estimation. Hence, by monotonically varying θ in a one-dimensional search within a possible range, the minimum of (2.61) can be uniquely determined, leading to consistent parameter estimation for model identification. This completes the proof. $\square$

In practical applications, the true model structure of the process to be identified is often not known exactly, especially for high-order processes. Although a low-order model of FOPDT or SOPDT is widely used in control system design and tuning (Ogunnaike and Ray 1994; Shinskey 1996; Seborg et al. 2004), a higher-order model may be preferred if it can result in a significantly smaller fitting error of time or frequency domain response. It is, therefore, desirable to determine the optimal model for representing the process dynamic response characteristics from a step test. A guideline is given accordingly as follows:

Theorem 2.2. *Given a general model structure as shown in (2.38), despite the presence of Gaussian white measurement noise, $\zeta(t) \sim N(0, \sigma_\zeta^2)$, in the step response tests for identification, the model order for obtaining optimal fitting can be uniquely determined using Algorithm-SS-IV and a hypothesis testing condition,*

$$H_0 : 1 - \frac{\sum_{k=1}^{N_s} \left[\hat{y}(kT_s) - \widehat{y}(kT_s)\right]^2 \Big|_{n=n_2}}{\sum_{k=1}^{N_s} \left[\hat{y}(kT_s) - \widehat{y}(kT_s)\right]^2 \Big|_{n=n_1}} \geq \beta \quad (2.64)$$

where $\hat{y}(kT_s)$ and $\widehat{\overline{y}}(kT_s)$ denotes, respectively, the measured process output and the model output in such a step test, n_1 is the current model order, n_2 is a higher model order to be verified, and β is the significance level.

Proof. Owing to $E(\hat{y} - \overline{y})^2 \geq 0$ for models of any order, it follows from (2.63) that

$$\min(err) \geq E(y - \hat{y}_{op})^2 + \sigma_\zeta^2 \qquad (2.65)$$

where $\hat{y}_{op}(t)$ denotes the step response of the optimal model. It is obvious that $\min(err) = \sigma_\zeta^2$ if $y = \hat{y}_{op}$, corresponding to a perfect match of the identified model and the process.

Given the order of the model structure shown in (2.38), it follows from Corollary 2.1 that the optimal parameter estimation can be uniquely obtained using Algorithm-SS-IV. Therefore, by monotonically increasing the model order for parameter estimation using Algorithm-SS-IV, the optimal model order that gives the lower bound of (2.65) can be uniquely determined using the statistical hypothesis testing condition given in (2.64). This completes the proof. □

For implementation, it is generally suggested that $\beta = 0.9$ be taken for verification. It can be seen from (2.64) that this choice corresponds to the acceptance of a higher-order model if the resulting err shown in (2.61) is no larger than one tenth of that of the current model.

2.2.7 Illustrative Examples

Four examples studied in the recent literature are used here to illustrate the effectiveness of the presented algorithms for frequency response estimation and model identification. Examples 2.1–2.3 are given to show the good accuracy of the presented algorithms in identifying low-order processes in terms of the exact model structures, together with measurement noise tests to demonstrate identification robustness. Example 2.4 is used to show the effectiveness of the presented algorithms for the identification of higher-order processes. The measurement noise level is evaluated in terms of the noise-to-signal ratio (NSR),

$$\text{NSR} = \frac{\text{mean(abs(noise))}}{\text{mean(abs(signal))}} \qquad (2.66)$$

or reciprocally, SNR, SNR = $20\log_{10}(1/\text{NSR})$(dB). The time domain fitting criterion in (2.42) is used to assess identification accuracy.

Example 2.1. Consider the FOPDT process studied by Bi et al. (1999),

$$G_1 = \frac{1}{s+1}e^{-s}.$$

2.2 Model Identification from an Open-Loop Step Test

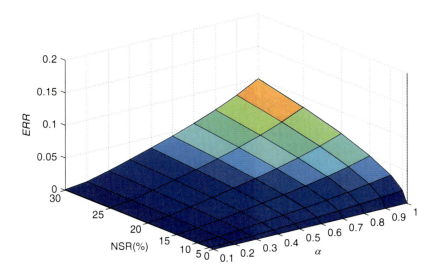

Fig. 2.2 The frequency response estimation error for Example 2.1 regarding different choices of α

Based on a unity step response test, Bi et al. (1999) gave an FOPDT model, $G_m = 1.00e^{-1.00s}/(0.997s + 1)$, by using a time domain LS fitting algorithm. For illustration, the same step test is performed here with a sampling period $T_s = 0.01$ (s). According to the guidelines given in (2.8) and (2.10), $\alpha = 0.5$ and $t_N = 30$ (s) are chosen to apply Algorithm-SS-I, resulting in the exact FOPDT model.

To demonstrate identification robustness regarding different choices of α, the frequency response estimation errors (*ERR*) resulting from taking $\alpha = 0.1, 0.2, \ldots, 1.0$ under a variety of random measurement noise levels (NSR = 0, 5%, 10%, 15%, 20%, 25%, 30%) are plotted in Fig. 2.2. It is seen that different choices of α result in *ERR* = 0 for no presence of measurement noise (NSR = 0), leading to the exact identification of the FOPDT model. When NSR $\neq 0$, a smaller value of α gives a smaller *ERR* for the same noise level, which indicates that a better statistical averaging effect against measurement noise can thus be obtained for frequency response estimation in terms of (2.6) (or (2.12)). In particular for $\alpha = 0.1$, it is seen that *ERR* becomes almost zero for different noise levels.

To demonstrate consistent estimation against measurement noise, the results of frequency response estimation for $G_1(0.5) = 0.4044$ under a variety of random noise levels (NSR = 5%, 10%, 20%, 30%) are listed in Table 2.1 based on a number of step tests ($M_s = 1, 5, 10, 20$). Note that different step tests are simulated by randomly varying the "seed" of the noise generator. The last row in Table 2.1 shows the identified models for $M_s = 20$, where the model parameters are shown by the mean of identification results, along with the sample standard deviation in parentheses. It is seen that good identification accuracy is obtained against different

Table 2.1 Frequency response estimation of $G_1(0.5) = 0.4044$ for Example 2.1 under different measurement noise levels

M_s	NSR = 5%	ERR	NSR = 10%	ERR	NSR = 20%	ERR	NSR = 30%	ERR
1	0.4066	0.55%	0.4088	1.09%	0.4133	2.21%	0.4179	3.32%
5	0.4031	0.32%	0.4019	0.61%	0.3995	1.22%	0.3971	1.82%
10	0.4034	0.26%	0.4024	0.49%	0.4005	0.97%	0.3985	1.46%
20	0.4038	0.15%	0.4033	0.28%	0.4022	0.54%	0.4011	0.81%
Model	$\dfrac{0.9986(\pm 0.008)e^{-1.0024(\pm 0.024)s}}{1.0059(\pm 0.085)s + 1}$		$\dfrac{0.9973(\pm 0.015)e^{-1.0051(\pm 0.048)s}}{1.0087(\pm 0.167)s + 1}$		$\dfrac{0.9947(\pm 0.031)e^{-1.0115(\pm 0.096)s}}{1.0034(\pm 0.347)s + 1}$		$\dfrac{1.0016(\pm 0.037)e^{-0.9887(\pm 0.117)s}}{1.0799(\pm 0.479)s + 1}$	

2.2 Model Identification from an Open-Loop Step Test

noise levels, and consistent estimation of the frequency response can gradually be reached as the number of step tests increase. Moreover, it can easily be verified that, given a random noise level, the sample standard deviation of frequency response estimation error is much smaller than that of the measurement noise, as clarified by Theorem 2.1.

Example 2.2. Consider the SOPDT process studied by Ahmed et al. (2006) and Liu et al. (2007),

$$G_2 = \frac{1.25}{0.25s^2 + 0.7s + 1} e^{-0.234s}.$$

For illustration, a unity step change is used for the step response test. The output response is like the one shown in Fig. 2.1. The sampling period is taken as $T_s = 0.01$ (s), which in fact is not an integer multiple of the process time delay. According to the guidelines given in (2.8) and (2.10), $\alpha = 0.2$ and $t_N = 100$ (s) are chosen to apply Algorithm-SS-I, resulting in an FOPDT model,

$$G_{m-1} = \frac{1.2505}{0.232s + 1} e^{-0.708s}.$$

Correspondingly, a referential cutoff angular frequency of the process frequency response can be derived as $\omega_{rc} = 3.4786$ (rad/s). Taking $\alpha = 0.2, 0.4, 0.6, 0.8, 1.0$, it can be easily verified that Algorithm-SS-II and Algorithm-SS-III result in the exact model. Also, it can be verified that Algorithm-SS-IV gives the exact model based on 11 points of frequency response estimation, $G_2(0.2 + j\omega_k)$, where $\omega_k = k\omega_{rc}/10$ and $k = 0, 1, 2, \ldots, 10$.

To demonstrate identification robustness against measurement noise, assume that there exists a random measurement noise of $N(0, \sigma_\xi^2 = 0.024)$, causing NSR = 10%. By performing 200 Monte Carlo tests in terms of varying the "seed" of the noise generator from 1 to 200, Algorithm-SS-IV using the above frequency response estimation gives

$$G_{m-2} = \frac{1.25(\pm 0.006)}{0.25(\pm 0.03)s^2 + 0.7(\pm 0.03)s + 1} e^{-0.234(\pm 0.04)s}.$$

It is seen that further improved identification robustness is thus obtained, compared to the results shown by Ahmed et al. (2006) and Liu et al. (2007).

Moreover, to demonstrate the achievable identification robustness of Algorithm-SS-IV without using the unbiased frequency response estimation in the presence of measurement noise, i.e., model identification based on frequency response estimation for each test, Fig. 2.3 shows the identification results for a variety of random noise levels (NSR = 1%, 5%, 10%, 15%, 20%, 25%, 30%), where the result for each model parameter to a given noise level is shown as a vertical linear segment along with the sample standard deviation in parentheses for 200 Monte

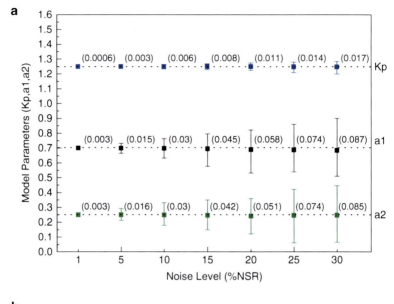
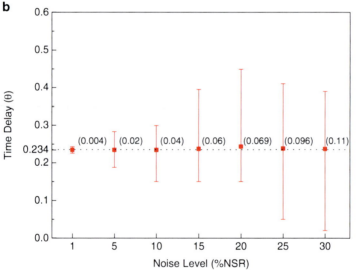

Fig. 2.3 The results of 200 Monte Carlo tests for the identification of Example 2.2 against measurement noise

Carlo tests. The square solid point in each linear segment denotes the mean of the 200 identification results, and the upper and lower bars correspond to the maximum and the minimum of parameter estimation, respectively. Note that a possible range of the time delay is estimated from the worst case of NSR = 30% as $\theta \in (0, 0.5]$ for a one-dimensional search using a search step size that equals the sampling

2.2 Model Identification from an Open-Loop Step Test

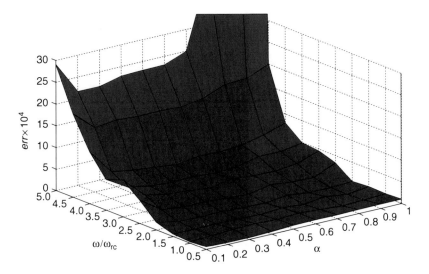

Fig. 2.4 The step response fitting error for Example 2.2 regarding different choices of α

period, together with the parameter estimation constraints: $a_1 > 0$ and $a_2 > 0$. It is, therefore, demonstrated that Algorithm-SS-IV can maintain good identification robustness, even without using the unbiased frequency response estimation.

In addition, to demonstrate identification robustness with respect to different choices of α, the step response fitting error for $\alpha = 0.1, 0.2, \ldots, 1.0$ and the frequency range of $(0.5 \sim 5.0)\omega_{rc}$ under the noise level of NSR = 10% for applying Algorithm-SS-IV is plotted in Fig. 2.4. It is seen that for the same frequency range for model fitting, different choices of α result in almost the same *err*. Note that *err* increases gradually with respect to the frequency range for model fitting, due to lower estimation accuracy for the process frequency responses at higher frequencies corresponding to lower SNR. This fact demonstrates why frequency response estimation in the low-frequency range is of primary use for model fitting from a step response test.

Example 2.3. Consider the SOPDT process with a right-half-plane (RHP) zero, widely studied in the references (Wang et al. 2001; Ahmed et al. 2006; Liu et al. 2007),

$$G_3 = \frac{-4s + 1}{9s^2 + 2.4s + 1} e^{-s}.$$

By performing a step test as in Example 2.1, Algorithm-SS-III gives the process model,

$$G_m = \frac{-3.9989s + 0.9998}{9.0183s^2 + 2.3951s + 1} e^{-s},$$

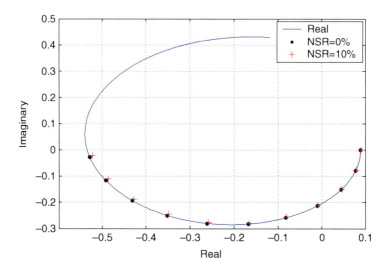

Fig. 2.5 Frequency response estimation for Example 2.3

by taking $\alpha = 0.05, 0.1, 0.15, 0.2$, $t_N = 200$ (s), and an initial estimation of the time delay as zero for a one-dimensional search with the fitting threshold of $err \leq 1 \times 10^{-4}$. Correspondingly, a referential cutoff angular frequency of the process frequency response can be derived as $\omega_{rc} = 0.366$ (rad/s). Note that using these choices of α, Algorithm-SS-I and Algorithm-SS-II cannot give an acceptable FOPDT or SOPDT model due to severe model mismatch. This fact was also found by Wang et al. (2001).

Figure 2.5 shows the frequency response estimation of $G_3(0.2 + j\omega_k)$, where $\omega_k = k\omega_{rc}/10$ and $k = 0, 1, 2, \ldots, 10$, demonstrating good accuracy. Accordingly, Algorithm-SS-IV results in the exact model.

To demonstrate identification robustness against measurement noise, assume that there exists a random measurement noise of $N(0, \sigma_\xi^2 = 0.015)$ in the step test, causing NSR $= 10\%$. The frequency response estimation of the above points is also plotted in Fig. 2.5 for comparison, which indicates good identification robustness. Accordingly, the proposed Algorithm-SS-IV gives a model,

$$G_m = \frac{-3.96945s + 0.9978}{9.0537s^2 + 2.3906s + 1} e^{-1.03s}$$

which results in $err = 1.93 \times 10^{-5}$ in terms of the settling time of $t_{set} = 50$ (s), compared to that of Wang et al. (2001), which gave $err = 6.25 \times 10^{-4}$. It can be verified that an unbiased frequency response estimation can be obtained using $5 \sim 10$ tests for a precise identification of the process model under this noise level.

2.2 Model Identification from an Open-Loop Step Test

Example 2.4. Consider the high-order process studied by Wang et al. (2001),

$$G_4 = \frac{2.15(-2.7s + 1)(158.5s^2 + 6s + 1)}{(17.5s + 1)^4(20s + 1)} e^{-14s}.$$

By performing the same step test as these two references, Algorithm-SS-I with $\alpha = 0.01$ and $t_N = 800$ (s) gives an FOPDT model,

$$G_{m-1} = \frac{2.1434}{53.2173s + 1} e^{-49.2712s}$$

which corresponds to $err = 4.54 \times 10^{-3}$ in terms of the settling time of $t_{set} = 500$ (s).

Note that improved fitting accuracy can be obtained using Algorithm-SS-IV based on the frequency response estimation of $G_4(0.01 + j\omega_k)$, where $\omega_k = k\omega_{rc}/10$, $k = 0, 1, 2, \ldots, 10$, and $\omega_{rc} = 0.0407$ (rad/s), that is derived from the above FOPDT model. The resulting FOPDT model is

$$G_{m-1} = \frac{2.1771}{55.0668s + 1} e^{-51.57s}$$

which corresponds to $err = 3.41 \times 10^{-3}$.

For comparison, Algorithm-SS-IV is also used to yield an SOPDT model,

$$G_{m-2} = \frac{2.1413 e^{-27.96s}}{1903.0013s^2 + 70.9754s + 1}$$

which corresponds to $err = 2.74 \times 10^{-4}$, and a third-order model,

$$G_{m-3} = \frac{244.0559s^2 + 9.0939s + 2.1507}{27587.1611s^3 + 2297.2191s^2 + 79.6136s + 1} e^{-25.48s}$$

which corresponds to $err = 3.67 \times 10^{-6}$, and a fourth-order model,

$$G_{m-4} = \frac{33.1113s^3 + 340.8036s^2 + 13.7927s + 2.1486}{320502.235s^4 + 45289.5023s^3 + 2879.9862s^2 + 85.7314s + 1} e^{-21.28s}$$

which corresponds to $err = 6.81 \times 10^{-7}$.

According to the model structure selection guideline in (2.64) with $\beta = 0.9$, the third-order model can be determined as the optimal model for the step response fitting.

It should be noted that Wang et al. (2001a) derived an SOPDT model corresponding to $err = 4.53 \times 10^{-3}$, and a third-order model corresponding to $err = 3.86 \times 10^{-5}$. Wang et al. (2001b) gave an SOPDT model corresponding

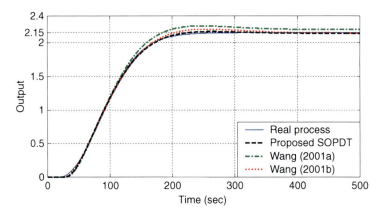

Fig. 2.6 Step response fitting of identified SOPDT models for Example 2.4

to $err = 5.68 \times 10^{-4}$. The step responses of these SOPDT models are plotted in Fig. 2.6. Note that the step response of the proposed third-order model almost overlaps with that of the process and thus is omitted.

2.3 Robust Identification Under Unsteady Initial Process Conditions and Load Disturbance

To cope with unsteady initial process conditions and unexpected load disturbance in practical applications, a modified implementation of the step response test is proposed here. Consequently, three robust identification algorithms are presented for obtaining the widely used low-order models of FOPDT and SOPDT (with or without a zero).

2.3.1 Implementation of a Step-Like Test

It is widely recognized that only the process transient response data to a step change are useful for model identification (Rake 1980; Shinskey 1996; Åström and Hägglund 2005). To facilitate robust identification under unsteady initial process conditions and unexpected load disturbance, a modified implementation of the step response test is proposed as illustrated in Figs. 2.7 or 2.8, where t_0 denotes the time when the step change is added. The time when the step change is subsequently removed, denoted by t_{ss}, is determined in terms of the transient response to the step change being nearly completed. Note that the proposed step-like test differs from the conventional step test in that the process transient response after removing the

2.3 Robust Identification Under Unsteady Initial Process Conditions and Load Disturbance

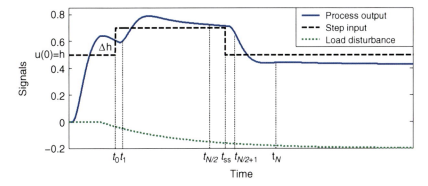

Fig. 2.7 A modified step test under unsteady initial process conditions and load disturbance

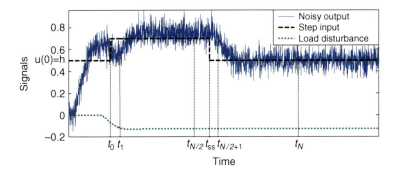

Fig. 2.8 A modified step test under unsteady initial process conditions, measurement noise and load disturbance

step change is also observed for identification. Also the proposed test differs from a standard pulse or rectangular wave test, where the duration of the excitation signal is predetermined (Söderström and Stoica 1989; Hwang and Lai 2004).

To effectively extract transient response data from the proposed step-like test, the step change should be added to initiate an observable dynamic response in contrast to measurement noise, and then be removed when the process response is observed to have almost entered into a steady state, as illustrated in Figs. 2.7 or 2.8. Note that both segments of the step response data – one corresponding to the addition of the step change (i.e., $t_1 \sim t_{N/2}$) and the other to the removal (i.e., $t_{N/2+1} \sim t_N$) – will be used for model identification. An important merit of such a choice is that the excitation sequence and the corresponding output sequence, composed of the two segments of data, are no longer correlated to the time sequence or any other time-oriented sequences like the transient response sequence resulting from unsteady initial process conditions or load disturbance. The uncorrelated relationship will be used in the identification algorithms to be presented later.

When adding the step change in the proposed test, the process output trend should be monitored beforehand. If the initial process output, $y(t_0)$, has a decreasing trend, i.e., $\dot{y}(t_0) < 0$, a positive step change to the process input (u) is suggested, as shown in Fig. 2.7, so that an obvious turning point in the output response can be observed. The starting point (t_1) of the first segment of the step response data can then be taken for identification. If the initial process output has an increasing trend, a negative step change should be introduced accordingly. If the trend of the initial process output cannot be observed clearly, especially in the presence of severe measurement noise as shown in Fig. 2.8, either a positive or a negative step change may be used, the magnitude of which should be set reasonably large to create an observable and admissible fluctuation of the process output. The final point ($t_{N/2}$) of the first segment of data can be taken as the time when the step change is removed or slightly earlier, where N is the number of step response data points used for identification.

It is, of course, ideal to remove the step change after the process response has completely moved into a steady state. However, this may be difficult to observe in the presence of measurement noise and load disturbance. It is, therefore, suggested to remove the step change when the process response is observed to have almost moved into a steady state. Correspondingly, the first obvious turning point of the output response after the step change is removed may be taken as the starting point ($t_{N/2+1}$) of the second segment of data for identification. The final point (t_N) of the second segment of data can be taken as the time after which the process response has almost recovered to a steady state, as shown in Figs. 2.7 or 2.8. Note that if there surely exists no load disturbance, it will not be needed to wait for such a steady state before removing the step input. The reason can be seen from the identification algorithms presented later.

The total number of data points (N) making up the two segments used for identification is suggested to be in the range of $50 \sim 200$ for a good trade-off between identification accuracy and computational efficiency.

2.3.2 Model Identification

For the identification of a low-order process model to facilitate control design, a representative model structure is

$$G = \frac{b_1 s + b_0}{a_2 s^2 + a_1 s + 1} e^{-\theta s} \qquad (2.67)$$

which corresponds to the time domain response of

$$a_2 \ddot{y}(t) + a_1 \dot{y}(t) + y(t) = b_1 \dot{u}(t-\theta) + b_0 u(t-\theta) + l(t) \qquad (2.68)$$

2.3 Robust Identification Under Unsteady Initial Process Conditions and Load Disturbance

where $y(t)$ denotes the process output, $u(t)$ the process input, $l(t)$ the output deviation caused by load disturbance, θ the process time delay, and b_0 is customarily named as the process static gain. For $b_1 < 0$, (2.67) or (2.68) indicates an SOPDT model with an RHP zero. $b_1 = 0$ corresponds to an SOPDT model of either one of three response types – underdamped, critically damped, and overdamped. The combination of $b_1 = 0$ and $a_2 = 0$ leads to an FOPDT model.

For clarity, the development of the identification algorithm for obtaining an SOPDT model with a zero (Case 1) is presented in detail. Then, the development of the identification algorithms for obtaining an SOPDT model without zeros (Case 2) and an FOPDT model (Case 3) is briefly described.

Case 1 SOPDT model with a zero. Given the nonzero initial conditions of $y(t_0) \neq 0$ and $u(t_0) = h$, $h \in \mathfrak{R}$, by using a time shift of t_0, i.e., letting $t_0 = 0$, it follows from implementation of the modified step test that

$$u(t) = \begin{cases} h, & 0 \leq t \leq \theta; \\ h + \Delta h, & \theta < t \leq t_{ss} + \theta; \\ h, & t > t_{ss} + \theta. \end{cases} \quad (2.69)$$

Note that for $t_1 \leq t \leq t_{ss} + \theta$, where $t_1 \geq t_0 + \theta$, it can be derived in terms of $t_0 = 0$ that

$$\int_0^t u(t)dt = \int_0^\theta h\,dt + \int_\theta^t (h + \Delta h)dt = (h + \Delta h)t - \Delta h \theta \quad (2.70)$$

$$\int_0^t \dot{u}(t)dt = \int_0^{\theta_-} 0\,dt + \int_{\theta_-}^{\theta_+} \dot{u}(t)dt + \int_{\theta_+}^t 0\,dt = \Delta h. \quad (2.71)$$

Similarly, it can be derived for $t_{ss} + \theta < t \leq t_N$ that

$$\int_0^t u(t)dt = \int_0^\theta h\,dt + \int_\theta^{t_{ss}+\theta} (h + \Delta h)dt + \int_{t_{ss}+\theta}^t (h + \Delta h)dt = \Delta h t_{ss} + ht \quad (2.72)$$

$$\int_0^t \dot{u}(t)dt = \int_0^{\theta_-} 0\,dt + \int_{\theta_-}^{\theta_+} \dot{u}(t)dt + \int_{\theta_+}^{t_{ss}+\theta_-} 0\,dt + \int_{t_{ss}+\theta_-}^{t_{ss}+\theta_+} \dot{u}(t)dt + \int_{t_{ss}+\theta_+}^t 0\,dt = 0. \quad (2.73)$$

For clarity, denote multiple integrals for a time function of $f(t)$ as

$$\int_{[0,t]}^{(m)} f(t) = \int_0^t \int_0^{\tau_{m-1}} \cdots \int_0^{\tau_1} f(\tau_0)d\tau_0 d\tau_1 \cdots d\tau_{m-1}, \quad m \geq 2. \quad (2.74)$$

For $t_1 \leq t \leq t_{ss} + \theta$, by using (2.70) it follows that

$$\int_{[0,t]}^{(2)} u(t) = \int_0^\theta \int_0^{\tau_1} h\, d\tau_0 d\tau_1 + \int_\theta^t [(h + \Delta h)\tau_1 - \Delta h\theta]\, d\tau_1$$

$$= \frac{h + \Delta h}{2} t^2 - \Delta h t\theta + \frac{\Delta h}{2}\theta^2. \tag{2.75}$$

Analogously, it can be derived that

$$\int_{[0,t]}^{(2)} u(t) = \begin{cases} \dfrac{h + \Delta h}{2} t^2 - \Delta h t\theta + \dfrac{\Delta h}{2}\theta^2, & \theta < t \leq t_{ss} + \theta; \\ \dfrac{h}{2} t^2 + \Delta h t_{ss} t - \dfrac{\Delta h}{2} t_{ss}^2 - \Delta h t_{ss}\theta, & t > t_{ss} + \theta. \end{cases} \tag{2.76}$$

$$\int_{[0,t]}^{(3)} u(t) = \begin{cases} \dfrac{h + \Delta h}{6} t^3 - \dfrac{\Delta h}{2} t^2\theta + \dfrac{\Delta h}{2} t\theta^2 - \dfrac{\Delta h}{6}\theta^3, & \theta < t \leq t_{ss} + \theta; \\ \dfrac{h}{6} t^3 + \dfrac{\Delta h}{2} t_{ss} t^2 - \dfrac{\Delta h}{2} t_{ss}^2 t + \dfrac{\Delta h}{6} t_{ss}^3 + \left(\dfrac{\Delta h}{2} t_{ss}^2 - \Delta h t_{ss} t\right)\theta + \dfrac{\Delta h}{2} t_{ss}\theta^2, & t > t_{ss} + \theta. \end{cases} \tag{2.77}$$

$$\int_{[0,t]}^{(2)} \dot{u}(t) = \begin{cases} \Delta h(t - \theta), & \theta < t \leq t_{ss} + \theta; \\ \Delta h t_{ss}, & t > t_{ss} + \theta. \end{cases} \tag{2.78}$$

$$\int_{[0,t]}^{(3)} \dot{u}(t) = \begin{cases} \dfrac{\Delta h}{2} t^2 - \Delta h t\theta + \dfrac{\Delta h}{2}\theta^2, & \theta < t \leq t_{ss} + \theta; \\ \Delta h t_{ss} t - \dfrac{\Delta h}{2} t_{ss}^2 - \Delta h t_{ss}\theta, & t > t_{ss} + \theta. \end{cases} \tag{2.79}$$

$$\int_{[0,t]}^{(3)} \dot{y}(t) = \int_{[0,t]}^{(2)} y(t) - \frac{1}{2} y(0) t^2 \tag{2.80}$$

$$\int_{[0,t]}^{(3)} \ddot{y}(t) = \int_{[0,t]}^{(1)} y(t) - y(0)t - \frac{1}{2}\dot{y}(0) t^2. \tag{2.81}$$

With the description of (2.69), one can reformulate the time domain response of (2.68) as

$$a_2 \ddot{y}(t) + a_1 \dot{y}(t) + y(t) = b_1 \dot{u}(t) + b_0 u(t) + l(t). \tag{2.82}$$

By integrating both sides of (2.82) three times and rearranging the resulting equation using (2.77), (2.79)–(2.81), one can obtain the LS form of parameter estimation,

$$\psi(t) = \phi^T(t)\gamma + \varepsilon(t) \tag{2.83}$$

2.3 Robust Identification Under Unsteady Initial Process Conditions and Load Disturbance

where $\varepsilon(t)$ denotes the residual error, and

$$\begin{cases} \psi(t) = \int_{[0,t]}^{(3)} y(t), \\ \phi(t) = \left[-\int_{[0,t]}^{(1)} y(t), -\int_{[0,t]}^{(2)} y(t), F_0(t), F_1(t), F_2(t), F_3(t), 1, t, t^2, \ldots, t^q \right]^T, \\ \gamma = \left[a_2, a_1, b_0, b_0\theta - b_1, b_0\theta^2 - 2b_1\theta, b_0\theta^3 - 3b_1\theta^2, \eta_0, \eta_1, \eta_2, \ldots, \eta_q \right]^T. \end{cases} \quad (2.84)$$

$$F_0(t) = \begin{cases} \dfrac{h + \Delta h}{6} t^3, & \theta < t \leq t_{ss} + \theta; \\ \dfrac{h}{6} t^3 + \dfrac{\Delta h}{2} t_{ss} t^2 - \dfrac{\Delta h}{2} t_{ss}^2 t + \dfrac{\Delta h}{6} t_{ss}^3, & t > t_{ss} + \theta. \end{cases} \quad (2.85)$$

$$F_1(t) = \begin{cases} -\dfrac{\Delta h}{2} t^2, & \theta < t \leq t_{ss} + \theta; \\ \dfrac{\Delta h}{2} t_{ss}^2 - \Delta h t_{ss} t, & t > t_{ss} + \theta. \end{cases} \quad (2.86)$$

$$F_2(t) = \begin{cases} \dfrac{\Delta h}{2} t, & \theta < t \leq t_{ss} + \theta; \\ \dfrac{\Delta h}{2} t_{ss}, & t > t_{ss} + \theta. \end{cases} \quad (2.87)$$

$$F_3(t) = \begin{cases} -\dfrac{\Delta h}{6}, & \theta < t \leq t_{ss} + \theta; \\ 0, & t > t_{ss} + \theta. \end{cases} \quad (2.88)$$

Note that in the parameter vector γ to be estimated, η_i ($i = 0, 1, 2, \ldots, q$) are used for the Maclaurin series approximation of the unforced output response resulting from a nonzero initial state of the process and the load disturbance response, according to the linear superposition principle for analysis of system response. That is, the effect of $y(0)$, $\dot{y}(0)$ and $l(t)$ on the LS fitting of (2.83) is approximately quantified as

$$\hat{Q}(t) = \sum_{j=0}^{q} \eta_j t^j \quad (2.89)$$

which is based on the fact that the time domain response arising from nonzero initial process conditions and load disturbance can generally be expressed as follows:

$$Q(t) = \sum_{j=0}^{m} \alpha_j e^{-\beta_j t} \quad (2.90)$$

where $\alpha_j = \sum_{i=0}^{r} c_i t^i$, $c_i \in \Re$, $\text{Re}(\beta_j) \in \Re_+$, r is limited by the number of repeated poles of the process and the input type, m is the number of different poles of the process, and there exist

$$e^{-\text{Re}(\beta_j)t} = 1 - \text{Re}(\beta_j)t + \frac{[\text{Re}(\beta_j)]^2}{2!}t^2 + \frac{[-\text{Re}(\beta_j)]^3}{3!}t^3 + \cdots + \frac{[-\text{Re}(\beta_j)]^n}{n!}t^n + R_n(t) \quad (2.91)$$

$$\lim_{t \to \infty} R_n(t) = \lim_{t \to \infty} \frac{[-\text{Re}(\beta_j)t]^{n+1}}{(n+1)!} e^{-\delta \text{Re}(\beta_j)t} = 0, \quad \delta \in (0, 1). \quad (2.92)$$

For instance, if there exists no load disturbance, i.e., $l(t) = 0$, it can be concluded using (2.80) and (2.81) that $q = 2$ for (2.89) is sufficient to solve the LS fitting of (2.83), and $\eta_0 = 0$, $\eta_1 = a_2 y(0)$, and $\eta_2 = [a_1 y(0) + a_2 \dot{y}(0)]/2$ can be explicitly derived from the formulation of (2.83). In the presence of a static load disturbance, i.e., $l(t) = c, c \in \Re$, it can be similarly verified that $q = 3$ is sufficient for parameter estimation, if the model structure matches the process.

In cases where a nonstatic load disturbance occurs in such a modified step test, the order of q for representing the corresponding output response with respect to the time used in the test depends on the achievable approximation of (2.89). This explains why it is necessary to remove the step change after the process response has almost moved into a steady state. Thereby, the transient response resulting from nonzero initial process conditions and load disturbance can be sufficiently represented by (2.89).

Note that the influence of a time-varying load disturbance that cannot be effectively approximated by (2.89) may distort the LS fitting of (2.83). However, the influence is likely to be observed according to the guideline for removing the step change. In such a case, the modified step test may simply be repeated for verification, owing to the fact that the proposed algorithms are developed specifically for practical application subject to nonzero initial process conditions.

To determine a suitable order of q for solving the LS fitting of (2.83), a statistical hypothesis test can be used. That is, with an initial estimation of $q = 3$ for iteration using the same transient response data collected for identification, the updating law, $q(k+1) = q(k) + 1$, can be used if the following convergent condition is not satisfied at the kth iteration step,

$$J_p = \sqrt{\frac{1}{p} \sum_{i=1}^{p} \left[\gamma^{[k]}(i) - \gamma^{[k-1]}(i) \right]^2} < \varepsilon \quad (2.93)$$

where ε is a user-specified threshold for assessing the fitting variance of the model parameters, and p is the number of model parameters to be identified. It can be seen from (2.84) that $\gamma(i)$ ($i = 1, 2, \ldots, p$) correspond to the model parameters, which are relatively independent of the Maclaurin series coefficients ($\eta_j, j = 0, 1, 2, \ldots, q$) for approximation.

2.3 Robust Identification Under Unsteady Initial Process Conditions and Load Disturbance   45

Hence, by using the two proposed segments of step response data and letting $\Psi = [\psi(t_1), \psi(t_2), \ldots, \psi(t_N)]^T$ and $\Phi = [\phi(t_1), \phi(t_2), \ldots, \phi(t_N)]^T$, one can establish an LS algorithm for parameter estimation,

$$\Psi = \Phi \gamma. \tag{2.94}$$

Accordingly, the parameter vector can be solved as

$$\gamma = (\Phi^T \Phi)^{-1} \Phi^T \Psi. \tag{2.95}$$

To clarify the invertibility of $\Phi^T \Phi$ in solving (2.95), the following theorem is given.

Theorem 2.3. $\Phi^T \Phi$ *is guaranteed invertible by a modified step test as shown in Fig. 2.7.*

Proof. Note that the last q columns in Φ are all time vectors with different integral exponents, so they are linearly independent of each other. The first two columns in Φ, sequences of single and double integrals for the process output, are obviously linearly independent of the last q columns. Owing to the modified step test, $F_j(t)$ for $j = 0, 1, 2, 3$ are all piecewise continuous-time functions, so the corresponding columns in Φ are not only linearly independent of each other, but also linearly independent of the remaining columns in Φ. Therefore, Φ is guaranteed to have full column rank. Since $\text{rank}(\Phi^T \Phi) = \text{rank}(\Phi)$, one ensures that $\Phi^T \Phi$ is invertible for the computation of (2.95). This completes the proof. □

Remark 2.4. If a conventional step test is used for identification, the transient response data correspond only to the first segment of data of the modified step test. It can be seen from (2.85)–(2.88) that the resulting $F_j(t)$ for $j = 0, 1, 2$ will all be time functions, while $F_3(t)$ becomes a constant. The corresponding columns will definitely be linearly dependent of the last $q + 1$ columns in Φ, causing $\Phi^T \Phi$ to be not invertible for computation. It is, thus, demonstrated that a conventional step test is not suitable for parameter estimation under unsteady initial process conditions or load disturbance. ◇

In the sequel, the model parameters can be retrieved from $\gamma(i)$ ($i = 1, 2, \ldots, p$) as

$$[a_2, a_1, b_0, b_1, \theta]^T = \left[\gamma(1), \gamma(2), \gamma(3), \pm \sqrt{\gamma^2(4) - \gamma(3)\gamma(5)}, [b_1 + \gamma(4)]/\gamma(3) \right]^T. \tag{2.96}$$

Obviously, a positive b_1 indicates a minimum-phase process, and a negative b_1 corresponds to a nonminimum-phase process with an initial inverse response, i.e., the output response to a step change will initially be in the opposite direction to its final value, which can evidently be observed in practice. Hence, there can only be one unique choice of b_1.

It should be noted that the redundant fitting condition, $\gamma(6)$, can surely be satisfied if the model structure adopted fits the real process well. In case there exists model mismatch, the condition may be used to establish an LS fitting of the model parameters to enhance the model fitting accuracy, as introduced in Sect. 2.2.3.

In the presence of measurement noise, $\zeta(t)$, there exists $\hat{y}(t) = y(t) + \zeta(t)$, where $\hat{y}(t)$ denotes the measured process output and $y(t)$ the true process output. Substituting $\hat{y}(t)$ into (2.82) and then triply integrating both sides of (2.82), one can obtain

$$\psi(t) = \phi^T(t)\gamma + v(t) \tag{2.97}$$

where

$$\begin{cases} \psi(t) = \int_{[0,t]}^{(3)} \hat{y}(t), \\ \phi(t) = \left[-\int_{[0,t]}^{(1)} \hat{y}(t), -\int_{[0,t]}^{(2)} \hat{y}(t), F_0(t), F_1(t), F_2(t), F_3(t), 1, t, t^2, \ldots, t^q \right]^T, \\ \gamma = \left[a_2, a_1, b_0, b_0\theta - b_1, b_0\theta^2 - 2b_1\theta, b_0\theta^3 - 3b_1\theta^2, \eta_0, \eta_1, \eta_2, \ldots, \eta_q \right]^T, \\ v(t) = a_2 \int_{[0,t]}^{(1)} \zeta(t) + a_1 \int_{[0,t]}^{(2)} \zeta(t) + \int_{[0,t]}^{(3)} \zeta(t). \end{cases} \tag{2.98}$$

It is seen from (2.97) and (2.98) that $\phi(t)$ is now correlated with $v(t)$, which is the influence of measurement noise. Therefore, parameter estimation from (2.95) may not be consistent. To circumvent this issue, the instrumental variable (IV) method (Söderström and Stoica 1989) can be used. There is, however, no unified choice of the IV matrix for consistent estimation. A feasible choice is proposed in the following theorem.

Theorem 2.4. *With* $Z = [z_1, z_2, \ldots, z_N]^T$ *chosen as the IV matrix where* $z_i = [1/t_i^{p+1}, 1/t_i^p, \ldots, 1/t_i, 1, t_i, t_i^2, \ldots, t_i^q]^T$, *which satisfies the two limiting conditions: 1. the inverse of* $\lim_{N \to \infty} (Z^T \Phi)/N$ *exists; 2.* $\lim_{N \to \infty} (Z^T v)/N = 0$, *where* $v = [v(t_1), v(t_2), \ldots, v(t_N)]^T$, *a consistent parameter estimation is given by* $\gamma = (Z^T \Phi)^{-1} Z^T \Psi$.

Proof. The two limiting conditions in Theorem 2.4 are sufficient for consistent estimation (Söderström and Stoica 1989), which, in fact, indicate that the instruments must be correlated with the regression variables but uncorrelated with the measurement noise. It is, thus, required to demonstrate that the proposed IV matrix complies with these two limiting conditions.

2.3 Robust Identification Under Unsteady Initial Process Conditions and Load Disturbance

For the first limiting condition, which is equivalent to saying that $(Z^T\Phi)/N$ is guaranteed nonsingular for $N \to \infty$, it has been shown that both Φ^T and Z^T have full row rank for $N \geq \dim(\gamma)$. Note that

$$\operatorname{rank}(Z^T) = \operatorname{rank}(\Phi^T) \tag{2.99}$$

$$\operatorname{rank}(Z^T) = \operatorname{rank}(Z) = \operatorname{rank}(Z^T Z) \tag{2.100}$$

$$\operatorname{rank}(\Phi^T) = \operatorname{rank}(\Phi) = \operatorname{rank}(\Phi^T \Phi) \tag{2.101}$$

Besides, there exists

$$\left(\frac{1}{N}Z^T\Phi\right)^T \left(\frac{1}{N}Z^T\Phi\right) = \Phi^T\left(\frac{1}{N^2}ZZ^T\right)\Phi. \tag{2.102}$$

Here, denote

$$A = \frac{1}{N^2}ZZ^T. \tag{2.103}$$

For $0 < t_1 < t_2 < \cdots < t_N$, it can easily be verified that

$$X^T A X \geq 0, \quad \forall X \in \Re^{N \times 1}, \quad \text{and} \quad X^T A X = 0 \Leftrightarrow X = 0 \tag{2.104}$$

which indicates that A is positive definite. Accordingly, it follows that

$$\operatorname{rank}(\Phi^T A \Phi) = \operatorname{rank}(\Phi^T). \tag{2.105}$$

Hence, one can conclude from (2.102) and (2.105) that $(Z^T\Phi)/N$ is nonsingular. For the second limiting condition, it follows for $N \to \infty$ that the measurement noise sequence, $\zeta(t_i)(i = 1, 2, \ldots, N)$, can be viewed as a Gaussian white noise sequence with zero mean. The corresponding $v(t_i)(i = 1, 2, \ldots, N)$ can, thus, be determined as a random distribution with zero mean, which is independent of the time origin or any other time such as t_0 used in the modified step test. Therefore, the random vector v is uncorrelated with the rows of Z^T that are time vectors with different integral exponents. For the row of Z^T containing constants only, i.e., $[1, 1, \ldots, 1]$, its inner product with v can be derived as

$$\lim_{N \to \infty} \frac{1}{N} \sum_{i=1}^{N} v(t_i) = 0. \tag{2.106}$$

Hence, the second condition is also satisfied. This completes the proof. □

Case 2 SOPDT model without zeroes. By letting $b_1 = 0$, the above identification algorithm can directly be applied to obtain an SOPDT model without zeroes. The only difference is that the parameter vector in (2.84) should be amended to $\gamma = [a_2, a_1, b_0, b_0\theta, b_0\theta^2, b_0\theta^3, \eta_0, \eta_1, \eta_2, \ldots, \eta_q]^T$. Accordingly, the model parameters can be retrieved as

$$[a_2, a_1, b_0, \theta]^T = [\gamma(1), \gamma(2), \gamma(3), \gamma(4)/\gamma(3)]^T. \quad (2.107)$$

Case 3 FOPDT model. Since $b_1 = 0$ and $a_2 = 0$, by doubly integrating both sides of (2.82) and rearranging the resulting equation using (2.76) and (2.78), one can formulate an LS fitting similar to (2.83), for which

$$\begin{cases} \psi(t) = \int_{[0,t]}^{(2)} y(t), \\ \phi(t) = \left[-\int_{[0,t]}^{(1)} y(t), F_0(t), F_1(t), F_2(t), 1, t, t^2, \ldots, t^q\right]^T, \\ \gamma = [a_1, b_0, b_0\theta, b_0\theta^2, \eta_0, \eta_1, \eta_2, \ldots, \eta_q]^T. \end{cases} \quad (2.108)$$

$$F_0(t) = \begin{cases} \dfrac{h + \Delta h}{2} t^2, & \theta < t \leq t_{ss} + \theta; \\ \dfrac{h}{2} t^2 + \Delta h t_{ss} t - \dfrac{\Delta h}{2} t_{ss}^2, & t > t_{ss} + \theta. \end{cases} \quad (2.109)$$

$$F_1(t) = \begin{cases} -\Delta h t, & \theta < t \leq t_{ss} + \theta; \\ -\Delta h t_{ss}, & t > t_{ss} + \theta. \end{cases} \quad (2.110)$$

$$F_2(t) = \begin{cases} \dfrac{\Delta h}{2}, & \theta < t \leq t_{ss} + \theta; \\ 0, & t > t_{ss} + \theta. \end{cases} \quad (2.111)$$

It can easily be verified that $q = 1$ for (2.89) is sufficient to solve the LS fitting if there exists no load disturbance, for which $\eta_0 = 0$ and $\eta_1 = a_1 y(0)$ can be derived accordingly. In the presence of a static load disturbance, it can also be verified that $q = 2$ is sufficient for parameter estimation, if the model structure matches the process.

Accordingly, the model parameters can be derived using (2.95) and (2.108) as

$$[a_1, b_0, \theta]^T = [\gamma(1), \gamma(2), \gamma(3)/\gamma(2)]^T. \quad (2.112)$$

For consistent estimation against measurement noise, the corresponding IV can be taken according to Theorem 2.4 as $z_i = [1/t_i^4, 1/t_i^3, 1/t_i^2, 1/t_i, 1, t_i, t_i^2, \ldots, t_i^q]^T$.

Remark 2.5. By letting $u(0) = 0$ (i.e., $h = 0$ for $t < t_0 = 0$), all the above identification algorithms can be directly applied for zero or steady nonzero initial process conditions, regardless of whether load disturbance exists or not. In such a

2.3 Robust Identification Under Unsteady Initial Process Conditions and Load Disturbance

case, $q = 0$ can be used if there is no load disturbance. It should be noted that the above identification algorithms can be extended transparently to the case where the initial process input is also time varying, because the variation in such a case is usually under control in practice and, thus, is known and can be expressed explicitly for computation. ◇

2.3.3 Illustrative Examples

Three examples from existing literature are used here to illustrate the effectiveness and accuracy of the above robust identification algorithms. For assessing the achievable identification accuracy, the transient response error criterion shown in (2.42) is adopted for reference.

Example 2.5. Consider the SOPDT process studied by Ahmed et al. (2006) and Liu et al. (2007),

$$G_5 = \frac{1.25e^{-0.234s}}{0.25s^2 + 0.7s + 1}.$$

The corresponding time domain response is in the form of

$$0.25\ddot{y}(t) + 0.7\dot{y}(t) + y(t) = 1.25u(t - 0.234) + l(t).$$

Based on zero initial process conditions and $l(t) = 0$, Liu et al. (2007) derived an SOPDT model using multiple step response tests, and Ahmed et al. (2006) developed an SOPDT algorithm using a random binary signal (RBS) for excitation. For illustration, assume that the initial conditions are $y(t_0) = 0.6$, $\dot{y}(t_0) = -0.06$, and $u(t_0) = 0.5$, as shown in Fig. 2.7. In view of the fact that the initial process output has a decreasing trend, a step change of $\Delta h = 0.2$ is added at $t_0 = 3$ (s), ahead of which a load disturbance with a slow dynamic of $G_d = 0.2/(0.5s + 1)$ is injected into the process output at $t = 2$ (s). The step change is removed at $t_{ss} = 10$ (s) before the load disturbance becomes steady.

Using the process output data in the time intervals of [3.5, 10](s) and [10.5, 20](s), the identification algorithm for Case 2 taking $N = 100$ and $q = 8$ gives a model listed in Table 2.2 (NSR = 0), which indicates good accuracy. Note that the Maclaurin approximation for the effect of the nonzero initial process state and the load disturbance converges at $q = 8$ with a prescribed threshold of $\varepsilon = 5\%$.

To demonstrate identification robustness in the presence of measurement noise, assume that a random noise of $N(0, \sigma_N^2 = 0.45\%)$, causing NSR = 10%, is added to the process output measurement, while a step-type load disturbance with a magnitude of 0.1 is added to the process input at $t = 2$ (s), as shown in Fig. 2.8. Repeating the above step test and using the process output data in the time intervals of [4, 10](s) and [11, 25](s) that are obvious for observing the process dynamic

Table 2.2 Identification results under a variety of measurement noise levels

NSR	Example	Identified model	Fitting error (err)
0	2.5	$\dfrac{1.2507e^{-0.2536s}}{0.2503s^2 + 0.6994s + 1}$	3.88×10^{-5}
	2.6	$\dfrac{(-4.000s + 1.0000)e^{-1.0187s}}{9.0000s^2 + 2.4000s + 1}$	4.04×10^{-6}
1%	2.5	$\dfrac{1.247e^{-0.2501s}}{0.2548s^2 + 0.7021s + 1}$	6.19×10^{-5}
	2.6	$\dfrac{(-4.0039s + 0.9971)e^{-1.0189s}}{9.0459s^2 + 2.4125s + 1}$	2.76×10^{-5}
10%	2.5	$\dfrac{1.2202e^{-0.2234s}}{0.2998s^2 + 0.7153s + 1}$	9.22×10^{-4}
	2.6	$\dfrac{(-4.0425s + 0.9726)e^{-0.9307s}}{9.421s^2 + 2.5255s + 1}$	1.08×10^{-3}
20%	2.5	$\dfrac{1.1914e^{-0.2338s}}{0.3209s^2 + 0.6886s + 1}$	3.11×10^{-3}
	2.6	$\dfrac{(-4.0968s + 0.9437)e^{-0.823s}}{9.8338s^2 + 2.676s + 1}$	4.4×10^{-3}

response to the addition and removal of the step change, the above identification algorithm with $q = 3$ gives a model listed in Table 2.2, which demonstrates good identification robustness. The results for NSR = 1% and 20% are also listed in Table 2.2 to show the achievable accuracy.

Example 2.6. Consider the SOPDT process with a RHP zero studied by Wang et al. (2001), Ahmed et al. (2008), and Liu et al. (2007),

$$G_6 = \frac{(-4s + 1)e^{-s}}{9s^2 + 2.4s + 1}.$$

The corresponding time domain response is in the form of

$$9\ddot{y}(t) + 2.4\dot{y}(t) + y(t) = -4\dot{u}(t-1) + u(t-1) + l(t).$$

Zero initial process conditions were assumed by Wang et al. (2001) and Liu et al. (2007), and nonzero initial conditions but with no load disturbance were assumed by Ahmed et al. (2008). Here, the initial conditions are considered to be $y(0) = 0.5$, $\dot{y}(0) = -0.026$, and $u(0) = 0.5$, which are extracted from the process transient response to a step change with a magnitude of 0.5, subject to a slowly changing load disturbance as in Example 2.5.

By performing a modified step test as in Example 2.5, the identification algorithm for Case 1 taking $N = 100$ and $q = 7$ in terms of the output response data in the time intervals of [6.5, 10](s) and [15, 35](s) gives a model listed in Table 2.2

2.3 Robust Identification Under Unsteady Initial Process Conditions and Load Disturbance

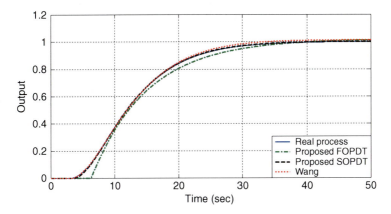

Fig. 2.9 Step response fitting for Example 2.7

(NSR = 0), which once again demonstrates good accuracy. Also it is shown that using a longer time length to collect the second segment of data reduces the Maclaurin approximation order of q with respect to $\varepsilon = 5\%$ as used in Example 2.5. Repeating the measurement noise tests in Example 2.5 and taking $t_{ss} = 35$ (s), the corresponding identification results obtained using $q = 3$ in terms of the output response data in the time intervals of [8, 35](s) and [40, 65](s) are also presented in Table 2.2, which again demonstrates good identification robustness.

Example 2.7. Consider the high-order process studied by Wang et al. (2006),

$$25\dddot{y}(t) + 35\ddot{y}(t) + 11\dot{y}(t) + y(t) = u(t - 2.5) + l(t)$$

with the initial conditions of $y(0) = 2$, $\dot{y}(0) = 0.4$, $\ddot{y}(0) = -0.4$, and $u(0) = 0$, together with a static load disturbance of $l(t) = 1$. For illustration, the same initial conditions but with a nonzero input of $u(0) = 1.5$ are considered here. In view of the fact that the initial process output has an increasing trend, a step change of $\Delta h = -0.5$ is added at $t_0 = 0$ (s) and then is removed at $t_{ss} = 10$ (s).

Using the process output data in the time intervals of [5, 10](s) and [15, 45](s), the identification algorithm for Case 3 taking $N = 100$ and $q = 3$ gives an FOPDT model, $G_{m-1} = 1.0192e^{-6.2227s}/(8.8339s + 1)$, corresponding to *err* = 7.42×10^{-4}. In contrast, the identification algorithm for Case 2 gives an SOPDT model, $G_{m-2} = 1.0025e^{-3.4681s}/(25.2131s^2 + 10.0171s + 1)$, corresponding to *err* = 1.32×10^{-5}. It should be noted that Wang et al. (2006) derived an SOPDT model, $G_m = 0.03411e^{-3.11s}/(s^2 + 0.3486s + 0.03366)$, corresponding to *err* = 1.46×10^{-4}. For comparison, the unity step responses of these models are shown in Fig. 2.9. It can be seen that both SOPDT models offer better fitting than the FOPDT model, while the proposed SOPDT model shows further improved fitting accuracy.

2.4 Piecewise Model Identification Under Inherent Load Disturbance

In the presence of a deterministic load disturbance, which is common in various repetitive industrial control systems and batch processes that are frequently or periodically initiated by a step change to the set-point, the resulting step response may be viewed as the pure process response plus load disturbance response according to the linear superposition principle. Such load disturbance is herein named inherent-type load disturbance. Modeling the pure process response only might not be sufficient to describe the overall dynamic response characteristics for control system design or controller tuning. For instance, a water pump in an air conditioning system obviously gives different step responses under different loads. Modeling both the pure pump response to the set-point without load and the disturbance response of load can facilitate the control design for the pump operation under a number of load levels. For the injection velocity (IVE) control of an industrial injection molding machine (Tian and Gao 1999; Tan et al. 2001), the open-loop response of the injection velocity to a step change of the valve opening will gradually decrease during the filling process because of the presence of mold cavity pressure that gradually increases until the mold is filled. Modeling the influence of the mold cavity pressure can facilitate advanced control design for maintaining the injection velocity during the filling process to guarantee product consistency and quality.

To facilitate step response identification subject to an inherent-type load disturbance, a piecewise model identification method is therefore presented for practical application. Both the process model and the inherent-type load disturbance model can be simultaneously derived from a step test.

2.4.1 Partition of Step Response Data

When performing a step response test for model identification, if there exists an inherent-type load disturbance, the corresponding turning point in the output response can be observed, as illustrated in Fig. 2.10. This turning point should, therefore, be taken as the starting point (t_d) of the load disturbance response for identification. Practically, it is suggested to take t_d as the time slightly before the observed tuning point to ensure identification effectiveness against measurement noise, in consideration of time delay usually existing in the process response.

Hence, the transient step response data in the time interval $[t_1, t_d)$, as shown in Fig. 2.10, can be used to identify a model of the pure process response. Correspondingly, the step response data in the time interval $[t_d, t_N]$ subtracted by the obtained process model response can be used to identify the load disturbance model, where t_N may be taken as roughly the time after the output response has recovered to a steady state.

2.4 Piecewise Model Identification Under Inherent Load Disturbance

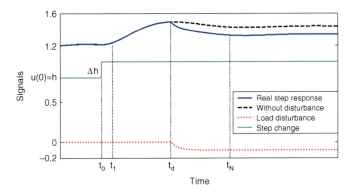

Fig. 2.10 Step response test under an inherent-type load disturbance

In the case where the inherent-type load disturbance occurs at the early stage of a step test, a reasonable partition of the observed transient response data is needed to identify the pure process response model and the load disturbance model separately. That is, the number of transient response data points (M) chosen in terms of the obvious turning point in the step response should be large enough for the identification of the pure process model. In general, it is suggested that M should at least be twice as large as the number of parameters to be estimated in such an LS fitting algorithm to guarantee effectiveness of the model identification. This will be illustrated later with an experimental example in Sect. 2.4.4. To ensure a sufficient number of transient response data points for model fitting, a compromise can be made between the two segments of data chosen, respectively, for the identification of the pure process model and the inherent-type load disturbance model. For the worst case, where the obvious turning point appears at a very early stage of a step test, causing insufficient data for the identification of the pure process model, the standard step test cannot allow for independent identification of the process model against the influence of load disturbance, as discussed in Liu and Gao (2008). A modified step test can, therefore, be performed to apply a robust identification algorithm for the identification of the pure process model, as presented in Sect. 2.3. Then, the inherent-type load disturbance model can be identified using the piecewise identification method to be presented in the next section.

In the case where the overall transient response to a step change can be effectively described by a single model structure or the modeling aims at controller tuning after the inherent-type disturbance response has become steady, the identification effort can be correspondingly reduced to the determination of only a single low-order model for describing the dynamic response characteristics of interest to the control design. In other words, there is no further need to choose a t_d for model identification.

2.4.2 Model Identification

Under a step test, the process output response to the step change and the inherent-type load disturbance can be generally expressed as

$$Y(s) = G(s)U(s) + G_d(s)U_d(s) \tag{2.113}$$

where $G(s)$ and $G_d(s)$ denote, respectively, the process model and the disturbance model, and $U(s)$ and $U_d(s)$ denote, respectively, the step change and the inherent-type load disturbance. For the convenience of model identification, here $U_d(s)$ is normalized as a unity step signal, and correspondingly, the static gain of $G_d(s)$ reflects the magnitude of the inherent-type load disturbance.

As the low-order process models of FOPDT and SOPDT are two of the most widely used in control system design, the following model structures are used here,

$$G(s) = \frac{k_p}{a_2 s^2 + a_1 s + 1} e^{-\theta s} \tag{2.114}$$

$$G_d(s) = \frac{k_d}{\tau_d s + 1} e^{-\theta_d s} \tag{2.115}$$

where k_p denotes the process static gain, θ the process time delay, a_1 and a_2 are positive coefficients that reflect the process dynamic characteristics; k_d denotes the magnitude of the inherent-type load disturbance, θ_d the time delay of the disturbance response, and τ_d the time constant of the disturbance response.

Note that the time delay in (2.115) can be decomposed as $\theta_d = \tilde{\theta}_d + t_d$, where t_d is the observed turning point of the process response to the inherent-type load disturbance. By using a time shift of t_d (i.e., letting $t_d = 0$), $\tilde{\theta}_d$ can be separately derived for fitting the load disturbance response in the time interval $[t_d, t_N]$.

Using the linear superposition principle, the time domain response of (2.113) can be decomposed as

$$y = y_r + y_d \tag{2.116}$$

$$a_2 \ddot{y}_r(t) + a_1 \dot{y}_r(t) + y_r(t) = k_p u(t - \theta) \tag{2.117}$$

$$\tau_d \dot{y}_d(t) + y_d(t) = k_d u_d(t - \theta_d) \tag{2.118}$$

where y_r denotes the pure process response to the input change and y_d is the inherent-type load disturbance response.

Note that, by letting $a_2 = 0$, the expression of (2.114) is reduced to an FOPDT model similar to (2.115). The development of an identification algorithm for obtaining an SOPDT model (Case 1) is described in detail and then the development of another identification algorithm for obtaining an FOPDT model (Case 2) is briefly summarized.

2.4 Piecewise Model Identification Under Inherent Load Disturbance

To allow for practical identification with unsteady or nonzero initial process conditions, here the initial process state for a step test is considered as $u(t_0) = h$, $h \in \mathfrak{R}$, and $y(t_0) \neq 0$, as illustrated in Fig. 2.10. Accordingly, the following identification algorithms are developed based on the raw step response data to facilitate practical applications.

Case 1 SOPDT model. With initial process conditions as described above, by using a time shift of t_0 (i.e., letting $t_0 = 0$), the process input under a step test can be expressed as

$$u(t) = \begin{cases} h, & 0 \le t < \theta; \\ h + \Delta h, & t \ge \theta. \end{cases} \qquad (2.119)$$

where θ denotes the process time delay to be identified.

For $t > \theta$, it can be derived that

$$\int_0^t u(t)dt = \int_0^\theta h\,dt + \int_\theta^t (h + \Delta h)dt = (h + \Delta h)t - \Delta h\theta. \qquad (2.120)$$

Denote the multiple integrals for a time function of $f(t)$ as in (2.74). It can be derived for $t > \theta$ that

$$\int_{[0,t]}^{(2)} u(t) = \frac{h + \Delta h}{2}t^2 - \Delta h t\theta + \frac{\Delta h}{2}\theta^2 \qquad (2.121)$$

$$\int_{[0,t]}^{(3)} u(t) = \frac{h + \Delta h}{6}t^3 - \frac{\Delta h}{2}t^2\theta + \frac{\Delta h}{2}t\theta^2 - \frac{\Delta h}{6}\theta^3 \qquad (2.122)$$

$$\int_{[0,t]}^{(2)} \dot{y}(t) = \int_0^t y(t) - y(0)t \qquad (2.123)$$

$$\int_{[0,t]}^{(3)} \dot{y}(t) = \int_{[0,t]}^{(2)} y(t) - \frac{1}{2}y(0)t^2 \qquad (2.124)$$

$$\int_{[0,t]}^{(3)} \ddot{y}(t) = \int_0^t y(t) - y(0)t - \frac{1}{2}\dot{y}(0)t^2 \qquad (2.125)$$

Note that $y(t) = y_r(t)$ for $t < t_d$ as shown in Fig. 2.10. With the input description of (2.119), the time domain response shown in (2.117) can be equivalently expressed as

$$a_2\ddot{y}_r(t) + a_1\dot{y}_r(t) + y_r(t) = k_p u(t). \qquad (2.126)$$

By triply integrating both sides of (2.126) and rearranging the resulting equation using (2.122), (2.124), and (2.125), one can obtain

$$\psi(t) = \phi^T(t)\gamma \tag{2.127}$$

where

$$\begin{cases} \psi(t) = \int_{[0,t]}^{(3)} y(t), \\ \phi(t) = \left[-\int_0^t y(t), -\int_{[0,t]}^{(2)} y(t), (h+\Delta h)t^3/6, t^2/2, t/2, -\Delta h/6 \right]^T, \\ \gamma = \left[a_2, a_1, k_p, -\Delta h k_p \theta + a_2 \dot{y}(0) + a_1 y(0), \Delta h k_p \theta^2 + 2a_2 y(0), k_p \theta^3 \right]^T. \end{cases} \tag{2.128}$$

Hence, using the step response data in the time interval $[t_1, t_d)$ as shown in Fig. 2.10 (i.e., $t_0 < t_1 < t_2 < \cdots < t_M < t_d$), and letting $\Psi = [\psi(t_1), \psi(t_2), \ldots, \psi(t_M)]^T$ and $\Phi = [\phi(t_1), \phi(t_2), \ldots, \phi(t_M)]^T$, one can obtain a linear LS algorithm for parameter estimation,

$$\Psi = \Phi \gamma. \tag{2.129}$$

Accordingly, the parameter vector can be solved as

$$\gamma = (\Phi^T \Phi)^{-1} \Phi^T \Psi. \tag{2.130}$$

Note that the first two columns of Φ, vectors of single and double integrals for the process output, are obviously independent of the remaining columns of Φ, while the third to fifth columns of Φ are all time vectors with different integral exponents. All columns of Φ are, therefore, linearly independent of each other. Correspondingly, Φ is guaranteed to have full column rank. Owing to the matrix property, $\text{rank}(\Phi^T \Phi) = \text{rank}(\Phi)$, one can ensure that $\Phi^T \Phi$ is invertible for the computation of (2.130), corresponding to a unique solution of γ.

Consequently, the model parameters can be retrieved from (2.130) as

$$\begin{bmatrix} a_2 \\ a_1 \\ k_p \\ \theta \end{bmatrix} = \begin{bmatrix} \gamma(1) \\ \gamma(2) \\ \gamma(3) \\ \sqrt[3]{\gamma(6)/\gamma(3)} \end{bmatrix}. \tag{2.131}$$

Note that the initial state of the process output can be estimated from (2.128) and (2.130) as

2.4 Piecewise Model Identification Under Inherent Load Disturbance

$$y(0) = \frac{y(5) - \Delta h k_p \theta^2}{2a_2} \tag{2.132}$$

$$\dot{y}(0) = \frac{1}{a_2}[y(4) + \Delta h k_p \theta - a_1 y(0)]. \tag{2.133}$$

The above estimation of the initial process state can, therefore, be used to represent the step response test for evaluating the fitting accuracy of the identified model under unsteady or nonzero initial process conditions.

Moreover, in the case of model mismatch in the identification of a higher-order process, enhanced fitting accuracy may be achieved by using the preliminary knowledge of initial process conditions. That is, apart from the other model parameters, the process time delay can alternatively be estimated from the redundant fitting conditions in (2.128) as

$$\theta = \frac{1}{\Delta h k_p}[a_2 \dot{y}(0) + a_1 y(0) - y(4)] \tag{2.134}$$

or

$$\theta = \sqrt{\frac{y(5) - 2a_2 y(0)}{\Delta h k_p}} \tag{2.135}$$

or the mean of the three values computed from (2.131), (2.134), and (2.135). The best choice can be determined from the model fitting of the step response.

Remark 2.6. By doubly integrating both sides of (2.126), an identification algorithm with less computation effort can be obtained in a similar way, but its identification robustness against measurement noise may be inferior to the above algorithm. It can be seen from (2.128) that, rather than using individual output response points measured from the step test, single to triple integrals of the output response points are used for parameter estimation, which reduces the influence of measurement errors according to the statistical averaging principle. ◇

To guarantee the parameter estimation consistency in the presence of measurement noise, a feasible choice of the IV matrix for consistent estimation is given according to Theorem 2.4 as

$$Z = [z_1, z_2, \ldots, z_M]^T \tag{2.136}$$

where $z_i = [1/t_i, 1/t_i^2, t_i^3, t_i^2, t_i, 1]^T$.

Correspondingly, the consistent parameter estimation can be obtained as

$$\gamma = (Z^T \Phi)^{-1} Z^T \Psi \tag{2.137}$$

Case 2 FOPDT model. For the identification of an FOPDT model for the inherent-type load disturbance in the time interval $(t_d, t_N]$, as shown in Fig. 2.10, the load disturbance response can be computed as

$$y_d = y - \hat{y}_r \qquad (2.138)$$

where $\hat{y}_r$ denotes the above identified process model response of the step test.

Because the inherent-type load disturbance is normalized as a unity step signal, it follows by letting $t_d = 0$ in $\theta_d = \tilde{\theta}_d + t_d$ that

$$u_d(t) = \begin{cases} 0, & 0 \leq t < \tilde{\theta}_d; \\ 1, & t \geq \tilde{\theta}_d. \end{cases} \qquad (2.139)$$

With the disturbance description of (2.139), the time domain response shown in (2.118) can be equivalently expressed as

$$\tau_d \dot{y}_d(t) + y_d(t) = k_d u_d(t). \qquad (2.140)$$

By doubly integrating both sides of (2.140) and rearranging the resulting equation using (2.121), (2.123), and (2.139), one can formulate a linear LS fitting in the form of (2.127), for which

$$\begin{cases} \psi(t) = \int_{[0,t]}^{(2)} y_d(t), \\ \phi(t) = \left[-\int_0^t y_d(t), t^2/2, t, 1/2 \right]^T, \\ \gamma = \left[\tau_d, k_d, -k_d \tilde{\theta}_d + \tau_d y_d(0), k_d \tilde{\theta}_d^2 \right]^T. \end{cases} \qquad (2.141)$$

Accordingly, the model parameters can be derived using (2.130) and (2.141) as

$$\begin{bmatrix} \tau_d \\ k_d \\ \tilde{\theta}_d \end{bmatrix} = \begin{bmatrix} \gamma(1) \\ \gamma(2) \\ \sqrt{\gamma(4)/\gamma(2)} \end{bmatrix}, \qquad (2.142)$$

To guarantee consistent estimation against measurement noise, the corresponding IV is suggested to be

$$z_i = [1/t_i, t_i^2, t_i, 1]^T. \qquad (2.143)$$

Remark 2.7. In case different step tests are used to verify the identification effectiveness of the load disturbance model, particularly in the presence of a high

2.4 Piecewise Model Identification Under Inherent Load Disturbance

noise level, different choices of t_d will result in different $\tilde{\theta}_d$. Therefore, verification of the time delay of the load disturbance model should be made in terms of $\theta_d = \tilde{\theta}_d + t_d$. That is, the model verification should be made in terms of a time domain response fitting criterion, $\sum_{k=1}^{N_d} \left[\hat{y}_d(kT_s) - y_d(kT_s) \right]^2 / N_d < \varepsilon$, where $\hat{y}_d(kT_s)$ is the disturbance model response obtained by adding a unity step change to the disturbance model as shown in (2.115), $y_d(kT_s)$ is the disturbance response obtained from (2.138), T_s the sampling period, $N_d T_s$ the transient response time, and ε a user-specified threshold for model fitting. ◇

For the identification of an FOPDT process model in the time interval $[t_1, t_d)$, by doubly integrating both sides of (2.126) with $a_2 = 0$ and rearranging the resulting equation using (2.121) and (2.123), one can formulate a linear LS fitting in the form of (2.127), for which

$$\begin{cases} \psi(t) = \int_{[0,t]}^{(2)} y(t), \\ \phi(t) = \left[-\int_0^t y(t), (h + \Delta h)t^2/2, t, \Delta h/2 \right]^T, \\ \gamma = \left[a_1, k_p, -\Delta h k_p \theta + a_1 y(0), k_p \theta^2 \right]^T. \end{cases} \quad (2.144)$$

Hence, the model parameters can be retrieved the same as (2.142). Also, consistent parameter estimation against measurement noise can be obtained using the IV given in (2.143).

2.4.3 Illustrative Examples

Two examples from the literature are used to demonstrate the effectiveness and accuracy of the above identification algorithms. Example 2.8 is given to demonstrate the good accuracy of these algorithms in identifying first- and second-order processes, together with measurement noise tests for demonstrating identification robustness. Example 2.9 is given to demonstrate the identification effectiveness for higher-order processes. In all tests, the number of transient response data points is taken to be $M = 100$ for computation, together with the sampling period of $T_s = 0.01$ (s). In assessing the model fitting error, the transient response error criterion shown in (2.42) is adopted for reference.

Example 2.8. Consider the second-order process studied by Ahmed et al. (2008),

$$Y(s) = \frac{1.2 e^{-6s}}{9s^2 + 2.4s + 1} U(s) + \frac{e^{-s}}{s+1} U_d(s).$$

Note that an FOPDT model studied by Bi et al. (1999) is used here to represent the dynamics of an inherent-type load disturbance. For illustration, assume that the initial process conditions are $y(t_0) = 1.2$, $\dot{y}(t_0) = 0.01$, and $u(t_0) = 1.0$. A step change of $\Delta h = 0.2$ is added to the set-point at $t_0 = 3$ (s) for model identification, and an inherent-type load disturbance with a magnitude of -0.1 is added through the above FOPDT model to the process at $t = 18$ (s).

For illustration, $t_d = 19$ (s), corresponding to an obvious turning point in the step response as shown in Fig. 2.10, is chosen for the inherent-type load disturbance identification. The transient response data points in the time interval [10, 19](s) are used to derive the process model. The result is presented in Table 2.3, along with the fitting error of the transient response used for identification. It is seen that applying the algorithm to Case 1 results in good accuracy.

Subsequently, using the load disturbance response estimated in the time interval [19, 26](s), that is, subtracting the resulting SOPDT process model response from the step response, an FOPDT disturbance model is, therefore, derived as presented in Table 2.3, which also indicates good accuracy. Note that the static gain of the FOPDT disturbance model is reduced by 10 times, as the magnitude of the load disturbance is normalized to unity for model identification. The fitting error of the load disturbance response is correspondingly evaluated in terms of the transient response in the time interval [19, 26](s).

To demonstrate identification robustness in the presence of measurement noise, assume that a random noise $N(0, \sigma_\zeta^2 = 0.0012\%)$, causing NSR $= 2\%$, is added to the output measurement. By performing 100 Monte Carlo tests in terms of varying the "seed" of noise generator from 1 to 100, the proposed algorithms based on the above time intervals of the step response data give the results presented in Table 2.3, where the model parameters are shown by the mean of the 100 Monte Carlo tests, along with the sample standard deviation in parentheses. The results for NSR $= 10\%$ and 30% are also given in Table 2.3 to show the achievable identification accuracy and robustness.

Note that, for NSR $= 10\%$ and in the absence of load disturbance but with nonzero initial conditions of $y(t_0) = 0.2$, $\dot{y}(t_0) = 0.01$, and $u(t_0) = 0$, Ahmed et al. (2008) gave the identification results based on 100 Monte Carlo tests and using the settling time length of $t_N = 50$ (s) and $M = 500$ for computation,

$$G_m(s) = \frac{1.2(\pm 0.007)e^{-5.95(\pm 0.36)s}}{9.1(\pm 0.8)s^2 + 2.41(\pm 0.16)s + 1}$$

which obviously achieved better accuracy than the results presented in Table 2.3. Nevertheless, if the same time length of the transient response to the step change is used for the identification of the process model, the algorithm for Case 1 using $M = 100$ for computation gives the following results,

$$G_m(s) = \frac{1.2(\pm 0.0003)e^{-6.065(\pm 0.22)s}}{8.98(\pm 0.21)s^2 + 2.39(\pm 0.09)s + 1}.$$

2.4 Piecewise Model Identification Under Inherent Load Disturbance

Table 2.3 Step response identification for Example 2.8 under different measurement noise levels

NSR	Process model	err	Load disturbance model	err
0	$\dfrac{1.2000e^{-6.0466 s}}{8.9999 s^2 + 2.4001 s + 1}$	1.2×10^{-5}	$\dfrac{0.1000 e^{-1.0192 s}}{1.0001 s + 1}$	2.65×10^{-7}
2%	$\dfrac{1.2007(\pm 0.013)e^{-6.0458(\pm 0.26)s}}{8.9946(\pm 0.35)s^2 + 2.4212(\pm 0.36)s + 1}$	5.52×10^{-5}	$\dfrac{0.1019(\pm 0.015)e^{-1.0306(\pm 0.13)s}}{1.0011(\pm 0.36)s + 1}$	2.25×10^{-6}
10%	$\dfrac{1.2184(\pm 0.047)e^{-6.2913(\pm 0.89)s}}{8.4433(\pm 1.25)s^2 + 2.9135(\pm 1.38)s + 1}$	3.81×10^{-3}	$\dfrac{0.1123(\pm 0.046)e^{-0.9539(\pm 0.24)s}}{1.1894(\pm 0.79)s + 1}$	6.04×10^{-5}
30%	$\dfrac{1.2531(\pm 0.055)e^{-7.0594(\pm 0.75)s}}{8.4726(\pm 2.09)s^2 + 3.8558(\pm 1.71)s + 1}$	4.02×10^{-2}	$\dfrac{0.1405(\pm 0.052)e^{-0.8529(\pm 0.19)s}}{1.2489(\pm 0.71)s + 1}$	9.01×10^{-4}

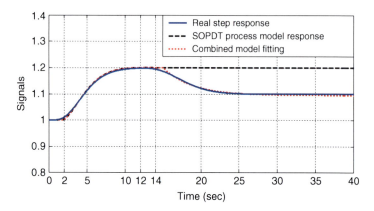

Fig. 2.11 Step response fitting for Example 2.9

It is seen that the proposed algorithm can also give very good accuracy in the absence of load disturbance but with nonzero initial process conditions, thus demonstrating good applicability for various step tests in practice.

Example 2.9. Consider the high order process studied by Wang et al. (2001),

$$Y(s) = \frac{1}{(s+1)^5} U(s) + \frac{1}{(s+1)^8} U_d(s).$$

Note that the eighth-order process model studied by Bi et al. (1999) is used to describe the dynamics of an inherent-type load disturbance. Assume that the initial process conditions are $y(t_0) = 1.0$, $\dot{y}(t_0) = 0$, and $u(t_0) = 1.0$. A step change of $\Delta h = 0.2$ is added to the set-point at $t_0 = 0$ (s) and an inherent-type load disturbance with a magnitude of -0.1 is added to the process at $t = 10$ (s). The corresponding step response is shown in Fig. 2.11.

Using the process transient response in the time interval $[2, 12]$(s), the identification algorithm for Case 1 gives an SOPDT process model, $G_m = 0.9976 e^{-2.06s}/(3.4825 s^2 + 3.1803 s + 1)$, corresponding to $err = 1.74 \times 10^{-3}$ in the time interval $[0, 12]$(s). Note that further enhanced fitting accuracy can be obtained using the known initial process conditions and the corresponding identification formula of (2.134) as $G_m = 0.9976 e^{-1.72s}/(3.4825 s^2 + 3.1803 s + 1)$, which corresponds to $err = 3.42 \times 10^{-4}$. Then, using the load disturbance response estimated in the time interval $[14, 30]$(s) with $t_d = 12$ (s), which is slightly ahead of the observed turning point at $t = 14$ (s) (i.e., subtracting the above SOPDT model response from the step response), an FOPDT disturbance model is, therefore, derived as $G_d = 0.1045 e^{-3.03s}/(3.7663 s + 1)$, corresponding to $err = 2.14 \times 10^{-4}$ for the transient response in the time interval $[12, 30]$(s). The combined step

2.4 Piecewise Model Identification Under Inherent Load Disturbance

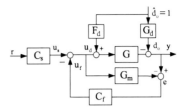

Fig. 2.12 A block diagram of the 2DOF IMC plus feedforward control scheme for inherent-type disturbance rejection

response of the above SOPDT process model and the FOPDT disturbance model for representing the step response are also plotted in Fig. 2.11 for comparison, which demonstrates good fitting accuracy.

To demonstrate identification robustness to different choices of the time length of the transient response associated with the turning point chosen for the piecewise model identification, Table 2.4 lists the identification results when different time lengths of the transient response are used. The results indicate that the proposed piecewise model identification method is not sensitive to different choices of turning point and transient response data for model fitting.

To demonstrate the achievable control effect based on the identified models, a two-degrees-of-freedom (2DOF) internal model control (IMC) plus feedforward control scheme is correspondingly proposed for load disturbance rejection, which is shown in Fig. 2.12, where C_s is the set-point tracking controller, C_f is the closed-loop feedback controller, and F_d is the feedforward controller. According to the aforementioned normalization of the inherent-type load disturbance as a unity step signal, the process input can be derived as

$$U(s) = C_s(s)R(s) + C_f(s)E(s) + \frac{F_d(s)}{s} \tag{2.145}$$

where $R(s)$ and $E(s)$ denote the Laplace transforms of r and e, respectively.

To combat the inherent-type load disturbance, it is ideal to let

$$G_d(s) = G(s)F_d(s). \tag{2.146}$$

Substituting (2.114) and (2.115) into (2.146), one obtains

$$F_{d-\text{ideal}}(s) = \frac{k_d(a_2 s^2 + a_1 s + 1)}{k_p(\tau_d s + 1)} e^{-(\theta_d - \theta)s}. \tag{2.147}$$

It is seen that the ideal feedforward controller is not physically proper. A first-order low-pass filter, $1/(\lambda_d s + 1)$, is therefore introduced for implementation, resulting in

$$F_d(s) = \frac{k_d(a_2 s^2 + a_1 s + 1)}{k_p(\tau_d s + 1)(\lambda_d s + 1)} e^{-(\theta_d - \theta)s} \tag{2.148}$$

Table 2.4 Identification results for Example 2.9 using different time lengths of the transient response

Time interval (s)	Process model	err	Time interval (s)	Inherent disturbance model	err
[2, 12]	$\dfrac{0.9976e^{-1.72s}}{3.4825s^2 + 3.1803s + 1}$	3.42×10^{-4}	[14, 30]	$\dfrac{0.1045e^{-3.03s}}{3.7663s + 1}$	2.14×10^{-4}
[2, 13]	$\dfrac{0.9981e^{-1.74s}}{3.4021s^2 + 3.1862s + 1}$	3.22×10^{-4}	[15, 30]	$\dfrac{0.1027e^{-3.39s}}{3.2801s + 1}$	2.01×10^{-4}
[2, 11]	$\dfrac{0.9963e^{-1.68s}}{3.6297s^2 + 3.1551s + 1}$	4.28×10^{-4}	[16, 30]	$\dfrac{0.1015e^{-3.69s}}{2.9068s + 1}$	1.91×10^{-4}

2.4 Piecewise Model Identification Under Inherent Load Disturbance

where the filter time constant is suggested to be $\lambda_d = (0.1 \sim 1.0)\tau_d$ to adjust the feedforward control action. When λ_d is zero, the feedforward control is ideal, but tends to be sensitive to model mismatch and measurement noise. On the contrary, increasing λ_d mitigates the control action so that unexpected model mismatch can be accommodated.

In the nominal case, i.e., when $G = G_m$ and so is for the feedforward control, there is an "open-loop" control for the set-point tracking. Based on the process model of (2.114), the desired system transfer function according to the IMC theory (Morari and Zafiriou 1989) to be presented in Sect. 7.3 is in the form of

$$T_r(s) = \frac{e^{-\theta s}}{(\lambda_s s + 1)^2} \qquad (2.149)$$

where λ_s is an adjustable time constant for obtaining desirable set-point tracking performance. Using the nominal relationship of $T_r(s) = G(s)C_s(s)$, the set-point tracking controller can be inversely derived as

$$C_s = \frac{a_2 s^2 + a_1 s + 1}{k_p(\lambda_s s + 1)^2}. \qquad (2.150)$$

The closed-loop structure set between the process input and output is used for eliminating the output error in the presence of model mismatch and other process uncertainties. Note that if there exists model mismatch in the feedforward control, the redundant control signal (Δu_d) may be viewed as a load disturbance (denoted as d_i) that enters into the process input. If $G = G_m$, the transfer function from u_d to u_f can be derived as

$$H_{d_i}(s) = G(s)C_f(s) \qquad (2.151)$$

which is exactly equivalent to the nominal "open-loop" system transfer function for set-point tracking. According to the optimal load disturbance rejection strategy developed in Liu et al. (2005), the desired closed-loop transfer function is in the form of

$$T_f(s) = H_{d_i}(s) = \frac{e^{-\theta s}}{(\lambda_f s + 1)^2} \qquad (2.152)$$

where λ_f is an adjustable time constant used to tune the closed-loop performance for disturbance rejection.

Substituting (2.114) and (2.152) into (2.151) yields

$$C_f = \frac{a_2 s^2 + a_1 s + 1}{k_p(\lambda_f s + 1)^2} \qquad (2.153)$$

which is similar to the form of C_s in (2.150). However, the tuning of C_f is subject to a stability constraint of the closed-loop structure. According to the IMC theory (Morari and Zafiriou 1989), tuning C_f aims at a good trade-off between the disturbance rejection performance of the closed-loop structure and its robust stability, i.e.,

$$|\Delta_m(s)T_f(s)| + |W(s)[1 - T_f(s)]| < 1 \quad (2.154)$$

where $\Delta_m(s) = [G(s) - G_m(s)]/G(s)$ defines the process multiplicative uncertainty, and $W(s)$ is a weighting function of the closed-loop sensitivity function, $S_f(s) = 1 - T_f(s)$. For instance, $W(s) = 1/s$ can be taken for a step change in the load disturbance that enters into the process input. Decreasing λ_f can improve the disturbance rejection performance of the closed-loop structure, but degrades its robust stability in the presence of process uncertainties. In contrast, increasing λ_f can strengthen the robust stability of the closed-loop structure, but in exchange for a degradation in its disturbance rejection performance.

According to the small gain theorem (Zhou et al. 1996), the closed-loop structure for disturbance rejection holds robust stability if and only if

$$\|\Delta_m(s)T_f(s)\|_\infty < 1 \quad (2.155)$$

Substituting (2.152) into (2.155), one can obtain the robust stability constraint for tuning λ_f,

$$\sqrt{\lambda_f^2 \omega^2 + 1} > \Delta_m(j\omega), \quad \forall \omega \geq 0 \quad (2.156)$$

which can be intuitively checked by observing whether the magnitude plot of the left-hand side of (2.156) is larger than the right-hand side for $\omega \in [0, \infty)$. Therefore, given an upper bound of Δ_m as usually specified in practice (e.g., the maximal range of the model parameters), the admissible tuning range of λ_f can be numerically ascertained from (2.156).

The above control scheme is applied in comparison with the standard 2DOF IMC control structure (see Fig. 8.1). For the above initial process conditions and the set-point change with an inherent-type load disturbance as in the above step test, the control results are shown in Fig. 2.13. It is seen that given the same control parameters (i.e., $\lambda_s = 0.5$ and $\lambda_f = 1.0$), the standard 2DOF IMC control structure that uses the identified SOPDT process model gives similar set-point tracking performance but better load disturbance rejection compared to the structure that uses the real process model. To obtain the same disturbance rejection performance, the tuning parameter λ_f of C_f should be increased to 1.6 if the identified SOPDT process model is used, which, in fact, facilitates better closed-loop stability according to the IMC theory. Note that, based on the identified FOPDT disturbance model, the proposed control scheme with $\lambda_s = 0.5$, $\lambda_f = 1.6$, and $\lambda_d = 2.0$ obviously gives an

2.4 Piecewise Model Identification Under Inherent Load Disturbance

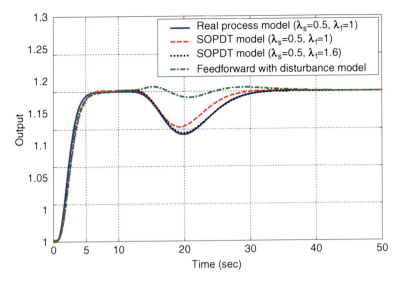

Fig. 2.13 A comparison of control effects for Example 2.9

improved load disturbance response. This goes to demonstrate that identifying both the model of the process and the model of the inherent-type load disturbance from a step test facilitates advanced control design and performance.

2.4.4 Application to Injection Velocity Control

Consider the injection velocity control of an industrial reciprocating screw injection molding machine (Chen-Hsong, model no. JM88-MKIII-C), the schematic of which is shown in Fig. 2.14. The injection velocity is regulated by a proportional valve (4WRP-10–63S-1X/G24Z24/W), denoted as PV1 in Fig. 2.14, and is measured with an MTS Temposonics III displacement and velocity transducer (RH-N-0200M-RG0–1-V0–1). A 16-bit data acquisition card (PCL-816) from ADVANTECH is used for analog-to-digital (A/D) and digital-to-analog (D/A) conversions. For illustration, a rectangular mold of length 150(mm), width 200(mm) and thickness 2(mm), corresponding to a weight of 27.8(g), is used for the injection molding experiments. The plastic material is higher-density polyethylene (HDPE).

For an open-loop step test in which the valve opening of PV1 suddenly goes through a change from 0% to 40% (as a percentage), the injection velocity response measured in a sampling period of 0.005(s) is plotted in Fig. 2.15. It is seen that the injection velocity response starting from an initial value of about −2(m/s) has an obvious overshoot, and then drops to a roughly steady value of about 28(m/s) in the time interval [0.1, 0.2](s). Because of the presence of mold cavity pressure, which increases gradually during the filling process, the injection velocity decreases

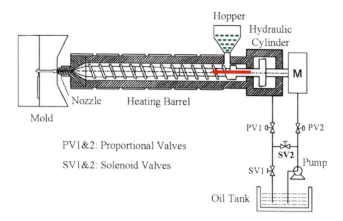

Fig. 2.14 Schematic of an injection molding machine

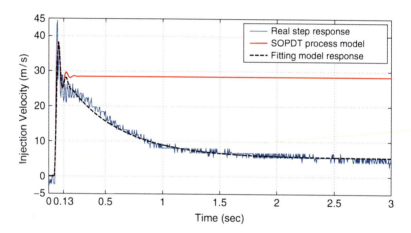

Fig. 2.15 Open-loop step test for injection velocity response

continuously until the end of the filling process. In view of the obvious overshoot in the step response, an SOPDT model structure is adopted for identification. It can be seen from the corresponding LS fitting algorithm shown in (2.128) that the number of parameters to be estimated ($\dim(\gamma)$) is 6. To ensure identification effectiveness, the turning point is chosen as $t = 0.13$ (s) for determining the starting point of the load disturbance response. Accordingly, the transient response data points in the time interval $[0.05, 0.13]$(s) are used to identify the dynamic response of injection velocity to the valve opening, corresponding to $M = 17$, which complies with the guideline given in Sect. 2.4.1 for a piecewise model identification.

By using a low-pass third-order Butterworth filter with a cutoff frequency of $f_c = 20$(HZ) for predenoising the measured data, the identification algorithm for Case 1 gives the SOPDT model,

2.4 Piecewise Model Identification Under Inherent Load Disturbance

$$G_m = \frac{71.1198e^{-0.05s}}{1.0792 \times 10^{-4}s^2 + 6.6425 \times 10^{-3}s + 1}.$$

Then, by letting $t_d = 0.12$ (s) in consideration of the process time delay, the load disturbance response in the time interval [0.13, 2.5](s), estimated by subtracting the above SOPDT model response from the step response, is used for modeling the influence of mold cavity pressure. An FOPDT disturbance model is, therefore, derived as

$$G_d = \frac{23.025e^{-0.13s}}{0.4726s + 1}.$$

For comparison, the SOPDT model response and the combined model response for representing the step test are also plotted in Fig. 2.15, which shows good fitting effect.

Based on the identified models, the 2DOF IMC plus feedforward control scheme shown in Fig. 2.12 is applied for closed-loop control of the injection velocity (IVE) at a desired constant value, IVE $= 30$ (m/s),for the filling process of injection molding. For implementation, the control sampling period is taken as $T_s = 0.01$ (s), and the one-step backward discretization operator, $\dot{e}(kT_s) = [e(kT_s) - e((k-1)T_s)]/T_s$, is used for computational simplicity. Experimental results based on the tuning parameters of $\lambda_s = 0.2$, $\lambda_f = 0.5$, and $\lambda_d = 0.1$ are plotted in Fig. 2.16. It is seen from Fig. 2.16a that fast set-point tracking without overshoot is obtained using the proposed control method. A slight drop of the injection velocity during the time interval [0.6, 1](s) is due to model mismatch in describing the influence of the mold cavity pressure, which, however, is quickly compensated by the feedback controller (C_f). Figure 2.16b shows the valve opening (as a percentage) and the controller outputs. For comparison, the control result obtained by using the standard 2DOF IMC scheme is also plotted in Fig. 2.16a, which indicates that the set-point tracking is obviously slower and the filling time is longer, when the feedforward control based on the identified disturbance model is not used.

To further demonstrate the achievable control performance, assume that the injection velocity profile shown in Fig. 2.17a (dash line) is prescribed for molding a product of the convex shape. The proposed control scheme with the above tuning parameters gives the result shown in Fig. 2.17a, b. Note that, for the set-point change from 30 to 40(m/s) or the reverse, the output of the set-point tracking controller C_s is implemented one sampling step ahead of the set-point change to compensate for the identified time delay of the injection velocity response. Correspondingly, the tuning parameters of C_s and C_f are adjusted to $\lambda_s = 0.05$ and $\lambda_f = 0.1$ to deal with the step change, so that an implementation constraint of moderating a step change of the set-point into a ramp type (Tian and Gao 1999; Tan et al. 2001) is no longer necessary.

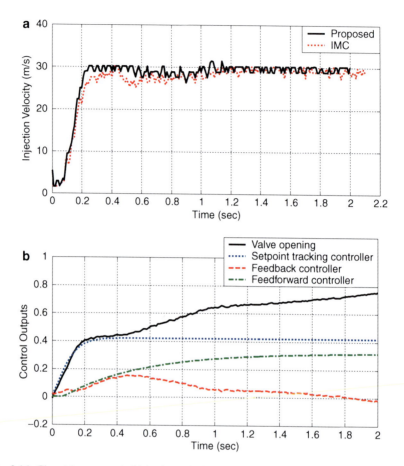

Fig. 2.16 Closed-loop control of injection velocity during the filling process

2.5 Model Identification from a Closed-Loop Step Test

Generally, a closed-loop step test is performed by adding a step change to the setpoint and then collecting the transient output response data for model identification. The closed-loop test is based on a simple low-order controller like PID for the closed-loop configuration, as shown in Fig. 2.18, where G denotes the process to be identified and C is the closed-loop controller, r denotes the set-point, u the process output, and y the process output.

With a prescribed closed-loop controller, one can ensure the closed-loop system has entered into a steady state before adding a step change to the set-point for a closed-loop step test (Jin et al. 1998; Li et al. 2005). This can facilitate model identification around the set-point for online autotuning, in contrast with an open-loop step test.

2.5 Model Identification from a Closed-Loop Step Test

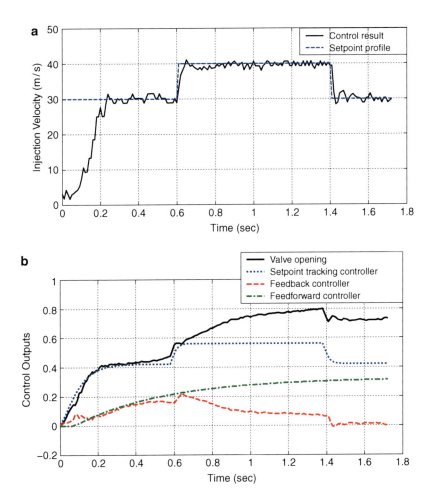

Fig. 2.17 Tracking an injection velocity profile during the filling process

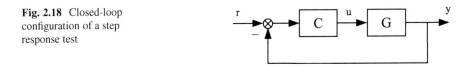

Fig. 2.18 Closed-loop configuration of a step response test

2.5.1 Frequency Response Estimation

A typical closed-loop step response is shown in Fig. 2.19. It is seen that the Fourier transform of the closed-loop step response does not exist because $\Delta y(t) \neq 0$ for $t \to \infty$, where $\Delta y(t) = y(t) - y(t_0)$, $y(t_0)$ denotes the initial steady output value corresponding to the set-point value. However, by letting $s = \alpha + j\omega$ in the Laplace

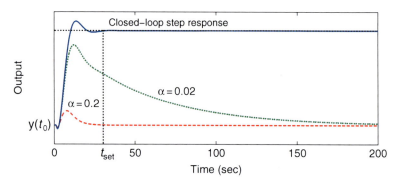

Fig. 2.19 Illustration of choosing α in a closed-loop step test

transform for the step response,

$$\Delta Y(s) = \int_0^\infty \Delta y(t) e^{-st} dt \qquad (2.157)$$

one can obtain

$$\Delta Y(\alpha + j\omega) = \int_0^\infty \left[\Delta y(t) e^{-\alpha t}\right] e^{-j\omega t} dt. \qquad (2.158)$$

Therefore, similar to the frequency response estimation introduced in Sect. 2.2.1, by regarding α as a damping factor of the closed-loop step response in the Laplace transform, one can compute $\Delta Y(\alpha + j\omega)$ from a finite number of step response points as

$$\Delta Y(\alpha + j\omega) = \int_0^{t_N} \left[\Delta y(t) e^{-\alpha t}\right] e^{-j\omega t} dt. \qquad (2.159)$$

For a closed-loop step test in a steady state initially, i.e., $y(t) = r(t) = c$ for $t \leq t_0$, where c is a constant and t_0 is the moment for the step test, one can formulate the step change of the set-point by using a time shift of t_0 (i.e., letting $t_0 = 0$) as

$$\Delta r(t) = \begin{cases} 0, & t \leq 0; \\ h, & t > 0. \end{cases} \qquad (2.160)$$

where h is the magnitude of the step change. Its Laplace transform for $s = \alpha + j\omega$ with $\alpha > 0$ can be explicitly derived as

$$\Delta R(\alpha + j\omega) = \int_0^\infty h e^{-(\alpha + j\omega)t} dt = \frac{h}{\alpha + j\omega}. \qquad (2.161)$$

2.5 Model Identification from a Closed-Loop Step Test

Correspondingly, the closed-loop frequency response can be derived using (2.159) and (2.161) as

$$T(\alpha + j\omega) = \frac{\alpha + j\omega}{h}\Delta Y(\alpha + j\omega), \quad \alpha > 0. \tag{2.162}$$

Note that the guideline for choosing α and t_N is the same as that presented in (2.8) and (2.10).

Also, in the presence of measurement noise, enhanced identification robustness can be obtained by computing the closed-loop frequency response through

$$T(\alpha + j\omega) = \frac{\frac{\Delta Y(\alpha+j\omega)}{\alpha+j\omega}}{\frac{\Delta R(\alpha+j\omega)}{\alpha+j\omega}} = \frac{(\alpha+j\omega)^2}{h}\int_0^{t_N}\left[\int_0^t \Delta y(\tau)d\tau\right]e^{-\alpha t}e^{-j\omega t}dt. \tag{2.163}$$

According to the definition of the nth order derivative for a complex function of $F(s)$ with respect to the Laplace operator, s, as shown in (2.15), it follows from (2.157) and (2.162) that

$$T^{(1)}(s) = \frac{1}{h}\int_0^\infty (1-st)\Delta y(t)e^{-st}dt \tag{2.164}$$

$$T^{(2)}(s) = \frac{1}{h}\int_0^\infty t(st-2)\Delta y(t)e^{-st}dt \tag{2.165}$$

Hence, by letting $s = \alpha$ and choosing α according to the guideline for computing (2.159), the time integral in (2.164) and (2.165) can be numerically computed using a finite number of step response points. The corresponding time lengths t_N can be determined using the numerical constraints,

$$|(1-\alpha t_N)\Delta y(t_N)|\,T_s e^{-\alpha t_N} < \delta \tag{2.166}$$

$$|t_N(\alpha t_N - 2)\Delta y(t_N)|\,T_s e^{-\alpha t_N} < \delta. \tag{2.167}$$

Note that the closed-loop transfer function can be derived from Fig. 2.18 as

$$T(s) = \frac{G(s)C(s)}{1+G(s)C(s)}. \tag{2.168}$$

Hence, with a known form of $C(s)$ for closed-loop step test, the process frequency response can be inversely derived from (2.168) as

$$G(s) = \frac{T(s)}{C(s)[1-T(s)]}. \tag{2.169}$$

Accordingly, the first derivative of $G(s)$ can be derived from (2.169) as

$$G^{(1)} = \frac{T^{(1)}C + C^{(1)}T(T-1)}{C^2(1-T)^2} \qquad (2.170)$$

and the second derivative of $G(s)$ can be derived as

$$G^{(2)} = \frac{CT^{(2)} + 2C^{(1)}T^{(1)}T + C^{(2)}T(T-1)}{C^2(1-T)^2}$$
$$- \frac{2[CT^{(1)} + C^{(1)}T(T-1)][C^{(1)}(1-T) - CT^{(1)}]}{C^3(1-T)^3}. \qquad (2.171)$$

For example, assume that a conventional PID controller is used in the closed-loop configuration as shown in Fig. 2.18 for the step test, which is generally in the form of

$$C(s) = k_C + \frac{1}{\tau_I s} + \frac{\tau_D s}{\tau_F s + 1} \qquad (2.172)$$

where k_C denotes the controller gain, τ_I the integral constant, τ_D the derivative constant, and τ_F a filter constant that is usually taken as $\tau_F = (0.01 \sim 0.1)\tau_D$ for implementation. It can be derived that

$$C^{(1)}(s) = -\frac{1}{\tau_I s^2} + \frac{\tau_D}{(\tau_F s + 1)^2} \qquad (2.173)$$

$$C^{(2)}(s) = \frac{2}{\tau_I s^3} - \frac{2\tau_D \tau_F}{(\tau_F s + 1)^3} \qquad (2.174)$$

Hence, by substituting $s = \alpha + j\omega_k$ ($k = 1, 2, \ldots, M$) into (2.169), where M is the number of representative frequency response points in a user-specified frequency range for identification, the process frequency response, $G(\alpha + j\omega_k)$, can thus be estimated for model fitting.

2.5.2 Model Identification

Based on the above frequency response estimation of the process, the identification algorithms presented in Sect. 2.2 can be used for model identification. If the model structure adopted matches the process, all the identification algorithms presented in Sect. 2.2 will give good fitting accuracy. For simplicity, the analytical identification algorithms – Algorithm-SS-I and Algorithm-SS-II – presented in Sects. 2.2.2 and 2.2.3 are preferred for application. When there exists a model mismatch, in particular for the identification of higher-order processes, the Algorithm-SS-IV

2.5 Model Identification from a Closed-Loop Step Test

presented in Sect. 2.2.5 is preferred to improve the fitting accuracy over a specified frequency range of interest to control design and online autotuning.

In practical applications, low-order process models of FOPDT and SOPDT are most widely used for online controller autotuning such as the PID autotuning for improving system performance of load disturbance rejection. When there exists a model mismatch, especially in the identification of higher-order processes, the following identification algorithm for improving fitting accuracy over the low frequency, which is similar to Algorithm-SS-IV presented in Sect. 2.2.5, can be used.

Substituting the process frequency response in (2.169) with $s = \alpha + j\omega_k$ ($k = 1, 2, \ldots, M$) and the process model of FOPDT in (2.13) or SOPDT in (2.27) into the left-hand side of (2.43) and letting it equal zero, a weighted LS solution for parameter estimation can be derived accordingly as

$$\gamma = (\bar{\Phi}^T W \bar{\Phi})^{-1} \bar{\Phi}^T W \bar{\Psi} \tag{2.175}$$

where $W = diag\{\rho_1, \ldots, \rho_M, \rho_1, \ldots, \rho_M\}$,

$$\bar{\Psi} = \begin{bmatrix} \text{Re}[\Psi] \\ \text{Im}[\Psi] \end{bmatrix}, \quad \bar{\Phi} = \begin{bmatrix} \text{Re}[\Phi] \\ \text{Im}[\Phi] \end{bmatrix},$$

where $\Psi = [\psi(\alpha + j\omega_1), \psi(\alpha + j\omega_2), \ldots, \psi(\alpha + j\omega_M)]^T$, $\Phi = [\phi(\alpha + j\omega_1), \phi(\alpha + j\omega_2), \ldots, \phi(\alpha + j\omega_M)]^T$. For obtaining an FOPDT model shown in (2.13), $\psi(\alpha + j\omega_k)$, $\phi(\alpha + j\omega_k)$, and γ assume the form of

$$\begin{cases} \psi(\alpha + j\omega_k) = G_1(\alpha + j\omega_k), \\ \phi(\alpha + j\omega_k) = [-(\alpha + j\omega_k)G_1(\alpha + j\omega_k), e^{-(\alpha + j\omega_k)\theta}]^T, \\ \gamma = [\tau_p, k_p]^T. \end{cases} \tag{2.176}$$

For obtaining an SOPDT model shown in (2.27), $\psi(\alpha + j\omega_k)$, $\phi(\alpha + j\omega_k)$, and γ assume the form of

$$\begin{cases} \psi(\alpha + j\omega_k) = G_2(\alpha + j\omega_k), \\ \phi(\alpha + j\omega_k) = [-(\alpha + j\omega_k)G_2(\alpha + j\omega_k), \\ \quad -(\alpha + j\omega_k)^2 G_2(\alpha + j\omega_k), e^{-(\alpha + j\omega_k)\theta}]^T, \\ \gamma = [a_1, a_2, k_p]^T. \end{cases} \tag{2.177}$$

It can easily be verified that all the columns of $\bar{\Phi}$ are linearly independent of each other, so $(\bar{\Phi}^T W \bar{\Phi})^{-1}$ is guaranteed to be nonsingular for computation. Accordingly, there exists a unique solution for parameter estimation. For implementation, $M \in [10, 50]$ is suggested for a good trade-off between the fitting accuracy and computational efficiency.

Accordingly, the FOPDT model parameters can be retrieved as

$$\begin{cases} \tau_p = \gamma(1) \\ k_p = \gamma(2) \end{cases} \qquad (2.178)$$

and the SOPDT model parameters can be retrieved as

$$\begin{cases} a_1 = \gamma(1) \\ a_2 = \gamma(2) \\ k_p = \gamma(3) \end{cases} \qquad (2.179)$$

Note that a preliminary value of the process time delay (θ) is needed to derive the remaining model parameters from (2.175). In fact, an approximate estimate of the process time delay can be obtained from the step response or a time delay model derived from Algorithm-SS-I or Algorithm-SS-II, which may be used as an initial value in a one-dimensional search that uses a convergent condition for fitting the closed-loop step response,

$$err = \frac{1}{N_s} \sum_{k=1}^{N_s} \left[\Delta y(kT_s) - \Delta \widehat{y}(kT_s) \right]^2 < \varepsilon \qquad (2.180)$$

where $\Delta y(kT_s)$ and $\Delta \widehat{y}(kT_s)$ denote, respectively, the process and model outputs to the closed-loop step test, and $N_s T_s$ is the settling time. Therefore, given a specified threshold of ε, a suitable solution of the model parameters can be obtained by monotonically varying θ in a possible range for computation. The one-dimensional search step size may be taken as a small multiple of the sampling period for implementation. The optimal fitting can be determined by deriving such a model that yields the smallest value of err.

It is obvious that minimization of the time domain fitting condition in (2.180) can lead to the minimum of the frequency domain objective function in (2.43), if the model structure adopted matches the process. When there exists a model mismatch, the combination of (2.43) and (2.180) for deriving the model parameters can guarantee a good compromise between the time domain response fitting and the frequency response fitting, but may not realize the global minimization of (2.43) or (2.180).

2.5.3 Illustrative Examples

Three examples studied in the recent literature are used here to illustrate the effectiveness of the presented algorithms for closed-loop frequency response estimation and model identification. In all identification tests, the sampling period is taken to be $T_s = 0.01$ (s) for implementation.

2.5 Model Identification from a Closed-Loop Step Test

Example 2.10. Consider the FOPDT stable process studied by Padhy and Majhi 2006,

$$G(s) = \frac{1}{s+1} e^{-0.5s}.$$

Based on a closed-loop relay feedback test with two P controllers, Padhy and Majhi (2006) derived an FOPDT model, $G_m = 1.0e^{-0.5s}/(0.9996s + 1)$. For illustration, the unity feedback control structure with a proportional controller of $k_c = 3.5$, which is equivalent to that in Padhy and Majhi (2006), is used here for a closed-loop step test. By adding a step change with a magnitude of $h = 0.5$ to the set-point, the closed-loop step response is like the one shown in Fig. 2.19. According to the guidelines given in (2.8) and (2.10), $\alpha = 0.1$ and $t_N = 200$ (s) are chosen to apply Algorithm-SS-I, resulting in an FOPDT model, $G_m = 1.0e^{-0.5002s}/(0.9998s + 1)$, which indicates good accuracy for no model mismatch or measurement noise. Also, the identification algorithm given in Sect. 2.5.2 based on the frequency response estimation of $G(0.1 + j\omega_k)$, where $\omega_k = k\omega_{rc}/10$, $k = 0, 1, 2, \ldots, 10$, and $\omega_{rc} = 3.6718$ (rad/s) that is estimated from the above FOPDT model, can result in the exact process model.

To demonstrate identification robustness with regard to different choices of α ($\alpha = 0.1, 0.2, 0.5$), the results of frequency response estimation for the choices of $\alpha = 0.1, 0.2, 0.5$ under a variety of noise levels (NSR = 0, 5, 20%) are listed in Table 2.5 based on a number of closed-loop step tests ($N = 1, 10, 20$). Note that different step tests are simulated by randomly varying the "seed" of the noise generator, and correspondingly, the results for $N = 10$ and $N = 20$ are denoted by a mean value along with the sample standard deviation in parentheses. It is seen that precise frequency response estimation is obtained with regard to different choices of α for NSR = 0. Given these noise levels, taking a smaller value of α facilitates better identification robustness, and so does using a larger number of tests, which is in accordance with the statistical averaging principle.

To demonstrate the consistent parameter estimation of the identification algorithm given in Sect. 2.5.2, which uses the time domain fitting criterion of (2.180), assume that a random noise $N(0, \sigma_\xi^2 = 0.94\%)$, causing NSR = 20%, is added to the process output measurement, which is then used for feedback control. By performing 100 Monte Carlo tests in terms of varying the "seed" of the noise generator from 1 to 100, the identified results are obtained as

$$G_m = \frac{1.0003(\pm 0.006)}{0.9904(\pm 0.25)s + 1} e^{-0.5012(\pm 0.042)s}$$

where the model parameters are shown by the mean of 100 Monte Carlo tests, along with the sample standard deviation in parentheses. It is thus demonstrated that this identification algorithm results in consistent parameter estimation under the severe noise level.

Table 2.5 Frequency response estimation for Example 2.10 under different measurement noise levels

Process Frequency Response	Number of Tests	NSR = 0	Relative Error	NSR = 5%	Relative Error	NSR = 20%	Relative Error
$G(\alpha = 0.1) = 0.8648$	$N = 1$	0.8648	0	0.8645	0.03%	0.8634	0.15%
	$N = 10$	0.8648	0	0.8647($\pm$ 0.0017)	0.01%	0.8637($\pm$ 0.0069)	0.12%
$G(\alpha = 0.2) = 0.754$	$N = 1$	0.754	0	0.7544	0.05%	0.7518	0.29%
	$N = 10$	0.754	0	0.7538($\pm$ 0.0019)	0.04%	0.7554($\pm$ 0.0078)	0.19%
$G(\alpha = 0.5) = 0.5192$	$N = 1$	0.5192	0	0.5204	0.24%	0.5242	0.97%
	$N = 10$	0.5192	0	0.5185($\pm$ 0.0021)	0.12%	0.5167($\pm$ 0.0085)	0.48%
	$N = 20$	0.5192	0	0.5189($\pm$ 0.0019)	0.06%	0.5179($\pm$ 0.0078)	0.24%

2.5 Model Identification from a Closed-Loop Step Test

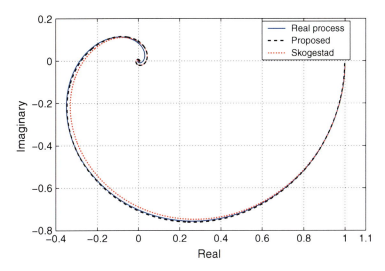

Fig. 2.20 Nyquist plots of identified SOPDT models for Example 2.11

Example 2.11. Consider a high-order process studied by Skogestad (2003),

$$G(s) = \frac{-s+1}{(6s+1)(2s+1)^2} e^{-s}.$$

By using an analytical model reduction method, Skogestad (2003) gave an SOPDT model, $G_m = 1.0e^{-3s}/[(6s+1)(3s+1)]$, to tune a PID controller, $C = (1+1/6s)(3s+1)/(0.03s+1)$, for the closed-loop control. By performing a closed-loop step test with a unity step change using the above PID controller, Algorithm-SS-I taking $\alpha = 0.01$ and $t_N = 1{,}500$ (s) gives an FOPDT model, $G_{m-1} = 0.9992e^{-6.6268s}/(5.2148s+1)$, corresponding to the closed-loop transient response error $err = 6.39 \times 10^{-2}$ in the time interval [0, 30]s. In contrast, Algorithm-SS-II based on $\alpha = 0.01, 0.11, 0.21, 0.31, 0.41$ and $t_N = 1{,}200$ (s) gives an SOPDT model, $G_{m-2} = 0.9107e^{-3.18s}/(23.6983s^2 + 5.4853s + 1)$, corresponding to $err = 3.09 \times 10^{-2}$. To improve model fitting accuracy over the low frequency range concerned for controller tuning, the identification algorithm given in Sect. 2.5.2 based on the frequency response estimation of $G(0.01 + j\omega_k)$, where $\omega_k = k\omega_{rc}/10$, $k = 0, 1, 2, \ldots, 10$, and $\omega_{rc} = 0.3188$(rad/s) that is estimated from the above FOPDT model, gives $G_{m-2} = 0.9986e^{-3.04s}/(18.6132s^2 + 8.7483s + 1)$, corresponding to $err = 6.04 \times 10^{-5}$. Note that the SOPDT model in Skogestad (2003) corresponds to $err = 1.98 \times 10^{-4}$. The Nyquist plots of the identified SOPDT models are shown in Fig. 2.20. It is seen that the Nyquist curve of the proposed SOPDT model almost entirely overlaps that of the real process, especially in the low frequency range.

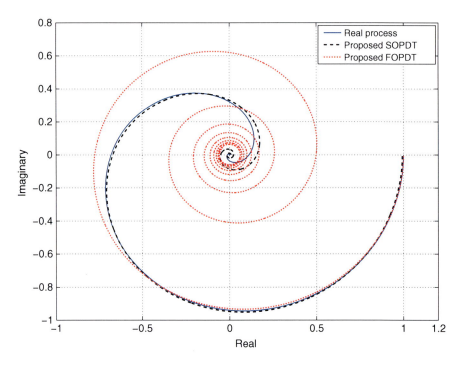

Fig. 2.21 Nyquist plots of identified FOPDT and SOPDT models for Example 2.12

Example 2.12 Consider another high-order process studied by Huang et al. (2005),

$$G(s) = \frac{1}{(4s^2 + 2.8s + 1)(s + 1)^2} e^{-2.2s}.$$

Based on a closed-loop relay feedback test for frequency response estimation, Huang et al. (2005) gave a PID tuning, $C = 0.314(1 + 1/2.59s + 2.103s)/(0.1s + 1)$. By performing a closed-loop step test with a unity step change using the above PID controller, Algorithm-SS-I taking $\alpha = 0.05$ and $t_N = 500$ (s) gives an FOPDT model, $G_{m-1} = 1.0005e^{-5.2748s}/(1.754s + 1)$, corresponding to $err = 1.09 \times 10^{-3}$ in terms of the closed-loop transient response in the time interval [0, 50]s. Based on the frequency response estimation of $G(0.05 + j\omega_k)$, where $\omega_k = k\omega_{rc}/10$, $k = 0, 1, 2, \ldots, 10$., and $\omega_{rc} = 0.4657$ (rad/s) that is estimated from the above FOPDT model, the identification algorithm given in Sect. 2.5.2 gives an SOPDT model, $G_{m-2} = 0.9934e^{-3.54s}/(5.5069s^2 + 3.4095s + 1)$, corresponding to $err = 1.51 \times 10^{-5}$. The Nyquist plots of the identified models are shown in Fig. 2.21. It is seen that the Nyquist curves of both the FOPDT and SOPDT models almost entirely overlap that of the real process in the low frequency range, while the SOPDT model gives improved fitting at higher frequencies.

2.6 Summary

The step response test has been widely practiced for model identification in various process industries. There are typically two types of step response test – open-loop and closed-loop. The guidelines for implementing these step tests have been presented in Sects. 2.2 and 2.5.

For the use of an open-loop step test, a frequency response estimation algorithm has been presented (Liu and Gao 2010a) by introducing a damping factor to the step response for the realization of Laplace transform. Based on the proposed frequency response estimation, four model identification algorithms (Algorithm-SS-I, Algorithm-SS-II, Algorithm-SS-III and Algorithm-SS-IV) have been presented to meet different requirements of identification accuracy and computation effort in practical applications. Note that these algorithms can reduce n-fold multiple integrals to a single integral for the identification of an nth order model ($n \geq 2$), while being insensitive to the time length of the step response, compared to most existing step identification methods using multiple integrals to establish time domain LS fitting. Moreover, Algorithm-SS-IV is able to obtain the optimal fitting accuracy for a given model structure over a specified frequency range of interest for control design. To deal with model mismatch in the identification of high-order processes, as encountered in engineering practice, a guideline for model structure selection to realize the optimal fitting has been given accordingly. Four examples from the literature have demonstrated that the proposed frequency response estimation algorithm maintains good robustness with respect to different choices of the damping factor for computation. All of these identification algorithms have been demonstrated to result in good accuracy if the model structure adopted matches the process, together with good identification robustness against measurement noise.

For practical application of step response identification subject to unsteady or unknown initial process conditions, or unexpected load disturbance, a modified implementation of the step response test is suggested in Sect. 2.3. Correspondingly, a robust step identification method (Liu and Gao 2008) has been detailed for obtaining the widely used low-order process models of FOPDT and SOPDT with or without zeroes, which can also be transparently extended to obtain a specific higher-order model. A distinguishing feature of the modified step test is that the process transient response from removing the step change is also included for model identification. It is, therefore, demonstrated that the conventional step test is not suitable for parameter estimation in the presence of unsteady initial process conditions or unexpected load disturbance. Multiple integrals of the differential system equation have been used to establish a linear regression, through which individual output response points that are likely to be subject to measurement noise are avoided for deriving the model parameters, thereby facilitating identification robustness. With a user-specified threshold for convergence, the Maclaurin series approximation for representing the influence of nonzero initial process conditions and load disturbance can be conveniently implemented in the proposed identification algorithms. As nonzero initial process conditions are considered, the proposed identification method can easily be repeated online, if unexpected load disturbance impedes convergence. A practical IV method has also been given to guarantee consistent estimation against measurement noise. The applications to three examples from the references have demonstrated that the proposed identification method can give good accuracy and robustness.

For step response identification subject to inherent-type load disturbance in practical applications, a piecewise model identification method (Liu et al. 2010) has been presented in Sect. 2.4 that allows the use of raw step-response data for simultaneously identifying the pure process model and the inherent-type load disturbance model from a step test, based on an intuitive partition of the transient step response data. Identification algorithms have been detailed for obtaining the widely used low-order models of FOPDT and SOPDT. Note that these identification algorithms can be transparently extended to obtain higher-order models for describing more complex dynamic response characteristics of the process and the inherent-type load disturbance. Two illustrative examples have demonstrated that good identification accuracy and robustness can be obtained through the use of the piecewise model identification method. Accordingly, a 2DOF IMC plus feedforward control scheme has been proposed for improving process operation against inherent-type load disturbance. The application to the velocity control of injection molding has demonstrated the effectiveness of the piecewise model identification method and the corresponding control method.

For the use of a closed-loop step test, a frequency response estimation algorithm similar to that given in Sect. 2.2.1 is presented to estimate the closed-loop frequency response (Liu and Gao 2010b). Accordingly, the process frequency response can be analytically derived from the closed-loop frequency response with knowledge of the loop controller. Based on the estimated frequency response points of the process, the identification algorithms given in Sect. 2.2 for open-loop step test can also be applied for the identification of models of any specific order. To cope with model mismatch in the identification of high-order processes, an identification algorithm is detailed for improving the fitting accuracy over a specified frequency range in terms of an FOPDT or SOPDT model structure that is most widely used for online controller tuning. This algorithm is based on using a weighted LS fitting for multiple frequency response points estimated in the specified frequency range. The applications to three examples from the literature have demonstrated the effectiveness and merits of the proposed closed-loop step response estimation and identification algorithms.

References

Ahmed S, Huang B, Shah SL (2006) Parameter and delay estimation of continuous-time models using a linear filter. J Process Control 16:323–331

Ahmed S, Huang B, Shah SL (2007) Novel identification method from step response. Control Eng Pract 15:545–556

References

Ahmed S, Huang B, Shah SL (2008) Identification from step responses with transient initial conditions. J Process Control 18:121–130

Åström KJ, Hägglund T (1995) PID controller: theory, design, and tuning, 2nd edn. ISA Society of America, Research Triangle Park

Åström KJ, Hägglund T (2005) Advanced PID control. ISA Society of America, Research Triangle Park

Bi Q, Cai WJ, Lee EL, Wang QG, Hang CC, Zhang Y (1999) Robust identification of first-order plus dead-time model from step response. Control Eng Pract 7:71–77

Goodwin GC, Graebe L, Salgado ME (2001) Control system design. Prentice Hall, Upper Saddle River

Huang HP, Jeng JC, Luo KY (2005) Auto-tune system using single-run relay feedback test and model-based controller design. J Process Control 15:713–727

Huang HP, Lee MW, Chen CL (2001) A system of procedures for identification of simple models using transient step response. Ind Eng Chem Res 40:1903–1915

Hwang SH, Lai ST (2004) Use of two-stage least-squares algorithms for identification of continuous systems with time delay based on pulse responses. Automatica 40:1561–1568

Jin HP, Heung IP, Lee I-B (1998) Closed-loop on-line process identification using a proportional controller. Chem Eng Sci 53(9):1713–1724

Li SY, Cai WJ, Mei H, Xiong Q (2005) Robust decentralized parameter identification for two-input two-output process from closed-loop step responses. Control Eng Pract 13:519–531

Liu M, Wang QG, Huang B, Hang CC (2007) Improved identification of continuous-time delay processes from piecewise step tests. J Process Control 17:51–57

Liu T, Gao F (2008) Robust step-like identification of low-order process model under nonzero initial conditions and disturbance. IEEE Trans Autom Control 53:2690–2695

Liu T, Gao F (2010a) A frequency domain step response identification method for continuous-time processes with time delay. J Process Control 20(7):800–809

Liu T, Gao F (2010b) Closed-loop step response identification of integrating and unstable processes. Chem Eng Sci 65(10):2884–2895

Liu T, Zhang WD, Gu DY (2005) Analytical two-degree-of-freedom tuning design for open-loop unstable processes with time delay. J Process Control 15:559–572

Liu T, Zhou F, Yang Y, Gao F (2010) Step response identification under inherent-type load disturbance with application to injection molding. Ind Eng Chem Res 49(22):11572–11581

Luyben WL (1990) Process modeling, simulation, and control for chemical engineers. McGraw Hill, New York

Morari M, Zafiriou E (1989) Robust process control. Prentice Hall, Englewood Cliff

Ogunnaike BA, Ray WH (1994) Process dynamics, modeling, and control. Oxford University Press, New York

Padhy PK, Majhi S (2006) Relay based PI-PD design for stable and unstable FOPDT processes. Comput Chem Eng 30:790–796

Pintelon R, Schoukens J (2001) System identification: a frequency domain approach. IEEE Press, New York

Rake H (1980) Step response and frequency response methods. Automatica 16:519–526

Rangaiah GP, Krishnaswamy PR (1996) Estimating second-order dead time parameters from underdamped process transients. Chem Eng Sci 51:1149–1155

Seborg DE, Edgar TF, Mellichamp DA (2004) Process dynamics and control, 2nd edn. Wiley, Hoboken

Shinskey FG (1996) Process control system, 4th edn. McGraw Hill, New York

Skogestad S (2003) Simple analytical rules for model reduction and PID controller tuning. J Process Control 13:291–309

Söderström T, Stoica P (1989) System identification. Prentice Hall, New York

Tan KK, Huang SN, Jiang X (2001) Adaptive control of ram velocity for the injection moulding machine. IEEE Trans Control Syst Technol 9:663–671

Tian YC, Gao F (1999) Injection velocity control of thermoplastic injection molding via a double controller scheme. Ind Eng Chem Res 38:3396–3406

Wang QG, Guo X, Zhang Y (2001a) Direct identification of continuous time delay systems from step responses. J Process Control 11:531–542

Wang QG, Zhang Y (2001b) Robust identification of continuous systems with dead-time from step responses. Automatica 37:377–390

Wang QG, Lee TH, Lin C (2003) Relay feedback: analysis, identification and control. Springer, London

Wang QG, Liu M, Hang CC, Tang W (2006) Robust process identification from relay tests in the presence of nonzero initial conditions and disturbance. Ind Eng Chem Res 47:4063–4070

Wang QG, Liu M, Hang CC, Zhang Y, Zheng WX (2008) Integral identification of continuous-time delay systems in the presence of unknown initial conditions and disturbances from step tests. Ind Eng Chem Res 47:4929–4936

Yu CC (2006) Autotuning of PID controllers: a relay feedback approach, 2nd edn. Springer, London

Zheng WX (1996) Identification of closed-loop systems with low-order controllers. Automatica 32(12):1753–1757

Zhou KM, Doyle JC, Glover K (1996) Robust and optimal control. Prentice Hall, Englewood Cliff

Chapter 3
Step Response Identification of Integrating and Unstable Processes

3.1 Practical Implementation Issues

Some industrial and chemical processes, e.g., heating boilers and continuous-stirred-tank-reactors (CSTRs), show integrating or unstable dynamics to the input change, such as a step or ramp signal that often occurs in process operations. Such a dynamic response can be intuitively observed from the phenomenon: Given a step change or a load disturbance to the process input, the process output will not recover the previous state or move into another steady state, but keep on varying beyond the output limit of system operation. A monotonically increasing or decreasing response case is usually indicative of a process of an integrating type, and in contrast, other cases with irregular behavior that exceed the output limits are classified as unstable processes.

For safety and economic reasons, integrating and unstable processes are usually not allowed to be operated in an open-loop manner (Morari and Zafiriou 1989; Shinskey 1996; Seborg et al. 2004). A closed-loop step test is therefore necessary for model identification. For the convenience of implementation, a low-order controller such as a P-, PI-, or PID-type controller may be initially used for closed-loop stabilization, based on a preliminary knowledge or experiences of the process operation.

Generally, a closed-loop identification test is performed in terms of the unity feedback control structure as shown in Fig. 3.1, where G denotes the process to be identified and C is the closed-loop controller, r denotes the set-point, u the process input, y the process output, and d indicates external signal such as load disturbance entering into the process.

There are two alternative choices for adding an excitation signal to conduct a closed-loop identification test (Söderström and Stoica 1989; Pintelon and Schoukens 2001), i.e., adding it to the process input or the set-point. For the

Fig. 3.1 The closed-loop configuration for an identification test

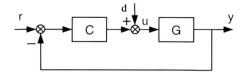

identification of an integrating or unstable process, it is suggested to add the excitation signal to the set-point of the closed-loop system. The reason is given below:

From Fig. 3.1, the transfer function relationship relating r and d to y and u can be derived as

$$\begin{bmatrix} y \\ u \end{bmatrix} = \begin{bmatrix} \dfrac{G(s)C(s)}{1+G(s)C(s)} & \dfrac{G(s)}{1+G(s)C(s)} \\ \dfrac{C(s)}{1+G(s)C(s)} & \dfrac{1}{1+G(s)C(s)} \end{bmatrix} \begin{bmatrix} r \\ d \end{bmatrix}. \tag{3.1}$$

Denote the closed-loop transfer function by

$$T(s) = \frac{G(s)C(s)}{1+G(s)C(s)}. \tag{3.2}$$

Substituting (3.2) into (3.1) yields

$$\begin{bmatrix} y \\ u \end{bmatrix} = \begin{bmatrix} T(s) & G(s)[1-T(s)] \\ C(s)[1-T(s)] & 1-T(s) \end{bmatrix} \begin{bmatrix} r \\ d \end{bmatrix}. \tag{3.3}$$

Note that the closed-loop system holds internal stability if and only if all the four transfer functions shown in (3.3) are maintained to be stable, according to the robust control theory (Zhou et al. 1996). If $C(s)$ is tuned to keep $T(s)$ stable, the closed-loop sensitivity function, $S(s) = 1 - T(s)$, will satisfy the asymptotic tracking constraint,

$$\lim_{s \to 0} S(s) = 0 \tag{3.4}$$

which implies that, besides a P-type controller, a low-order controller of PI- or PID-type can also maintain $C(s)[1 - T(s)]$ stable, if such a controller has been tuned to keep $T(s)$ stable.

For a stable process, it can be seen from (3.3) and (3.4) that the excitation signal for a closed-loop identification test in terms of a low-order controller of P-, PI-, or PID- type may be added to the set-point or the process input (through d), without affecting the closed-loop system stability.

However, for an integrating or unstable process, even if $T(s)$ is maintained to be stable, the closed-loop system may not hold internal stability, because the transfer function from d to y, $G(s)[1 - T(s)]$, may not be stable due to the unstable pole(s) in $G(s)$. That is, the closed-loop controller is required not only to keep $T(s)$ stable but also to let $1 - T(s)$ contain the corresponding zero(s) to cancel out the unstable pole(s) in $G(s)$, if the excitation signal is added to the process input (through d). Since the process is in fact to be identified, such excitation will evidently bring more difficulties to the tuning of $C(s)$ for the closed-loop stabilization, compared to adding the excitation signal to the set-point.

Based on the above analysis, it is suggested to add the excitation signal of a step change to the set-point for closed-loop identification of an integrating or unstable process, for the convenience of tuning the closed-loop controller for stabilization while preventing the process output from drifting too far way from its working range.

For integrating processes, it has been clarified that using the standard relay feedback control structure can undoubtedly result in steady oscillation for model identification (Liu and Gao 2008). In fact, the standard relay feedback structure can be reduced to the unity feedback control structure with a P-type controller, by letting the relay hysteresis and the negative switch of the relay output to be zero. Therefore, it is suggested for simplicity to use a P-type controller for stabilization to conduct a closed-loop step test. The gain of the P-type controller may be initially taken as a small value, and then be gradually increased online to facilitate the observation of the transient step response for model identification.

For unstable processes, the above closed-loop step test in terms of a P-type controller may be used for identifying such processes with no or small time delays, in consideration of that the standard relay feedback structure can only result in steady oscillation for such cases (Tan et al. 1998; Liu and Gao 2008). For unstable processes with longer time delays, knowledge or experience of the process operation should be used to tune a low-order controller (e.g., PI or PID) for closed-loop stabilization, including the trial-and-error methods (Seborg et al. 2004; Åström and Hägglund 2005). Accordingly, a closed-loop step test can be conducted as above, together with a precaution of the controller windup to guarantee identification effectiveness.

It should be noted that the closed-loop step test may be performed iteratively to facilitate a better observation of the closed-loop transient response for model identification, based on an improved tuning of the closed-loop controller with a model identified from the initial closed-loop step test.

Since the pioneering work of using a simple P-type controller to stabilize an unstable process for a closed-loop step test (Deshpande 1980), only a few closed-loop step response identification methods have been reported for integrating and unstable processes, in particular in the presence of time delay. Following an early idea of considering the process response delay as the closed-loop system delay, Ananth and Chidambaram (1999) proposed an FOPDT model identification method for unstable processes by assigning the dominant closed-loop system pole for model fitting. This work was further extended in the references (Sree and Chidambaram 2006; Cheres 2006) by using the peak values of a closed-loop step response.

Using the so-called pseudo-derivative feedback (PDF) for closed-loop stabilization, Paraskevopoulos et al. (2004) developed another FOPDT identification algorithm by establishing amplitude fitting conditions of a closed-loop step response. For integrating processes, closed-loop step identification algorithms with a P-type controller for stabilization were simultaneously developed to obtain an FOPDT or SOPDT model for model-based controller tuning on-line (Sung and Lee 1996; Kwak et al. 1997; Jin et al. 1998).

Note that monotonically increasing or decreasing the process output in a certain range around the set-point may be allowed for operating some industrial integrating processes involving the regulations of temperature, pressure, and liquid level etc. An open-loop step response test is therefore possible for model identification around the set-point, for the control design of set-point tracking and load disturbance rejection (Luyben 1990). Moreover, a stable model structure with a large time constant may be considered for describing the dynamic response characteristics of a slowly integrating process, to facilitate the closed-loop controller tuning in practice (Åström and Hägglund 1995).

Accordingly, an open-loop step response identification method (Liu et al. 2009) is first presented here for identifying an integrating process. Then, a closed-loop step response identification method (Liu and Gao 2010) is presented for the identification of integrating and unstable processes.

3.2 Open-Loop Step Response Identification of Integrating Processes

Generally, zero initial process state or nonzero steady state is required for the implementation of an open-loop step test. To identify an integrating process, zero initial process state is preferred for the test, since a nonzero steady state of such a process usually is not available in practice. In the case where the process dynamic response characteristics around a nonzero set-point value is required for control design or on-line controller tuning, a reasonable zero initialization of the nonzero initial process conditions is needed to perform an open-loop step test, based on a preliminary knowledge of the process operation conditions. This will be illustrated by the experimental example in Sect. 3.4.

To describe the fundamental dynamic response characteristics of an integrating process, a low-order model structure of FOPDT or SOPDT is mostly adopted for control-oriented identification in engineering practice (Luyben 1990; Åström and Hägglund 1995), not only for simplicity but also for effectiveness to accommodate process uncertainties. Correspondingly, a low-order model identification method is presented in the following subsection.

3.2 Open-Loop Step Response Identification of Integrating Processes

3.2.1 Low-Order Model Identification

A low-order integrating process model is usually expressed as

$$G_m = \frac{k_p e^{-\theta s}}{s(\tau_p s + 1)} \tag{3.5}$$

where k_p is a proportional gain, θ is the process time delay, and τ_p is a time constant reflecting the process inertial characteristics. Note that if $\tau_p \to 0$, the above model is reduced to describe a purely integrating process.

In an open-loop step test, by letting $u = h$ for $t > 0$ ($h \in \mathfrak{R}$) and $u = 0$ for $t \leq 0$ in terms of a zero initialization of the process state, the time domain response of an integrating process described in (3.5) can be derived as

$$\begin{cases} y(t) = 0, & 0 < t \leq \theta; \\ \tau_p \ddot{y}(t) + \dot{y}(t) = k_p h(t - \theta), & t > \theta. \end{cases} \tag{3.6}$$

Triply integrating both sides of (3.6) for $t > \theta$ yields

$$\int_0^t \int_0^{\tau_2} y(\tau_1) d\tau_1 d\tau_2 = -\tau_p \int_0^t y(\tau) d\tau + \frac{1}{6} k_p h(t - \theta)^3 \tag{3.7}$$

which can be rewritten as

$$\psi(t) = \phi^{\mathrm{T}}(t)\gamma \tag{3.8}$$

where $\psi(t) = \int_0^t \int_0^{\tau_2} y(\tau_1) d\tau_1 d\tau_2$, $\phi(t) = [ht^3/6, -ht^2/2, ht/2, -h/6, -\int_0^t y(\tau) d\tau]^{\mathrm{T}}$, and $\gamma = [k_p, k_p\theta, k_p\theta^2, k_p\theta^3, \tau_p]^{\mathrm{T}}$.

In view of that $\psi(t) = y(t) = 0$ for $t \leq \theta$, it is suggested to choose the time sequence, t_i ($i = 1, 2, \ldots, N$), with a limitation of $\theta \leq t_1 < t_2 < \cdots < t_N$ for model fitting. In fact, the process time delay may not be explicitly known beforehand, especially in the presence of measurement noise. It is therefore suggested to choose t_1 slightly larger than the observed response delay from the step test.

Then by letting $\Psi = [\psi(t_1), \psi(t_2), \cdots, \psi(t_N)]^{\mathrm{T}}$ and $\Phi = [\phi(t_1), \phi(t_2), \cdots, \phi(t_N)]^{\mathrm{T}}$, an LS algorithm for parameter estimation can be established as

$$\gamma = (\Phi^{\mathrm{T}}\Phi)^{-1}\Phi\Psi \tag{3.9}$$

It can easily be verified that all the columns of Φ are linearly independent with each other, such that Φ is guaranteed non-singular for computing (3.9). Accordingly, there exists a unique solution of γ for the time sequence.

For practical application, the number of data points (N) for fitting may be taken in a range of 50–200, for a good trade-off between the fitting accuracy and computational efficiency.

In the sequel, the process time constant can be retrieved as

$$\tau_p = \gamma(5) \tag{3.10}$$

Note that, besides $k_p = \gamma(1)$ and $k_p\theta = \gamma(2)$, there exist two redundant fitting conditions, $k_p\theta^2 = \gamma(3)$ and $k_p\theta^3 = \gamma(4)$, which can certainly be satisfied if the model structure matches the process. To procure the fitting accuracy for a higher order process, one may take the natural logarithm for both sides of these fitting conditions to obtain

$$\begin{bmatrix} 1 & 0 \\ 1 & 1 \\ 1 & 2 \\ 1 & 3 \end{bmatrix} \begin{bmatrix} \ln k_p \\ \ln \theta \end{bmatrix} = \begin{bmatrix} \ln \gamma(1) \\ \ln \gamma(2) \\ \ln \gamma(3) \\ \ln \gamma(4) \end{bmatrix} \tag{3.11}$$

Thus, k_p and θ can be retrieved from (3.11) through an LS fitting algorithm.

By letting $\tau_p = 0$, the above algorithm can be transparently applied to identify an FOPDT integrating model, $G_m = k_p e^{-\theta s}/s$, which, however, is inferior to an SOPDT model for describing a higher order process in practice, due to the deficiency of representing the inertial characteristics in the transient response, as demonstrated by Liu and Gao (2008).

Besides, it should be noted that the step response of an integrating process will increase infinitely as $t \to \infty$, and therefore, should be limited in an admissible range around the set-point. Obviously, a larger range of the step response corresponding to a longer time sequence can facilitate a better identification accuracy.

In the presence of measurement noise, $\zeta(t)$, during the step response test, there is $\hat{y}(t) = y(t) + \zeta(t)$, where $\hat{y}(t)$ denotes the measured output and $y(t)$ is the true output. It follows from (3.7) that

$$\Psi = \Phi\gamma + \upsilon \tag{3.12}$$

where $\upsilon = [\delta(t_1), \delta(t_2), \ldots, \delta(t_N)]^T$ and $\delta(t) = \int_0^t \int_0^{\tau_2} \zeta(\tau_1) d\tau_1 d\tau_2 + \tau_p \int_0^t \zeta(\tau) d\tau$. In view of that Φ is now correlated with υ, the LS estimation given in (3.9) may not be consistent. A solution to this problem is the use of the IV method (Söderström and Stoica 1989). A feasible choice of such an IV matrix is given in the following theorem:

Theorem 3.1. *With* $Z = [z_1, z_2, \ldots, z_N]^T$ *chosen as the IV matrix where* $z_i = [1/t_i^2, 1/t_i, 1, t_i, t_i^2]^T$ *for* $i = 1, 2, \ldots, N$, *which satisfies the two limiting conditions:* (1). *the inverse of* $\lim_{N \to \infty} (Z^T \Phi)/N$ *exists*; (2). $\lim_{N \to \infty} (Z^T \upsilon)/N = 0$, *a consistent estimation is given by* $\gamma = (Z^T \Phi)^{-1} Z \Psi$.

3.2 Open-Loop Step Response Identification of Integrating Processes

Proof. The two limiting conditions in Theorem 3.1 are sufficient for consistent estimation (Söderström and Stoica 1989). It is therefore required to demonstrate that the proposed IV matrix satisfies them. For the first condition, one can derive that

$$\lim_{N\to\infty} \frac{Z^T \Phi}{N}$$

$$= \begin{bmatrix} \frac{h}{6}\lim_{N\to\infty}\frac{1}{N}\sum_{i=1}^{N}t_i - \frac{h}{2} & 0 & 0 & 0 \\ \frac{h}{6}\lim_{N\to\infty}\frac{1}{N}\sum_{i=1}^{N}t_i^2 - \frac{h}{2}\lim_{N\to\infty}\frac{1}{N}\sum_{i=1}^{N}t_i & \frac{h}{2} & 0 & 0 \\ \frac{h}{6}\lim_{N\to\infty}\frac{1}{N}\sum_{i=1}^{N}t_i^3 - \frac{h}{2}\lim_{N\to\infty}\frac{1}{N}\sum_{i=1}^{N}t_i^2 & \frac{h}{2}\lim_{N\to\infty}\frac{1}{N}\sum_{i=1}^{N}t_i - \frac{h}{6} & -\lim_{N\to\infty}\frac{1}{N}\int_0^{t_i}y(\tau)d\tau \\ \frac{h}{6}\lim_{N\to\infty}\frac{1}{N}\sum_{i=1}^{N}t_i^4 - \frac{h}{2}\lim_{N\to\infty}\frac{1}{N}\sum_{i=1}^{N}t_i^3 & \frac{h}{2}\lim_{N\to\infty}\frac{1}{N}\sum_{i=1}^{N}t_i^2 - \frac{h}{6}\lim_{N\to\infty}\frac{1}{N}\sum_{i=1}^{N}t_i & -\lim_{N\to\infty}\frac{t_i}{N}\int_0^{t_i}y(\tau)d\tau \\ \frac{h}{6}\lim_{N\to\infty}\frac{1}{N}\sum_{i=1}^{N}t_i^5 - \frac{h}{2}\lim_{N\to\infty}\frac{1}{N}\sum_{i=1}^{N}t_i^4 & \frac{h}{2}\lim_{N\to\infty}\frac{1}{N}\sum_{i=1}^{N}t_i^3 - \frac{h}{6}\lim_{N\to\infty}\frac{1}{N}\sum_{i=1}^{N}t_i^2 & -\lim_{N\to\infty}\frac{t_i^2}{N}\int_0^{t_i}y(\tau)d\tau \end{bmatrix}$$

(3.13)

Note that

$$\lim_{N\to\infty} \frac{t_N}{N} = c \tag{3.14}$$

where $c \in \Re_+$. For instance, $c = T_s$ is for the case where $t_i = t_1 + T_s(i-1)$ ($i = 1, 2, \ldots, N$) and T_s is the sampling period for identification. Therefore, all rows or columns in the square matrix of (3.13) are nonzero vectors and linearly independent of each other, which guarantee

$$\det\left(\lim_{N\to\infty} \frac{Z^T \Phi}{N}\right) \neq 0. \tag{3.15}$$

Hence, the first condition in Theorem 3.1 is satisfied.

For the second condition, it can be derived that

$$\lim_{N\to\infty} \frac{Z^T \sigma}{N} = \left[0\; 0\; \lim_{N\to\infty}\frac{1}{N}\sum_{i=1}^{N}\delta(t_i)\; \lim_{N\to\infty}\frac{1}{N}\sum_{i=1}^{N}t_i\delta(t_i)\; \lim_{N\to\infty}\frac{1}{N}\sum_{i=1}^{N}t_i^2\delta(t_i)\right]^T. \tag{3.16}$$

Note that the measurement noise sequence, $\zeta(t_i)$ ($i = 1, 2, \ldots, N$), may be viewed as white noise for $N \to \infty$, and correspondingly, its mean tends to zero. So, it follows from $\delta(t) = \int_0^t \int_0^{\tau_2} \zeta(\tau_1) d\tau_1 d\tau_2 + \tau_p \int_0^t \zeta(\tau) d\tau$ that

$$\lim_{N\to\infty} \sum_{i=1}^{N} \delta(t_i) = 0 \tag{3.17}$$

which indicates that $\delta(t_i)$ $(i = 1, 2, \ldots, N)$ is also a random noise sequence for $N \to \infty$ and thus is uncorrelated with the time sequence, t_i or t_i^2 $(i = 1, 2, \ldots, N)$. Hence, the last two elements in the vector at the right-hand side of (3.16) also become zero for $N \to \infty$. This completes the proof. □

3.2.2 Illustrative Examples

Two examples from the existing literature are used here to illustrate the effectiveness of the above identification algorithm. Example 3.1 is given to demonstrate good accuracy of this algorithm for identifying a low-order integrating process in terms of the exact model structure, together with measurement noise tests to demonstrate identification robustness. Example 3.2 is used to show the effectiveness of this algorithm for identifying higher order integrating processes. In all the step tests, the sampling period is taken as $T_s = 0.01(s)$ for computation.

Example 3.1. Consider the SOPDT integrating process studied by Kaya (2006),

$$G = \frac{e^{-10s}}{s(20s + 1)}$$

Based on a unity step response test with zero initial process conditions, the proposed identification algorithm using the measured output data in the time interval [8, 30](s) gives a SOPDT model, $G_m = 1.0000e^{-10.01s}/s(20.0000s + 1)$, which demonstrates good accuracy.

Now, suppose that a random noise $N(0, \sigma_\xi^2 = 0.0127)$ is added to the process output measurement, causing NSR = 5%. Using the measured output data in the above time interval, the proposed IV identification method results in a model, $G_m = 1.0362e^{-9.8205s}/s(21.1995s + 1)$. According to the time domain fitting criterion shown in (2.42), the identified model corresponds to $err = 1.15 \times 10^{-3}$, thus demonstrating good identification robustness. Then assume that the noise level is increased to NSR = 30% by introducing a random noise $N(0, \sigma_\xi^2 = 0.572)$, the proposed IV identification method based on the above time interval of output data gives a model, $G_m = 1.0541e^{-9.2686s}/s(22.2082s + 1)$, corresponding to $err = 2.82 \times 10^{-2}$.

Example 3.2. Consider the high-order integrating process studied by Ingimundarson and Hägglund (2001),

$$G = \frac{(-s + 1)e^{-5s}}{s(s + 1)^5}$$

Based on the unity step response data, Ingimundarson and Hägglund (2001) gave an FOPDT model, $G_m = 1.0000e^{-11.0000s}/s$. Using the measured data in the time

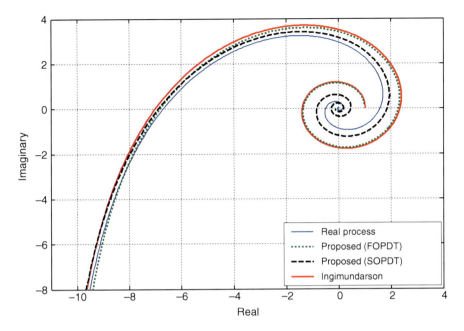

Fig. 3.2 Nyquist plots of identified models for Example 3.2

interval [10, 30](s) of the unity step response, the proposed identification algorithm gives a SOPDT model,

$$G_m = \frac{0.9936 e^{-9.4145s}}{s(1.7364s + 1)}$$

For comparison, by letting $\tau_p = 0$ the proposed algorithm can also be used to obtain an FOPDT model, $G_m = 0.9811 e^{-10.9029s}/s$. The Nyquist plots of these models are shown in Fig. 3.2. It can be seen that the proposed FOPDT model exhibits improved fitting over that of Ingimundarson and Hägglund (2001), while the proposed SOPDT model obtains apparently better fitting.

3.3 Closed-Loop Step Response Identification of Integrating and Unstable Processes

Based on a closed-loop step response test, as shown in Fig. 3.1, the closed-loop frequency response can be computed using the frequency response estimation algorithm presented in Sect. 2.5.1. Also, the process frequency response can be inversely derived from the closed-loop transfer function, with a known form of the closed-loop controller. The corresponding computation formulae are the same as given in Sect. 2.5.1 and thus are omitted.

3.3.1 The FOPDT/SOPDT Model for Integrating Processes

Based on the frequency response estimation of the process, identification algorithms are analytically developed here for obtaining a low-order integrating model of FOPDT or SOPDT. Consider the following SOPDT integrating model,

$$G_I = \frac{k_p e^{-\theta s}}{s(\tau_p s + 1)} \tag{3.18}$$

where k_p denotes a proportional gain, θ is the process time delay, and τ_p is a time constant reflecting the process inertial characteristics. Note that letting $\tau_p = 0$ in (3.18) leads to an FOPDT model.

Regarding $s \in \Re_+$ and taking the natural logarithm for both sides of (3.18), one can obtain

$$\ln[G_I(s)] = \ln(k_p) - \ln(s) - \ln(\tau_p s + 1) - \theta s. \tag{3.19}$$

According to the definition of a derivative for a complex function with respect to the Laplace operator, s, as shown in (2.15), one may take the first derivative for both sides of (3.19) with respect to s, obtaining

$$\frac{G_I^{(1)}(s)}{G_I(s)} = -\frac{1}{s} - \frac{\tau_p}{\tau_p s + 1} - \theta. \tag{3.20}$$

Also, the second derivative for both sides of (3.19) can be obtained as

$$\frac{G_I^{(2)}(s)G_I(s) - [G_I^{(1)}(s)]^2}{G_I^2(s)} = \frac{1}{s^2} + \frac{\tau_p^2}{(\tau_p s + 1)^2}. \tag{3.21}$$

For simplicity, the left-hand side of (3.20) is denoted by $Q_1(s)$ and the left-hand side of (3.21) is denoted by $Q_2(s)$.

By substituting $s = \alpha$ into (3.21), it can be derived that

$$\tau_p = \begin{cases} \dfrac{1 - \alpha^2 Q_2(\alpha) + \sqrt{\alpha^2 Q_2(\alpha) - 1}}{\alpha^3 Q_2(\alpha) - 2\alpha}, & \text{if } Q_2(\alpha) > \dfrac{2}{\alpha^2}; \\[2mm] \dfrac{\alpha^2 Q_2(\alpha) - 1 + \sqrt{\alpha^2 Q_2(\alpha) - 1}}{2\alpha - \alpha^3 Q_2(\alpha)}, & \text{if } Q_2(\alpha) < \dfrac{2}{\alpha^2}. \end{cases} \tag{3.22}$$

Note that there exists a limiting condition of $Q_2(\alpha) \geq 1/\alpha^2$.

Consequently, the remaining model parameters can be derived from (3.20) and (3.18) as

$$\theta = -Q_1(\alpha) - \frac{1}{\alpha} - \frac{\tau_p}{\tau_p \alpha + 1} \tag{3.23}$$

3.3 Closed-Loop Step Response Identification of Integrating and Unstable Processes

$$k_p = \alpha(\tau_p\alpha + 1)G_1(\alpha)e^{\alpha\theta} \tag{3.24}$$

Hence, the above algorithm named Algorithm-SI-I for obtaining an SOPDT model for an integrating process can be summarized as follows.

Algorithm-SI-I

(i) Choose α and t_N in terms of (2.8), (2.10), (2.166), and (2.167) to compute $T(\alpha)$, $T^{(1)}(\alpha)$, and $T^{(2)}(\alpha)$ from (2.162) (or (2.163)), (2.164), and (2.165).
(ii) Compute $C(\alpha)$, $C^{(1)}(\alpha)$, and $C^{(2)}(\alpha)$ such as from (2.172), (2.173), and (2.174).
(iii) Compute $G_1(\alpha)$, $G_1^{(1)}(\alpha)$, and $G_1^{(2)}(\alpha)$ from (2.169), (2.170), and (2.171).
(iv) Compute $Q_1(\alpha)$ and $Q_2(\alpha)$ from the left-hand sides of (3.20) and (3.21).
(v) Compute the process time constant, τ_p, from (3.22).
(vi) Compute the process time delay, θ, from (3.23).
(vii) Compute the proportional gain, k_p, from (3.24).

Note that by letting $\tau_p = 0$ Algorithm-SI-I can be applied to obtain an FOPDT model. Accordingly, a referential cutoff angular frequency of the process, ω_{rc}, can be derived in terms of the critical phase condition,

$$-\frac{\pi}{2} - \arctan(\tau_p\omega_{rc}) = -\pi \tag{3.25}$$

In practice, ω_{rc} can be used to determine the low-frequency range that is of primary concern in control design for the model fitting.

For implementation, it is suggested to check if $Q_2(\alpha) \geq 1/\alpha^2$ when choosing α for applying Algorithm-SI-I to obtain an SOPDT model, because a too large α might not result in a positive solution of τ_p due to the computation loss on the transient response for the Laplace transform.

3.3.2 The FOPDT Model for Unstable Processes

An FOPDT model for an unstable process is generally in the form of

$$G_{U-1} = \frac{k_p e^{-\theta s}}{\tau_p s - 1} \tag{3.26}$$

where k_p denotes the proportional gain, θ the process time delay, and τ_p a positive coefficient that reflects fundamental dynamic response characteristics of the process.

By taking the natural logarithm for both sides of (3.26) with regard to $0 < s < 1/\tau_p$, one can obtain

$$\ln[-G_{U-1}(s)] = \ln(k_p) - \ln(1 - \tau_p s) - \theta s. \tag{3.27}$$

Taking the first and second derivatives for both sides of (3.27) with respect to s yields

$$\frac{G^{(1)}_{U-1}(s)}{G_{U-1}(s)} = \frac{\tau_p}{1-\tau_p s} - \theta \qquad (3.28)$$

$$Q_2(s) = \frac{\tau_p^2}{(1-\tau_p s)^2} \qquad (3.29)$$

where $Q_2(s) = d[Q_1(s)]/ds$ and $Q_1(s)$ denotes the left-hand side of (3.28). By substituting $s = \alpha$ into (3.29), it can be derived that

$$\tau_p = \begin{cases} \dfrac{\sqrt{Q_2(\alpha)}}{\alpha\sqrt{Q_2(\alpha)} - 1} \text{ or } \dfrac{\sqrt{Q_2(\alpha)}}{\alpha\sqrt{Q_2(\alpha)} + 1}, & \text{if } Q_2(\alpha) > \dfrac{1}{\alpha^2}; \\[2ex] \dfrac{\sqrt{Q_2(\alpha)}}{\alpha\sqrt{Q_2(\alpha)} + 1}, & \text{if } Q_2(\alpha) \leq \dfrac{1}{\alpha^2}. \end{cases} \qquad (3.30)$$

Note that there exists a limiting condition of $Q_2(\alpha) > 0$. For $Q_2(\alpha) > 1/\alpha^2$, one may determine a suitable solution based on the model fitting accuracy for the closed-loop step response.

Remark 3.1. The use of $0 < s < 1/\tau_p$ for deriving (3.27)–(3.30) is implied by the guideline of choosing the damping factor (α) for frequency response estimation. If the resulting τ_p does not satisfy $\alpha < 1/\tau_p$, an inverse assumption of $s > 1/\tau_p$ may be used to derive τ_p, following a similar procedure as above. ◇

Consequently, the remaining model parameters can be derived from (3.28) and (3.26) as

$$\theta = -Q_1(\alpha) + \frac{\tau_p}{1-\tau_p \alpha} \qquad (3.31)$$

$$k_p = (\tau_p \alpha - 1)G_{U-1}(\alpha)e^{\alpha\theta} \qquad (3.32)$$

Hence, the above algorithm named Algorithm-SU-I for obtaining an FOPDT model for an unstable process can be summarized as follows.

Algorithm-SU-I

(i) Choose α and t_N in terms of (2.8), (2.10), (2.166), and (2.167) to compute $T(\alpha)$, $T^{(1)}(\alpha)$, and $T^{(2)}(\alpha)$ from (2.162) (or (2.163)), (2.164), and (2.165).
(ii) Compute $C(\alpha)$, $C^{(1)}(\alpha)$, and $C^{(2)}(\alpha)$ such as from (2.172), (2.173), and (2.174).
(iii) Compute $G_{U-1}(\alpha)$, $G^{(1)}_{U-1}(\alpha)$, and $G^{(2)}_{U-1}(\alpha)$ from (2.169), (2.170), and (2.171).
(iv) Compute $Q_1(\alpha)$ and $Q_2(\alpha)$ from the left-hand sides of (3.28) and (3.29).
(v) Compute the positive coefficient, τ_p, from (3.30).
(vi) Compute the process time delay, θ, from (3.31).
(vii) Compute the proportional gain, k_p, from (3.32).

3.3.3 The SOPDT Model for Unstable Processes

A SOPDT model for an unstable process is generally in the form of

$$G_{U-2} = \frac{k_p e^{-\theta s}}{(\tau_1 s - 1)(\tau_2 s + 1)}, \quad (3.33)$$

where k_p denotes the proportional gain, θ is the process time delay, and τ_1 and τ_2 are positive coefficients that reflect fundamental dynamic response characteristics of the process.

By taking the natural logarithm for both sides of (3.33) with regard to $0 < s < 1/\tau_1$, one can obtain

$$\ln[-G_{U-2}(s)] = \ln(k_p) - \ln(1 - \tau_1 s) - \ln(\tau_2 s + 1) - \theta s \quad (3.34)$$

Taking the first and second derivatives for both sides of (3.34) with respect to s yields

$$\frac{G^{(1)}_{U-2}(s)}{G_{U-2}(s)} = \frac{\tau_1}{1 - \tau_1 s} - \frac{\tau_2}{\tau_2 s + 1} - \theta \quad (3.35)$$

$$Q_2(s) = \frac{\tau_1^2}{(1 - \tau_1 s)^2} + \frac{\tau_2^2}{(\tau_2 s + 1)^2}, \quad (3.36)$$

where $Q_2(s) = d[Q_1(s)]/ds$ and $Q_1(s)$ denotes the left-hand side of (3.35).
Substituting $s = \alpha$ into (3.36) yields

$$Q_2(\alpha) = \left[2\alpha^2 - \alpha^4 Q_2(\alpha)\right] \tau_1^2 \tau_2^2 + \left[2\alpha - 2\alpha^3 Q_2(\alpha)\right](\tau_1^2 \tau_2 - \tau_1 \tau_2^2)$$
$$+ 4\alpha^2 Q_2(\alpha) \tau_1 \tau_2 + \left[1 - \alpha^2 Q_2(\alpha)\right](\tau_1^2 + \tau_2^2) + 2\alpha Q_2(\alpha)(\tau_1 - \tau_2)$$
$$(3.37)$$

To solve τ_1 and τ_2 from (3.37), one can reformulate (3.37) into

$$\psi(\alpha) = \phi(\alpha)^T \gamma, \quad (3.38)$$

where

$$\begin{cases} \psi(\alpha) = Q_2(\alpha), \\ \phi(\alpha) = [2\alpha^2 - \alpha^4 Q_2(\alpha), \ 2\alpha - 2\alpha^3 Q_2(\alpha), \ -\alpha^2 Q_2(\alpha), \ 1, \ 2\alpha Q_2(\alpha)]^T, \\ \gamma = \left[\tau_1^2 \tau_2^2, \ \tau_1^2 \tau_2 - \tau_1 \tau_2^2, \ \tau_1^2 + \tau_2^2 - 4\tau_1 \tau_2, \ \tau_1^2 + \tau_2^2, \ \tau_1 - \tau_2\right]^T. \end{cases}$$
$$(3.39)$$

By choosing five different values of α in terms of the guideline given in (2.8) and denoting $\Psi = [\psi(\alpha_1), \psi(\alpha_2), \ldots, \psi(\alpha_5)]^T$ and $\Phi = [\phi(\alpha_1), \phi(\alpha_2), \ldots, \phi(\alpha_5)]^T$, the following LS solution can be obtained:

$$\gamma = (\Phi^T \Phi)^{-1} \Phi^T \Psi. \tag{3.40}$$

It is obvious that all the columns of Φ are linearly independent with each other, so that Φ is guaranteed to be non-singular for computation, corresponding to a unique solution of γ.

In the sequel, the model parameters can be retrieved from the relationship,

$$\begin{cases} \tau_1 - \tau_2 = \gamma(5) \\ \tau_1 \tau_2 = \dfrac{\gamma(2)}{\gamma(5)} \end{cases} \tag{3.41}$$

which can be solved as

$$\begin{cases} \tau_1 = \dfrac{\gamma(5)}{2} + \dfrac{1}{2}\sqrt{\gamma^2(5) + 4\dfrac{\gamma(2)}{\gamma(5)}} \\ \tau_2 = \tau_1 - \gamma(5) \end{cases} \tag{3.42}$$

Note that there exist three redundant fitting conditions in the parameter estimation of γ, which can certainly be satisfied if the model structure matches the process. To procure fitting accuracy for a higher order process, one can use $\gamma(1)$, $\gamma(3)$, and $\gamma(4)$ together with $\gamma(2)$ and $\gamma(5)$ to establish an LS fitting solution by taking the natural logarithm of τ_1 and τ_2,

$$\begin{bmatrix} 2 & 2 \\ 1 & 1 \\ 1 & 0 \\ 0 & 1 \end{bmatrix} \begin{bmatrix} \ln \tau_1 \\ \ln \tau_2 \end{bmatrix} = \begin{bmatrix} \ln \gamma(1) \\ \ln \left[\dfrac{\gamma(4) - \gamma(3)}{4} \right] \\ \ln \left[\dfrac{\gamma(5)}{2} + \dfrac{1}{2}\sqrt{\gamma^2(5) + 4\dfrac{\gamma(2)}{\gamma(5)}} \right] \\ \ln \left[\dfrac{\gamma(5)}{2} + \dfrac{1}{2}\sqrt{\gamma^2(5) + 4\dfrac{\gamma(2)}{\gamma(5)}} \right] - \gamma(5) \end{bmatrix} \tag{3.43}$$

Consequently, the remaining model parameters can be derived from (3.35) and (3.33) as

$$\theta = -Q_1(\alpha) + \dfrac{\tau_1}{1 - \tau_1 \alpha} - \dfrac{\tau_2}{\tau_2 \alpha + 1} \tag{3.44}$$

$$k_p = (\tau_1 \alpha - 1)(\tau_2 \alpha + 1) G_{U-2}(\alpha) e^{\alpha \theta} \tag{3.45}$$

3.3 Closed-Loop Step Response Identification of Integrating and Unstable Processes   99

Hence, the above algorithm named Algorithm-SU-II for obtaining an SOPDT model for an unstable process can be summarized as follows.

Algorithm-SU-II

(i) Choose α and t_N in terms of (2.8), (2.10), (2.166), and (2.167) to compute $T(\alpha)$, $T^{(1)}(\alpha)$, and $T^{(2)}(\alpha)$ from (2.162) (or (2.163)), (2.164), and (2.165).
(ii) Compute $C(\alpha)$, $C^{(1)}(\alpha)$, and $C^{(2)}(\alpha)$ such as from (2.172), (2.173), and (2.174).
(iii) Compute $G_{U-2}(\alpha)$, $G_{U-2}^{(1)}(\alpha)$, and $G_{U-2}^{(2)}(\alpha)$ from (2.169), (2.170), and (2.171).
(iv) Compute $Q_1(\alpha)$ and $Q_2(\alpha)$ from the left-hand sides of (3.35) and (3.36).
(v) Compute the positive coefficients, τ_1 and τ_2, from (3.42) (or (3.43)).
(vi) Compute the process time delay, θ, from (3.44).
(vii) Compute the proportional gain, k_p, from (3.45).

3.3.4 Improving Fitting Accuracy Against Model Mismatch

The above Algorithm-SI-I, Algorithm-SU-I and Algorithm-SU-II can give good fitting accuracy if the model structure adopted matches the process to be identified. In case there exists a model mismatch for the identification of a higher order process, the resulting fitting accuracy might not be optimal since these algorithms establish frequency response fitting only around the zero frequency ($\omega = 0$). To improve the fitting accuracy over a user specified frequency range, e.g., the aforementioned low frequency range for control design and on-line tuning, another identification algorithm named Algorithm-SU-III is derived as follows.

Given a user-specified frequency range for model fitting, one can let $s = \alpha + j\omega_k$ ($k = 1, 2, \ldots, M$), where M is the number of representative frequency response points in the specified frequency range. The corresponding objective function for model identification can therefore be determined as (2.43), which has been used for improving the fitting accuracy for the identification of a high-order stable process in Sect. 2.2.5.

To improve the fitting accuracy over the low-frequency range that is mostly concerned for control design, it is suggested to choose $\omega_M = (1.0 \sim 2.0)\omega_{rc}$ and $\rho_k = \eta^k / \sum_{k=1}^{M} \eta^k$, $\eta \in [0.9, 0.99]$, where ω_{rc} is a referential cutoff angular frequency of the process that can be estimated from an FOPDT or SOPDT model derived from the above algorithms.

Substituting $s = \alpha + j\omega_k$ ($k = 1, 2, \ldots, M$), the process frequency response estimated by (2.169), and the process model of (3.18), (3.26), or (3.33) into (2.43), a weighted LS solution for parameter estimation can be derived accordingly

$$\gamma = (\bar{\Phi}^T W \bar{\Phi})^{-1} \bar{\Phi}^T W \bar{\Psi} \tag{3.46}$$

where $W = diag\{\rho_1, \ldots, \rho_M, \rho_1, \ldots, \rho_M\}$,

$$\bar{\Psi} = \begin{bmatrix} Re[\Psi] \\ Im[\Psi] \end{bmatrix}, \quad \bar{\Phi} = \begin{bmatrix} Re[\Phi] \\ Im[\Phi] \end{bmatrix},$$

$$\Psi = [\psi(\alpha + j\omega_1), \psi(\alpha + j\omega_2), \ldots, \psi(\alpha + j\omega_M)]^T,$$
$$\Phi = [\phi(\alpha + j\omega_1), \phi(\alpha + j\omega_2), \ldots, \phi(\alpha + j\omega_M)]^T.$$

To obtain an SOPDT model for an integrating process, $\psi(\alpha + j\omega_k)$, $\phi(\alpha + j\omega_k)$, and γ assume the form of

$$\begin{cases} \psi(\alpha + j\omega_k) = (\alpha + j\omega_k)G_I(\alpha + j\omega_k), \\ \phi(\alpha + j\omega_k) = \left[-(\alpha + j\omega_k)^2 G_I(\alpha + j\omega_k), e^{-(\alpha+j\omega_k)\theta}\right]^T, \\ \gamma = [\tau_p, k_p]^T. \end{cases} \quad (3.47)$$

To obtain an FOPDT model for an unstable process, $\psi(\alpha + j\omega_k)$, $\phi(\alpha + j\omega_k)$, and γ assume the form of

$$\begin{cases} \psi(\alpha + j\omega_k) = G_{U-1}(\alpha + j\omega_k), \\ \phi(\alpha + j\omega_k) = \left[(\alpha + j\omega_k)G_{U-1}(\alpha + j\omega_k), -e^{-(\alpha+j\omega_k)\theta}\right]^T, \\ \gamma = [\tau_p, k_p]^T. \end{cases} \quad (3.48)$$

To obtain an SOPDT model for an unstable process, $\psi(\alpha + j\omega_k)$, $\phi(\alpha + j\omega_k)$, and γ assume the form of

$$\begin{cases} \psi(\alpha + j\omega_k) = G_{U-2}(\alpha + j\omega_k), \\ \phi(\alpha + j\omega_k) = \Big[(\alpha + j\omega_k)^2 G_{U-2}(\alpha + j\omega_k), (\alpha + j\omega_k)G_{U-2}(\alpha + j\omega_k), \\ \qquad\qquad -e^{-(\alpha+j\omega_k)\theta}\Big]^T, \\ \gamma = [\tau_1\tau_2, \tau_1 - \tau_2, k_p]^T. \end{cases}$$
(3.49)

It can easily be verified that all the columns of $\bar{\Phi}$ for each case are linearly independent with each other, such that $(\bar{\Phi}^T W \bar{\Phi})^{-1}$ is guaranteed to be non-singular for computation. Accordingly, there exists a unique solution for parameter estimation. Generally, $M \in [10, 50]$ is suggested for practical application to meet a good trade-off between the computational efficiency and fitting accuracy.

In the sequel, the SOPDT model parameters for an integrating process can be retrieved as

$$\begin{cases} \tau_p = \gamma(1) \\ k_p = \gamma(2) \end{cases}, \quad (3.50)$$

which is the same for obtaining the FOPDT model parameters for an unstable process.

3.3 Closed-Loop Step Response Identification of Integrating and Unstable Processes

The SOPDT model parameters for an unstable process can be retrieved as

$$\begin{cases} \tau_1 = \dfrac{\gamma(2)}{2} + \dfrac{1}{2}\sqrt{\gamma^2(2) + 4\gamma(1)} \\ \tau_2 = \tau_1 - \gamma(2) \\ k_p = \gamma(3) \end{cases} \quad (3.51)$$

Note that a preliminary value of the process time delay (θ) is needed to derive the remaining model parameters from (3.46). In fact, a rough estimation of the process time delay can be obtained from the step response or an FOPDT or SOPDT model derived from Algorithm-SI-I, Algorithm-SU-I, or Algorithm-SU-II, which may be taken as an initial value in an one-dimensional search that uses a convergence condition for fitting the closed-loop step response,

$$err = \frac{1}{N_s}\sum_{k=1}^{N_s}\left[\Delta y(kT_s) - \Delta \widehat{y}(kT_s)\right]^2 < \varepsilon, \quad (3.52)$$

where $\Delta y(kT_s)$ and $\Delta \widehat{y}(kT_s)$ denotes respectively the process and model outputs to the closed-loop step test, and $N_s T_s$ is the settling time.

Given a user specified threshold of ε, a suitable solution of the model parameters can be derived by monotonically varying θ in a possible range for computation. The one-dimensional search step size can be taken as a small multiple of the sampling period for implementation. The optimal fitting can be determined by deriving such a model that yields the smallest value of *err*.

Hence, the above algorithm named Algorithm-SU-III for improving fitting accuracy in a user specified frequency range can be summarized.

Algorithm-SU-III

(i) Choose α and t_N in terms of (2.8) and (2.10) to compute $T(\alpha + j\omega_k)$ ($k = 1, 2, \ldots, M$) from (2.162) (or (2.163)), where ω_{rc} for the choice of ω_M can be determined from an initial model obtained using Algorithm-SI-I, Algorithm-SU-I or Algorithm-SU-II.
(ii) Compute $C(\alpha + j\omega_k)$ ($k = 1, 2, \ldots, M$) such as from (2.172).
(iii) Compute $G(\alpha + j\omega_k)$ ($k = 1, 2, \ldots, M$) from (2.169).
(iv) Obtain an initial estimate of the process time delay, θ, from the step test, or a model obtained using Algorithm-SI-I, Algorithm-SU-I, or Algorithm-SU-II.
(v) Solve the remaining model parameters from (3.46) to (3.51).
(vi) End the algorithm if the fitting condition in (3.52) is satisfied. Otherwise, go back to Step (v) by monotonically varying θ for a one-dimensional search within a possible range as observed from the step test.

Note that in the presence of measurement noise, Algorithm-SI-I, Algorithm-SU-I, and Algorithm-SU-II can give convergent parameter estimation owing to the development of analytical formulae. Whether a consistent estimation can be reached or not depends in essence on the unbiasedness of the closed-loop frequency response estimation algorithm as presented in Sect. 2.5.1. Regarding Algorithm-SU-III, the use of time domain fitting criterion in (3.52) needs to be clarified for consistent parameter estimation, even if unbiased frequency response estimation is used for model fitting. Following the proof of Corollary 2.1 in Sect. 2.2.6, the below corollary can be given:

Corollary 3.1. *Given Gaussian white measurement noise,* $\zeta(t) \sim N(0, \sigma_\zeta^2)$, *in a closed-loop step test, consistent parameter estimation can be obtained by Algorithm-SU-III in terms of using the time domain fitting criterion in (3.52).*

3.3.5 Illustrative Examples

Five examples studied in the existing literature are used to demonstrate the effectiveness of the presented identification algorithms. Examples 3.3–3.5 are given to demonstrate the accuracy of the presented algorithms for identifying first- and second-order integrating and unstable processes, respectively. Examples 3.6 and 3.7 are given to show the effectiveness of the presented algorithms for the identification of higher order integrating and unstable processes. Measurement noise tests are also included in Examples 3.3–3.5 to demonstrate the identification robustness of the presented algorithms. In all identification tests, the sampling period is taken as $T_s = 0.01$ (s) for computation.

Example 3.3. Consider the SOPDT integrating process studied by Sung and Lee (1996),

$$G = \frac{e^{-0.2s}}{s(0.1s+1)}$$

Based on a closed-loop step test using a P-type controller ($k_C = 3$) by adding a unity step change to the set-point, Sung and Lee (1996) derived an SOPDT model, $G = e^{-0.176s}/[s(0.122s + 1)]$. For illustration, the same test is performed here. According to the guideline given in (2.8) and (2.10), $\alpha = 1.0$ and $t_N = 20$(s) are chosen to apply Algorithm-SI-I, resulting in an FOPDT model listed in Table 3.1, which indicates good accuracy. The fitting error is given in terms of the closed-loop transient response in the time interval [0, 5]s. Note that taking $\alpha = 0.5$ and $t_N = 40$ (s), or $\alpha = 2.0$ and $t_N = 10$ (s), can result in the exact model as well, which demonstrates that the closed-loop frequency response estimation algorithm presented in Sect. 2.5.1 maintains good robustness with regard to different choices of α.

To demonstrate identification robustness against measurement noise, assume that a random noise $N(0, \sigma_N^2 = 0.36\%)$, causing NSR $= 5\%$, is added to the process output measurement which is then used for feedback control. By performing 100 Monte Carlo tests in terms of varying the "seed" of the noise generator from 1

3.3 Closed-Loop Step Response Identification of Integrating and Unstable Processes 103

Table 3.1 Closed-loop step response identification results under different noise levels

NSR	Example 3.3	err	Example 3.4	err	Example 3.5	err
0	$\dfrac{1.0000e^{-0.2000s}}{s(0.1001s+1)}$	1.46×10^{-8}	$\dfrac{1.0000e^{-0.8008s}}{1.0006s-1}$	8.46×10^{-8}	$\dfrac{0.9999e^{-0.5000s}}{(2.0000s-1)(0.5000s+1)}$	1.46×10^{-8}
5%	$\dfrac{1.0000(\pm 0.0027)e^{-0.2026(\pm 0.025)s}}{s[0.0969(\pm 0.0026)s+1]}$	2.09×10^{-6}	$\dfrac{1.0001(\pm 0.0011)e^{-0.8019(\pm 0.045)s}}{1.0015(\pm 0.033)s-1}$	4.15×10^{-7}	$\dfrac{0.9999(\pm 0.0021)e^{-0.4997(\pm 0.0079)s}}{[2.0002(\pm 0.0029)s-1][0.5002(\pm 0.0095)s-1]}$	1.67×10^{-7}
10%	$\dfrac{1.0001(\pm 0.0056)e^{-0.1956(\pm 0.041)s}}{s[0.1051(\pm 0.042)s+1]}$	5.97×10^{-6}	$\dfrac{1.0003(\pm 0.0021)e^{-0.8018(\pm 0.089)s}}{1.0013(\pm 0.066)s-1}$	5.08×10^{-7}	$\dfrac{0.9999(\pm 0.0041)e^{-0.4991(\pm 0.016)s}}{[2.0003(\pm 0.0058)s-1][0.5007(\pm 0.019)s-1]}$	7.57×10^{-7}
20%	$\dfrac{1.0002(\pm 0.011)e^{-0.1866(\pm 0.061)s}}{s[0.1261(\pm 0.063)s+1]}$	2.71×10^{-4}	$\dfrac{1.0006(\pm 0.0041)e^{-0.7971(\pm 0.183)s}}{0.9973(\pm 0.136)s-1}$	2.86×10^{-6}	$\dfrac{0.9998(\pm 0.0082)e^{-0.4982(\pm 0.031)s}}{[2.0005(\pm 0.012)s-1][0.5016(\pm 0.037)s-1]}$	2.75×10^{-6}

to 100, the identified results are shown in Table 3.1, where the model parameters are respectively the mean of 100 Monte Carlo tests, and the values in the adjacent parentheses are the sample standard deviation. The results for the noise levels of NSR = 10% and 20% are also listed in Table 3.1 to show the achievable identification accuracy and robustness.

Example 3.4. Consider the FOPDT unstable process studied by Padhy and Majhi (2006),

$$G = \frac{1}{s-1}e^{-0.8s}$$

Based on relay feedback tests with two P-type controllers, Padhy and Majhi (2006) derived an FOPDT model, $G_m = 1.0e^{-0.8033s}/(1.0007s-1)$. For illustration, a closed-loop step test by adding a step change of $h = 0.05$ to the set-point and using a P-type controller of $k_C = 1.2$, which is equivalent to that of Padhy and Majhi (2006), is performed here. Algorithm-SU-I taking $\alpha = 0.1$ and $t_N = 150(s)$ gives an FOPDT model listed in Table 3.1, again demonstrating good accuracy. The fitting error is given in terms of the closed-loop transient response in the time interval [0, 50]s. Identification results for the measurement noise levels of NSR = 5%, 10%, and 20% are also shown in Table 3.1, which demonstrates good identification robustness.

Example 3.5. Consider the SOPDT unstable process studied in the literature (Cheres 2006; Sree and Chidambaram 2006; Liu and Gao 2008),

$$G = \frac{1}{(2s-1)(0.5s+1)}e^{-0.5s}$$

Based on a closed-loop step test using a PID controller ($k_C = 2.71, \tau_I = 4.43$ and $\tau_D = 0.319$ in (2.172)) and adding a unity step change to the set-point, Cheres (2006) derived an FOPDT model for controller tuning, as was also done in Sree and Chidambaram (2006). By performing the same closed-loop step test, Algorithm-SU-II based on the choice of $\alpha = 0.1, 0.15, 0.2, 0.25, 0.3$ and $t_N = 300(s)$ gives $G_{m-2} = 0.9999e^{-0.4996s}/(2.0000s-1)(0.5000s+1)$, and Algorithm-SU-III results in almost the exact SOPDT model, as listed in Table 3.1. The fitting error is given in terms of the closed-loop transient response in the time interval [0, 30]s.

To demonstrate consistent parameter estimation of Algorithm-SU-III against measurement noise, 100 Monte Carlo tests are performed under the noise levels of NSR = 5%, 10% and 20%, respectively. Identification results obtained using a search range of [0.3, 0.7](s) for the time delay and a computational constraint of positive solution for the remaining model parameters are also shown in Table 3.1, which demonstrate good identification accuracy and robustness.

Example 3.6. Consider the high-order integrating process studied by Ingimundarson and Hägglund (2001),

$$G = \frac{64}{s(s+1)(s+2)(s+4)(s+8)}e^{-5s}$$

3.3 Closed-Loop Step Response Identification of Integrating and Unstable Processes

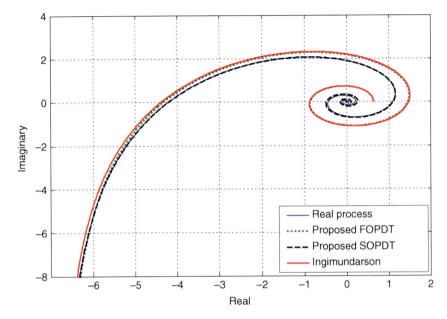

Fig. 3.3 Nyquist plots of identified models for Example 3.6

Based on an open-loop step test, Ingimundarson and Hägglund (2001) derived an FOPDT model, $G_m = 1.0e^{-6.9s}/s$. For illustration, assume that the process is operated in a closed-loop manner with a PI controller ($k_C = 0.101$ and $\tau_I = 39.6$) tuned by the so-called SIMC method (Skogestad 2003, taking $\tau_C = 3$) in terms of the above model. By adding a unity step change to the set-point in a closed-loop step test, Algorithm-SI-I taking $\alpha = 0.05$ and $t_N = 500$ (s) gives an FOPDT model, $G_{m-1} = 0.9984e^{-6.8113s}/s$, corresponding to $err = 2.71 \times 10^{-5}$ in terms of the closed-loop transient response in the time interval [0, 200]s, and a SOPDT model, $G_{m-2} = 1.0000e^{-5.7061s}/[s(1.1698s + 1)]$, corresponding to $err = 9.32 \times 10^{-8}$. Algorithm-SU-III based on the frequency response estimation of $G(0.05 + j\omega_k)$, where $\omega_k = k\omega_{rc}/10$, $k = 0, 1, 2, \ldots, 10$, and $\omega_{rc} = 0.2306$ (rad/s) derived from the above FOPDT model, gives a SOPDT model, $G_{m-2} = 1.0000e^{-5.717s}/[s(1.1616s + 1)]$, corresponding to $err = 7.12 \times 10^{-8}$. The Nyquist plots of the above FOPDT models and the proposed LS-based SOPDT model are shown in Fig. 3.3.

To demonstrate the achievable control performance, assume that a unity step change is added to the set-point and a load disturbance with a magnitude of -0.07 is added at $t = 200$ (s). The closed-loop output responses resulting from the above PI controller and a PID controller ($k_C = 0.1029$, $\tau_I = 38.868$, and $\tau_D = 1.1616$) tuned by the SIMC method (taking $\tau_C = 4$) based on the proposed LS-based SOPDT model, are shown in Fig. 3.4, in terms of the same set-point tracking speed for comparison. It can be seen that the proposed SOPDT model facilitates obtaining improved control performance.

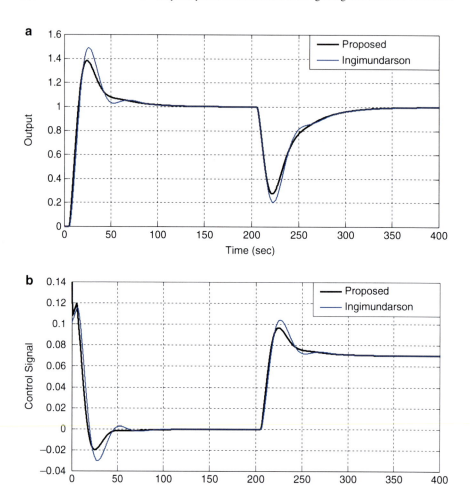

Fig. 3.4 Comparison of closed-loop control performance for Example 3.6

Subsequently, assume that the process time delay is perturbed to be 30% larger. The perturbed output responses are shown in Fig. 3.5, which demonstrates that the proposed SOPDT model also facilitates control robustness.

Example 3.7. Consider the high-order unstable process studied by Majhi (2007),

$$G = \frac{e^{-0.5s}}{(5s-1)(2s+1)(0.5s+1)}$$

Using a relay feedback test, Majhi (2007) derived an SOPDT model, $G_m = 1.001e^{-0.939s}/(10.354s^2 + 2.932s - 1)$, from a state-space analysis on the process relay response. For illustration, by performing a closed-loop step test as in

3.3 Closed-Loop Step Response Identification of Integrating and Unstable Processes

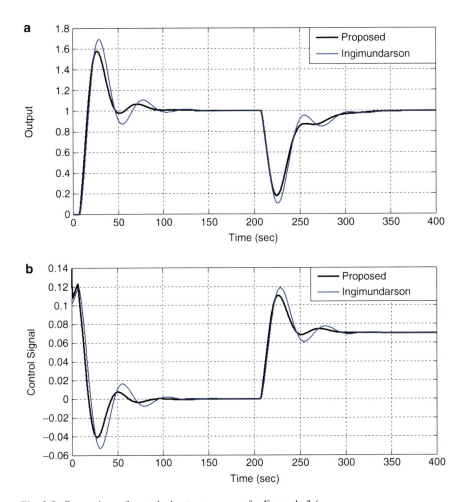

Fig. 3.5 Comparison of perturbed output response for Example 3.6

Example 3.3 but with $h = 0.1$ and $k_C = 2.0$ for the closed-loop stabilization, Algorithm-SU-I, taking $\alpha = 0.1$ and $t_N = 150$ (s), gives an FOPDT model, $G_{m-1} = 0.9907e^{-2.792s}/(5.037s - 1)$, corresponding to $err = 1.53 \times 10^{-3}$ in terms of the closed-loop transient response in the time interval $[0, 150]$s.

To improve fitting accuracy over the low frequency range, Algorithm-SU-III based on the frequency response estimation of $G(0.1 + j\omega_k)$, where $\omega_k = k\omega_{rc}/10$, $k = 0, 1, 2, \ldots, 10$, and $\omega_{rc} = 0.3963$ (rad/s) estimated from the above FOPDT model, gives $G_{m-1} = 0.9492e^{-2.774s}/(5.2644s - 1)$, corresponding to $err = 4.12 \times 10^{-4}$.

Note that Algorithm-SU-II, taking $\alpha = 0.02, 0.04, 0.06, 0.08, 0.1$ and $t_N = 1,000$ (s), gives an SOPDT model, $G_{m-2} = 0.9808e^{-0.5221s}/(4.8996s - 1)(2.0874s + 1)$, corresponding to $err = 1.89 \times 10^{-4}$. Further improved fitting

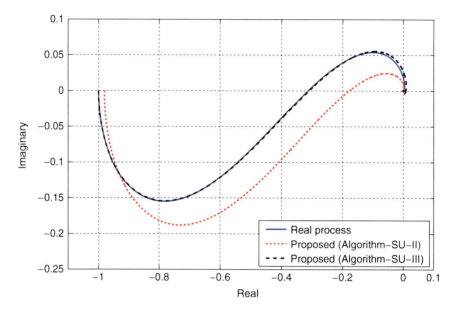

Fig. 3.6 Nyquist plots of identified SOPDT models for Example 3.7

accuracy can be obtained by Algorithm-SU-III based on the above frequency response estimation, which results in an SOPDT model, $G_{m-2} = 1.0000e^{-0.938s}/(5.0107s - 1)(2.0773s + 1)$, corresponding to $err = 1.31 \times 10^{-7}$. The Nyquist plots of these identified SOPDT models are shown in Fig. 3.6. It should be noted that the SOPDT model of Majhi (2007) corresponds to $err = 3.92 \times 10^{-7}$.

3.4 Application to the Heating-Up Control of Barrel Temperature in Injection Molding

Consider the barrel temperature control of an industrial Chen-Hsong reciprocating screw injection molding machine (model no. JM88-MKIII-C), the schematic of which is shown in Fig. 3.7. There are 6 temperature zones in the heating barrel. Each zone is equipped with an electric heater with a capacity of 1,040(W), which is regulated via a zero-crossing solid state relay (SSR) with pulse-width modulation (PWM). Each zone temperature is measured using a K-type thermocouple. A 16-bit data acquisition card (AT-MIO-16X) from National Instruments (NI) is used for analog-to-digital (A/D) and digital-to-analog (D/A) conversions. The semi-crystalline material of high-density polyethylene (HDPE) is used for the injection molding experiment. According to a ramp-up heating technique for the raw materials, the rear three zones, 4–6, are primarily used for heating the raw materials below a temperature of 200°C, while the front three zones, 1–3, are required to

3.4 Application to the Heating-Up Control of Barrel Temperature in Injection Molding

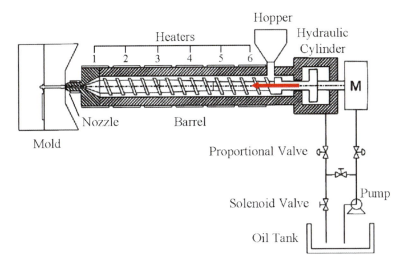

Fig. 3.7 Schematic of an injection molding machine

reach the melting temperature of 220°C and tightly maintain the temperature, e.g., within ±0.5°C, for injection molding. Therefore, identification and control tuning are herein focused on the front zones 1–3 for illustration. It should be noted that all of these heating zones have positive correlation with each other for both the heating-up process and the injection molding process of cycling, the individual loop control structure (i.e., multiloop) is therefore adopted for simplicity. The positive correlation between these heating zones is also considered in the following identification procedure to ensure identification-effectiveness.

To identify an integrating SOPDT model for representing the heating-up response characteristics for each of the three zones, an open-loop step response test is conducted for each zone: First, the three zones are heated to a temperature over 210°C, and then, the corresponding heaters are shut off for cooling down by air convection to the environment, in order to have a zero initialization of the temperature response for each zone; When the temperature of the zone to be identified drops to 210°C, the three zones are simultaneously heated with the corresponding heaters turned fully on, until the temperature of the zone to be identified reaches the set-point temperature of 220°C. The experimental results with a sampling period of 200(ms) are shown in Fig. 3.8. It can be seen that due to air convection, the temperatures of zones 1–3 drop slowly before increasing to 220°C. The time of temperature drop can be viewed as time delay of the heating-up response, which may be referenced to choose the initial point of the transient response for model identification. To reduce the influence of measurement noise, a third-order Butterworth filter with a cutoff frequency, $f_c = 0.5$ (Hz), is used in both the forward and reverse directions to recover the noisy data for identification, as shown by the thick dashed lines in Fig. 3.8. Using the open-loop step response identification algorithm presented in

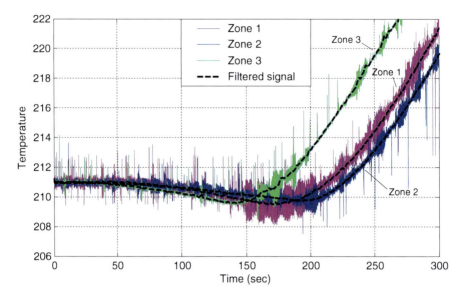

Fig. 3.8 Open-loop step response tests for zones 1–3

Table 3.2 Heating-up response models for zones 1–3

Zone	Proposed method	Z-N method
1	$\dfrac{0.2842e^{-33.23s}}{s(147.1585s + 1)}$	$\dfrac{0.1337e^{-75s}}{s}$
2	$\dfrac{0.1719e^{-25.01s}}{s(56.7819s + 1)}$	$\dfrac{0.1463e^{-63s}}{s}$
3	$\dfrac{0.197e^{-25.04s}}{s(78.52s + 1)}$	$\dfrac{0.1431e^{-68s}}{s}$

Sect. 3.2.1, SOPDT integrating models of these three zones are obtained as listed in Table 3.2 together with the FOPDT integrating models obtained using the early Ziegler-Nichols (Z-N) step response method (Åström and Hägglund 1995).

Accordingly, an internal model control (IMC) based scheme is constructed for the heating-up control of zones 1–3, which is shown in Fig. 3.9, where the heating zone to be controlled plus the heater are treated together as the process (denoted by G) to be controlled, G_m is the process model as above identified, C is the controller, r is the set-point temperature, y is the measured temperature of the heating zone to be controlled, and d_0 denotes load disturbance with a transfer function of G_d, including air convection and heat absorption from the raw materials.

From Fig. 3.9, the closed-loop transfer function can be derived as

$$T_r = \frac{GC}{1 + (G - G_m)C} \tag{3.53}$$

3.4 Application to the Heating-Up Control of Barrel Temperature in Injection Molding

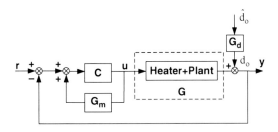

Fig. 3.9 Block diagram of the IMC-based temperature control structure

In the nominal case ($G = G_m$), the transfer function can be simplified to $T_r = GC$.

Based on the above identified SOPDT integrating model for each zone, the desired closed-loop transfer function can be determined according to the IMC theory (Morari and Zafiriou 1989), as will be introduced in the later Sect. 7.3, in the form of

$$T_r = \frac{e^{-\theta s}}{(\lambda_h s + 1)^2}, \tag{3.54}$$

where λ_h is an adjustable closed-loop time constant.

The corresponding closed-loop controller, C, can then be derived from the nominal relationship, $T_r = G_m C$, as

$$C = \frac{s(\tau_p s + 1)}{k_p (\lambda_h s + 1)^2} \tag{3.55}$$

It is seen that λ_h becomes the adjustable parameter of C, which can be tuned to obtain a desirable heating-up performance.

Note that due to the heat loss from air convection or radiation to the environment, which is viewed as load disturbance, the control output should actually be augmented with a certain value to prevent temperature drop during or after the heating-up stage, i.e.,

$$u = u_{r-h} + u_0, \tag{3.56}$$

where u_0 corresponds to the heating power required for balancing the heat loss from air convection or radiation, which can be ascertained from an open-loop test of maintaining the set-point temperature.

It should be noted that there exists an implemental constraint of $0 \leq u \leq 1$, corresponding to 0–100% of the heater power.

The load disturbance transfer function from $\hat{d}_o$ to y shown in Fig. 3.9 can be derived as

$$H_d(s) = G_d(s)(1 - T_r(s)), \tag{3.57}$$

where $G_d(s)$ is a stable transfer function that reflects the fundamental dynamics of the load disturbance. It can be verified using the above $T_r(s)$ in (3.54) that

$$\lim_{s \to 0} H_d(s) = 0, \tag{3.58}$$

which indicates that the influence of a step-type load disturbance entering into the process from its output side can be eliminated completely. Note that no step-type load disturbance could enter into the process input since it is purely manipulated by the controller as shown in Fig. 3.9, so the influence of such load disturbance does not need to be considered here, though it may cause control system instability as discussed by Liu et al. (2005).

Based on the above control system design, a heating-up test is performed, starting from an initial temperature of 210°C. That is, zones 1–3 are heated to slightly over 210°C, and then, the control scheme shown in Fig. 3.9 is switched over after the temperature of any one of the three zones drops to 210°C via shutting off the corresponding heaters. The sampling period for control implementation is taken as $T_s = 0.5(s)$, in view of that the maximal temperature increment (ΔT) is actually no larger than 0.1°C/s, as can be verified from the filtered signals shown in Fig. 3.8. Accordingly, a first-order backward discretization operator, $\dot{e}(kT_s) = [e(kT_s) - e((k-1)T_s)]/T_s$, is used for the differential computation.

To reduce the influence of measurement noise for computing the control output, an on-line noise spike filtering strategy (Seborg et al. 2004) is adopted for feedback control,

$$y(kT_s) = \begin{cases} \hat{y}((k-1)T_s) + \Delta T, & \hat{y}(kT_s) - \hat{y}((k-1)T_s) \geq \Delta T; \\ \hat{y}((k-1)T_s) - \Delta T, & \hat{y}(kT_s) - \hat{y}((k-1)T_s) \leq -\Delta T; \\ \hat{y}(kT_s), & \text{else.} \end{cases} \tag{3.59}$$

where $y(kT_s)$ denotes the filtered temperature for feedback control, $\hat{y}(kT_s)$ the measured temperature, and $\Delta T = 0.005°C$ a specified threshold for filtering measurement noise.

By setting the control parameters as $\lambda_{h-1} = \lambda_{h-2} = \lambda_{h-3} = 20$, respectively for zones 1–3, the experimental results are shown in Fig. 3.10. It is seen that fast heating-up without the temperature overshoot has been obtained for all these three zones, compared to the Z-N PID tuning method (Åström and Hägglund 1995) based on using the FOPDT integrating models listed in Table 3.2. For comparison, using the identified SOPDT integrating models, the control results obtained from the IMC-based PID tuning method (Skogestad 2003) with the parameter settings of $\tau_{c-1} = \tau_{c-2} = \tau_{c-3} = 30$ are also shown in Fig. 3.10. It is seen that apparently improved heating up performance is obtained by using the identified SOPDT integrating models for these three zones, while the proposed IMC design gives better control performance. Note that the control signals have a practical range of [0, 100], corresponding to the heating power in a range of 0–100%. This implies that no negative control output can be used to bring down the temperature. In the case of

3.4 Application to the Heating-Up Control of Barrel Temperature in Injection Molding

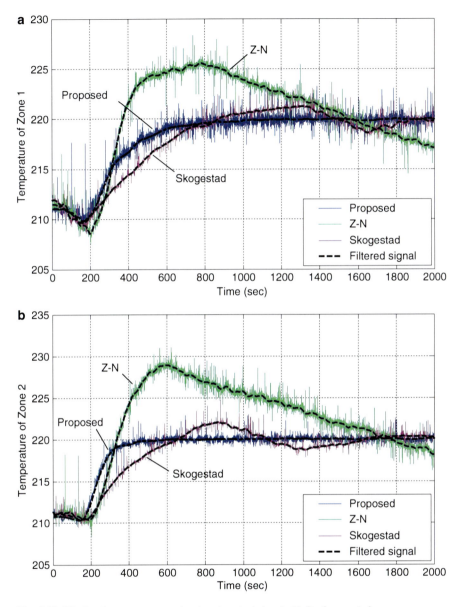

Fig. 3.10 The heating-up responses (**a**–**c**) and control signals (**d**–**f**) of zones 1–3

overheating, only load disturbance from the air convection can help to drop down the temperature, unavoidably leading to a prolonged settling time (Yao and Gao 2007). Besides, it can be seen from Fig. 3.10d–f that after the heating-up process, each of the control outputs maintains a constant value to balance the heat loss from air convection.

114   3 Step Response Identification of Integrating and Unstable Processes

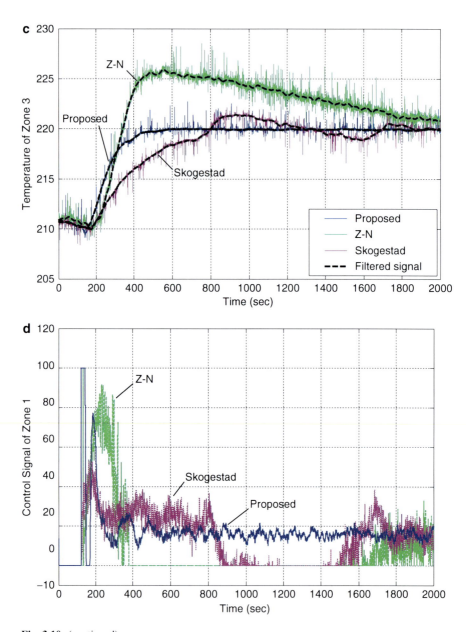

Fig. 3.10 (continued)

3.4 Application to the Heating-Up Control of Barrel Temperature in Injection Molding 115

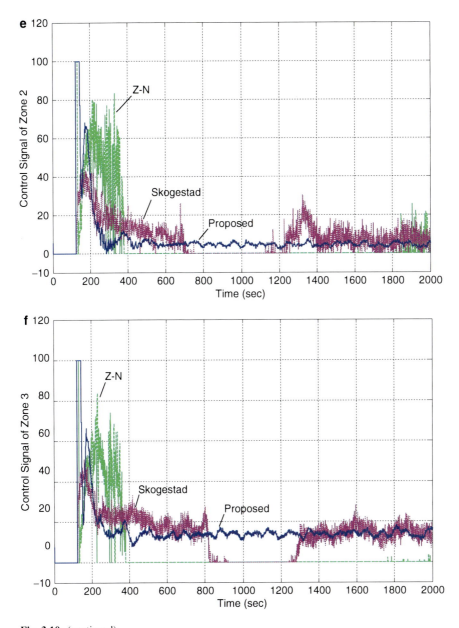

Fig. 3.10 (continued)

3.5 Summary

For the identification of integrating and unstable processes to facilitate control design in engineering practice, a step response test is generally preferred to be performed in a closed-loop structure, such as the unity feedback control loop with a prescribed low-order controller of PID for stabilization. It is clarified that the excitation signal like a step change should be added to the set-point, rather than the process input, to facilitate the implementation of a closed-loop identification test. Guidelines of choosing a low-order controller for implementing a closed-loop step response test have been given accordingly.

For the identification of integrating processes, the response feature of monotonically increasing or decreasing to a step change can be used for the implementation of an open-loop step test from zero initial process state, or around a nonzero setpoint for operation but subject to a limit of the output range. Correspondingly, an open-loop step response identification algorithm (Liu et al. 2009) has been presented to obtain the widely used low-order models of FOPDT and SOPDT for controller tuning, based on an LS fitting of the transient response data in terms of a time integral approach to the linearly differential expression of the model structure for identification. At the same time, a practical IV method has also been given to guarantee consistent parameter estimation against measurement noise. The applications to two examples from existing references have demonstrated that the proposed identification method can give good accuracy and robustness.

For the use of a closed-loop step test, the closed-loop frequency response estimation algorithm presented in Sect. 2.5.1 can be adopted to estimate the process frequency response. Note that this algorithm can also be applied in the case where the closed-loop output response has not yet moved into a steady state before stopping the identification test. This advantage can obviate the requirement of controller tuning for stabilization to perform a closed-loop identification test. Correspondingly, three model identification algorithms, Algorithm-SI-I, Algorithm-SU-I, and Algorithm-SU-II, have been analytically developed (Liu and Gao 2010) in the frequency domain to obtain the low-order models of FOPDT and SOPDT, for the identification and control of integrating and unstable processes. To deal with model mismatch, as encountered for the identification of a higher order integrating or unstable process, a weighted LS fitting algorithm, Algorithm-SU-III, has also been presented based on the frequency response estimation in the low frequency range that is mostly concerned with control design, which in fact can be extended to obtain a very specific model with more descriptive parameters. The applications to five examples from the existing literature have demonstrated the effectiveness and merits of these identification algorithms.

An experimental application to the barrel temperature control of an injection molding machine has been detailed, including practical implementation issues such as the denoising strategies for identification. The experiment results well demonstrate the effectiveness of the above open-loop step response identification method for integrating processes, and the good performance of a model-based IMC scheme for the heating-up control.

References

Ananth I, Chidambaram M (1999) Closed loop identification to transfer function model for unstable systems. J Franklin Inst 336(7):1055–1061

Åström KJ, Hägglund T (1995) PID controller: theory, design, and tuning, 2nd edn. ISA Society of America, Research Triangle Park

Åström KJ, Hägglund T (2005) Advanced PID control. ISA Society of America, Research Triangle Park

Cheres E (2006) Parameter estimation of an unstable system with a PID controller in a closed loop configuration. J Franklin Inst 343(2):204–209

Deshpande PB (1980) Process identification of open-loop unstable systems. AIChE J 26(2):305–308

Ingimundarson A, Hägglund T (2001) Robust tuning procedures of dead-time compensating controllers. Control Eng Pract 9:1195–1208

Jin HP, Heung IP, Lee I-B (1998) Closed-loop on-line process identification using a proportional controller. Chem Eng Sci 53(9):1713–1724

Kaya I (2006) Parameter estimation for integrating processes using relay feedback control under static load disturbances. Ind Eng Chem Res 45:4726–4731

Kwak HJ, Sung SW, Lee I-B (1997) On-line process identification and autotuning for integrating processes. Ind Eng Chem Res 36(12):5329–5338

Liu T, Gao F (2008) Identification of integrating and unstable processes from relay feedback. Comput Chem Eng 32(12):3038–3056

Liu T, Gao F (2010) Closed-loop step response identification of integrating and unstable processes. Chem Eng Sci 65(10):2884–2895

Liu T, Zhang WD, Gu DY (2005) Analytical two-degree-of-freedom tuning design for open-loop unstable processes with time delay. J Process Control 15:559–572

Liu T, Yao K, Gao F (2009) Identification and autotuning of temperature control system with application to injection molding. IEEE Trans Control Syst Technol 17(6):1282–1294

Luyben WL (1990) Process modeling, simulation, and control for chemical engineers. McGraw Hill, New York

Majhi S (2007) Relay based identification of processes with time delay. J Process Control 17(2):93–101

Morari M, Zafiriou E (1989) Robust process control. Prentice Hall, Englewood Cliff, NJ

Padhy PK, Majhi S (2006) Relay based PI-PD design for stable and unstable FOPDT processes. Comput Chem Eng 30(5):790–796

Paraskevopoulos PN, Pasgianos GD, Arvanitis KG (2004) New tuning and identification methods for unstable first order plus dead-time. IEEE Trans Control Syst Technol 12(3):455–464

Pintelon R, Schoukens J (2001) System identification: a frequency domain approach. IEEE Press, New York

Seborg DE, Edgar TF, Mellichamp DA (2004) Process dynamics and control, 2nd edn. Wiley, Hoboken

Shinskey FG (1996) Process control system, 4th Edition, McGraw Hill, New York

Skogestad S (2003) Simple analytical rules for model reduction and PID controller tuning. J Process Control 13(4):291–309

Söderström T, Stoica P (1989) System identification. Prentice Hall, New York

Sree RP, Chidambaram M (2006) Improved closed loop identification of transfer function model for unstable systems. J Franklin Inst 343(2):152–160

Sung SW, Lee I-B (1996) Limitations and countermeasures of PID controllers. Ind Eng Chem Res 35(8):2596–2610

Tan KK, Wang QG, Lee TH (1998) Finite spectrum assignment control of unstable time delay processes with relay tuning. Ind Eng Chem Res 37:1351–1357

Yao K, Gao F (2007) Optimal start-up control of injection molding barrel temperature. Polym Eng Sci 47(3):254–261

Zhou KM, Doyle JC, Glover K (1996) Robust and optimal control. Prentice Hall, Englewood Cliff

Chapter 4
Relay Feedback Identification of Stable Processes

4.1 Implementation of a Relay Feedback Test

Owing to the fact that a relay feedback test can generate sustained oscillations of the controlled output for closed-loop identification, model identification from the relay feedback has attracted significantly increasing attentions in the process control community, since the pioneering works of the 1980s, see Atherton (1982), Tsypkin (1984), Åström and Hägglund (1984), and Luyben (1987). Compared to an open-loop identification test such as using a step or pulse signal, the fundamental dynamic response characteristics of the process may be better observed from the sustained oscillations under relay feedback, especially in the presence of measurement noise. Moreover, a relay feedback test will not cause the process to drift too far away from its set-point. This is very necessary in many practical applications, in particular for the control-oriented piecewise identification of a highly nonlinear process that is subject to rigorously operating conditions (Luyben 1987).

Generally, there are two types of relay feedback tests – unbiased (symmetrical) and biased (asymmetrical). A biased relay function, which is depicted in Fig. 4.1, can be specified as

$$u(t) = \begin{cases} u_+ \text{ for } \{e(t) > \varepsilon_+\} \text{ or } \{e(t) \geq \varepsilon_- \text{ and } u(t_-) = u_+\} \\ u_- \text{ for } \{e(t) < \varepsilon_-\} \text{ or } \{e(t) \leq \varepsilon_+ \text{ and } u(t_-) = u_-\} \end{cases} \quad (4.1)$$

where $u_+ = \Delta\mu + \mu_0$ and $u_- = \Delta\mu - \mu_0$ denote, respectively, the positive and negative relay magnitudes; ε_+ and ε_- denote, respectively, the positive and negative switch hystereses. The initial output of the relay is assumed as u_- for zero input, as commonly set in a commercial relay module in industrial applications. Note that letting $\Delta\mu = 0$ and $|\varepsilon_+| = |\varepsilon_-|$ leads to an unbiased relay function.

To capture the process dynamic response characteristics around the set-point for online tuning, a closed-loop relay test is usually performed in terms of nonzero steady initial process conditions. A typical closed-loop configuration for a relay test

Fig. 4.1 Depiction of a relay function

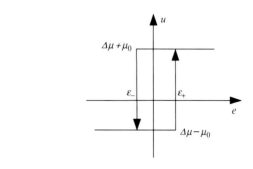

Fig. 4.2 Block diagram of a relay feedback test

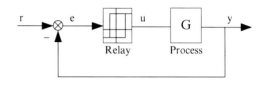

Fig. 4.3 Schematic of a relay feedback test on-line

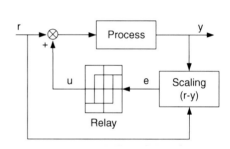

is shown in Fig. 4.2, where r denotes the set-point, y the process output, and u the relay output. Correspondingly, an online relay identification test can be constructed as shown in Fig. 4.3, where the scaling unit $(r - y)$ is set to normalize r and y for computing the output error, e.

To implement a relay feedback test online, the magnitudes of u_+ and u_- should be set based on an admissible fluctuation range of the process output around the set-point. It is obvious that a larger magnitude of the relay output will facilitate the observation of more details in the process dynamic response. To avoid measurement noise causing incorrect relay switches, it is suggested that the magnitudes of ε_+ and ε_- should be set at least twice larger than the noise band, together with an upper limit almost equal to 0.95 times of the absolute minimum of u_+ and u_- (Wang et al. 2003). After the process has moved into the operating range, a short "listening period" (e.g., 20–100 samples) should be referenced to set ε_+ and ε_- properly.

A scenario of the relay feedback response online is shown in Fig. 4.4a, where the process output response and the relay output are plotted together for illustration. It can be seen that the relay feedback test begins at $t = 20(s)$ while the set-point is taken as $r = 5$. Then, the process output response moves into a steady oscillation after several relay switches. For the convenience of analysis, the initial relay

4.1 Implementation of a Relay Feedback Test 121

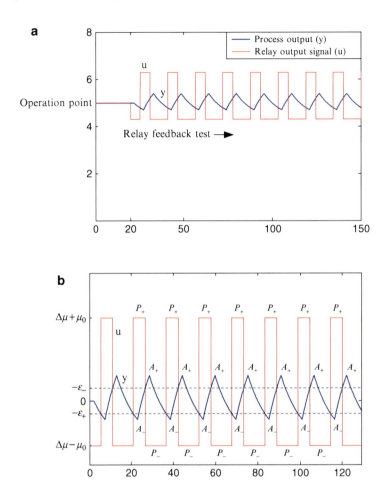

Fig. 4.4 On-line relay feedback response (**a**) and the shifted version for analysis (**b**)

response of the process output can be shifted to the origin in the reference coordinate plane, as shown in Fig. 4.4b, where A_+ denotes the positive amplitude in the steady oscillation that is specifically named the "limit cycle," and A_- indicates the negative amplitude. The steady oscillation period is computed as $P_u = P_+ + P_-$, where P_+ denotes the half period corresponding to the relay output u_+, and P_- is the other half period corresponding to u_-. Thus, by decomposing the process dynamic response from the set-point value, the resulting limit cycle can be separately studied for model fitting of the dynamic response.

In the presence of measurement noise, if the noise level is low (e.g., NSR < 10%), the statistical averaging method can be used in terms of 5 ∼ 20 steady oscillation periods to recover the limit cycle for model identification. To cope with a higher noise level, a low-pass Butterworth filter is suggested to recover the limit cycle for a relay feedback test or offline denoising. The Butterworth filter can be determined by specifying the filter order, n_f, and the cutoff angular frequency, ω_c, i.e.,

$$\text{Butter}(n_f, \omega_c) = \frac{b_1 + b_2 z^{-1} + b_3 z^{-2} + \cdots + b_{n_f+1} z^{-n_f}}{1 + a_2 z^{-1} + a_3 z^{-2} + \cdots + a_{n_f+1} z^{-n_f}} \quad (4.2)$$

where Butter(n_f, ω_c) denotes the filtering function with two input parameters of n_f and ω_c. In view of the fact that measurement noise is mainly composed of high-frequency components, a guideline for choosing the cutoff angular frequency is given as

$$\omega_c \geq 5\omega_u = 10\pi/P_u \quad (4.3)$$

where $\omega_u = 2\pi/P_u$ is the angular frequency of the limit cycle that can be computed from the measurement of P_u.

Thus, in an online relay feedback test, the measured output components only within the frequency band around ω_u can be passed through for feedback control. Note that the phase lag caused by the low-pass filter almost does not affect the measurement of the oscillation period and the amplitude of the limit cycle, because the relay output has the same phase lag under the filtered feedback control.

Moreover, a further improved denoising effect can be obtained by using an offline denoising strategy, i.e., filtering the noisy limit cycle data in both the forward and reverse directions with the same low-pass Butterworth filter.

4.2 Guidelines for Model Structure Selection

Owing to the fact that low-order models of FOPDT and SOPDT have been most widely used in practical applications, relay identification methods have been mainly devoted for identifying these low-order process models. Generally, FOPDT and SOPDT models are expressed as

$$G_{m\text{-}1} = \frac{k_p e^{-\theta s}}{\tau_p s + 1} \quad (4.4)$$

$$G_{m\text{-}2} = \frac{k_p e^{-\theta s}}{\tau_p^2 s^2 + 2\xi \tau_p s + 1} \quad (4.5)$$

where k_p denotes the process static gain and θ is the process time delay. Note that τ_p denotes the process time constant in an FOPDT model, while in an SOPDT model, it is a positive coefficient equal to the reciprocal of ω_p that is named the natural frequency of the process. The positive coefficient, ξ, in the denominator of

4.2 Guidelines for Model Structure Selection

Table 4.1 Relay response shapes of FOPDT models with different parameter settings

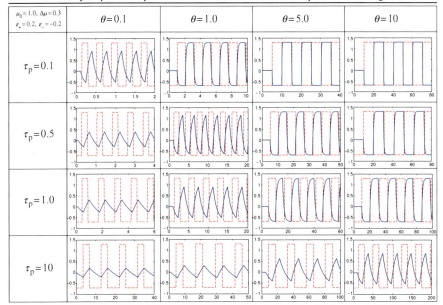

an SOPDT model is named the damping ratio of the process response. Typically, a classification of three types – underdamped ($0 < \xi < 1$), critically damped ($\xi = 1$), and overdamped ($\xi > 1$), has been widely used to describe different process dynamic response characteristics with respect to a step change in engineering practice (Ogunnaike and Ray 1994; Shinskey 1996; Seborg et al. 2004).

Given different output response shapes in relay identification tests, it is desired to choose a suitable low-order model structure for fitting the relay response shape of a process to be identified. For reference, Tables 4.1 and 4.2 show the relay response shapes of FOPDT and SOPDT models with different parameter settings under a biased relay test of $\mu_0 = 1.0$, $\Delta\mu = 0.3$, and $\varepsilon_+ = -\varepsilon_- = 0.2$. All of the model static gains have been fixed as the unity for benchmark comparison.

From these relay response shapes given in Tables 4.1 and 4.2, the guidelines for model structure selection are deduced as follows:

1. For FOPDT models, similar ratios of θ/τ result in the similar relay response shapes. This is also true for SOPDT models with a fixed damping ratio (ξ).
2. For FOPDT models and overdamped SOPDT models, the relay response shapes are similar to a triangular waveform with sharp edges and peaks. When $\theta/\tau \gg 1$, the relay response shapes of FOPDT models and critically damped SOPDT models tend to be a rectangular waveform following the relay output shape.
3. For underdamped and critically damped SOPDT models, the relay response shapes are similar to a sinusoidal waveform with smooth curvature and rounded peaks, and such characteristics become more apparent when ξ becomes smaller.

Table 4.2 Relay response shapes of SOPDT models with different parameter settings

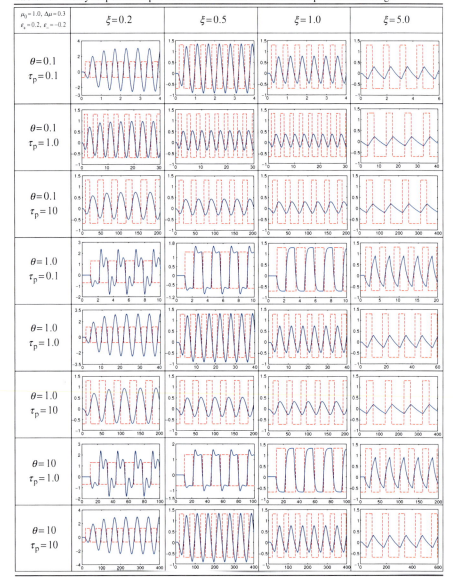

4. For FOPDT models and critically damped and overdamped ($\xi > 1$) SOPDT models, the relay responses move into a limit cycle almost after a single period of the relay output. The first relay response period is very similar (or even the same) in magnitude and shape to the following periods.

5. For an underdamped SOPDT model, when ξ becomes smaller, more switches of the relay output are needed for the relay response to move into a limit cycle, and multiple peaks are likely to occur in the limit cycle when $\theta/\tau \gg 1$.
6. For an FOPDT model, the amplitude of the limit cycle is inversely proportional to the process time constant and directly proportional to the process time delay, until the relay response becomes a rectangular waveform following the relay output shape.
7. For all these three types of SOPDT models, the amplitude of the limit cycle is proportional to the ratio of θ/τ with respect to a fixed ξ, and proportional to ξ with respect to a fixed ratio of θ/τ.

With the above guidelines, a suitable model structure can be intuitively chosen by comparing the relay response shape of the process to be identified with the reference relay response shapes shown in Tables 4.1 and 4.2. Note that the use of a biased or unbiased relay test or the height of the relay does not change the relay response shape of the process or any one of the above models. A reasonable estimation of ξ and the ratio of θ/τ can also be made with reference to the model parameter settings in Tables 4.1 and 4.2. Accordingly, parameter identification for the chosen model structure can now proceed to capture good fitting accuracy.

4.3 Low-Order Model Fitting Algorithms

Relay feedback identification methods for obtaining low-order process models of FOPDT and SOPDT have been increasingly reported in the past two decades. It is a trend to use only one relay feedback test for model identification to facilitate practical applications, as surveyed by Hang et al. (2002) and Atherton (2006). Based on an unbiased relay test, Luyben (2001) developed an FOPDT model identification method by defining a curvature factor for the relay response shape of the process; Vivek and Chidambaram (2005) derived another FOPDT algorithm based on the Fourier analysis of the process response; Panda and Yu (2005) gave an FOPDT algorithm by deriving an analytical expression for the unbiased relay response, which was then extended to obtain an underdamped SOPDT model (Panda 2006); Huang et al. (2005) developed an alternative identification method for obtaining FOPDT and underdamped SOPDT models based on the analysis of ultimate frequency; Majhi (2007a, b) derived relay response expressions and low-order model identification algorithms in terms of a state-space description of the relay control system. To resolve the difficulty associated with the derivation of the process static gain under an unbiased relay feedback test, biased relay identification methods have been developed (Li et al. 1991; Shen et al. 1996). Based on a biased relay test, Wang et al. (1997) derived an FOPDT algorithm by using the algebraic properties of periodic oscillations; Sung and Lee (1997) suggested a frequency response estimation algorithm for model fitting from the Fourier analysis; Srinivasan and Chidambaram (2003) also proposed such an FOPDT identification algorithm to effectively represent the process response over the low-frequency range, which was further extended to obtain an overdamped SOPDT model for improving fitting effect of the process frequency response (Ramakrishnan and Chidambaram 2003). Using complex function analysis, Kaya and Atherton (2001) developed a so-called A-locus identification algorithm for obtaining an FOPDT model or an overdamped SOPDT model.

It is well known that under a relay test the process response moves into the limit cycle. Following an early idea of deriving the relay response expression of the limit cycle for a linear time-invariant system (Atherton 1982; Tsypkin 1984), relay response expressions based on an unbiased relay test had been derived for FOPDT and SOPDT models (Thyagarajan and Yu 2003; Panda and Yu 2003). For generality, analytical relay response expressions (Liu and Gao 2008; Liu et al. 2008) for assessing the process response under an unbiased or biased relay test are presented here, based on a low-order model structure of FOPDT or SOPDT that is chosen for fitting the relay response. Accordingly, quantitative relationships between the measured limit cycle information and the model parameters are established from these relay response expressions for parameter estimation.

For clarity, these relay response expressions and identification algorithms are detailed in the following two subsections for obtaining FOPDT and SOPDT models, respectively.

4.3.1 The FOPDT Model

Consider an FOPDT model shown in (4.4), the following proposition gives the corresponding relay response expression under a biased relay test:

Proposition 4.1. *For a first-order stable process modeled by (4.4) under a biased relay test as shown in Fig. 4.5, the resulting limit cycle of the process output response is characterized by*

$$y_+(t) = k_p (\Delta \mu + \mu_0) - 2k_p \mu_0 E e^{-\frac{t}{\tau_p}}, \quad t \in [0, P_+] \quad (4.6)$$

$$y_-(t) = k_p (\Delta \mu - \mu_0) - 2k_p \mu_0 F e^{-\frac{t}{\tau_p}}, \quad t \in [0, P_-] \quad (4.7)$$

where $y_+(t)$ is the monotonically ascending part for $t \in [0, P_+]$, $y_-(t)$ is the monotonically descending part for $t \in [0, P_-]$ that corresponds to $t \in [P_+, P_u]$ in the limit cycle, $P_u = P_+ + P_-$ is the oscillation period, and

$$E = \frac{1 - e^{-\frac{P_-}{\tau_p}}}{1 - e^{-\frac{P_u}{\tau_p}}} \quad (4.8)$$

$$F = -\frac{1 - e^{-\frac{P_+}{\tau_p}}}{1 - e^{-\frac{P_u}{\tau_p}}} \quad (4.9)$$

4.3 Low-Order Model Fitting Algorithms

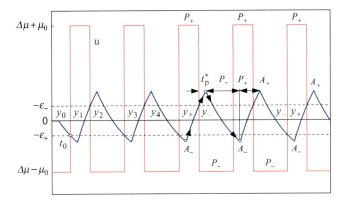

Fig. 4.5 Limit cycle analysis for an FOPDT model

Proof. The initial step response of an FOPDT stable process in (4.4) arising from the relay output, $u(t) = \Delta\mu - \mu_0$, can be derived as

$$y_0(t) = k_p (\Delta\mu - \mu_0) \left(1 - e^{-\frac{t-\theta}{\tau_p}}\right) \quad (4.10)$$

When it comes to the first relay switch point denoted by t_0 as shown in Fig. 4.5, the relay output changes to $\Delta\mu + \mu_0$, indicating that a step change of $2\mu_0$ is added to the process input. According to the linear superposition principle, the process output response can be derived as

$$y_1(t) = y_0(t + t_0) + 2k_p\mu_0 \left(1 - e^{-\frac{t-\theta}{\tau_p}}\right) \quad (4.11)$$

By using a time shift of $t_0 + \theta$, (4.11) can be rewritten as

$$y_1(t)|_{\text{shift}} = y_0(t + t_0 + \theta) + 2k_p\mu_0 - 2k_p\mu_0 e^{-\frac{t}{\tau_p}} \quad (4.12)$$

When it comes to the second relay switch point, the relay output changes to $\Delta\mu - \mu_0$, indicating that a step change of $-2\mu_0$ is added to the process input. According to the linear superposition principle, the process output response can be derived as

$$y_2(t) = y_1(t + P_+) - 2k_p\mu_0 \left(1 - e^{-\frac{t-\theta}{\tau_p}}\right) \quad (4.13)$$

By using a time shift of $t_0 + \theta + P_+$, (4.13) can be rewritten as

$$y_2(t)|_{\text{shift}} = y_0(t + t_0 + \theta + P_+) + 2k_p\mu_0 (1 - 1) - 2k_p\mu_0 e^{-\frac{t}{\tau_p}} \left(e^{-\frac{P_+}{\tau_p}} - 1\right) \quad (4.14)$$

At the third relay switch point, the relay output changes back to $\Delta\mu + \mu_0$, indicating that a step change of $2\mu_0$ is once again added to the process input. The process output response with a time shift of $t_0 + \theta + P_u$ can be derived as

$$y_3(t)|_{\text{shift}} = y_0(t + t_0 + \theta + P_u) + 2k_p\mu_0(1 - 1 + 1)$$
$$- 2k_p\mu_0 e^{-\frac{L}{\tau_p}}\left(e^{-\frac{P_u}{\tau_p}} - e^{-\frac{P_-}{\tau_p}} + 1\right) \tag{4.15}$$

The process output response following the fourth relay switch point is the result of four interlaced step changes, respectively, with a magnitude of $2\mu_0$. The process output response with a time shift of $t_0 + \theta + P_u + P_+$ can thus be derived as

$$y_4(t)|_{\text{shift}} = y_0(t + t_0 + \theta + P_u + P_+) + 2k_p\mu_0(1 - 1 + 1 - 1)$$
$$- 2k_p\mu_0 e^{-\frac{L}{\tau_p}}\left(e^{-\frac{P_u+P_+}{\tau_p}} - e^{-\frac{P_u}{\tau_p}} + e^{-\frac{P_+}{\tau_p}} - 1\right) \tag{4.16}$$

Hence, the time shifted process output response after each relay switch point can be summarized as

$$y_{2n+1}(t)|_{\text{shift}} = y_0(t + t_0 + \theta + nP_u) + 2k_p\mu_0 - 2k_p\mu_0 E e^{-\frac{L}{\tau_p}} \tag{4.17}$$

$$y_{2n+2}(t)|_{\text{shift}} = y_0(t + t_0 + \theta + nP_u + P_+) - 2k_p\mu_0 F e^{-\frac{L}{\tau_p}} \tag{4.18}$$

where $n = 0, 1, 2, \ldots$, and

$$E = 1 + \sum_{k=1}^{n}\left(e^{-\frac{kP_u}{\tau_p}} - e^{-\frac{(k-1)P_u+P_-}{\tau_p}}\right) \tag{4.19}$$

$$F = \sum_{k=0}^{n}\left(e^{-\frac{kP_u+P_+}{\tau_p}} - e^{-\frac{kP_u}{\tau_p}}\right) \tag{4.20}$$

Note that $0 < e^{-P_u/\tau_p} < 1$. It follows for $n \to \infty$ that

$$\sum_{k=0}^{n} e^{-\frac{kP_u}{\tau_p}} = \frac{1}{1 - e^{-\frac{P_u}{\tau_p}}} \tag{4.21}$$

Substituting (4.21) into (4.19) and (4.20), respectively, one can obtain

$$E = \frac{1}{1 - e^{-\frac{P_u}{\tau_p}}} - \frac{e^{-\frac{P_-}{\tau_p}}}{1 - e^{-\frac{P_u}{\tau_p}}} = \frac{1 - e^{-\frac{P_-}{\tau_p}}}{1 - e^{-\frac{P_u}{\tau_p}}} \tag{4.22}$$

4.3 Low-Order Model Fitting Algorithms

$$F = \left(e^{-\frac{P_+}{\tau_p}} - 1\right) \cdot \frac{1}{1 - e^{-\frac{P_u}{\tau_p}}} = -\frac{1 - e^{-\frac{P_+}{\tau_p}}}{1 - e^{-\frac{P_u}{\tau_p}}} \quad (4.23)$$

It can be seen from (4.10) that $y_0(t + t_0 + \theta + nP_u) = y_0(t + t_0 + \theta + nP_u + P_+) = k_p(\Delta\mu - \mu_0)$ for $n \to \infty$. Hence, in the limit cycle it follows that

$$y_+(t) = \lim_{n\to\infty} y_{2n+1}(t)|_{\text{shift}} = k_p(\Delta\mu + \mu_0) - 2k_p\mu_0 E e^{-\frac{t}{\tau_p}}, \quad t \in [0, P_+] \quad (4.24)$$

$$y_-(t) = \lim_{n\to\infty} y_{2n+2}(t)|_{\text{shift}} = k_p(\Delta\mu - \mu_0) - 2k_p\mu_0 F e^{-\frac{t}{\tau_p}}, \quad t \in [0, P_-] \quad (4.25)$$

In view of the fact that $E > 0$, $F < 0$, and e^{-t/τ_p} decreases monotonically with respect to t, one can conclude from (4.24) and (4.25) that $y_+(t)$ increases monotonically for $t \in [0, P_+]$, while $y_-(t)$ decreases monotonically for $t \in [0, P_-]$ that corresponds to $t \in [P_+, P_u]$ in the limit cycle.

From the above derivation for (4.24) and (4.25), it can be seen that the limit cycle can be definitely formed for an FOPDT stable process. It should be noted that before the process output response moves into the limit cycle, the time intervals between the relay switch points may not be equal to the corresponding half periods in the limit cycle. Nevertheless, one can equalize them as done in (4.13)–(4.16) to derive the exact expression of the limit cycle, in view of the fact that the limit cycle does not reflect these initial response differences. In other words, the process response in steady oscillation has the same limit cycle with that of the corresponding ideal oscillation, which has identical time intervals between the sequential relay switch points from beginning to end. This completes the proof. □

If an unbiased relay feedback test is used, there exist $\Delta\mu = 0$ and $P_+ = P_- = P_u/2$. By substituting them into (4.6)–(4.9), one can obtain the corresponding relay response expression,

$$y_+(t) = -y_-(t) = k_p\mu_0 - 2k_p\mu_0 E e^{-\frac{t}{\tau_p}} \quad (4.26)$$

where $y_+(t)$ is for $t \in [0, P_u/2]$, $y_-(t)$ is for $t \in (P_u/2, P_u]$, and

$$E = \frac{1}{1 + e^{-\frac{P_u}{2\tau_p}}} \quad (4.27)$$

Note that in the limit cycle, the process output is a periodic function with respect to the oscillation angular frequency, $\omega_u = 2\pi/P_u$. Using the idea of a time shift, one may view it as a periodic signal from the very beginning, so its Fourier transform can be derived as

$$Y(j\omega_u) = \lim_{N\to\infty} N \int_0^{P_u} y_{os}(t)e^{-j\omega_u t}dt = \lim_{N\to\infty} N \int_{t_{os}}^{t_{os}+P_u} y(t)e^{-j\omega_u t}dt \quad (4.28)$$

where $y_{os}(t) = y(t)$ for $t \in [t_{os}, \infty)$ and t_{os} can be taken as any relay switch point in steady oscillation, such that the influence from the initial process response to the above periodic integral can be excluded.

Similarly, it follows that

$$U(j\omega_u) = \lim_{N\to\infty} N \int_{t_{os}}^{t_{os}+P_u} u(t)e^{-j\omega_u t}dt \quad (4.29)$$

Thereby, the process frequency response at ω_u can be obtained as

$$G(j\omega_u) = \frac{Y(j\omega_u)}{U(j\omega_u)} = \frac{\int_{t_{os}}^{t_{os}+P_u} y(t)e^{-j\omega_u t}dt}{\int_{t_{os}}^{t_{os}+P_u} u(t)e^{-j\omega_u t}dt} = A_u e^{j\varphi_u} \quad (4.30)$$

In fact, the numerical integral in (4.30) can be computed using the trapezoidal rule or the fast Fourier transform (FFT) for $Y(j\omega_u)$ and $U(j\omega_u)$.

It should be noted that $\angle G(j\omega_u) = -\pi$ had been used for parameter identification in many existing references (e.g., Atherton 1982; Åström and Hägglund 1984; Luyben 1987; Li et al. 1991; Shen et al. 1996; Huang et al. 2005; Yu 2006), based on the describing function analysis. Such exercise will result in degraded identification accuracy since $\angle G(j\omega_u)$ is actually larger than $-\pi$ due to the phase lag caused by the relay, as will be illustrated by the later identification algorithms and examples.

When a biased relay test is used, the process static gain can be derived from (4.30) as

$$k_p = G(0) = \frac{\int_{t_{os}}^{t_{os}+P_u} y(t)dt}{\int_{t_{os}}^{t_{os}+P_u} u(t)dt} \quad (4.31)$$

Note that Proposition 4.1 indicates that the process time delay can be intuitively measured as the time taken to reach the peak of the process output response from the initial relay switch point in a half period of the relay, which is denoted as t_p^* as shown in Fig. 4.5. Correspondingly, it follows from (4.6) and (4.7) that

$$y_+(P_+) = k_p(\Delta\mu + \mu_0) - 2k_p\mu_0 E e^{-\frac{P_+}{\tau_p}} = A_+ \quad (4.32)$$

$$y_-(P_-) = k_p(\Delta\mu - \mu_0) - 2k_p\mu_0 F e^{-\frac{P_-}{\tau_p}} = A_- \quad (4.33)$$

4.3 Low-Order Model Fitting Algorithms

$$y_+ (P_+ - \theta) = k_p (\Delta\mu + \mu_0) - 2k_p\mu_0 E e^{-\frac{P_+ - \theta}{\tau_p}} = -\varepsilon_- \quad (4.34)$$

$$y_- (P_- - \theta) = k_p (\Delta\mu - \mu_0) - 2k_p\mu_0 F e^{-\frac{P_- - \theta}{\tau_p}} = -\varepsilon_+ \quad (4.35)$$

Substituting (4.32) into (4.34) yields

$$\tau_p = \frac{\theta}{\ln \frac{k_p(\Delta\mu+\mu_0)+\varepsilon_-}{k_p(\Delta\mu+\mu_0)-A_+}} \quad (4.36)$$

Alternatively, substituting (4.33) into (4.35) yields

$$\tau_p = \frac{\theta}{\ln \frac{k_p(\Delta\mu-\mu_0)+\varepsilon_+}{k_p(\Delta\mu-\mu_0)-A_-}} \quad (4.37)$$

Therefore, the process time constant can be derived from (4.36) or (4.37). It is preferred to use (4.36) for better accuracy in the presence of model mismatch, in view of that the positive fitting part in a half period of the limit cycle occupies a larger percentage compared to the negative fitting part in the other half period, and vice versa.

Hence, the above identification algorithm named Algorithm-RS-FA1 for obtaining an FOPDT model under a biased relay test can be summarized.

Algorithm-RS-FA1

(i) Measure P_u and A_+ from the limit cycle.
(ii) Measure the process time delay as $\theta = t_p^*$.
(iii) Compute the process static gain, k_p, from (4.31).
(iv) Compute the process time constant, τ_p, from (4.36).

Note that by substituting the FOPDT model in (4.4) into (4.30), one can obtain the process response fitting conditions at the oscillation frequency,

$$\frac{k_p}{\sqrt{\tau_p^2\omega_u^2 + 1}} = A_u \quad (4.38)$$

$$-\theta\omega_u - \arctan(\tau_p\omega_u) = \varphi_u \quad (4.39)$$

It can be seen from (4.39) that $\varphi_u \in (-\pi, -\pi/2)$, rather than $\varphi_u = -\pi$ that had been conventionally used in the describing function analysis (Atherton 1982).

Accordingly, the process time constant and time delay can be derived from (4.38) and (4.39) as

$$\tau_p = \frac{1}{\omega_u}\sqrt{\frac{k_p^2}{A_u^2} - 1} \tag{4.40}$$

$$\theta = -\frac{1}{\omega_u}\left[\varphi_u + \arctan(\tau_p\omega_u)\right] \tag{4.41}$$

Hence, an alternative identification algorithm named Algorithm-RS-FA2 for obtaining an FOPDT model under a biased relay test can be summarized.

Algorithm-RS-FA2

(i) Measure P_u from the limit cycle.
(i) Compute $G(j\omega_u)$ from (4.30).
(ii) Compute the process static gain, k_p, from (4.31).
(iii) Compute the process time constant, τ_p, from (4.40).
(iv) Compute the process time delay, θ, from (4.41).

Owing to the fact that the process response at the oscillation frequency is precisely represented by the FOPDT model derived from Algorithm-RS-FA2, it can be used to obtain enhanced identification accuracy for a higher order process compared to Algorithm-RS-FA1, but at the cost of more computation effort.

When an unbiased relay test is used, it can also be verified by substituting $\Delta\mu = 0$ and $P_+ = P_- = P_u/2$ into Proposition 4.1 that the process time delay can be directly measured as the time taken to reach the peak of the process output response from the initial relay switch point in a half period of the relay, i.e., t_p^*, as shown in Fig. 4.5. Accordingly, solving (4.32) and (4.34) together to eliminate k_p yields

$$\varepsilon_+\left(1 - e^{-\frac{P_u}{2\tau_p}}\right) = A_+\left(1 + e^{-\frac{P_u}{2\tau_p}} - 2e^{-\frac{P_u - 2\theta}{2\tau_p}}\right) \tag{4.42}$$

Note that the left-hand side of (4.42) is monotonically decreasing with respect to τ_p, subject to a practical constraint of $0 < \tau_p < P_u$. Only finite solutions of τ_p can, therefore, exist for (4.42), which may be derived using any iterative algorithm such as the Newton-Raphson method. The initial estimation for iteration may be taken as $\widehat{\tau}_p = P_u/2 - \theta$, in consideration of the fact that the influence from the process time constant to the relay response corresponds to this time interval.

The process static gain can then be derived from (4.32) as

$$k_p = \frac{A_+\left(1 + e^{-\frac{P_u}{2\tau_p}}\right)}{\mu_0\left(1 - e^{-\frac{P_u}{2\tau_p}}\right)} \tag{4.43}$$

4.3 Low-Order Model Fitting Algorithms

Note that in the case where multiple solutions of τ_p and k_p are obtained from the above computation, a suitable solution pair of τ_p and k_p can be determined by comparing the relay response of the resulting model with that of the process, or by using the critical oscillation conditions from the describing function analysis (Atherton 1982; Yu 2006).

Hence, the above identification algorithm named Algorithm-RS-FB1 for obtaining an FOPDT model under an unbiased relay test can be summarized.

Algorithm-RS-FB1

(i) Measure P_u and A_+ from the limit cycle.
(ii) Measure the process time delay as $\theta = t_p^*$.
(iii) Compute the process time constant, τ_p, from (4.42) by using the Newton-Raphson iteration method. The initial estimation for iteration may be taken as $\hat{\tau}_p = P_u/2 - \theta$.
(iv) Compute the process static gain, k_p, from (4.43).
(v) Determine the suitable solution pair of τ_p and k_p by comparing the relay response of the resulting model with that of the process or by checking if
$|N(A_+)\hat{G}(j\omega_u)| \to 1$ and $\angle N(A_+) + \angle \hat{G}(j\omega_u) \to -\pi$ are satisfied, where $N(A_+) = 4\mu_0 e^{-j\arcsin(\varepsilon_+/A_+)}/(\pi A_+)$ denotes the describing function of the unbiased relay and $\hat{G}(j\omega_u)$ is the FOPDT model response at the oscillation frequency.

With the process time delay directly measured from the limit cycle, the process static gain and time constant can also be derived from the fitting conditions at the oscillation frequency as shown in (4.38) and (4.39),

$$\tau_p = \frac{1}{\omega_u}\tan(-\varphi_u - \theta\omega_u), \quad \varphi_u \in (-\pi, -\pi/2) \tag{4.44}$$

$$k_p = A_u\sqrt{\tau_p^2\omega_u^2 + 1} \tag{4.45}$$

To procure fitting accuracy for the identification of a higher order process as encountered in practice, a one-dimensional search of θ can be implemented in terms of the following time domain fitting condition of the relay response,

$$err = \frac{1}{N_p}\sum_{k=1}^{N_p}\left[y(kT_s + t_{os}) - \hat{y}(kT_s + t_{os})\right]^2 < \varepsilon \tag{4.46}$$

where $y(kT_s + t_{os})$ and $\hat{y}(kT_s + t_{os})$ denote the relay responses of the process and the model in the limit cycle, respectively, T_s is the sampling period corresponding to $N_p = P_u/T_s$, and ε is a user-specified fitting threshold for computation. The optimal fitting can be determined by deriving such a model that yields the smallest value of *err*. Improved fitting accuracy can thus be obtained for the identification of a higher order process, compared to Algorithm-RS-FB1.

Hence, an alternative identification named Algorithm-RS-FB2 for obtaining an FOPDT model under an unbiased relay test can be summarized.

Algorithm-RS-FB2

(i) Measure P_u and the process time delay ($\theta = t_p^*$) from the limit cycle.
(ii) Compute $G(j\omega_u)$ from (4.30).
(iii) Compute the process time constant, τ_p, from (4.44).
(iv) Compute the process static gain, k_p, from (4.45).
(v) End the algorithm if the fitting condition of relay response shown in (4.46) is satisfied. Otherwise, go back to Step (iii) by monotonically varying θ for a one-dimensional search within a possible range as observed from the initial step response in the relay test.

4.3.2 The SOPDT Model

According to the classification of the three types – underdamped, critically damped, and overdamped – for identifying different process response characteristics, the corresponding model identification algorithms are presented as follows:

Type 1 The Critically Damped SOPDT Model

A critically damped second-order process is generally in the form of

$$G_{sc} = \frac{k_p e^{-\theta s}}{(\tau_p s + 1)^2} \qquad (4.47)$$

The following proposition gives the exact relay response expression under a biased relay test.

Proposition 4.2. *For a critically damped second-order process modeled by (4.47) under a biased relay test as shown in Fig. 4.6a, the resulting limit cycle of the process output response is characterized by*

$$y_+(t) = k_p(\Delta\mu + \mu_0) - 2k_p\mu_0 e^{-\frac{t}{\tau_p}}\left(E_1 + \frac{E_1 t + E_2}{\tau_p}\right), \quad t \in [0, P_+] \quad (4.48)$$

$$y_-(t) = k_p(\Delta\mu - \mu_0) - 2k_p\mu_0 e^{-\frac{t}{\tau_p}}\left(F_1 + \frac{F_1 t + F_2}{\tau_p}\right), \quad t \in [0, P_-] \quad (4.49)$$

4.3 Low-Order Model Fitting Algorithms

where $y_+(t)$ is for $t \in [0, P_+]$ while $y_-(t)$ is for $t \in (P_+, P_u]$, $E_2 < 0 < E_1$, $F_1 < 0 < F_2$, and

$$E_1 = \frac{1 - e^{-\frac{p_-}{\tau_p}}}{1 - e^{-\frac{p_u}{\tau_p}}} \tag{4.50}$$

$$E_2 = \frac{p_u e^{-\frac{p_u}{\tau_p}}\left(1 - e^{-\frac{p_-}{\tau_p}}\right)}{\left(1 - e^{-\frac{p_u}{\tau_p}}\right)^2} - \frac{p_- e^{-\frac{p_-}{\tau_p}}}{1 - e^{-\frac{p_u}{\tau_p}}} \tag{4.51}$$

$$F_1 = -\frac{1 - e^{-\frac{p_+}{\tau_p}}}{1 - e^{-\frac{p_u}{\tau_p}}} \tag{4.52}$$

$$F_2 = -\frac{p_u e^{-\frac{p_u}{\tau_p}}\left(1 - e^{-\frac{p_+}{\tau_p}}\right)}{\left(1 - e^{-\frac{p_u}{\tau_p}}\right)^2} + \frac{p_+ e^{-\frac{p_+}{\tau_p}}}{1 - e^{-\frac{p_u}{\tau_p}}} \tag{4.53}$$

Proof. The initial step response of a critically damped SOPDT process in (4.47) arising from the relay output, $u(t) = \Delta\mu - \mu_0$, can be derived as

$$y_0(t) = k_p (\Delta\mu - \mu_0)\left[1 - \left(1 + \frac{t - \theta}{\tau_p}\right)e^{-\frac{t-\theta}{\tau_p}}\right] \tag{4.54}$$

When it comes to the first relay switch point denoted by t_0 as shown in Fig. 4.6a, the relay output changes to $\Delta\mu + \mu_0$, indicating that a step change of $2\mu_0$ is added to the process input. According to the linear superposition principle, the process output response can be derived as

$$y_1(t) = y_0(t + t_0) + 2k_p\mu_0 - 2k_p\mu_0\left(1 + \frac{t - \theta}{\tau_p}\right)e^{-\frac{t-\theta}{\tau_p}} \tag{4.55}$$

By using a time shift of $t_0 + \theta$, (4.55) can be rewritten as

$$y_1(t)|_{\text{shift}} = y_0(t + t_0 + \theta) + 2k_p\mu_0 - 2k_p\mu_0\left(1 + \frac{t}{\tau_p}\right)e^{-\frac{t}{\tau_p}} \tag{4.56}$$

When it comes to the second relay switch point, the relay output changes to $\Delta\mu - \mu_0$, indicating that a step change of $-2\mu_0$ is added to the process input. According to the linear superposition principle, the process output response can be derived as

$$y_2(t) = y_1(t + P_+) - 2k_p\mu_0 + 2k_p\mu_0\left(1 + \frac{t - \theta}{\tau_p}\right)e^{-\frac{t-\theta}{\tau_p}} \tag{4.57}$$

136　　　　　　　　　　4　Relay Feedback Identification of Stable Processes

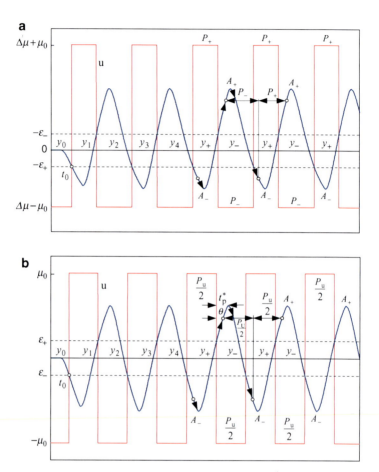

Fig. 4.6 Limit cycle analysis for a critically damped SOPDT model

Using a time shift of $t_0 + \theta + P_+$, (4.57) can be rewritten as

$$y_2(t)|_{\text{shift}} = y_0\left(t + t_0 + \theta + P_+\right) + 2k_p\mu_0\left(1 - 1\right)$$

$$- 2k_p\mu_0 e^{-\frac{t}{\tau_p}} \left\{ e^{-\frac{P_+}{\tau_p}} - 1 + \frac{1}{\tau_p}\left[t\left(e^{-\frac{P_+}{\tau_p}} - 1\right) + P_+ e^{-\frac{P_+}{\tau_p}}\right]\right\}$$
(4.58)

At the third relay switch point, the relay output changes back to $\Delta\mu + \mu_0$, indicating that a step change of $2\mu_0$ is once again added to the process input. The process output response with a time shift of $t_0 + \theta + P_u$ can be derived accordingly as

4.3 Low-Order Model Fitting Algorithms

$$y_3(t)|_{\text{shift}} = y_0(t + t_0 + \theta + P_u) + 2k_p\mu_0(1 - 1 + 1)$$

$$- 2k_p\mu_0 e^{-\frac{L}{\tau_p}} \left\{ e^{-\frac{P_u}{\tau_p}} - e^{-\frac{P_-}{\tau_p}} + 1 + \frac{1}{\tau_p}\left[t\left(e^{-\frac{P_u}{\tau_p}} - e^{-\frac{P_-}{\tau_p}} + 1\right) + P_u e^{-\frac{P_u}{\tau_p}} - P_- e^{-\frac{P_-}{\tau_p}}\right]\right\}$$
(4.59)

where $P_u = P_+ + P_-$.

The process output response following the fourth relay switch point is the result of four interlaced step changes, respectively, with a magnitude of $2\mu_0$. The process output response with a time shift of $t_0 + \theta + P_u + P_+$ can thus be derived as

$$y_4(t)|_{\text{shift}} = y_0(t + t_0 + \theta + P_u + P_+) + 2k_p\mu_0(1 - 1 + 1 - 1)$$

$$- 2k_p\mu_0 e^{-\frac{L}{\tau_p}} \left\{ e^{-\frac{P_u+P_+}{\tau_p}} - e^{-\frac{P_u}{\tau_p}} + e^{-\frac{P_+}{\tau_p}} - 1 + \frac{1}{\tau_p}\left[t\left(e^{-\frac{P_u+P_+}{\tau_p}} - e^{-\frac{P_u}{\tau_p}} + e^{-\frac{P_+}{\tau_p}} - 1\right)\right.\right.$$

$$\left.\left. + (P_u + P_+)e^{-\frac{P_u+P_+}{\tau_p}} - P_u e^{-\frac{P_u}{\tau_p}} + P_+ e^{-\frac{P_+}{\tau_p}}\right]\right\}$$
(4.60)

It can be summarized from (4.56)–(4.60) that

$$y_{2n+1}(t)|_{\text{shift}} = y_0(t + t_0 + \theta + nP_u) + 2k_p\mu_0 - 2k_p\mu_0 e^{-\frac{L}{\tau_p}} \left[E_1 + \frac{(E_1 t + E_2)}{\tau_p}\right]$$
(4.61)

$$y_{2n+2}(t)|_{\text{shift}} = y_0(t + t_0 + \theta + nP_u + P_+) - 2k_p\mu_0 e^{-\frac{L}{\tau_p}} \left[F_1 + \frac{(F_1 t + F_2)}{\tau_p}\right]$$
(4.62)

where $n = 0, 1, 2, \ldots$, and

$$E_1 = 1 + \sum_{k=1}^{n}\left(e^{-\frac{kP_u}{\tau_p}} - e^{-\frac{(k-1)P_u + P_-}{\tau_p}}\right)$$
(4.63)

$$E_2 = \sum_{k=1}^{n} kP_u e^{-\frac{kP_u}{\tau_p}} - \sum_{k=1}^{n}[(k-1)P_u + P_-]e^{-\frac{(k-1)P_u + P_-}{\tau_p}}$$
(4.64)

$$F_1 = \sum_{k=0}^{n}\left(e^{-\frac{kP_u+P_+}{\tau_p}} - e^{-\frac{kP_u}{\tau_p}}\right)$$
(4.65)

$$F_2 = \sum_{k=0}^{n}\left[(kP_u + P_+)e^{-\frac{kP_u+P_+}{\tau_p}} - kP_u e^{-\frac{kP_u}{\tau_p}}\right]$$
(4.66)

In view of the fact that $0 < e^{-P_u/\tau_p} < 1$, it follows for $n \to \infty$ that

$$\sum_{k=0}^{n} e^{-\frac{kP_u}{\tau_p}} = \frac{1}{1 - e^{-\frac{P_u}{\tau_p}}} \qquad (4.67)$$

Note that

$$\lim_{n \to \infty} \sum_{k=1}^{n} k P_u e^{-\frac{kP_u}{\tau_p}} = \lim_{n \to \infty} P_u \left(\sum_{k=1}^{n} e^{-\frac{kP_u}{\tau_p}} + \sum_{k=2}^{n} e^{-\frac{kP_u}{\tau_p}} + \cdots + \sum_{k=n-1}^{n} e^{-\frac{kP_u}{\tau_p}} + e^{-\frac{nP_u}{\tau_p}} \right)$$

$$= \lim_{n \to \infty} \frac{P_u}{1 - e^{-\frac{P_u}{\tau_p}}} \left(e^{-\frac{P_u}{\tau_p}} + e^{-\frac{2P_u}{\tau_p}} + \cdots + e^{-\frac{(n-1)P_u}{\tau_p}} + e^{-\frac{nP_u}{\tau_p}} \right)$$

$$= \frac{P_u e^{-\frac{P_u}{\tau_p}}}{\left(1 - e^{-\frac{P_u}{\tau_p}}\right)^2}$$

(4.68)

By substituting (4.67) and (4.68) into (4.63)–(4.66), one can obtain the simplified forms of E_1, E_2, F_1, and F_2, as shown in (4.50)–(4.53).

Note that $y_0(t + t_0 + \theta + n P_u) = y_0(t + t_0 + \theta + n P_u + P_+) = k_p(\Delta \mu - \mu_0)$ for $n \to \infty$. Hence, in the limit cycle it follows that

$$y_+(t) = \lim_{n \to \infty} y_{2n+1}(t)|_{\text{shift}} = k_p(\Delta \mu + \mu_0)$$

$$- 2k_p \mu_0 e^{-\frac{t}{\tau_p}} \left[E_1 + \frac{(E_1 t + E_2)}{\tau_p} \right], \quad t \in [0, P_+] \qquad (4.69)$$

$$y_-(t) = \lim_{n \to \infty} y_{2n+2}(t)|_{\text{shift}} = k_p(\Delta \mu - \mu_0)$$

$$- 2k_p \mu_0 e^{-\frac{t}{\tau_p}} \left[F_1 + \frac{(F_1 t + F_2)}{\tau_p} \right], \quad t \in [0, P_-] \qquad (4.70)$$

where $y_+(t)$ denotes the ascending output response in a half period, P_+, corresponding to a positive step change of the relay output, while $y_-(t)$ denotes the descending output response in the other half period, P_-, corresponding to a negative step change of the relay output.

Note that E_2 shown in (4.51) can be reformulated as

$$E_2 = \frac{P_+ e^{-\frac{P_u}{\tau_p}} \left(1 - e^{-\frac{P_-}{\tau_p}}\right) + P_- e^{-\frac{P_-}{\tau_p}} \left(e^{-\frac{P_+}{\tau_p}} - 1\right)}{\left(1 - e^{-\frac{P_u}{\tau_p}}\right)^2} \qquad (4.71)$$

4.3 Low-Order Model Fitting Algorithms

Let

$$g(x) = \frac{1 - e^{-\frac{x}{\tau_p}}}{x}, \quad x \in (0, \infty) \tag{4.72}$$

It follows that

$$\frac{dg(x)}{dx} = \frac{\left(\frac{x}{\tau_p} + 1\right) e^{-\frac{x}{\tau_p}} - 1}{x^2} < 0 \tag{4.73}$$

Thus, one can conclude that $g(P_+) > g(P_-)$ since $P_+ < P_- < P_u$. Accordingly, it can be concluded from (4.71) that $E_2 < 0$. In a similar way, one can ensure that $F_2 > 0$. This completes the proof. □

If an unbiased relay test is used, as shown in Fig. 4.6b, there exist $\Delta\mu = 0$ and $P_+ = P_- = P_u/2$. By substituting them into (4.48)–(4.53), one can obtain the corresponding expression,

$$y_+(t) = -y_-(t) = k_p \mu_0 - 2k_p \mu_0 e^{-\frac{t}{\tau_p}} \left(E_1 + \frac{E_1 t + E_2}{\tau_p} \right) \tag{4.74}$$

where $y_+(t)$ is for $t \in [0, P_u/2]$, $y_-(t)$ for $t \in (P_u/2, P_u]$, and

$$E_1 = \frac{1}{1 + e^{-\frac{P_u}{2\tau_p}}} \tag{4.75}$$

$$E_2 = -\frac{P_u e^{-\frac{P_u}{2\tau_p}}}{2\left(1 + e^{-\frac{P_u}{2\tau_p}}\right)^2} \tag{4.76}$$

When a biased relay test is used, the process static gain can be derived from (4.31). It can be derived from (4.48) and (4.49) that

$$\frac{dy_+(t)}{dt} = \frac{2k_p \mu_0 (E_1 t + E_2)}{\tau_p^2} e^{-\frac{t}{\tau_p}} \tag{4.77}$$

$$\frac{dy_-(t)}{dt} = \frac{2k_p \mu_0 (F_1 t + F_2)}{\tau_p^2} e^{-\frac{t}{\tau_p}} \tag{4.78}$$

It can, therefore, be concluded that $y_+(t)$ does not increase monotonically for $t \in [0, P_+]$, and correspondingly, $y_-(t)$ does not decrease monotonically for $t \in (P_+, P_u]$.

By substituting the process model in (4.47) into the frequency response fitting condition shown in (4.30), the process time constant can be derived as

$$\tau_p = \frac{1}{\omega_u} \sqrt{\frac{k_p}{A_u} - 1} \tag{4.79}$$

and then using the phase fitting condition,

$$-\theta\omega_u - 2\arctan(\tau_p\omega_u) = \varphi_u, \quad \varphi_u \in (-\pi, -\pi/2) \tag{4.80}$$

the process time delay can be derived as

$$\theta = -\frac{1}{\omega_u}\left[\varphi_u + 2\arctan(\tau_p\omega_u)\right] \tag{4.81}$$

Hence, the above algorithm named Algorithm-RS-SC1 for the identification of a critically damped SOPDT model under a biased relay test can be summarized.

Algorithm-RS-SC1

(i) Measure P_+ and P_- from the limit cycle.
(ii) Compute $G(j\omega_u)$ from (4.30).
(iii) Compute the process static gain, k_p, from (4.31).
(iv) Compute the process time constant, τ_p, from (4.79).
(v) Compute the process time delay, θ, from (4.81).

When an unbiased relay test is used, by letting (4.77) equal zero one can compute the time to reach the peak value of $y_+(t)$ (or $y_-(t)$) as

$$t_p = -\frac{E_2}{E_1} = \frac{p_u}{2\left(1+e^{\frac{p_u}{2\tau_p}}\right)} \tag{4.82}$$

In fact, the time to reach the output response peak from the initial relay switch point in a half period of the relay, t_p^*, can be measured as shown in Fig. 4.6b. It follows that

$$t_p = t_p^* - \theta \tag{4.83}$$

Substituting (4.81) and (4.82) into (4.83) yields

$$t_p^* + \frac{1}{\omega_u}\left[\varphi_u + 2\arctan(\tau_p\omega_u)\right] = \frac{p_u}{2\left(1+e^{\frac{p_u}{2\tau_p}}\right)} \tag{4.84}$$

It can be seen that both sides of (4.84) increase monotonically with respect to τ_p. For $\tau_p \in (0,\infty)$, the left-hand side gives a value in the range of $(t_p^* + \varphi_u/\omega_u, t_p^* + (\pi + \varphi_u)/\omega_u)$, while the right-hand side is in the range of $(0, p_u/4)$. Therefore, there exists only finite solutions of τ_p for (4.84), which can be derived using any iterative algorithm such as the Newton-Raphson method. The initial estimation of τ_p for iteration may be taken as $\hat{\tau}_p = P_u/2 - t_p^*$, in view of the fact that the influence from the process time constant to the relay response corresponds to this time interval.

4.3 Low-Order Model Fitting Algorithms

The process time delay can then be derived from (4.81), and the process static gain can be inversely derived from the magnitude fitting condition as used in (4.79), i.e.,

$$k_p = A_u \left(\tau_p^2 \omega_u^2 + 1 \right) \qquad (4.85)$$

In the case where multiple solutions of τ_p are obtained for (4.84), the time domain fitting condition shown in (4.46) may be used to screen out the most suitable solution.

Alternatively, with a prespecified value of the process time delay (θ), the process time constant can be derived from the phase fitting condition shown in (4.80) as

$$\tau_p = \frac{1}{\omega_u} \tan\left[(-\varphi_u - \theta \omega_u)/2 \right], \quad \varphi_u \in (-\pi, -\pi/2) \qquad (4.86)$$

Correspondingly, a one-dimensional search of θ is needed in combination with the fitting condition of relay response shown in (4.46) to determine the optimal fitting.

Hence, the above algorithm named Algorithm-RS-SC2 for the identification of a critically damped SOPDT model under an unbiased relay test can be summarized.

Algorithm-RS-SC2

(i) Measure P_+, P_-, and t_p^* from the limit cycle.
(ii) Compute $G(j\omega_u)$ from (4.30).
(iii) Compute the process time constant, τ_p, from (4.84) using the Newton-Raphson iteration method, or from (4.86) based on a prespecified time delay (θ). The initial estimation of τ_p for iteration may be taken as $\hat{\tau}_p = P_u/2 - t_p^*$, or alternatively, a one-dimensional search of θ can be implemented within a possible range as observed from the initial step response in the relay test.
(iv) Compute the process time delay, θ, from (4.81) if τ_p has been computed from (4.84).
(v) Compute the process static gain, k_p, from (4.85).
(vi) End the algorithm if the fitting condition of relay response shown in (4.46) is satisfied. Otherwise, go back to Step (iii) by changing the initial estimation of τ_p, or monotonically varying θ for a one-dimensional search.

Note that for a critically damped second-order processes with a dominant time delay ($\theta/\tau \gg 1$), $y_+(t)$ increases monotonically while $y_-(t)$ decreases monotonically, owing to that $E_2 \to 0$ and $F_2 \to 0$. Accordingly, the process time delay can be directly measured as the time to reach the positive peak (A_+) of the output response from the initial relay switch point in a negative half period of the relay, so that the process time constant and static gain can then be derived from (4.86) and (4.85) for simplicity.

Type 2 The Overdamped SOPDT Model

An overdamped second-order process is generally in the form of

$$G_{so} = \frac{k_p e^{-\theta s}}{(\tau_1 s + 1)(\tau_2 s + 1)} \tag{4.87}$$

where $\tau_1 > \tau_2 > 0$ is assumed without loss of generality. The following proposition gives the exact relay response expression under a biased relay test:

Proposition 4.3. *For an overdamped second-order process modeled by (4.87) under a biased relay test as shown in Fig. 4.7a, the resulting limit cycle of the process output response is characterized by*

$$y_+(t) = k_p(\Delta\mu + \mu_0) - 2k_p\mu_0 \left(\frac{\tau_1 E_1}{\tau_1 - \tau_2} e^{-\frac{t}{\tau_1}} - \frac{\tau_2 E_2}{\tau_1 - \tau_2} e^{-\frac{t}{\tau_2}} \right), \quad t \in [0, P_+] \tag{4.88}$$

$$y_-(t) = k_p(\Delta\mu - \mu_0) - 2k_p\mu_0 \left(\frac{\tau_1 F_1}{\tau_1 - \tau_2} e^{-\frac{t}{\tau_1}} - \frac{\tau_2 F_2}{\tau_1 - \tau_2} e^{-\frac{t}{\tau_2}} \right), \quad t \in [0, P_-] \tag{4.89}$$

where $0 < E_1 < E_2$, $F_2 < F_1 < 0$, and

$$E_1 = \frac{1 - e^{-\frac{P_-}{\tau_1}}}{1 - e^{-\frac{P_u}{\tau_1}}} \tag{4.90}$$

$$E_2 = \frac{1 - e^{-\frac{P_-}{\tau_2}}}{1 - e^{-\frac{P_u}{\tau_2}}} \tag{4.91}$$

$$F_1 = -\frac{1 - e^{-\frac{P_+}{\tau_1}}}{1 - e^{-\frac{P_u}{\tau_1}}} \tag{4.92}$$

$$F_2 = -\frac{1 - e^{-\frac{P_+}{\tau_2}}}{1 - e^{-\frac{P_u}{\tau_2}}} \tag{4.93}$$

Proof. The initial step response of an overdamped SOPDT process in (4.87) arising from the relay output, $u(t) = \Delta\mu - \mu_0$, can be derived as

$$y_0(t) = k_p(\Delta\mu - \mu_0) \left(1 - \frac{\tau_1}{\tau_1 - \tau_2} e^{-\frac{t-\theta}{\tau_1}} + \frac{\tau_2}{\tau_1 - \tau_2} e^{-\frac{t-\theta}{\tau_2}} \right) \tag{4.94}$$

Following a similar analysis as in the proof for Proposition 4.2, the time shifted output response from the initial to the fourth relay switch point can be derived, respectively, as

4.3 Low-Order Model Fitting Algorithms

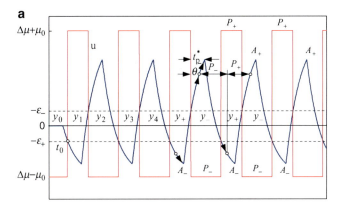

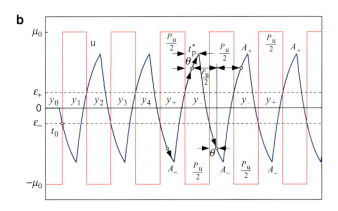

Fig. 4.7 Limit cycle analysis for an overdamped SOPDT model

$$y_1(t)|_{\text{shift}} = y_0(t + t_0 + \theta) + 2k_p\mu_0 - 2k_p\mu_0 \left(\frac{\tau_1}{\tau_1 - \tau_2} e^{-\frac{t}{\tau_1}} - \frac{\tau_2}{\tau_1 - \tau_2} e^{-\frac{t}{\tau_2}} \right)$$
(4.95)

$$y_2(t)|_{\text{shift}} = y_0(t + t_0 + \theta + P_+) + 2k_p\mu_0 (1 - 1)$$
$$- 2k_p\mu_0 \left[\frac{\tau_1}{\tau_1 - \tau_2} e^{-\frac{t}{\tau_1}} \left(e^{-\frac{P_+}{\tau_1}} - 1 \right) - \frac{\tau_2}{\tau_1 - \tau_2} e^{-\frac{t}{\tau_2}} \left(e^{-\frac{P_+}{\tau_2}} - 1 \right) \right]$$
(4.96)

$$y_3(t)|_{\text{shift}} = y_0(t + t_0 + \theta + P_u) + 2k_p\mu_0 (1 - 1 + 1)$$
$$- 2k_p\mu_0 \left[\frac{\tau_1}{\tau_1 - \tau_2} e^{-\frac{t}{\tau_1}} \left(e^{-\frac{P_u}{\tau_1}} - e^{-\frac{P_-}{\tau_1}} + 1 \right) - \frac{\tau_2}{\tau_1 - \tau_2} e^{-\frac{t}{\tau_2}} \left(e^{-\frac{P_u}{\tau_2}} - e^{-\frac{P_-}{\tau_2}} + 1 \right) \right]$$
(4.97)

$$y_4(t)|_{\text{shift}} = y_0(t + t_0 + \theta + P_\text{u} + P_+) + 2k_\text{p}\mu_0(1 - 1 + 1 - 1)$$

$$- 2k_\text{p}\mu_0 \left[\frac{\tau_1}{\tau_1 - \tau_2} e^{-\frac{t}{\tau_1}} \left(e^{-\frac{P_\text{u}+P_+}{\tau_1}} - e^{-\frac{P_\text{u}}{\tau_1}} + e^{-\frac{P_+}{\tau_1}} - 1 \right) \right.$$

$$\left. - \frac{\tau_2}{\tau_1 - \tau_2} e^{-\frac{t}{\tau_2}} \left(e^{-\frac{P_\text{u}+P_+}{\tau_2}} - e^{-\frac{P_\text{u}}{\tau_2}} + e^{-\frac{P_+}{\tau_2}} - 1 \right) \right] \quad (4.98)$$

The general relay response can, therefore, be summarized as

$$y_{2n+1}(t)|_{\text{shift}} = y_0(t + t_0 + \theta + nP_\text{u}) + 2k_\text{p}\mu_0 - 2k_\text{p}\mu_0 \left(\frac{\tau_1 E_1}{\tau_1 - \tau_2} e^{-\frac{t}{\tau_1}} - \frac{\tau_2 E_2}{\tau_1 - \tau_2} e^{-\frac{t}{\tau_2}} \right) \quad (4.99)$$

$$y_{2n+2}(t)|_{\text{shift}} = y_0(t + t_0 + \theta + nP_\text{u} + P_+) - 2k_\text{p}\mu_0 \left(\frac{\tau_1 F_1}{\tau_1 - \tau_2} e^{-\frac{t}{\tau_1}} - \frac{\tau_2 F_2}{\tau_1 - \tau_2} e^{-\frac{t}{\tau_2}} \right) \quad (4.100)$$

where $n = 0, 1, 2, \ldots$, and

$$E_1 = 1 + \sum_{k=1}^{n} \left(e^{-\frac{kP_\text{u}}{\tau_1}} - e^{-\frac{(k-1)P_\text{u}+P_-}{\tau_1}} \right) \quad (4.101)$$

$$E_2 = 1 + \sum_{k=1}^{n} \left(e^{-\frac{kP_\text{u}}{\tau_2}} - e^{-\frac{(k-1)P_\text{u}+P_-}{\tau_2}} \right) \quad (4.102)$$

$$F_1 = \sum_{k=0}^{n} \left(e^{-\frac{kP_\text{u}+P_+}{\tau_1}} - e^{-\frac{kP_\text{u}}{\tau_1}} \right) \quad (4.103)$$

$$F_2 = \sum_{k=0}^{n} \left(e^{-\frac{kP_\text{u}+P_+}{\tau_2}} - e^{-\frac{kP_\text{u}}{\tau_2}} \right) \quad (4.104)$$

By substituting (4.67) into (4.101)–(4.104), one can obtain the simplified forms of E_1, E_2, F_1, and F_2, as shown in (4.90)–(4.93).

Note that $y_0(t + t_0 + \theta + nP_\text{u}) = y_0(t + t_0 + \theta + nP_\text{u} + P_+) = k_\text{p}(\Delta\mu - \mu_0)$ for $n \to \infty$. Hence, in the limit cycle it follows that

$$y_+(t) = \lim_{n \to \infty} y_{2n+1}(t)|_{\text{shift}} = k_\text{p}(\Delta\mu + \mu_0)$$

$$- 2k_\text{p}\mu_0 \left(\frac{\tau_1 E_1}{\tau_1 - \tau_2} e^{-\frac{t}{\tau_1}} - \frac{\tau_2 E_2}{\tau_1 - \tau_2} e^{-\frac{t}{\tau_2}} \right), \quad t \in [0, P_+] \quad (4.105)$$

4.3 Low-Order Model Fitting Algorithms

$$y_-(t) = \lim_{n\to\infty} y_{2n+2}(t)|_{\text{shift}} = k_p(\Delta\mu - \mu_0)$$
$$- 2k_p\mu_0 \left(\frac{\tau_1 F_1}{\tau_1 - \tau_2} e^{-\frac{t}{\tau_1}} - \frac{\tau_2 F_2}{\tau_1 - \tau_2} e^{-\frac{t}{\tau_2}} \right), \quad t \in [0, P_-] \quad (4.106)$$

It can be derived from (4.105) and (4.106) that

$$\frac{dy_+(t)}{dt} = \frac{2k_p\mu_0}{\tau_1 - \tau_2} \left(E_1 e^{-\frac{t}{\tau_1}} - E_2 e^{-\frac{t}{\tau_2}} \right) \quad (4.107)$$

$$\frac{dy_-(t)}{dt} = \frac{2k_p\mu_0}{\tau_1 - \tau_2} \left(F_1 e^{-\frac{t}{\tau_1}} - F_2 e^{-\frac{t}{\tau_2}} \right) \quad (4.108)$$

To see if (4.105) and (4.106) have any positive solution for the time variable, t, here, define $z = x/P_-$, $a = P_-/P_u$, and a fractional function,

$$f(x) = \frac{1 - e^{-\frac{P_-}{x}}}{1 - e^{-\frac{P_u}{x}}}, \quad x \in (0, \infty) \quad (4.109)$$

It follows that

$$f(z) = \frac{1 - e^{-\frac{1}{z}}}{1 - e^{-\frac{1}{az}}} \quad (4.110)$$

$$\frac{df(z)}{dz} = \frac{e^{-\frac{a+1}{az}} \left(e^{\frac{1}{z}} - ae^{\frac{1}{az}} + a - 1 \right)}{az^2 \left(1 - e^{-\frac{1}{az}} \right)^2} \quad (4.111)$$

Let

$$g(z) = e^{\frac{1}{z}} - ae^{\frac{1}{az}} + a - 1 \quad (4.112)$$

It can be verified that

$$\frac{dg(z)}{dz} = \frac{1}{z^2} \left(e^{\frac{1}{az}} - e^{\frac{1}{z}} \right) > 0 \quad (4.113)$$

$$\lim_{z\to\infty} g(z) = 0 \quad (4.114)$$

It can, thus, be concluded that $g(z) < 0$ for $z \in (0, \infty)$, and correspondingly,

$$\frac{df(z)}{dz} < 0 \quad (4.115)$$

Hence, $f(z)$ decreases monotonically with respect to z and so does for $f(x)$ with respect to x. One can then conclude from $f(\tau_1) < f(\tau_2)$ that $E_2 > E_1 > 0$.

Following a similar analysis, it can be concluded that $F_2 < F_1 < 0$. This completes the proof. □

If an unbiased relay test is used as shown in Fig. 4.7b, by substituting $\Delta\mu = 0$ and $P_+ = P_- = P_u/2$ into (4.88)–(4.93) one can obtain the corresponding relay response expression,

$$y_+(t) = -y_-(t) = k_p\mu_0 - 2k_p\mu_0 \left(\frac{\tau_1 E_1}{\tau_1 - \tau_2} e^{-\frac{t}{\tau_1}} - \frac{\tau_2 E_2}{\tau_1 - \tau_2} e^{-\frac{t}{\tau_2}} \right) \quad (4.116)$$

where

$$E_1 = \frac{1}{1 + e^{-\frac{P_u}{2\tau_1}}} \quad (4.117)$$

$$E_2 = \frac{1}{1 + e^{-\frac{P_u}{2\tau_2}}} \quad (4.118)$$

When a biased relay test is used, the process static gain can be derived from (4.31). Substituting (4.87) into the frequency response fitting condition shown in (4.30), one can obtain

$$\frac{k_p}{\sqrt{\left(\tau_1^2 \omega_u^2 + 1\right)\left(\tau_2^2 \omega_u^2 + 1\right)}} = A_u \quad (4.119)$$

$$-\theta\omega_u - \arctan(\tau_1\omega_u) - \arctan(\tau_2\omega_u) = \varphi_u, \quad \varphi_u \in (-\pi, -\pi/2) \quad (4.120)$$

By letting $\frac{dy_-(t)}{dt} = 0$, the time to reach the single extreme value of $y_-(t)$ can be derived as

$$t_{p_-} = \frac{\tau_1 \tau_2}{\tau_1 - \tau_2} \ln \frac{F_2}{F_1} \quad (4.121)$$

Note that the time to reach the single extreme value of $y_-(t)$ from the initial relay switch point in a negative half period of the relay, $t_{p_-}^*$, can be measured as shown in Fig. 4.7a. It follows that

$$t_{p_-} = t_{p_-}^* - \theta \quad (4.122)$$

Substituting (4.119), (4.120), and (4.121) into (4.122) to eliminate τ_2 and θ yields

$$\frac{\tau_1 \tau_2}{\tau_1 - \tau_2} \ln \frac{F_2}{F_1} = t_{p_-}^* + \frac{1}{\omega_u} [\varphi_u + \arctan(\tau_1\omega_u) + \arctan(\tau_2\omega_u)] \quad (4.123)$$

4.3 Low-Order Model Fitting Algorithms

where

$$\tau_2 = \frac{1}{\omega_u} \sqrt{\frac{k_p^2}{A_u^2 (\tau_1^2 \omega_u^2 + 1)} - 1} \qquad (4.124)$$

It is seen that (4.123) is a transcendental equation with respect to τ_1. One can solve this equation with the numerical Newton-Raphson method. The initial estimation of τ_1 for iteration may be taken as

$$\widehat{\tau}_1 = \frac{1}{\omega_u} \sqrt{\frac{k_p}{A_u} - 1} \qquad (4.125)$$

which is estimated from (4.119) by letting $\tau_1 = \tau_2$.

The other time constant and the process time delay can then be derived from (4.124) and (4.120).

Alternatively, substituting (4.124) into (4.120) yields a transcendental equation with respect to τ_1, where the process time delay (θ) is prespecified for computation. Correspondingly, a one-dimensional search of θ is needed in combination with the fitting condition of relay response shown in (4.46) to determine the optimal fitting.

Hence, the above algorithm named Algorithm-RS-SO1 for the identification of an overdamped SOPDT model under a biased relay test can be summarized.

Algorithm-RS-SO1

(i) Measure P_+, P_-, and $t^*_{p_-}$ from the limit cycle.
(ii) Compute $G(j\omega_u)$ from (4.30).
(iii) Compute the process static gain, k_p, from (4.31).
(iv) Compute the process time constant, τ_1, from (4.123) using the Newton-Raphson iteration method, or from the equation resulting from substituting (4.124) into (4.120) based on a prespecified time delay (θ). The initial estimation of τ_1 for iteration may be taken from (4.125), or alternatively, a one-dimensional search of θ can be implemented within a possible range as observed from the initial step response in the relay test.
(v) Compute the process time constant, τ_2, from (4.124).
(vi) Compute the process time delay, θ, from (4.120) if τ_1 has been computed from (4.123).
(vii) End the algorithm if the fitting condition of relay response shown in (4.46) is satisfied. Otherwise, go back to Step (iii) by changing the initial estimation of τ_1, or monotonically varying θ in a one-dimensional search.

When an unbiased relay test is used, it follows from (4.116) that there exist two boundary conditions

$$y_+\left(\frac{P_u}{2} - \theta\right) = k_p \mu_0 - 2k_p \mu_0 \left(\frac{\tau_1 E_1}{\tau_1 - \tau_2} e^{-\frac{P_u - 2\theta}{2\tau_1}} - \frac{\tau_2 E_2}{\tau_1 - \tau_2} e^{-\frac{P_u - 2\theta}{2\tau_2}}\right) = \varepsilon_+ \qquad (4.126)$$

$$y_+(t_p) = k_p\mu_0 - 2k_p\mu_0\left(\frac{\tau_1 E_1}{\tau_1 - \tau_2}e^{-\frac{t_p}{\tau_1}} - \frac{\tau_2 E_2}{\tau_1 - \tau_2}e^{-\frac{t_p}{\tau_2}}\right) = A_- \quad (4.127)$$

where t_p is the time to reach the single extreme value of $y_+(t)$ (or $y_-(t)$), which can be derived from $\frac{dy_+(t)}{dt} = 0$ as

$$t_p = \frac{\tau_1\tau_2}{\tau_1 - \tau_2}\ln\frac{E_2}{E_1} \quad (4.128)$$

Note that

$$t_p = t_p^* - \theta \quad (4.129)$$

where t_p^* is the time to reach the single extreme value of $y_+(t)$ (or $y_-(t)$) from the initial relay switch point in a half period of the relay, which can be measured from the limit cycle.

It follows from dividing (4.126) by (4.127) that

$$\varepsilon + \left(1 - \frac{2\tau_1 E_1}{\tau_1 - \tau_2}e^{-\frac{t_p}{\tau_1}} + \frac{2\tau_2 E_2}{\tau_1 - \tau_2}e^{-\frac{t_p}{\tau_2}}\right) = A_-\left(1 - \frac{2\tau_1 E_1}{\tau_1 - \tau_2}e^{-\frac{P_u - 2\theta}{2\tau_1}} + \frac{2\tau_2 E_2}{\tau_1 - \tau_2}e^{-\frac{P_u - 2\theta}{2\tau_2}}\right) \quad (4.130)$$

Using the frequency response fitting condition as shown in (4.119) and (4.120), one can obtain

$$k_p = A_u\sqrt{(\tau_1^2\omega_u^2 + 1)(\tau_2^2\omega_u^2 + 1)} \quad (4.131)$$

$$\theta = -\frac{1}{\omega_u}[\varphi_u + \arctan(\tau_1\omega_u) + \arctan(\tau_2\omega_u)] \quad (4.132)$$

Substituting (4.128) and (4.132) into (4.129) yields an implicit equation with respect to τ_1 and τ_2,

$$F_1(\tau_1, \tau_2) = 0 \quad (4.133)$$

Similarly, substituting (4.128) and (4.132) into (4.130) yields another implicit equation with respect to τ_1 and τ_2,

$$F_2(\tau_1, \tau_2) = 0 \quad (4.134)$$

Therefore, one can solve (4.133) and (4.134) together to obtain the solution pair of τ_1 and τ_2. This can be performed by using a nonlinear programming algorithm. The initial values of τ_1 and τ_2 for iteration may be approximately estimated from the reference relay response shapes shown in Table 4.2. The search direction may

4.3 Low-Order Model Fitting Algorithms

be chosen as the gradients of $F_1(\tau_1, \tau_2)$ and $F_2(\tau_1, \tau_2)$. The iterative procedure can be programmed using a first-order Taylor expansion as

$$F_1(\tau_{1,k+1}, \tau_{2,k+1}) = F_1(\tau_{1,k}, \tau_{2,k}) + \left.\frac{\partial F_1}{\partial \tau_1}\right|_{\tau_{i,k}} \cdot \Delta\tau_1 + \left.\frac{\partial F_1}{\partial \tau_2}\right|_{\tau_{i,k}} \cdot \Delta\tau_2 \quad (4.135)$$

$$F_2(\tau_{1,k+1}, \tau_{2,k+1}) = F_2(\tau_{1,k}, \tau_{2,k}) + \left.\frac{\partial F_2}{\partial \tau_1}\right|_{\tau_{i,k}} \cdot \Delta\tau_1 + \left.\frac{\partial F_2}{\partial \tau_2}\right|_{\tau_{i,k}} \cdot \Delta\tau_2 \quad (4.136)$$

where $F_i(\tau_{1,k}, \tau_{2,k})$ denotes the value of $F_i(\tau_1, \tau_2)$ at the kth iteration step for $i = 1, 2$, while $\left.\frac{\partial F_i}{\partial \tau_i}\right|_{\tau_{i,k}}$ is the partial derivative of $F_i(\tau_1, \tau_2)$ with respect to τ_i at the kth step for $i = 1, 2$. The step size is denoted as $\Delta\tau_i$ ($i = 1, 2$), which may be practically chosen no larger than 0.01 to guarantee the computation accuracy. The optimal objective function for convergence can be specified as

$$J_{\min} = \sqrt{\frac{1}{2}\left[F_1^2(\tau_{1,k+1}, \tau_{2,k+1}) + F_2^2(\tau_{1,k+1}, \tau_{2,k+1})\right]} < \delta \quad (4.137)$$

where δ is a convergent threshold that can be practically set no larger than 1% for implementation. To avoid any local optimal solutions of τ_1 and τ_2, the relay response fitting condition shown in (4.46) can be used to screen out the most suitable solution pair.

The process static gain and time delay can then be derived from (4.131) and (4.132), respectively.

Alternatively, substituting (4.131) into (4.126) to eliminate k_p yields an implicit equation with respect to τ_1 and τ_2. By prespecifying a value of the process time delay (θ), this implicit equation together with (4.132) may be solved using the above nonlinear programming. A one-dimensional search of θ is, therefore, needed in combination with the fitting condition of relay response, shown in (4.46), to determine the optimal fitting.

Hence, the above algorithm named Algorithm-RS-SO2 for the identification of an overdamped SOPDT model under an unbiased relay test can be summarized.

Algorithm-RS-SO2

(i) Measure A_+(or A_-), P_u, and t_p^* from the limit cycle.
(ii) Compute $G(j\omega_u)$ from (4.30).
(iii) Compute the process time constants, τ_1 and τ_2, from (4.133) and (4.134) using the nonlinear programming algorithm given in (4.135)–(4.137), or from (4.132) and the implicit equation resulting from substituting (4.131) into (4.126) based on a prespecified time delay (θ). The initial values of τ_1 and τ_2 for iteration may be estimated from the reference relay response shapes shown in Table 4.2, or alternatively, a one-dimensional search of θ can be implemented within a possible range as observed from the initial step response in the relay test.
(iv) Compute the process static gain, k_p, from (4.131).

(v) Compute the process time delay, θ, from (4.132) if τ_1 and τ_2 have been computed from (4.133) and (4.134).
(vi) End the algorithm if the fitting condition of relay response shown in (4.46) is satisfied. Otherwise, go back to Step (iii) by changing the initial estimation of τ_1 and τ_2, or monotonically varying θ in a one-dimensional search.

It should be noted that for an overdamped second-order process with a dominant time delay ($\theta/\tau_1 \gg 1$ and $\theta/\tau_2 \gg 1$), $y_+(t)$ increases monotonically while $y_-(t)$ decreases monotonically, owing to that $E_1 \approx E_2 \approx 1$ and $F_1 \approx F_2 \approx -1$. Therefore, the process time delay can directly be measured as the time to reach the positive peak (A_+) of the process output response from the initial relay switch point in a negative half period of the relay. The process time constants τ_1 and τ_2 can then be derived by solving (4.130) and (4.132) together, or alternatively, from (4.132) and the implicit equation resulting from substituting (4.131) into (4.126), in terms of the fitting condition of relay response shown in (4.46) to determine the optimal solution.

Type 3 The Underdamped SOPDT Model

An underdamped second-order process is generally in the form of (4.5), where $\xi \in (0, 1)$ is named the damping ratio of the process response. The following proposition gives the exact relay response expression under a biased relay test.

Proposition 4.4. *For an underdamped second-order process modeled by (4.5) under a biased relay test as shown in Fig. 4.8a, the resulting limit cycle of the process output response is characterized by*

$$y_+(t) = k_p(\Delta\mu + \mu_0) - \frac{2k_p\mu_0\rho_1}{\eta\gamma} e^{-\frac{\xi t}{\tau_p}} \sin\left(\frac{\eta t}{\tau_p} + \psi_1\right), \quad t \in [0, P_+] \quad (4.138)$$

$$y_-(t) = k_p(\Delta\mu - \mu_0) + \frac{2k_p\mu_0\rho_2}{\eta\gamma} e^{-\frac{\xi t}{\tau_p}} \sin\left(\frac{\eta t}{\tau_p} + \psi_2\right), \quad t \in [0, P_-] \quad (4.139)$$

where $\eta = \sqrt{1-\xi^2}$, $\rho_1 = \sqrt{U_1^2 + V_1^2}$, $\rho_2 = \sqrt{U_2^2 + V_2^2}$, $\psi_1 = \arctan(V_1/U_1)$, $\psi_2 = \arctan(V_2/U_2)$, $\phi = \arctan(\sqrt{1-\xi^2}/\xi)$, *and*

$$\gamma = 1 - 2e^{-\frac{P_u\xi}{\tau_p}} \cos\frac{\eta P_u}{\tau_p} + e^{-\frac{2P_u\xi}{\tau_p}} \quad (4.140)$$

$$U_1 = \cos\phi - e^{-\frac{P_u\xi}{\tau_p}} \cos\left(\phi - \frac{\eta P_u}{\tau_p}\right) - e^{-\frac{P_-\xi}{\tau_p}} \left[\cos\left(\phi + \frac{\eta P_-}{\tau_p}\right) - e^{-\frac{P_u\xi}{\tau_p}} \cos\left(\phi - \frac{\eta P_+}{\tau_p}\right)\right] \quad (4.141)$$

4.3 Low-Order Model Fitting Algorithms

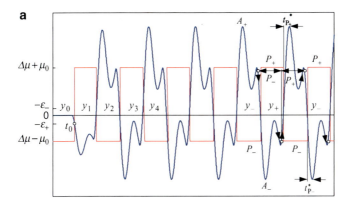

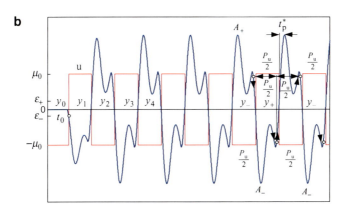

Fig. 4.8 Limit cycle analysis for an underdamped SOPDT model

$$V_1 = \sin\phi - e^{-\frac{P_u\xi}{\tau_p}} \sin\left(\phi - \frac{\eta P_u}{\tau_p}\right)$$
$$- e^{-\frac{P_-\xi}{\tau_p}} \left[\sin\left(\phi + \frac{\eta P_-}{\tau_p}\right) - e^{-\frac{P_u\xi}{\tau_p}} \sin\left(\phi - \frac{\eta P_+}{\tau_p}\right)\right] \quad (4.142)$$

$$U_2 = \cos\phi - e^{-\frac{P_u\xi}{\tau_p}} \cos\left(\phi - \frac{\eta P_u}{\tau_p}\right)$$
$$- e^{-\frac{P_+\xi}{\tau_p}} \left[\cos\left(\phi + \frac{\eta P_+}{\tau_p}\right) - e^{-\frac{P_u\xi}{\tau_p}} \cos\left(\phi - \frac{\eta P_-}{\tau_p}\right)\right] \quad (4.143)$$

$$V_2 = \sin\phi - e^{-\frac{P_u\xi}{\tau_p}} \sin\left(\phi - \frac{\eta P_u}{\tau_p}\right)$$
$$- e^{-\frac{P_+\xi}{\tau_p}} \left[\sin\left(\phi + \frac{\eta P_+}{\tau_p}\right) - e^{-\frac{P_u\xi}{\tau_p}} \sin\left(\phi - \frac{\eta P_-}{\tau_p}\right)\right] \quad (4.144)$$

Proof. The initial step response of an underdamped SOPDT process arising from the relay output, $u(t) = \Delta\mu - \mu_0$, can be derived as

$$y_0(t) = k_p(\Delta\mu - \mu_0)\left[1 - \frac{1}{\eta}e^{-\frac{\xi(t-\theta)}{\tau_p}}\sin\left(\frac{\eta(t-\theta)}{\tau_p} + \phi\right)\right] \quad (4.145)$$

Following a similar analysis as in the proof for Proposition 4.2, the time shifted output response from the initial to the fourth relay switch point can be derived, respectively, as

$$y_1(t)|_{\text{shift}} = y_0(t + t_0 + \theta) + 2k_p\mu_0 - \frac{2k_p\mu_0}{\eta}e^{-\frac{\xi t}{\tau_p}}\sin\left(\frac{\eta t}{\tau_p} + \phi\right) \quad (4.146)$$

$$y_2(t)|_{\text{shift}} = y_0(t + t_0 + \theta + P_+) + 2k_p\mu_0(1-1)$$
$$-\frac{2k_p\mu_0}{\eta}e^{-\frac{\xi t}{\tau_p}}\left[e^{-\frac{P_+\xi}{\tau_p}}\sin\left(\frac{\eta(t+P_+)}{\tau_p} + \phi\right) - \sin\left(\frac{\eta t}{\tau_p} + \phi\right)\right]$$
$$(4.147)$$

$$y_3(t)|_{\text{shift}} = y_0(t + t_0 + \theta + P_u) + 2k_p\mu_0(1 - 1 + 1)$$
$$-\frac{2k_p\mu_0}{\eta}e^{-\frac{\xi t}{\tau_p}}\left[e^{-\frac{P_u\xi}{\tau_p}}\sin\left(\frac{\eta(t+P_u)}{\tau_p} + \phi\right)\right.$$
$$\left. - e^{-\frac{P_-\xi}{\tau_p}}\sin\left(\frac{\eta(t+P_-)}{\tau_p} + \phi\right) + \sin\left(\frac{\eta t}{\tau_p} + \phi\right)\right] \quad (4.148)$$

$$y_4(t)|_{\text{shift}} = y_0(t + t_0 + \theta + P_u + P_+) + 2k_p\mu_0(1 - 1 + 1 - 1)$$
$$-\frac{2k_p\mu_0}{\eta}e^{-\frac{\xi t}{\tau_p}}\left[e^{-\frac{(P_u+P_+)\xi}{\tau_p}}\sin\left(\frac{\eta(t+P_u+P_+)}{\tau_p} + \phi\right)\right.$$
$$\left. -e^{-\frac{P_u\xi}{\tau_p}}\sin\left(\frac{\eta(t+P_u)}{\tau_p} + \phi\right) + e^{-\frac{P_+\xi}{\tau_p}}\sin\left(\frac{\eta(t+P_+)}{\tau_p} + \phi\right) - \sin\left(\frac{\eta t}{\tau_p} + \phi\right)\right]$$
$$(4.149)$$

The general relay response can therefore be summarized as

$$y_{2n+1}(t)|_{\text{shift}} = y_0(t + t_0 + \theta + nP_u) + 2k_p\mu_0 - \frac{2k_p\mu_0 E}{\eta}e^{-\frac{\xi t}{\tau_p}} \quad (4.150)$$

$$y_{2n+2}(t)|_{\text{shift}} = y_0(t + t_0 + \theta + nP_u + P_+) - \frac{2k_p\mu_0 F}{\eta}e^{-\frac{\xi t}{\tau_p}} \quad (4.151)$$

4.3 Low-Order Model Fitting Algorithms

where $n = 0, 1, 2, \ldots$, and

$$E = \sin\left(\frac{\eta t}{\tau_p} + \phi\right) + \sum_{k=1}^{n} e^{-\frac{k P_u \xi}{\tau_p}} \sin\left(\frac{\eta(t + k P_u)}{\tau_p} + \phi\right)$$
$$- \sum_{k=1}^{n} e^{-\frac{[(k-1)P_u + P_-]\xi}{\tau_p}} \sin\left(\frac{\eta[t + (k-1)P_u + P_-]}{\tau_p} + \phi\right) \quad (4.152)$$

$$F = \sum_{k=0}^{n} \left[e^{-\frac{(k P_u + P_+)\xi}{\tau_p}} \sin\left(\frac{\eta(t + k P_u + P_+)}{\tau_p} + \phi\right) - e^{-\frac{k P_u \xi}{\tau_p}} \sin\left(\frac{\eta(t + k P_u)}{\tau_p} + \phi\right) \right] \quad (4.153)$$

Using the Euler formula, it follows that

$$\sum_{k=1}^{n} e^{-\frac{k P_u \xi}{\tau_p}} \sin\left(\frac{\eta(t + k P_u)}{\tau_p} + \phi\right) = \sum_{k=1}^{n} \frac{e^{-\frac{k P_u \xi}{\tau_p}}}{2j} \left(e^{j[\frac{\eta(t+k P_u)}{\tau_p} + \phi]} - e^{-j[\frac{\eta(t+k P_u)}{\tau_p} + \phi]} \right) \quad (4.154)$$

Then, using the convergent formula shown in (4.67), one can obtain

$$\sum_{k=1}^{n} e^{-\frac{k P_u \xi}{\tau_p}} \sin\left(\frac{\eta(t + k P_u)}{\tau_p} + \phi\right)$$

$$= \frac{e^{j(\frac{\eta t}{\tau_p} + \phi)}}{2j} \sum_{k=1}^{n} e^{-\frac{k P_u(\xi - j\eta)}{\tau_p}} - \frac{e^{-j(\frac{\eta t}{\tau_p} + \phi)}}{2j} \sum_{k=1}^{n} e^{-\frac{k P_u(\xi + j\eta)}{\tau_p}}$$

$$= \frac{e^{j(\frac{\eta t}{\tau_p} + \phi)}}{2j} \cdot \frac{e^{-\frac{P_u(\xi - j\eta)}{\tau_p}}}{1 - e^{-\frac{P_u(\xi - j\eta)}{\tau_p}}} - \frac{e^{-j(\frac{\eta t}{\tau_p} + \phi)}}{2j} \cdot \frac{e^{-\frac{P_u(\xi + j\eta)}{\tau_p}}}{1 - e^{-\frac{P_u(\xi + j\eta)}{\tau_p}}}$$

$$= -\sin\left(\frac{\eta t}{\tau_p} + \phi\right) + \frac{\sin\left(\frac{\eta t}{\tau_p} + \phi\right) - e^{-\frac{P_u \xi}{\tau_p}} \sin\left(\frac{\eta(t - P_u)}{\tau_p} + \phi\right)}{1 - 2e^{-\frac{P_u \xi}{\tau_p}} \cos\frac{\eta P_u}{\tau_p} + e^{-\frac{2 P_u \xi}{\tau_p}}} \quad (4.155)$$

Similarly, it can be derived that

$$\sum_{k=0}^{n} e^{-\frac{(k P_u + P_-)\xi}{\tau_p}} \sin\left(\frac{\eta(t + k P_u + P_-)}{\tau_p} + \phi\right)$$

$$= \frac{e^{-\frac{P_-\xi}{\tau_p}} \left[\sin\left(\frac{\eta(t + P_-)}{\tau_p} + \phi\right) - e^{-\frac{P_u \xi}{\tau_p}} \sin\left(\frac{\eta(t - P_+)}{\tau_p} + \phi\right) \right]}{1 - 2e^{-\frac{P_u \xi}{\tau_p}} \cos\frac{\eta P_u}{\tau_p} + e^{-\frac{2 P_u \xi}{\tau_p}}} \quad (4.156)$$

Substituting (4.155) and (4.156) into (4.152) yields

$$E = \frac{\sin\left(\frac{\eta t}{\tau_p} + \phi\right) - e^{-\frac{P_u\xi}{\tau_p}} \sin\left(\frac{\eta(t-P_u)}{\tau_p} + \phi\right) - e^{-\frac{P-\xi}{\tau_p}}\left[\sin\left(\frac{\eta(t+P_-)}{\tau_p} + \phi\right) - e^{-\frac{P_u\xi}{\tau_p}} \sin\left(\frac{\eta(t-P_+)}{\tau_p} + \phi\right)\right]}{1 - 2e^{-\frac{P_u\xi}{\tau_p}} \cos\frac{\eta P_u}{\tau_p} + e^{-\frac{2P_u\xi}{\tau_p}}}$$

$$= \frac{\rho_1 \sin\left(\frac{\eta t}{\tau_p} + \psi_1\right)}{\gamma} \tag{4.157}$$

where $\rho_1 = \sqrt{U_1^2 + V_1^2}$, $\psi_1 = \arctan(V_1/U_1)$, and

$$\gamma = 1 - 2e^{-\frac{P_u\xi}{\tau_p}} \cos\frac{\eta P_u}{\tau_p} + e^{-\frac{2P_u\xi}{\tau_p}} \tag{4.158}$$

$$U_1 = \cos\phi - e^{-\frac{P_u\xi}{\tau_p}} \cos\left(\phi - \frac{\eta P_u}{\tau_p}\right)$$

$$- e^{-\frac{P-\xi}{\tau_p}}\left[\cos\left(\phi + \frac{\eta P_-}{\tau_p}\right) - e^{-\frac{P_u\xi}{\tau_p}} \cos\left(\phi - \frac{\eta P_+}{\tau_p}\right)\right] \tag{4.159}$$

$$V_1 = \sin\phi - e^{-\frac{P_u\xi}{\tau_p}} \sin\left(\phi - \frac{\eta P_u}{\tau_p}\right)$$

$$- e^{-\frac{P-\xi}{\tau_p}}\left[\sin\left(\phi + \frac{\eta P_-}{\tau_p}\right) - e^{-\frac{P_u\xi}{\tau_p}} \sin\left(\phi - \frac{\eta P_+}{\tau_p}\right)\right] \tag{4.160}$$

Following a similar derivation, one can obtain

$$F = -\frac{\rho_2 \sin\left(\frac{\eta t}{\tau_p} + \psi_2\right)}{\gamma} \tag{4.161}$$

where $\rho_2 = \sqrt{U_2^2 + V_2^2}$, $\psi_2 = \arctan(V_2/U_2)$, and

$$U_2 = \cos\phi - e^{-\frac{P_u\xi}{\tau_p}} \cos\left(\phi - \frac{\eta P_u}{\tau_p}\right)$$

$$- e^{-\frac{P_+\xi}{\tau_p}}\left[\cos\left(\phi + \frac{\eta P_+}{\tau_p}\right) - e^{-\frac{P_u\xi}{\tau_p}} \cos\left(\phi - \frac{\eta P_-}{\tau_p}\right)\right] \tag{4.162}$$

$$V_2 = \sin\phi - e^{-\frac{P_u\xi}{\tau_p}} \sin\left(\phi - \frac{\eta P_u}{\tau_p}\right)$$

$$- e^{-\frac{P_+\xi}{\tau_p}}\left[\sin\left(\phi + \frac{\eta P_+}{\tau_p}\right) - e^{-\frac{P_u\xi}{\tau_p}} \sin\left(\phi - \frac{\eta P_-}{\tau_p}\right)\right] \tag{4.163}$$

4.3 Low-Order Model Fitting Algorithms

Note that $y_0(t + t_0 + \theta + n P_u) = y_0(t + t_0 + \theta + n P_u + P_+) = k_p(\Delta \mu - \mu_0)$ for $n \to \infty$. Hence, in the limit cycle it follows that

$$y_+(t) = \lim_{n \to \infty} y_{2n+1}(t)|_{\text{shift}} = k_p(\Delta \mu + \mu_0) - \frac{2k_p\mu_0 E}{\eta} e^{-\frac{\xi t}{\tau_p}}, \quad t \in [0, P_+] \quad (4.164)$$

$$y_-(t) = y_{2n+2}(t)|_{\text{shift}} = k_p(\Delta \mu - \mu_0) - \frac{2k_p\mu_0 F}{\eta} e^{-\frac{\xi t}{\tau_p}}, \quad t \in [0, P_-] \quad (4.165)$$

This completes the proof. □

If an unbiased relay feedback test is used, as shown in Fig. 4.8b, by substituting $\Delta \mu = 0$ and $P_+ = P_- = P_u/2$ into (4.138)–(4.144) one can obtain the corresponding expression,

$$y_+(t) = -y_-(t) = k_p\mu_0 - \frac{2k_p\mu_0 \rho}{\eta \gamma} e^{-\frac{\xi t}{\tau_p}} \sin\left(\frac{\eta t}{\tau_p} + \psi\right) \quad (4.166)$$

where $\rho = \sqrt{U^2 + V^2}$, $\psi = \arctan(V/U)$, and

$$U = \cos\phi - e^{-\frac{P_u\xi}{\tau_p}} \cos\left(\phi - \frac{\eta P_u}{\tau_p}\right)$$

$$- e^{-\frac{P_u\xi}{2\tau_p}} \left[\cos\left(\phi + \frac{\eta P_u}{2\tau_p}\right) - e^{-\frac{P_u\xi}{\tau_p}} \cos\left(\phi - \frac{\eta P_u}{2\tau_p}\right)\right] \quad (4.167)$$

$$V = \sin\phi - e^{-\frac{P_u\xi}{\tau_p}} \sin\left(\phi - \frac{\eta P_u}{\tau_p}\right)$$

$$- e^{-\frac{P_u\xi}{2\tau_p}} \left[\sin\left(\phi + \frac{\eta P_u}{2\tau_p}\right) - e^{-\frac{P_u\xi}{\tau_p}} \sin\left(\phi - \frac{\eta P_u}{2\tau_p}\right)\right] \quad (4.168)$$

When a biased relay test is used, the process static gain can be derived from (4.31). By using the frequency response fitting condition shown in (4.30), one can obtain

$$\frac{k_p}{\sqrt{\left(1 - \tau_p^2\omega_u^2\right)^2 + 4\xi^2\tau_p^2\omega_u^2}} = A_u \quad (4.169)$$

$$-\theta\omega_u - \arctan\left(\frac{2\xi\tau_p\omega_u}{1 - \tau_p^2\omega_u^2}\right) = \varphi_u, \quad \varphi_u \in (-\pi, -\pi/2) \quad (4.170)$$

It follows from (4.169) that

$$\tau_p = \begin{cases} \frac{1}{\omega_u}\sqrt{1 - 2\xi^2 + \sqrt{\frac{k_p^2}{A_u^2} + 4\xi^2(\xi^2 - 1)}} \text{ or } \frac{1}{\omega_u}\sqrt{1 - 2\xi^2 - \sqrt{\frac{k_p^2}{A_u^2} + 4\xi^2(\xi^2 - 1)}}, & 0 < \xi < \sqrt{2}/2; \\ \frac{1}{\omega_u}\sqrt{1 - 2\xi^2 + \sqrt{\frac{k_p^2}{A_u^2} + 4\xi^2(\xi^2 - 1)}}, & \xi \geq \sqrt{2}/2. \end{cases}$$
(4.171)

$$\theta = -\frac{1}{\omega_u}\left[\varphi_u + \arctan\left(\frac{2\xi\tau_p\omega_u}{1 - \tau_p^2\omega_u^2}\right)\right]$$
(4.172)

It can be derived from (4.138) and (4.139) that

$$\frac{dy_+(t)}{dt} = -\frac{2k_p\mu_0\rho_1}{\eta\gamma\tau_p}e^{-\frac{\xi t}{\tau_p}}\left[\eta\cos\left(\frac{\eta t}{\tau_p} + \psi_1\right) - \xi\sin\left(\frac{\eta t}{\tau_p} + \psi_1\right)\right]$$
(4.173)

$$\frac{dy_-(t)}{dt} = \frac{2k_p\mu_0\rho_2}{\eta\gamma\tau_p}e^{-\frac{\xi t}{\tau_p}}\left[\eta\cos\left(\frac{\eta t}{\tau_p} + \psi_2\right) - \xi\sin\left(\frac{\eta t}{\tau_p} + \psi_2\right)\right]$$
(4.174)

Then, by letting $\frac{dy_+(t)}{dt} = 0$ and $\frac{dy_-(t)}{dt} = 0$, the time to reach the peak of $y_+(t)$ or $y_-(t)$ can be obtained, respectively, as

$$t_{p_+,k} = \frac{\tau_p}{\eta}(\phi - \psi_1 + k\pi), \quad k = 0, 1, 2, \ldots.$$
(4.175)

$$t_{p_-,k} = \frac{\tau_p}{\eta}(\phi - \psi_2 + k\pi), \quad k = 0, 1, 2, \ldots.$$
(4.176)

In fact, if there exist several peaks of $y_+(t)$ (or $y_-(t)$) as shown in Fig. 4.8a, the time to reach the largest peak denoted as A_+ (or A_-) from the terminal relay switch point in the corresponding half period of the relay, $t^*_{p_+}$ (or $t^*_{p_+}$), can be measured. It follows that

$$t_{p_+,k} - t_{p_-,k} = \frac{\tau_p}{\eta}(\psi_2 - \psi_1) = P_+ - \theta + t^*_{p_+} - \left(P_- - \theta + t^*_{p_-}\right)$$

$$= P_+ - P_- + t^*_{p_+} - t^*_{p_-}$$
(4.177)

It can be seen from (4.177) that $t_{p_+,k} - t_{p_-,k}$ gives the same value as long as the corresponding peaks of $y_+(t)$ and $y_-(t)$ are measured for computation.

Substituting (4.171) into (4.177) to eliminate τ_p yields a transcendental equation with respect to ξ. One can solve this equation using the Newton-Raphson iteration method. The initial value of ξ for iteration may be approximately estimated from the reference relay response shapes shown in Table 4.2. Note that the practical constraint, $0 < \xi < 1$, can be used to limit the search range and intuitively exclude those unsuitable solutions.

4.3 Low-Order Model Fitting Algorithms

Accordingly, the process time constant and time delay can then be derived from (4.171) and (4.172), respectively. Note that if $0 < \xi < \sqrt{2}/2$, the excessive solution of τ_p as shown in (4.171) can be excluded through the iterative algorithm for computing ξ. To relieve the computation effort, it is suggested to first use the search step size, $\xi = 0.01$, to locate θ and τ_p with a relaxed fitting constraint of the relay response, and then to reduce the step size such as $\xi = 0.001$ in conjunction with a much tighter fitting threshold for better computation accuracy.

Alternatively, substituting (4.171) into (4.170) to eliminate τ_p yields a transcendental equation with respect to ξ, where the process time delay (θ) is prespecified for computation. A one-dimensional search of θ is, therefore, needed in combination with the fitting condition of relay response shown in (4.46) to determine the optimal fitting.

Hence, the above algorithm named Algorithm-RS-SU1 for the identification of an underdamped SOPDT model under a biased relay test can be summarized.

Algorithm-RS-SU1

(i) Measure P_+, P_-, $t^*_{p_+}$, and $t^*_{p_-}$ from the limit cycle.
(ii) Compute $G(j\omega_u)$ from (4.30).
(iii) Compute the process static gain, k_p, from (4.31).
(iv) Compute the process damping ratio, ξ, from the equation resulting from substituting (4.171) into (4.177) by using the Newton-Raphson iteration method, or from the equation resulting from substituting (4.171) into (4.170) based on a prespecified time delay (θ). The initial value of ξ for iteration may be estimated from the reference relay response shapes shown in Table 4.2, or alternatively, a one-dimensional search of θ can be implemented within a possible range as observed from the initial step response in the relay test.
(v) Compute the process time constant, τ_p, from (4.171).
(vi) Compute the process time delay, θ, from (4.172) if ξ has been computed from the equation resulting from substituting (4.171) into (4.177).
(vii) End the algorithm if the fitting condition of relay response shown in (4.46) is satisfied. Otherwise, go back to Step (iv) by changing the initial estimation of ξ, or monotonically varying θ in a one-dimensional search.

It should be noted that although one can derive from (4.169) that

$$\xi = \frac{1}{2\tau_p \omega_u} \sqrt{\frac{k_p^2}{A_u^2} - \left(1 - \tau_p^2 \omega_u^2\right)^2} \qquad (4.178)$$

and then numerically solve τ_p from a transcendental equation resulting from substituting (4.178) into (4.177), the computation effort may be much larger than that of solving ξ, because the initial estimation of τ_p becomes more difficult and so is for determining its possible range. Hence, it is not recommended unless an approximate value or range of τ_p can be known from a prior knowledge of the process operation.

When an unbiased relay test is used, it follows from (4.166) that

$$y_+\left(\frac{P_u}{2}-\theta\right)=k_p\mu_0-\frac{2k_p\mu_0\rho}{\eta\gamma}e^{-\frac{\xi(P_u-2\theta)}{2\tau_p}}\sin\left(\frac{\eta(P_u-2\theta)}{2\tau_p}+\psi\right)=\varepsilon_+ \quad (4.179)$$

$$y_+(t_p)=k_p\mu_0-\frac{2k_p\mu_0\rho}{\eta\gamma}e^{-\frac{\xi t_p}{\tau_p}}\sin\left(\frac{\eta t_p}{\tau_p}+\psi\right)=A_+ \quad (4.180)$$

where t_p is the time to reach the largest peak of $y_+(t)$ if there exist several peaks as shown in Fig. 4.8b, which can be derived from $\frac{dy_+(t)}{dt}=0$ as

$$t_p=\frac{\tau_p}{\eta}(\phi-\psi+k\pi) \quad (4.181)$$

where k may be determined from the practical constraint,

$$t_p=\frac{P_u}{2}-\theta+t_p^* \quad (4.182)$$

where t_p^* is the time to reach the largest peak of $y_+(t)$ (or $y_-(t)$) from the terminal relay switch point in the corresponding half period of the relay; i.e., k should be taken to satisfy

$$t_p^*<t_p<\frac{P_u}{2}+t_p^* \quad (4.183)$$

It follows from dividing (4.179) by (4.180) that

$$\varepsilon_+\left[1-\frac{2\rho}{\eta\gamma}e^{-\frac{\xi t_p}{\tau_p}}\sin\left(\frac{\eta t_p}{\tau_p}+\psi\right)\right]=A_+\left[1-\frac{2\rho}{\eta\gamma}e^{-\frac{\xi(P_u-2\theta)}{2\tau_p}}\sin\left(\frac{\eta(P_u-2\theta)}{2\tau_p}+\psi\right)\right] \quad (4.184)$$

By using the frequency response fitting condition shown in (4.169) and (4.170), one can obtain

$$k_p=A_u\sqrt{\left(1-\tau_p^2\omega_u^2\right)^2+4\xi^2\tau_p^2\omega_u^2} \quad (4.185)$$

$$\theta=-\frac{1}{\omega_u}\left[\varphi_u+\arctan\left(\frac{2\xi\tau_p\omega_u}{1-\tau_p^2\omega_u^2}\right)\right] \quad (4.186)$$

Substituting (4.181) and (4.186) into (4.182) to eliminate t_p and θ yields an implicit equation with respect to τ_p and ξ,

$$H_1(\tau_p,\xi)=0 \quad (4.187)$$

4.3 Low-Order Model Fitting Algorithms

Similarly, substituting (4.181) and (4.186) into (4.184) to eliminate t_p and θ yields another implicit equation with respect to τ_p and ξ,

$$H_2\left(\tau_p, \xi\right) = 0 \tag{4.188}$$

Therefore, one can solve (4.187) and (4.188) together to find the solution pair of τ_p and ξ. This can be performed by using a nonlinear programming as introduced in Algorithm-RS-SO2. The initial values of τ_p and ξ for iteration may be estimated from the reference relay response shapes shown in Table 4.2. The practical constraint, $0 < \xi < 1$, may be used to limit the search range and intuitively exclude those unsuitable solution pairs.

The process static gain and time delay can then be derived from (4.185) and (4.186), respectively.

Alternatively, substituting (4.185) into (4.179) to eliminate k_p yields an implicit equation with respect to τ_p and ξ. By prespecifying a value of the process time delay (θ), this implicit equation and another implicit equation of (4.186) may be solved together by using a nonlinear programming as in Algorithm-RS-SO2. A one-dimensional search of θ is, therefore, needed in combination with the fitting condition of relay response shown in (4.46) to determine the optimal fitting.

Hence, the above algorithm named Algorithm-RS-SU2 for the identification of an underdamped SOPDT model under an unbiased relay test can be summarized.

Algorithm-RS-SU2

(i) Measure A_+ (or A_-), P_u, and t_p^* from the limit cycle.
(ii) Compute $G(j\omega_u)$ from (4.30).
(iii) Compute the process time constant, τ_p, and damping ratio, ξ, from (4.187) and (4.188) by using a nonlinear programming as in Algorithm-RS-SO2, or from (4.186) and the implicit equation resulting from substituting (4.185) into (4.179) based on a prespecified time delay (θ). The initial values of τ_p and ξ for iteration may be estimated from the reference relay response shapes shown in Table 4.2, or alternatively, a one-dimensional search of θ can be implemented within a possible range as observed from the initial step response in the relay test.
(iv) Compute the process static gain, k_p, from (4.185).
(v) Compute the process time delay, θ, from (4.186) if τ_p and ξ have been computed from (4.187) and (4.188).
(vi) End the algorithm if the fitting condition of relay response shown in (4.46) is satisfied. Otherwise, go back to Step (iii) by changing the initial estimation of τ_p and ξ, or monotonically varying θ in a one-dimensional search.

4.3.3 Illustrative Examples

Eight examples from existing literature are used here to illustrate the effectiveness and merits of the above identification algorithms. Example 4.1 is given to demonstrate the fitting accuracy of Algorithm-RS-FA1, Algorithm-RS-FA2, Algorithm-RS-FB1, and Algorithm-RS-FB2 for identifying a first-order process, based on a biased or unbiased relay test. Measurement noise tests are also included to illustrate identification robustness. Examples 4.2 and 4.3 are used to show the effectiveness of these algorithms for identifying higher order processes. Examples 4.4, 4.5, and 4.6 are given to demonstrate the fitting accuracy of Algorithm-RS-SC1, Algorithm-RS-SC2, Algorithm-RS-SO1, Algorithm-RS-SO2, Algorithm-RS-SU1, and Algorithm-RS-SU2 for identifying a second-order process in terms of the exact model structure, together with measurement noise tests to illustrate identification robustness. Examples 4.7 and 4.8 are used to show the effectiveness of these SOPDT algorithms for identifying higher order processes. In all the relay tests, the sampling period is taken as $T_s = 0.01(s)$ for computation.

To assess the model fitting error, the widely used step response fitting error, err, as shown in (2.42), is adopted for reference, together with the maximal frequency response error,

$$ERR = \max_{\omega \in [0, \omega_c]} \left\{ \left| \frac{\widehat{G}(j\omega) - G(j\omega)}{G(j\omega)} \right| \times 100\% \right\}$$

where $G(j\omega)$ and $\widehat{G}(j\omega)$ denote the frequency responses of the process and the model, respectively, and ω_c is the cutoff angular frequency corresponding to $\angle G(j\omega_c) = -\pi$. In consideration of the fact that ω_u can be intuitively measured from a relay test and is only slightly smaller than ω_c, it is adopted here to compute ERR for convenience.

Example 4.1. Consider the first-order process widely studied in the literature (Shen et al. 1996; Srinivasan and Chidambaram 2003; Vivek and Chidambaram 2005),

$$G = \frac{e^{-2s}}{10s + 1}$$

Based on a biased relay test, Shen et al. (1996) derived an FOPDT model, $G_m = 0.999e^{-2.005s}/(8.118s + 1)$, and Srinivasan and Chidambaram (2003) gave a model, $G_m = 1.03e^{-2.3s}/(10.3s + 1)$. Using an unbiased relay test, Vivek and Chidambaram (2005) derived a model, $G_m = 0.9467e^{-2s}/(9.5028s + 1)$.

For illustration, a biased relay test using $u_+ = 1.3$, $u_- = -0.7$, and $\varepsilon_+ = -\varepsilon_- = 0.2$ is performed for model identification. The measured limit cycle data are listed in Table 4.3. The corresponding algorithms, Algorithm-RS-FA1 and Algorithm-RS-FA2, are used to obtain the process model, respectively. The results are also shown in Table 4.3, which demonstrate good accuracy. For comparison, an unbiased relay test using $u_+ = -u_- = 1.0$ and $\varepsilon_+ = -\varepsilon_- = 0.2$ is also performed for model identification. Correspondingly, Algorithm-RS-FB1 and Algorithm-RS-

4.3 Low-Order Model Fitting Algorithms

Table 4.3 Identified FOPDT models for three examples under a relay test

Process	Relay	P_+	P_-	A_+	A_-	A_u	φ_u	FA1	FA2	FB1	FB2
$\dfrac{e^{-2s}}{10s+1}$	Biased	5.69	9.88	0.3995	−0.2906	0.2405	−2.137	$\dfrac{1.0001e^{-2s}}{9.9954s+1}$	$\dfrac{1.0001e^{-2.005s}}{10.001s+1}$	$\dfrac{1.0052e^{-2s}}{10.0561s+1}$	$\dfrac{1.0039e^{-2s}}{10.045s+1}$
	Unbiased	7.2	7.2	0.3452	−0.3452	0.2234	−2.2203				
$\dfrac{(-s+1)e^{-s}}{(s+1)^5}$	Biased	6.3	8.08	1.1509	−0.6918	0.7051	−2.9108	$\dfrac{1.0001e^{-3.53s}}{1.7664s+1}$	$\dfrac{1.0001e^{-4.8578s}}{2.3017s+1}$		
$\dfrac{e^{-s}}{8(s+1)^3}$	Unbiased	6.32	6.32	0.1143	−0.1143	0.0898	−1.8835			$\dfrac{0.1369e^{-1.28s}}{2.6253s+1}$	$\dfrac{0.12e^{-2.33s}}{1.7832s+1}$

Measured Limit Cycle Data / Algorithm-RS-

Table 4.4 Identification results for Example 4.1 under different measurement noise levels

NSR	Denoising	% Error in the Limit cycle data				Identified model	
		A_+	A_-	A_u	φ_u	Algorithm-RS-FA1	Algorithm-RS-FA2
5%	Averaging	1.5	−0.37	−3.89	−2.08	$\dfrac{0.9939e^{-1.994s}}{9.5672s+1}$	$\dfrac{0.9939e^{-2.0092s}}{9.9402s+1}$
10%	Averaging	2.74	−2.36	−9.28	−5.32	$\dfrac{1.0236e^{-1.928s}}{9.3637s+1}$	$\dfrac{1.0236e^{-1.9971s}}{10.2331s+1}$
10%	Filtering	1.72	−0.96	4.19	−2.05	$\dfrac{1.0108e^{-2.134s}}{10.4172s+1}$	$\dfrac{1.0108e^{-2.2417s}}{10.1769s+1}$
30%	Filtering	−0.91	2.78	2.71	−2.68	$\dfrac{1.0292e^{-2.222s}}{10.8582s+1}$	$\dfrac{1.0292e^{-2.2211s}}{10.3754s+1}$

FB2 are used, obtaining the results listed also in Table 4.3, together with the limit cycle data measured for computation. It is seen that these two algorithms also give good accuracy.

Now, suppose that a random noise $N\left(0, \sigma_N^2 = 0.0112\%\right)$ is added to the process output measurement, causing NSR = 5%. The corrupted output measurement is then used for relay feedback control. It can be seen from Table 4.4 that both Algorithm-RS-FA1 and Algorithm-RS-FA2 maintain good identification robustness, based on using the statistical averaging method to obtain the limit cycle data from 10 steady oscillation periods, e.g., in the time interval of [60, 220] (s). Note that similar identification results can also be obtained using Algorithm-RS-FB1 and Algorithm-RS-FB2, and thus are omitted. When the noise level is increased to NSR = 10%, it can be seen from Table 4.4 that the process time constant is somewhat underestimated by Algorithm-RS-FA1 in comparison with Algorithm-RS-FA2. To enhance identification robustness, a first-order Butterworth filter with a cutoff angular frequency, $\omega_c = 4.0$ (rad/s), according to the guideline given in (4.3), is used to recover the corrupted limit cycle and also for relay feedback control. The corresponding results are also listed in Table 4.4, which demonstrate that improved accuracy for the measurement of limit cycle data and model identification can thus be obtained. Even in the case where the noise causing NSR = 30% is added, the proposed filter together with the averaging for 10 oscillation periods can ensure good identification robustness, as illustrated by the resulting models listed in Table 4.4.

Example 4.2. Consider a high-order process studied in the literature (Wang et al. 1997; Kaya and Atherton 2001),

$$G = \frac{(-s+1)e^{-s}}{(s+1)^5}$$

Based on a biased relay test, Wang et al. (1997) derived an FOPDT model, $G_m = 1.00e^{-4.24s}/(2.99s+1)$, and Kaya and Atherton (2001) gave an FOPDT model,

4.3 Low-Order Model Fitting Algorithms

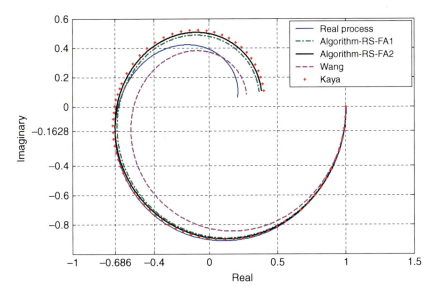

Fig. 4.9 Nyquist fitting of FOPDT models for Example 4.2

$G_m = 1.00e^{-5.082s}/(2.292s + 1)$. The biased relay test in Example 4.1 is performed for comparison. Correspondingly, Algorithm-RS-FA1 and Algorithm-RS-FA2 are used to derive the FOPDT models, which are shown in Table 4.3. The Nyquist plots of these FOPDT models are shown in Fig. 4.9. It is seen that Algorithm-RS-FA2 gives the best fit, owing to the use of the precise fitting condition of the process response at the oscillation frequency, i.e., $(-0.686, -j0.1628)$, as shown in Fig. 4.9. Note that the FOPDT model obtained from Algorithm-RS-FA2 corresponds to $ERR = 2.71\%$, while that of Kaya's method led to $ERR = 11\%$. In contrast, Algorithm-RS-FA1 gives slightly inferior fitting, but with a less computation effort.

Example 4.3. Consider a third-order process studied in the literature (Luyben 2001),

$$G = \frac{e^{-s}}{8(s+1)^3}$$

Based on an unbiased relay test, Luyben (2001) derived an FOPDT model, $G_m = 0.284e^{-1.96s}/(5.97s + 1)$. For illustration, an unbiased relay test using $u_+ = -u_- = 1.0$ and $\varepsilon_+ = -\varepsilon_- = 0.1$ is performed here for model identification. Correspondingly, Algorithm-RS-FB1 and Algorithm-RS-FB2 are used, obtaining the results listed in Table 4.3, together with the limit cycle data measured for computation. The Nyquist plots of these FOPDT models are shown in Fig. 4.10. It is seen that Algorithm-RS-FB2 yields the best fitting, in particular in the preferred low frequency range, owing to the fact that the model response coincides with the process at the oscillation frequency, i.e., $(-0.0276, -j0.0854)$.

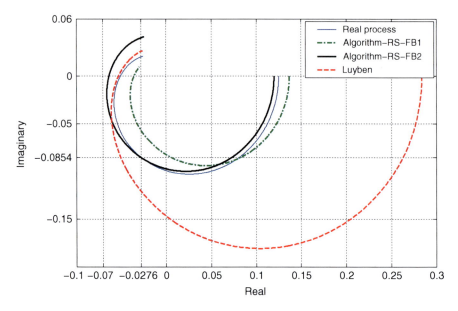

Fig. 4.10 Nyquist fitting of FOPDT models for Example 4.3

Example 4.4. Consider an overdamped second-order process widely studied in the literature (Li et al. 1991; Shen et al. 1996; Srinivasan and Chidambaram 2003; Ramakrishnan and Chidambaram 2003),

$$G = \frac{e^{-2s}}{(10s+1)(s+1)}$$

By using two different relay tests in terms of precise measurement of the process time delay, the so-called ATV method (Li et al. 1991) gave an SOPDT model, $G_m = 0.853e^{-2s}/(7.416s+1)(1.15s+1)$. Using a single biased relay test, Ramakrishnan and Chidambaram (2003) derived an SOPDT model, $G_m = 1.05e^{-1.814s}/(9.766s+1)(1.271s+1)$.

For illustration, a biased relay test with $u_+ = 1.3$ and $u_- = -0.7$ and an unbiased relay test with $u_+ = -u_- = 1.0$ are performed, respectively, together with $\varepsilon_+ = -\varepsilon_- = 0.2$ to avoid incorrect relay switches caused by measurement noise. Correspondingly, Algorithm-RS-SO1 and Algorithm-RS-SO2 are used to obtain the process models as listed in Table 4.5, together with the limit cycle data measured for computation. It is seen that good accuracy is obtained, corresponding to very small fitting errors.

Note that Ramakrishnan and Chidambaram (2003) derived an SOPDT model, $G_m = 1.06e^{-2.04s}/(10.73s+1)(0.92s+1)$, on condition that a random measurement noise $N(0, \sigma_N^2 = 0.5\%)$ is added to the relay feedback channel. It can be verified that this measurement noise causes NSR = 2%. By using the statistic

4.3 Low-Order Model Fitting Algorithms

Table 4.5 Identified SOPDT models for five processes under a relay test

Process	Relay	\multicolumn{6}{c}{Measured Limit Cycle Data}	\multicolumn{3}{c}{SOPDT Model Type}	\multicolumn{2}{c}{Fitting Error}
		P_+	P_-	A_+	A_-	A_u	φ_u	Overdamped	Critically damped	Underdamped	err	ERR
$\dfrac{e^{-2s}}{(10s+1)(s+1)}$	Biased	6.99	11.72	0.4222	-0.2953	0.2705	-2.2787	$\dfrac{1.0000e^{-1.9991s}}{(9.9921s+1)(1.0073s+1)}$			6.23×10^{-9}	0.17%
	Unbiased	8.73	8.73	0.3571	-0.3571	0.2519	-2.3667	$\dfrac{1.0122e^{-2.0037s}}{(10.1178s+1)(0.992s+1)}$			8.53×10^{-5}	1.22%
$\dfrac{e^{-10s}}{(s+1)^2}$	Biased	11.53	12.7	1.2998	-0.6999	0.937	-3.1019		$\dfrac{1.0000e^{-10s}}{(1.0001s+1)^2}$		2.5×10^{-10}	0.005%
	Unbiased	12.03	12.03	0.9998	-0.9998	0.9362	-3.1237		$\dfrac{1.0003e^{-9.9999s}}{(1.0027s+1)^2}$		1.54×10^{-7}	0.13%
$\dfrac{e^{-7s}}{s^2+0.4s+1}$	Biased	8.46	8.73	0.5823	-0.4728	1.1381	-2.7276		$\xi = 0.2001$, $\tau_p = 1.0001$	$\dfrac{1.0000e^{-7.004s}}{1.0004s^2+0.4009s+1}$	6.73×10^{-7}	0.19%
	Unbiased	8.66	8.66	0.4248	-0.4248	1.1359	-2.7084		$\xi = 0.2001$, $\tau_p = 1.0006$	$\dfrac{0.9999e^{-7.0044s}}{1.0012s^2+0.4004s+1}$	1.01×10^{-6}	0.18%
$\dfrac{e^{-4s}}{(0.5s+1)^3}$	Biased	5.24	5.96	-0.6989	1.2964	0.8926	-3.0672		$\dfrac{1.0000e^{-4.2771s}}{(0.6183s+1)^2}$		2.16×10^{-5}	0.36%
$\dfrac{e^{-0.5s}}{(s+1)(s^2+s+1)}$	Biased	2.71	3.53	1.0676	-0.7059	0.6996	-2.8818		$\xi = 0.719$, $\tau_p = 0.9871$	$\dfrac{1.0000e^{-1.3105s}}{0.9744s^2+1.4194s+1}$	7.12×10^{-4}	11.85%

averaging method on the measurement of 10 oscillation periods in the time interval [60, 260] (s), Algorithm-RS-SO1 results in the process model listed in Table 4.6, which indicates good identification robustness. Note that a similar result can be obtained using Algorithm-RS-SO2 and thus is omitted.

To further demonstrate identification robustness against measurement noise, a random noise $N(0, \sigma_N^2 = 0.055\%)$ is added to the process output measurement and relay feedback, causing NSR = 10%. It can be seen from Table 4.6 that the statistic averaging method results in notable identification errors for the model parameters, which, however, seem still acceptable from a view of control design in practice. To enhance identification robustness, a first-order Butterworth filter with the cutoff angular frequency, $\omega_c = 3.5$ (rad/s), according to the guideline given by (4.3), is used for the output measurement and relay feedback, leading to apparently improved identification robustness, as indicated by the resulting model shown in Table 4.6. Note that even in the case where a random noise causing NSR = 30% is added, using the proposed filter together with the averaging method on 10 oscillation periods, e.g., in the time interval [100, 320] (s), can also guarantee good identification robustness, as indicated by the resulting model and fitting error listed in Table 4.6.

Example 4.5. Consider a critically damped second-order process studied in the literature (Thyagarajan and Yu 2003),

$$G = \frac{e^{-10s}}{(s+1)^2}$$

Based on an unbiased relay test, Thyagarajan and Yu (2003) gave an approximate estimation of the corresponding SOPDT model. By performing a biased and an unbiased relay tests as in Example 4.4, respectively, the corresponding Algorithm-RS-SC1 and Algorithm-RS-SC2 give the process models as listed in Table 4.5, which once again demonstrate good accuracy.

To demonstrate identification robustness against measurement noise, the noise tests in Example 4.4 are performed. The corresponding identification results obtained using Algorithm-RS-SC1 are shown in Table 4.6, together with the limit cycle data measured for computation. It is seen that good identification robustness is, therefore, maintained.

Example 4.6. Consider an underdamped second-order process studied in the literature (Panda 2006),

$$G = \frac{e^{-7s}}{s^2 + 0.4s + 1}$$

Based on an unbiased relay test, Panda (2006) derived an SOPDT model, $G_m = 1.05e^{-6.8902s}/(1.03s^2 + 0.41s + 1)$. By performing a biased ($u_+ = 0.3$ and $u_- = -0.2$) and an unbiased ($u_+ = -u_- = 0.2$) relay tests with $\varepsilon_+ = -\varepsilon_- = 0.1$, respectively, Algorithm-RS-SU1 and Algorithm- RS-SU2 result in the SOPDT models shown in Table 4.5, which demonstrate that good identification accuracy can, therefore, be obtained.

4.3 Low-Order Model Fitting Algorithms

Table 4.6 Identification results for Examples 4.4–4.6 under different measurement noise levels

NSR	Denoising	Example	\% Error in the Limit Cycle Data A_+	A_-	A_u	φ_u	Identified Model	Fitting Error err	ERR
2%	Averaging	4.4	1.56	−2.07	−1.1	0.56	$\dfrac{0.9943 e^{-1.932 s}}{(9.8346 s+1)(1.098 s+1)}$	1.9×10^{-5}	0.57%
		4.5	−8.9	4.89	−0.02	0.29	$\dfrac{1.0003 e^{-10.0029 s}}{(1.0009 s+1)^2}$	1.54×10^{-7}	0.12%
		4.6	2.65	−2.47	0.01	−0.04	$\dfrac{1.0006 e^{-7.0521 s}}{0.9774 s^2+0.3595 s+1}$	1.53×10^{-4}	0.22%
10%	Averaging	4.4	−5.36	4.4	−9.64	−5.6	$\dfrac{1.0279 e^{-2.5707 s}}{(10.8686 s+1)(0.3197 s+1)}$	4.42×10^{-4}	3.27%
		4.5	−43.44	24.01	0.75	−2.93	$\dfrac{1.0025 e^{-10.1809 s}}{(1.0741 s+1)^2}$	6.9×10^{-4}	8.29%
		4.6	12.28	−14.67	0.24	−0.94	$\dfrac{1.0025 e^{-6.6759 s}}{1.1618 s^2+0.6823 s+1}$	2.86×10^{-3}	0.63%
10%	Filtering	4.4	−3.56	4.6	3.97	1.7	$\dfrac{1.0012 e^{-2.2686 s}}{(10.0433 s+1)(1.0216 s+1)}$	4.16×10^{-5}	9.38%
		4.5	−3.81	2.14	0.02	0.15	$\dfrac{1.001 e^{-10.2691 s}}{(1.0359 s+1)^2}$	5.08×10^{-4}	8.47%
		4.6	−0.02	−0.02	−1.54	−0.81	$\dfrac{1.0011 e^{-7.375 s}}{0.9296 s^2+0.3659 s+1}$	2.48×10^{-3}	9.46%
30%	Filtering	4.4	−3.8	7.96	3.2	−2.05	$\dfrac{1.0022 e^{-2.2091 s}}{(9.9832 s+1)(1.0878 s+1)}$	3.44×10^{-5}	9.27%
		4.5	−12.16	7.08	0.06	−0.01	$\dfrac{1.0041 e^{-10.2287 s}}{(1.05584 s+1)^2}$	7.68×10^{-4}	8.4%
		4.6	1.93	−1.03	−1.47	−0.78	$\dfrac{1.0023 e^{-7.4661 s}}{0.8943 s^2+0.2795 s+1}$	3.87×10^{-3}	11.51%

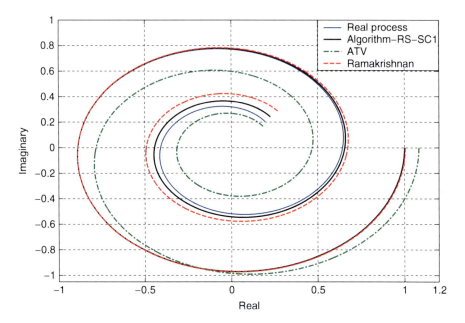

Fig. 4.11 Nyquist fitting effect for Example 4.7

To demonstrate identification robustness against measurement noise, the noise tests in Example 4.4 are also performed. The resulting models from Algorithm-RS-SU1 are shown in Table 4.6, which also demonstrate good identification robustness.

Example 4.7. Consider a high-order process studied in the literature (Li et al. 1991; Ramakrishnan and Chidambaram 2003),

$$G = \frac{e^{-4s}}{(0.5s+1)^3}$$

Based on a biased relay test, the ATV method of Li et al. (1991) gave an FOPDT model, $G_m = 1.0841e^{-4.04s}/(1.6362s+1)$, and Ramakrishnan and Chidambaram (2003) derived an overdamped SOPDT model, $G_m = 1.0046e^{-4.4239s}/(0.2286s+1)(0.8839s+1)$.

For illustration, a biased relay test as in Example 4.4 is performed for model identification. According to the guidelines given for model structure selection in terms of the reference relay response shapes shown in Table 4.2, Algorithm-RS-SC1 is chosen to derive a critically damped SOPDT model, which is shown in Table 4.5. For comparison, the Nyquist plots of the above models are shown in Fig. 4.11. It is seen that evidently improved fitting is obtained by the proposed SOPDT model.

4.3 Low-Order Model Fitting Algorithms

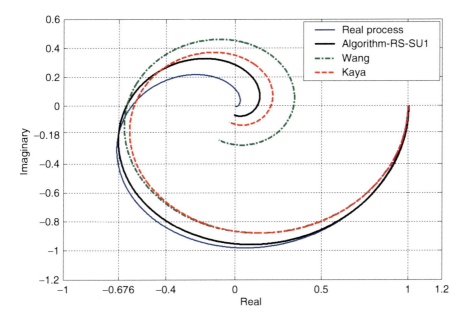

Fig. 4.12 Nyquist fitting effect for Example 4.8

Example 4.8. Consider another high-order process studied in the literature (Wang et al. 1997; Kaya and Atherton 2001),

$$G = \frac{e^{-0.5s}}{(s+1)(s^2+s+1)}$$

Based on a biased relay test, Wang et al. (1997) derived an FOPDT model, $G_m = 1.00e^{-2.1s}/(1.152s+1)$, and Kaya and Atherton (2001) gave a critically damped SOPDT model, $G_m = 1.00e^{-1.633s}/(0.785s+1)^2$.

For illustration, the biased relay test in Example 4.4 is performed for model identification. According to the guidelines given for model structure selection in terms of the reference relay response shapes shown in Table 4.2, Algorithm-RS-SU1 is chosen to derive an underdamped SOPDT model, resulting in the process model shown in Table 4.5. For comparison, the Nyquist plots of the above models are shown in Fig. 4.12, which once again demonstrate that obviously improved fitting is obtained using the proposed algorithm. Note that the proposed model response at $(-0.676, -j0.18)$ corresponding to the oscillation frequency coincides precisely with that of the process.

To demonstrate the achievable control performance in terms of the above identified models, the standard IMC method (Morari and Zafiriou 1989), as will be introduced in Chap. 7, is used here for benchmark comparison. If a perfect knowledge of the process is used, the IMC controller should be designed as

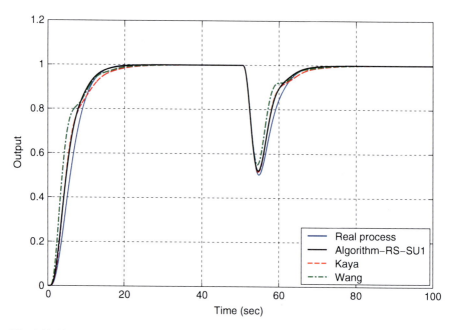

Fig. 4.13 Nominal system response for Example 4.8

$Q_P = (s + 1)(s^2 + s + 1)/(\lambda_P s + 1)^3$, where λ_P is an adjustable control parameter. Using the proposed SOPDT model, the corresponding IMC controller should be $Q_A = (0.9744s^2 + 1.4194s + 1)/(\lambda_A s + 1)^2$. If the SOPDT model of Kaya and Atherton (2001) is used, the corresponding IMC controller should be $Q_K = (0.785s + 1)^2/(\lambda_K s + 1)^2$. If the FOPDT model of Wang et al. (1997) is used, the corresponding IMC controller should be $Q_W = (1.152s + 1)/(\lambda_W s + 1)$.

By adding a unity step change to the set-point of the standard IMC system and then a step-type load disturbance with a magnitude of -0.5 to the process, and taking $\lambda_P = 2.0$, $\lambda_A = \lambda_K = 2.2$, and $\lambda_W = 3.0$ to obtain the same rising speed of the set-point response for comparison, the corresponding output responses are shown in Fig. 4.13, respectively. It is seen that the proposed SOPDT model facilitates good control performance. Then, assume that the process is perturbed to the form of $G_\Delta = e^{-2.5s}/(0.5s + 1)(s^2 + 0.5s + 1)$ due to practical uncertainties, the perturbed output responses are shown in Fig. 4.14, which demonstrate that the proposed SOPDT model also facilitates control robustness.

4.4 A Generalized Relay Identification Method

As introduced in the above section, under a biased relay test, the process static gain can be separately derived as the ratio of a periodic integral of the process output to that of the relay output, but the gain error may arise from unexpected

4.4 A Generalized Relay Identification Method

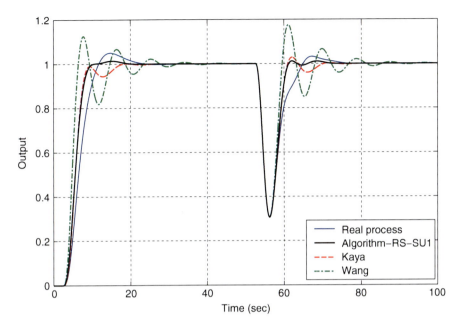

Fig. 4.14 Perturbed system response for Example 4.8

load disturbance. In contrast, under an unbiased relay test, the influence from load disturbance can be intuitively detected in terms of the symmetric shape of the limit cycle, but the process static gain cannot be derived like that under a biased relay test, because such a periodic integral results in zero.

A unified gain identification algorithm is, therefore, presented here for application under a biased or unbiased relay test. Moreover, a general model identification method is presented for identifying a model of any order, based on the frequency response estimation under a relay test.

Consider a general model structure that can describe both time delay and nonminimum phase (NMP) processes,

$$G(s) = \frac{k_p(-\tau_0 s + 1)}{(\tau_p s + 1)^m} e^{-\theta s} \quad (4.189)$$

where k_p denotes the process static gain, θ the time delay, $1/\tau_0$ the right-half-plane (RHP) zero, τ_p the time constant, and m is a positive integer indicating the model order. Note that $\tau_0 = 0$ and $\theta = 0$ corresponds to a linear and stable process; $\tau_0 = 0$ but $\theta \neq 0$ corresponds to a time delay process; $\tau_0 \neq 0$ but $\theta = 0$ corresponds to an inverse response process; $\tau_0 \neq 0$ and $\theta \neq 0$ corresponds to an inverse response process with time delay.

Hence, apart from identifying the process static gain, there are three specific cases for model identification: (i) identify τ_p and θ with $\tau_0 = 0$; (ii) identify τ_0 and

τ_p with $\theta = 0$; and (iii) identify τ_0, τ_p, and θ. Note that the model order, m, is user-specified, based on the process response characteristics, and is usually taken as 1, 2, or 3 for low-order modeling and controller tuning. Correspondingly, identification algorithms will be detailed for these three cases.

4.4.1 Relay Response Expression

For the convenience of assessing the fluctuation range of relay response or determining the model order for proper fitting, a general relay response expression for the model structure in (4.189) is given in the following proposition:

Proposition 4.5. *For a process described by (4.189) under a biased relay test, the resulting limit cycle is characterized by*

$$y_+(t) = k_p(\Delta\mu + \mu_0) - 2k_p\mu_0 e^{-t/\tau_p}$$
$$\times \left[E_0 + \frac{1}{\tau_p}E_1 + \frac{1}{2!\tau_p^2}E_2 + \cdots + \frac{1+\tau_0/\tau_p}{(m-1)!\tau_p^{m-1}}E_{m-1} \right], \quad t \in [0, P_+]$$
(4.190)

$$y_-(t) = k_p(\Delta\mu - \mu_0) - 2k_p\mu_0 e^{-t/\tau_p}$$
$$\times \left[F_0 + \frac{1}{\tau_p}F_1 + \frac{1}{2!\tau_p^2}F_2 + \cdots + \frac{1+\tau_0/\tau_p}{(m-1)!\tau_p^{m-1}}F_{m-1} \right], \quad t \in [0, P_-]$$
(4.191)

where $y_+(t)$ denotes the output response in the half period P_+, while $y_-(t)$ denotes the output response in the other half period P_- corresponding to $t \in (P_+, P_u]$, $P_u = P_+ + P_-$, $\rho_u = e^{-P_u/\tau_p}$, $\rho_+ = e^{-P_+/\tau_p}$, $\rho_- = e^{-P_-/\tau_p}$, and

$$E_i = t^i + \sum_{k=1}^{n\to\infty} \rho_u^k (t + k P_u)^i - \rho_- \sum_{k=1}^{n\to\infty} \rho_u^{k-1}[t + P_- + (k-1)P_u]^i,$$
$$i = 0, 1, 2, \ldots, m-1. \tag{4.192}$$

$$F_i = \rho_+ \sum_{k=0}^{n\to\infty} \rho_u^k(t + P_+ + k P_u)^i - \sum_{k=0}^{n\to\infty} \rho_u^k(t + k P_u)^i, \quad i = 0, 1, 2, \ldots, m-1.$$
(4.193)

4.4 A Generalized Relay Identification Method

Proof. The step response of the *m*th order process shown in (4.189) can be derived as

$$y(t) = k_p \left\{ 1 - \left[1 + \frac{t-\theta}{\tau_p} + \frac{(t-\theta)^2}{2!\tau_p^2} + \cdots + \frac{(1+\tau_0/\tau_p)(t-\theta)^{m-1}}{(m-1)!\tau_p^{m-1}} \right] e^{-(t-\theta)/\tau_p} \right\}$$

(4.194)

By using a time shift of θ, the initial process response arising from the relay output, $u(t) = \Delta\mu - \mu_0$, can be written as

$$y_0(t) = (\Delta\mu - \mu_0)y(t+\theta) = k_p(\Delta\mu - \mu_0)$$

$$\times \left\{ 1 - \left[1 + \frac{t}{\tau_p} + \frac{t^2}{2!\tau_p^2} + \cdots + \frac{(1+\tau_0/\tau_p)\,t^{m-1}}{(m-1)!\tau_p^{m-1}} \right] e^{-t/\tau_p} \right\} \quad (4.195)$$

After the first relay switch point, the relay output changes to be $\Delta\mu + \mu_0$, indicating that a step change of $2\mu_0$ is added to the process input. According to the linear superposition principle, one can obtain the process output response with a time shift of $t_0 + \theta$ as

$$y_1(t) = y_0(t+t_0) + 2k_p\mu_0 \left\{ 1 - \left[1 + \frac{t}{\tau_p} + \frac{t^2}{2!\tau_p^2} + \cdots + \frac{(1+\tau_0/\tau_p)\,t^{m-1}}{(m-1)!\tau_p^{m-1}} \right] e^{-t/\tau_p} \right\}$$

(4.196)

where t_0 is the time to reach the first relay switch point.

After the second relay switch point, the relay output changes to be $\Delta\mu - \mu_0$, indicating that a reverse step change of $2\mu_0$ is added to the process input. Accordingly, the process output response with a time shift of $t_0 + \theta + P_+$ can be derived as

$$y_2(t) = y_0(t+t_0+P_+) + 2k_p\mu_0(1-1) - 2k_p\mu_0 e^{-t/\tau_p}$$

$$\times \left\{ e^{-P_+/\tau_p} - 1 + \frac{1}{\tau_p}\left[(t+P_+)e^{-P_+/\tau_p} - t\right] + \frac{1}{2!\tau_p^2}\left[(t+P_+)^2 e^{-P_+/\tau_p} - t^2\right] \right.$$

$$\left. + \cdots + \frac{1+\tau_0/\tau_p}{(m-1)!\tau_p^{m-1}}\left[(t+P_+)^{m-1} e^{-P_+/\tau_p} - t^{m-1}\right] \right\}$$

(4.197)

After the third relay switch point, the relay output changes again to be $\Delta\mu + \mu_0$, indicating that a step change of $2\mu_0$ is once again added to the process input. Accordingly, the process output response with a time shift of $t_0 + \theta + P_u$ can be derived as

$$y_3(t) = y_0(t + t_0 + P_u) + 2k_p\mu_0(1 - 1 + 1) - 2k_p\mu_0 e^{-t/\tau_p}$$

$$\times \left\{ \rho_u - \rho_- + 1 + \frac{1}{\tau_p} [\rho_u(t + P_u) - \rho_-(t + P_-) + t] \right.$$

$$+ \frac{1}{2!\tau_p^2} \left[\rho_u(t + P_u)^2 - \rho_-(t + P_-)^2 + t^2 \right] + \cdots + \frac{1 + \tau_0/\tau_p}{(m-1)!\tau_p^{m-1}}$$

$$\left. \times \left[\rho_u(t + P_u)^{m-1} - \rho_-(t + P_-)^{m-1} + t^{m-1} \right] \right\} \qquad (4.198)$$

where $P_u = P_+ + P_-$, $\rho_u = e^{-P_u/\tau_p}$, $\rho_+ = e^{-P_+/\tau_p}$, and $\rho_- = e^{-P_-/\tau_p}$.

The process output response following the fourth relay switch point is the result of four interlaced step changes, respectively, with a magnitude of $2\mu_0$. Using a time shift of $t_0 + \theta + P_+ + P_u$, the process output response can be similarly derived as

$$y_4(t) = y_0(t + t_0 + P_+ + P_u) + 2k_p\mu_0(1 - 1 + 1 - 1) - 2k_p\mu_0 e^{-t/\tau_p}$$

$$\times \left\{ \rho_+\rho_u - \rho_u + \rho_+ - 1 + \frac{1}{\tau_p} [\rho_+\rho_u(t + P_+ + P_u) - \rho_u(t + P_u) \right.$$

$$+ \rho_+(t + P_+) - t] + \cdots + \frac{1 + \tau_0/\tau_p}{(m-1)!\tau_p^{m-1}} \left[\rho_+\rho_u(t + P_+ + P_u)^{m-1} \right.$$

$$\left. \left. - \rho_u(t + P_u)^{m-1} + \rho_+(t + P_+)^{m-1} - t^{m-1} \right] \right\} \qquad (4.199)$$

It is, therefore, summarized from (4.196)–(4.199) that

$$y_{2n+1}(t) = y_0(t + t_0 + nP_u) + 2k_p\mu_0 - 2k_p\mu_0 e^{-t/\tau_p}$$

$$\times \left[E_0 + \frac{1}{\tau_p} E_1 + \frac{1}{2!\tau_p^2} E_2 + \cdots + \frac{1 + \tau_0/\tau_p}{(m-1)!\tau_p^{m-1}} E_{m-1} \right] \qquad (4.200)$$

$$y_{2n+2}(t) = y_0(t + t_0 + P_+ + nP_u) - 2k_p\mu_0 e^{-t/\tau_p}$$

$$\times \left[F_0 + \frac{1}{\tau_p} F_1 + \frac{1}{2!\tau_p^2} F_2 + \cdots + \frac{1 + \tau_0/\tau_p}{(m-1)!\tau_p^{m-1}} F_{m-1} \right] \qquad (4.201)$$

where $n = 0, 1, 2, \ldots$, E_i and F_i ($i = 0, 1, 2, \ldots, m-1$) are shown in (4.192) and (4.193).

Owing to $0 < \rho_u < 1$, it follows for $n \to \infty$ that

$$\sum_{k=0}^{n} \rho_u^k = \frac{1}{1 - \rho_u} \qquad (4.202)$$

4.4 A Generalized Relay Identification Method

Note that there exists a recursive relationship,

$$\sum_{k=1}^{n} k^m \rho_u^k = \rho_u \frac{d}{d\rho_u}\left(\sum_{k=1}^{n} k^{m-1} \rho_u^k\right) \qquad (4.203)$$

Therefore, all the infinite series shown in (4.192) and (4.193) are uniformly convergent for $n \to \infty$, and the corresponding summations can be derived using (4.202) and (4.203).

It can be verified from (4.195) that $y_0(t + t_0 + nP_u) = y_0(t + t_0 + P_+ + nP_u) = k_p(\Delta\mu - \mu_0)$ for $n \to \infty$. Hence, in the limit cycle (4.200) and (4.201) converge to the forms shown in (4.190) and (4.191). This completes the proof. □

It is, therefore, concluded from the above derivation that the limit cycle can be definitely formed for a process that can be described by (4.189). The limiting condition to forming steady oscillation is only related to the choice of the relay function, e.g., $\varepsilon_+ = -\varepsilon_- = (0.1 \sim 0.95) \max\{|u_+|, |u_-|\}$ to avoid incorrect relay switches caused by measurement noise. A short "listening period," e.g., 20–100 samples, may be referenced to perform a relay test.

Note that, by letting $P_+ = P_- = P_u/2$ in (4.190) and (4.191), the relay response expression under an unbiased relay test can be obtained accordingly. If letting $t_0 = 0$ in the resulting relay response expression, it can be used to summarize the relay response expressions presented in the previous section for first- and second-order processes.

4.4.2 Frequency Response Estimation

In the limit cycle, the process output response is a periodic function with respect to the oscillation angular frequency, $\omega_u = 2\pi/P_u$. Using a time shift, it can be viewed as a periodic signal from the very beginning, so its Fourier transform can be derived as

$$\begin{aligned} Y(j\omega_u) &= \int_0^{P_u} y_{os}(t)e^{-j\omega_u t}\,dt + \int_{P_u}^{2P_u} y_{os}(t)e^{-j\omega_u t}\,dt + \cdots \\ &= \lim_{N\to\infty} N \int_0^{P_u} y_{os}(t)e^{-j\omega_u t}\,dt \\ &= \lim_{N\to\infty} N \int_{t_{os}}^{t_{os}+P_u} y(t)e^{-j\omega_u t}\,dt \end{aligned} \qquad (4.204)$$

where $y_{os}(t) = y(t)$ for $t \in [t_{os}, \infty)$, and t_{os} can be taken as any relay switch point in steady oscillation, such that the influence from the initial unsteady relay response can be excluded.

Similarly, it follows that

$$U(j\omega_u) = \lim_{N\to\infty} N \int_{t_{os}}^{t_{os}+P_u} u(t)e^{-j\omega_u t} dt \qquad (4.205)$$

The process frequency response at ω_u can, therefore, be derived as

$$G(j\omega_u) = \frac{Y(j\omega_u)}{U(j\omega_u)} = \frac{\int_{t_{os}}^{t_{os}+P_u} y(t)e^{-j\omega_u t} dt}{\int_{t_{os}}^{t_{os}+P_u} u(t)e^{-j\omega_u t} dt} = A_u e^{j\varphi_u} \qquad (4.206)$$

which can be numerically computed using the trapezoidal rule for numerical integral or the FFT series.

To estimate the process frequency response other than at ω_u, one can decompose the relay response as

$$Y(s) = \int_0^{t_{os1}} y(t)e^{-st} dt + \int_{t_{os1}}^{\infty} y(t)e^{-st} dt \qquad (4.207)$$

where t_{os1} denotes the time to reach steady oscillation under a relay test.

In view of the fact that $y(t)$ becomes a periodic signal after t_{os1}, the second integral in (4.207) can be derived as

$$\int_{t_{os1}}^{\infty} y(t)e^{-st} dt = \int_{t_{os1}}^{t_{os1}+P_u} y(t)e^{-st} dt + \int_{t_{os1}+P_u}^{t_{os1}+2P_u} y(t)e^{-st} dt + \cdots$$

$$= (1 + e^{-P_u s} + e^{-2P_u s} + \cdots) \int_{t_{os1}}^{t_{os1}+P_u} y(t)e^{-st} dt$$

$$= \lim_{n\to\infty} \frac{1 - e^{-nP_u s}}{1 - e^{-P_u s}} \int_{t_{os1}}^{t_{os1}+P_u} y(t)e^{-st} dt \qquad (4.208)$$

Note that if $\text{Re}(s) > 0$, there exists $e^{-nP_u s} \to 0$ for $n \to \infty$. One can thus obtain

$$Y(s) = \int_0^{t_{os1}} y(t)e^{-st} dt + \frac{1}{1 - e^{-P_u s}} \int_{t_{os1}}^{t_{os1}+P_u} y(t)e^{-st} dt \qquad (4.209)$$

Similarly, it follows for $\text{Re}(s) > 0$ that

$$U(s) = \int_0^{t_{os1}} u(t)e^{-st} dt + \frac{1}{1 - e^{-P_u s}} \int_{t_{os1}}^{t_{os1}+P_u} u(t)e^{-st} dt \qquad (4.210)$$

The process frequency response transfer function for $\text{Re}(s) > 0$ can be derived accordingly as

4.4 A Generalized Relay Identification Method

$$G(s) = \frac{Y(s)}{U(s)} = \frac{\left(1 - e^{-P_u s}\right) \int_0^{t_{os1}} y(t) e^{-st} dt + \int_{t_{os1}}^{t_{os1}+P_u} y(t) e^{-st} dt}{\left(1 - e^{-P_u s}\right) \int_0^{t_{os1}} u(t) e^{-st} dt + \int_{t_{os1}}^{t_{os1}+P_u} u(t) e^{-st} dt} \quad (4.211)$$

Substituting $s = \alpha + j\omega$ into (4.211) yields

$$G(\alpha + j\omega) = A_\alpha e^{j\varphi_\alpha}$$

$$= \frac{\left(1 - e^{-P_u(\alpha+j\omega)}\right) \int_0^{t_{os1}} [y(t) e^{-\alpha t}] e^{-j\omega t} dt + \int_{t_{os1}}^{t_{os1}+P_u} [y(t) e^{-\alpha t}] e^{-j\omega t} dt}{\left(1 - e^{-P_u(\alpha+j\omega)}\right) \int_0^{t_{os1}} [u(t) e^{-\alpha t}] e^{-j\omega t} dt + \int_{t_{os1}}^{t_{os1}+P_u} [u(t) e^{-\alpha t}] e^{-j\omega t} dt}$$

(4.212)

where $\alpha \in \Re_+$ may be viewed as a damping factor for the Laplace transform, as introduced in Sect. 2.2.1.

Note that $G(j\omega + \alpha) \to 0/0$ as $\alpha \to \infty$. The guideline for choosing α is, therefore, suggested as

$$\begin{cases} \alpha > \delta; \\ \min \left\{ \left| u(t_{os1} + P_u) e^{-\alpha(t_{os1}+P_u)} T_s \right|, \left| y(t_{os1} + P_u) e^{-\alpha(t_{os1}+P_u)} T_s \right| \right\} > \delta. \end{cases} \quad (4.213)$$

where δ denotes the computational precision that can be practically taken smaller than $10^{-6} \times \min \{|u(t_{os1} + P_u)T_s|, |y(t_{os1} + P_u)T_s|\}$ and T_s is the sampling period for computing the numerical integrals in (4.212).

Thereby, both the initial and steady oscillation responses of $y(t)$ and $u(t)$ can be effectively included in the computation of (4.212). Note that t_{os1} can be chosen as any moment in the steady oscillation, which can be verified from (4.208). For the convenience of computation, it is suggested to take t_{os1} as a small multiple of P_u.

Accordingly, one can compute multiple frequency response points of $G(\alpha + j\omega_k)$ from (4.212) for $\omega_k = k\omega_u/M_0 (k = 0, 1, 2, \ldots, N_0 - 1)$, where $N_0 = M_0 P_u/T_s$ and M_0 should be taken as an even integer for efficient computation of the FFT series.

Using the inverse Laplace transform of $g(t) e^{-\alpha t} = L^{-1} [G(\alpha + j\omega)]$, one can obtain

$$G(0) = \text{FFT}\{\text{FFT}^{-1} \{G(\alpha + j\omega_k)\} e^{\alpha k T_s}\}_{l=0} \quad (4.214)$$

where $\text{FFT}\{\cdot\}_{l=0}$ denotes the first element corresponding to $\omega_0 = 0$ in the resulting FFT series.

It is seen from (4.214) that the process static gain, k_p, can be separately derived from the other model parameters, regardless of whether the relay test is unbiased or biased.

Note that there exists $G(\alpha + j\omega_1) = G(\alpha + j\omega_2)$ in the case where $\omega_2 > \omega_1$ while $P_u(\omega_2 - \omega_1) = 2h\pi$ and $\omega_2 - \omega_1 = 2l\pi/T_s$ are satisfied, and both h and l are positive integers. Frequency response estimation from (4.214) is, therefore, limited not only by the sampling frequency for computing the numerical integrals in (4.212) but also by the frequency range of $\omega \in [0, 2\pi \cdot 10^r)$, where r is the minimal integer that satisfies $P_u \cdot 10^r = h$.

In the presence of measurement noise, to enhance identification robustness under a low noise level, e.g., NSR < 10%, it is suggested to use 10 ~ 20 periods in the steady oscillation for the frequency response estimation. Note that the relay output, $u(t)$, remains as a constant for each half period of the limit cycle, and thus can be used for measuring the averaged oscillation period, $\bar{P}_u$. Accordingly, the averaged frequency response estimation can be computed as

$$G(j\bar{\omega}_u) = \frac{\int_{t_{os}}^{t_{os}+N_s \bar{P}_u} y(t) e^{-j\bar{\omega}_u t} dt}{\int_{t_{os}}^{t_{os}+N_s \bar{P}_u} u(t) e^{-j\bar{\omega}_u t} dt} = \bar{A}_u e^{j\bar{\varphi}_u} \quad (4.215)$$

$G(\alpha + j\omega)$

$$= \frac{\left(1 - e^{-\bar{P}_u(\alpha+j\omega)}\right) \int_0^{t_{os1}+N_1 \bar{P}_u} [y(t) e^{-\alpha t}] e^{-j\omega t} dt + \int_{t_{os1}+N_1 \bar{P}_u}^{t_{os1}+(N_1+1)\bar{P}_u} [y(t) e^{-\alpha t}] e^{-j\omega t} dt}{\left(1 - e^{-\bar{P}_u(\alpha+j\omega)}\right) \int_0^{t_{os1}+N_1 \bar{P}_u} [u(t) e^{-\alpha t}] e^{-j\omega t} dt + \int_{t_{os1}+N_1 \bar{P}_u}^{t_{os1}+(N_1+1)\bar{P}_u} [u(t) e^{-\alpha t}] e^{-j\omega t} dt} \quad (4.216)$$

where $\bar{\omega}_u = 2\pi/\bar{P}_u$, N_s is the number of steady oscillation periods used for averaging, which may be taken in the range of 5–20, and N_1 may be taken as large as possible but subject to a numerical constraint similar to (4.213) for computation effectiveness.

To cope with the measurement noise causing a high NSR level, a low-pass Butterworth filter can be used for measuring the process output that is also used for relay feedback control, or offline denoising to improve identification accuracy, as introduced in Sect. 4.1.

4.4.3 Model Fitting Algorithms

Based on the process frequency response estimated in the above section, frequency response fitting conditions can be established in terms of the general model structure shown in (4.189).

For the process frequency response at ω_u, by substituting (4.189) into the left-hand side of (4.206), one can obtain

$$k_p \sqrt{\frac{\tau_0^2 \omega_u^2 + 1}{\left(\tau_p^2 \omega_u^2 + 1\right)^m}} = A_u \quad (4.217)$$

$$-\theta \omega_u - \arctan(\tau_0 \omega_u) - m \arctan(\tau_p \omega_u) = \varphi_u \quad (4.218)$$

4.4 A Generalized Relay Identification Method

Substituting the process model of (4.189) with $s = \alpha + j\omega_u$ into the leftmost side of (4.212), one can obtain the fitting conditions of the shifted process frequency response with respect to ω_u as

$$k_p e^{-\alpha\theta} \sqrt{\frac{(\alpha\tau_0 - 1)^2 + \tau_0^2 \omega_u^2}{\left[(\alpha\tau_p + 1)^2 + \tau_p^2 \omega_u^2\right]^m}} = A_\alpha \tag{4.219}$$

$$-\theta\omega_u - \arctan\left(\frac{\tau_0 \omega_u}{1 - \alpha\tau_0}\right) - m \arctan\left(\frac{\tau_p \omega_u}{\alpha\tau_p + 1}\right) = \varphi_\alpha \tag{4.220}$$

For the identification Case (i), with the process static gain separately derived from (4.214), the process time constant and time delay can then be derived from (4.217) and (4.218) as

$$\tau_p = \frac{1}{\omega_u} \sqrt{\left(\frac{k_p}{A_u}\right)^{\frac{2}{m}} - 1} \tag{4.221}$$

$$\theta = -\frac{1}{\omega_u}\left[\varphi_u + m \arctan\left(\tau_p \omega_u\right)\right] \tag{4.222}$$

For the identification Case (ii), it can be derived from (4.217) that

$$\tau_0 = \frac{1}{\omega_u} \sqrt{\left(\frac{A_u}{k_p}\right)^2 \left(\tau_p^2 \omega_u^2 + 1\right)^m - 1} \tag{4.223}$$

Substituting (4.223) into (4.218) yields a transcendental equation with respect to τ_p,

$$-\arctan\left(\sqrt{\left(\frac{A_u}{k_p}\right)^2 \left(\tau_p^2 \omega_u^2 + 1\right)^m - 1}\right) - m \arctan\left(\tau_p \omega_u\right) = \varphi_u \tag{4.224}$$

which can be numerically solved using any iterative algorithm such as the Newton-Raphson method. Since the left-hand side of (4.224) is monotonically decreasing with respect to τ_p, there exists only an unique solution of τ_p. Then, τ_0 can be derived from (4.223).

For the identification Case (iii), by subtracting (4.218) from (4.220) one can obtain

$$\varphi_\alpha - \varphi_u = \arctan\left(\tau_0 \omega_u\right) - \arctan\left(\frac{\tau_0 \omega_u}{1 - \alpha\tau_0}\right) + m\left[\arctan\left(\tau_p \omega_u\right) - \arctan\left(\frac{\tau_p \omega_u}{\alpha\tau_p + 1}\right)\right] \tag{4.225}$$

Substituting (4.223) into (4.225) yields a transcendental equation with respect to τ_p, which can be solved using the Newton-Raphson method. The other time constant, τ_0, can then be solved from (4.223). Consequently, θ can be derived from (4.218) as

$$\theta = -\frac{1}{\omega_u}\left[\varphi_u + \arctan(\tau_0\omega_u) + m\arctan(\tau_p\omega_u)\right] \quad (4.226)$$

Note that in the case where multiple solutions of τ_p are obtained from (4.225), a time domain fitting constraint of the relay response, as shown in (4.46), can be used to determine the most suitable solution.

To further enhance the fitting accuracy over a user specified frequency range rather than only at the oscillation frequency ($\omega = \omega_u$), e.g., the low-frequency range primarily concerned for controller tuning in practice, multiple frequency response fitting conditions can be established based on the frequency response estimation from (4.212). Such fitting conditions may also be used to obtain a very specific process model with more descriptive parameters, compared to the general form of (4.189). To minimize the fitting error, a weighted LS objective function similar to that introduced in Sect. 2.2.5 is proposed here,

$$J_{\text{opt}} = \sum_{k=0}^{M} \rho_k \left| G(\alpha + j\omega_k) - \widehat{G}(\alpha + j\omega_k) \right|^2 < \varepsilon^2 \quad (4.227)$$

where $G(\alpha + j\omega_k)$ and $\widehat{G}(\alpha + j\omega_k)$ denotes the frequency responses of the process and the model, respectively, ε is a user-specified threshold for convergence, M is the number of frequency response points for fitting, and ρ_k ($k = 0, 1, 2, \ldots, M$) are the weighting coefficients.

To emphasize frequency response fitting over the low frequency range, it is suggested to choose $\omega_M = (1.1 \sim 2.0)\omega_u$ and $\rho_k = \eta^k/\sum_{k=0}^{M}\eta^k$, where $M \in [5, 20]$ and $\eta \in [0.9, 0.99]$ may be taken for computation. It can be verified from (4.218) that $\omega_u < \omega_c$, where ω_c is the cutoff angular frequency corresponding to $\angle G(j\omega_c) = -\pi$. Therefore, the above choice of ω_M should suffice for computation.

To establish a linear regression for parameter estimation, assume that the process model is obtained as $\widehat{G}^{[i-1]}(s)$ at the $(i-1)$-th iteration, e.g., a model obtained from one of the above identification algorithms for Cases (i–iii), one can expand the model to be derived at the ith iteration using the multivariable Taylor series as

$$\widehat{G}^{[i]}(s) = \widehat{G}^{[i-1]}(s) - \frac{s\widehat{G}^{[i-1]}(s)}{\left(-\tau_0^{[i-1]}s + 1\right)}\left(\tau_0^{[i]} - \tau_0^{[i-1]}\right)$$

$$- \frac{ms\widehat{G}^{[i-1]}(s)}{\left(\tau_p^{[i-1]}s + 1\right)}\left(\tau_p^{[i]} - \tau_p^{[i-1]}\right) - s\widehat{G}^{[i-1]}(s)\left(\theta^{[i]} - \theta^{[i-1]}\right)$$

$$= \widehat{G}^{[i-1]}(s) + H^T\left(\gamma^{[i]} - \gamma^{[i-1]}\right) \quad (4.228)$$

4.4 A Generalized Relay Identification Method

where $\gamma = [\tau_0, \tau_p, \theta]^T$ and $H = [-s\widehat{G}^{[i-1]}(s)/(-\tau_0^{[i-1]}s+1), -ms\widehat{G}^{[i-1]}(s)/(\tau_p^{[i-1]}s+1), -s\widehat{G}^{[i-1]}(s)]^T$.

Let

$$Z(\alpha + j\omega_k) = G(\alpha + j\omega_k) + H^T(\alpha + j\omega_k)\gamma^{[i-1]} - \widehat{G}^{[i-1]}(\alpha + j\omega_k) \quad (4.229)$$

Substituting (4.228) and (4.229) into (4.227), one can obtain a recursive objective function,

$$J_{opt}^{[i]} = \sum_{k=1}^{M} \rho_k \left| H^T(\alpha + j\omega_k)\gamma^{[i]} - Z(\alpha + j\omega_k) \right|^2 \quad (4.230)$$

Denote that $\Phi = [H(\alpha + j\omega_1), \ldots, H(\alpha + j\omega_M)]^T$, $\Psi = [Z(\alpha + j\omega_1), \ldots, Z(\alpha + j\omega_M)]^T$, $W = \text{diag}\{\rho_1, \ldots, \rho_M, \rho_1, \ldots, \rho_M\}$, and

$$\bar{\Phi} = \begin{bmatrix} \text{Re}[\Phi] \\ \text{Im}[\Phi] \end{bmatrix}, \quad \bar{\Psi} = \begin{bmatrix} \text{Re}[\Psi] \\ \text{Im}[\Psi] \end{bmatrix}.$$

The recursive LS solution of (4.230) can be derived as

$$\gamma^{[i]} = \left(\bar{\Phi}^T W \bar{\Phi}\right)^{-1} \bar{\Phi}^T W \bar{\Psi} \quad (4.231)$$

It can be easily verified that all the columns of $\bar{\Phi}$ are linearly independent with each other, so $\left(\bar{\Phi}^T W \bar{\Phi}\right)^{-1}$ is guaranteed nonsingular for each iteration. Accordingly, the optimal solution can be derived to a given threshold of ε, unless it is prescribed too small to be satisfied.

4.4.4 Illustrative Examples

Four examples from existing literature are used here to illustrate the effectiveness and merits of the above identification algorithms. Examples 4.9 and 4.10 are given to demonstrate good accuracy for identifying a time delay and a NMP processes in terms of the exact model structure, together with measurement noise tests to demonstrate identification robustness. Examples 4.11 and 4.12 are used to illustrate the effectiveness of the presented algorithms for identifying high-order processes, in particular in the case of model mismatch. In all the relay tests, the sampling period is taken as $T_s = 0.01$ (s) for computation.

Example 4.9. Consider the first-order process with time delay widely studied in existing literature,

$$G = \frac{e^{-2s}}{10s + 1}$$

Based on an unbiased relay test, Vivek and Chidambaram (2005) derived an FOPDT model, $G_m = 0.9467e^{-2s}/(9.5028s + 1)$, and Majhi (2007a) gave an almost exact model by using a prior knowledge of the process static gain such as obtained from the ratio of the output response to the relay output using mutiple relay tests. Based on a biased relay test, Shen et al. (1996) derived a model, $G_m = 0.999e^{-2.005s}/(8.118s + 1)$, and Srinivasan and Chidambaram (2003) gave a model, $G_m = 1.03e^{-2.3s}/(10.3s + 1)$.

For illustration, an unbiased relay test using $u_+ = -u_- = 1.0$ and $\varepsilon_+ = -\varepsilon_- = 0.2$ is performed here. Taking $\alpha = 0.01$ for the frequency response estimation, the process static gain can be separately derived using the formula in (4.214) with a choice of $M_0 = 4$ as $k_p = 1.0062$. A very similar result can be obtained based on a biased relay test using $u_+ = 1.2$, $u_- = -0.8$, and $\varepsilon_+ = -\varepsilon_- = 0.2$. Accordingly, the identification algorithm for Case (i) results in the process model listed in Table 4.7, together with the measured limit cycle data for computation and the maximal frequency response error (ERR) for $\omega \in [0, \omega_u]$. It is seen that good identification accuracy is obtained.

Example 4.10. Consider the inverse response process studied in the literature (Majhi 2007a),

$$G = \frac{5(-3s + 1)}{(2s + 1)^3}$$

An unbiased relay test with $u_+ = -u_- = 1.0$ and $\varepsilon_+ = -\varepsilon_- = 0.5$ is performed similar to that in Majhi 2007a. Taking $\alpha = 0.01$ for the frequency response estimation, the process static gain can be separately derived using the formula in (4.214) with a choice of $M_0 = 2$ as $k_p = 5.0179$. Note that if k_p is scaled to the unity, it can be identified precisely as 1.0017. Accordingly, the identification algorithm for Case (ii) gives the other model parameters as shown in Table 4.7, which once again demonstrate good identification accuracy.

Now assume that a random noise $N(0, \sigma_N^2 = 0.03)$ is added to the process output measurement, which is then used for the relay feedback control, causing NSR = 5%. Using the statistical averaging method for 10 steady oscillation periods, e.g., in the time interval of [60, 230] (s), the estimation errors for $G(j\omega_u)$ and $G(\alpha + j\omega_u)$, the identified model, and the maximal frequency response error for $\omega \in [0, \omega_u]$ are listed in Table 4.8, which demonstrate good identification robustness. For comparison, a first-order Butterworth filter with a cutoff angular frequency, $\omega_c = 5.0$ (rad/s), according to the guideline given in (4.3), is used to recover the corrupted limit cycle and also for the relay feedback control. The corresponding results are listed in Table 4.8, which demonstrate that improved accuracy for frequency response estimation and model identification can, thus, be obtained.

Then, assume that a random noise $N(0, \sigma_N^2 = 1.104)$ as used in Majhi 2007a is added to the output measurement, which is then used for the relay feedback control, causing NSR = 30% (or SNR = 10dB). The relay test fails due to the

4.4 A Generalized Relay Identification Method

Table 4.7 Identification results for four examples under a relay test

Process	Measured Limit Cycle Data					Identified Model	ERR
	P_+	P_-	A_u	φ_u	A_α	φ_α		
$\dfrac{e^{-2s}}{10s+1}$	7.2	7.2	0.2234	-2.2204	0.2178	-2.1987	$\dfrac{1.0062e^{-2.0017s}}{10.0652s+1}$	0.62%
$\dfrac{5(-3s+1)}{(2s+1)^3}$	7.81	7.81	3.7066	-2.9133	3.5311	-2.8992	$\dfrac{5.0179(-2.9941s+1)}{(2.0041s+1)^3}$	0.36%
$\dfrac{1}{(s+1)^5}$	5.09	5.09	0.4462	-2.768	0.4304	-2.7458	$\dfrac{1.0076e^{-1.7557s}}{(1.8174s+1)^2}$	7.62%
$\dfrac{(-s+1)e^{-s}}{(s+1)^5}$	6.3	8.08	0.7051	-2.9108	0.6251	-2.8821	$\dfrac{0.9992e^{-4.8596s}}{2.298s+1}$	2.73%
							$\dfrac{0.9992e^{-4.037s}}{(1.4781s+1)^2}$	0.84%

Table 4.8 Identification results for example 4.10 under different measurement noise levels

Noise	NSR	Denoising	% Error in the Limit Cycle Data				Identified Model	ERR
			A_u	φ_u	A_α	φ_α		
$N(0, \sigma_N^2 = 0.03)$	5%	Averaging	−1.48	−1.66	1.34	−1.47	$\dfrac{4.7528(-3.1588s+1)}{(1.9691s+1)^3}$	4.94%
$N(0, \sigma_N^2 = 0.03)$	5%	Filtering	1.26	−0.74	1.25	−0.77	$\dfrac{4.8255(-3.2048s+1)}{(2.0078s+1)^3}$	3.88%
$N(0, \sigma_N^2 = 1.104)$	30%	Filtering	1.73	0.93	1.47	1.04	$\dfrac{5.7984(-2.2062s+1)}{(2.1772s+1)^3}$	15.96%
$N(0, \sigma_N^2 = 1.104)$	30%	Offline filtering	−0.12	0.13	0.55	0.01	$\dfrac{5.1522(-2.8428s+1)}{(2.0201s+1)^3}$	3.04%

4.4 A Generalized Relay Identification Method

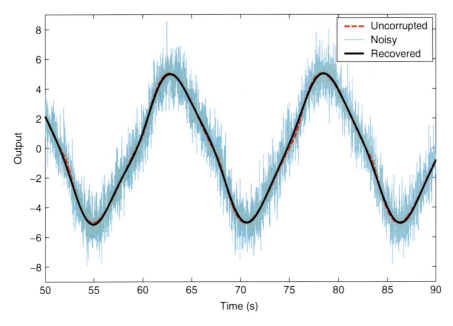

Fig. 4.15 Recovered limit cycle under a noise level of SNR = 10 dB

relay chattering. The above filter is, therefore, used to restore the relay test. The corresponding identification results are listed in Table 4.8, which indicate that the proposed denoising method works well under the high noise level. Note that if the noise affects only the process output measurement as assumed in Majhi 2007a, then offline denoising can be performed. Using a third-order Butterworth filter with a cutoff frequency of $f_c = \omega_c/2\pi = 0.25$ (Hz), the recovery effect is shown in Fig. 4.15 by performing the filtering in both the forward and reverse directions. It can be seen that the corrupted limit cycle has been well recovered. The corresponding identification results are listed also in Table 4.8, which demonstrate that evidently improved accuracy is thus obtained, compared to the offline denoising result given in Majhi 2007a based on the Fourier series fitting.

Example 4.11. Consider a high-order process studied in the literature (Majhi 2007a),

$$G = \frac{1}{(s+1)^5}$$

By performing an unbiased relay test as in Example 4.9, the frequency response estimation formula in (4.212) using $\alpha = 0.01$ together with the FFT formulae of $g(kT_s) = FFT^{-1}\{G(\alpha + j\omega)\} e^{\alpha k T_s}$ and $G(j\omega_k) = \text{FFT}\{g(kT_s)\}_{l=k}$, gives the frequency response estimation for 9 points ($\omega_k = k\omega_u/4$, $k = 0, 1, \ldots, 8$), which are shown in Fig. 4.16a. It is seen that good accuracy is obtained. Note that

$G(0)$ and $G(j2\omega_u)$ are evidently misestimated by the frequency response estimation algorithm given by Ma and Zhu (2006). The reason lies with the fact that the conventional Fourier transform of $s = j\omega$ was adopted by Ma and Zhu (2006) to develop a frequency response formula similar to (4.212), such that the first part in either the numerator or the denominator of (4.212) with $s = jk\omega_u$ becomes zero for the computation of $G(jk\omega_u)$ ($k = 0, 2, 3, \ldots$).

To obtain an SOPDT model for the PID controller tuning as done in Majhi 2007a, the identification algorithm for Case (i) is used accordingly, giving a model listed in Table 4.7. For comparison, the Nyquist plots of the proposed SOPDT model and that of Majhi 2007a (with the exactly known static gain, $k_p = 1.0$) are shown in Fig. 4.16b. It is seen that improved fitting over the low frequency range of interest to controller tuning is obtained by the proposed SOPDT model. The corresponding maximal frequency response error for $\omega \in [0, \omega_u]$ is $ERR = 7.62\%$, while the SOPDT model of Majhi 2007a yields $ERR = 11.26\%$. To further improve the fitting accuracy over the low frequency range of $\omega \in (0, \omega_u]$, the proposed recursive LS algorithm is then performed once based on the above nine frequency response points, resulting in $G_m = 1.0076e^{-1.689s}/(1.8123s + 1)^2$ that corresponds to $ERR = 5.68\%$. The corresponding Nyquist plot almost overlaps with that of the proposed SOPDT model listed in Table 4.7, except for better fitting over the low-frequency range, and thus is omitted.

Example 4.12. Consider a high-order process with inverse response studied in the literature (Wang et al. 1997; Kaya and Atherton 2001),

$$G = \frac{(-s+1)e^{-s}}{(s+1)^5}$$

It has been introduced in Example 4.2 that based on a biased relay test, Wang et al. (1997) derived an FOPDT model, $G_m = 1.0e^{-4.24s}/(2.99s+1)$, and by comparison, Kaya and Atherton (2001) gave an FOPDT model, $G_m = 1.0e^{-5.082s}/(2.292s+1)$, and an SOPDT model, $G_m = 1.0e^{-4.125s}/(1.4767s+1)^2$.

For illustration, the same biased relay test as in Example 4.2 is performed here. Taking $\alpha = 0.02$ for the frequency response computation, the process static gain can be separately derived as $k_p = 0.9992$ using the formula in (4.214) with a choice of $M_0 = 2$, which once again demonstrates good accuracy compared to the above use of an unbiased relay test in Examples 4.9, 4.10, and 4.11. Note that a similar result can also be obtained using an unbiased relay test as in Example 4.9. Accordingly, it can be verified that the identification algorithm for Case (iii) can give almost the exact process model. To make comparison with the referred references, low-order models of FOPDT and SOPDT are derived using the identification algorithm for Case (i), which are listed in Table 4.7. The Nyquist plots of these low-order models are shown in Fig. 4.17. It is seen that improved fitting is captured by the proposed FOPDT and SOPDT models, and the frequency response of the proposed models at $(-0.6864, -j0.1613)$, which corresponds to the oscillation frequency,

4.4 A Generalized Relay Identification Method

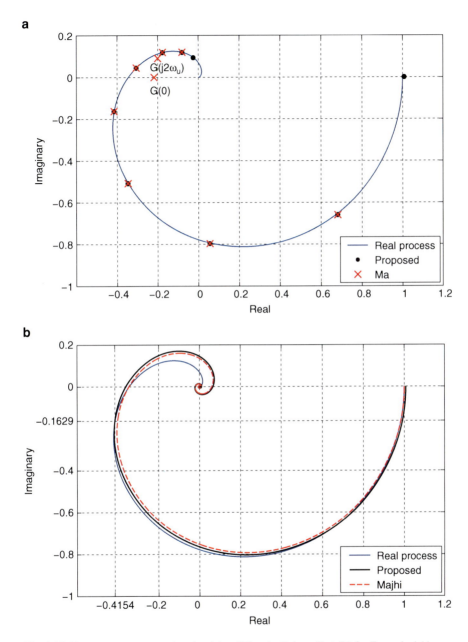

Fig. 4.16 Frequency response estimation (**a**) and Nyquist fitting effect (**b**) for Example 4.11

coincides precisely with that of the process. The maximal frequency response error for $\omega \in [0, \omega_u]$ is $ERR = 2.73\%$ for the proposed FOPDT model, compared to the FOPDT model given by Kaya and Atherton (2001) that yields $ERR = 9.82\%$.

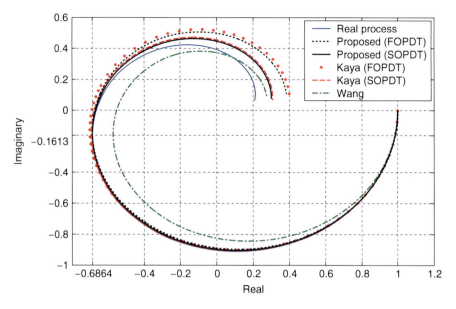

Fig. 4.17 Nyquist fitting effect for Example 4.12

The proposed SOPDT model corresponds to $ERR = 0.84\%$ compared to that of Kaya and Atherton (2001) yielding $ERR = 3.99\%$. The Nyquist plots shown in Fig. 4.17 indicate that improved fitting can be obtained using a higher order model like SOPDT in contrast to FOPDT.

4.5 Application to Barrel Temperature Maintenance in Injection Molding

Consider the barrel temperature control of an industrial injection molding machine as introduced in Sect. 3.4. To identify the temperature response models of zones 1–3 around the set-point temperature, 220°C, for injection molding, online relay tests are conducted for these three zones, respectively. The relay magnitudes are set as $u_+ = 20$ and $u_- = 0$, while the relay switch conditions are set as $\varepsilon_+ = 0.1$ and $\varepsilon_- = -0.2$, for implementation of a biased relay test. Note that $u_+ = 20$ corresponds to the use of 20% of the heating power and $u_- = 0$ corresponds to shutting off the heater, while the relay switch conditions of $\varepsilon_+ = 0.1$ and $\varepsilon_- = -0.2$ corresponds to the zone temperatures of 220.1°C and 219.8°C, respectively. A third-order Butterworth filter with a cutoff frequency, $f_c = 0.5$(HZ), is used to filter the temperature measurement for the relay feedback control, in view of that it can result in more regular oscillation periods compared to the use of a noise spike filter as introduced in Sect. 3.4.

4.5 Application to Barrel Temperature Maintenance in Injection Molding

Table 4.9 Temperature response models for zones 1–3

Zone	P_u	A_u	φ_u	Model for set-point operation
1	367.4	7.9775	−2.9144	$\dfrac{62.6212e^{-29.4s}}{(153.0371s+1)^2}$
2	374.4	9.0318	−2.8298	$\dfrac{62.7337e^{-27.8s}}{(145.2998s+1)^2}$
3	302.4	4.8005	−2.8246	$\dfrac{31.4213e^{-23.4s}}{(113.3369s+1)^2}$

(Limit cycle data)

To reduce the interaction between zones 1–3 under a relay test for each zone, the other two zones are similarly heated or cooled following the relay switch conditions for the zone to be identified. The experimental results are shown in Fig. 4.18. It is seen that the biased relay test for each zone has been effectively implemented, and meanwhile, there exist process uncertainties and unexpected load disturbance that cause slight irregularity between steady oscillation cycles of each zone.

To ensure identification effectiveness against process uncertainties and load disturbance, five similar oscillation periods are averaged for determining the limit cycle. According to the guidelines of model structure selection based on the reference relay response shapes shown in Table 4.2, a critically damped SOPDT model structure is chosen for model identification. Correspondingly, Algorithm-RS-SC1 is applied, obtaining the models of three zones as listed in Table 4.9, together with the limit cycle data measured for computation.

Based on these identified models, the control scheme shown in Fig. 3.9 is implemented to maintain the set-point temperature of 220°C for injection molding. According to the desired closed-loop transfer function shown in (3.54), the corresponding controller can be derived from the nominal relationship, $T_r = G_m C$, as

$$C = \frac{(\tau_p s + 1)^2}{k_p (\lambda_s s + 1)^2}$$

where λ_s is an adjustable parameter that corresponds to the closed-loop time constant, which can be monotonically tuned to obtain a desirable closed-loop performance specification.

The mold shape is rectangular with a length of 150 mm, width of 200 mm, and thickness of 2 mm. The cycle time for turning out a product with a weight of 27.8 g is nearly 40 s. For illustration, 20 cycles are run to test the control performance. The experimental results are shown in Fig. 4.19, by taking $\lambda_{s-1} = \lambda_{s-2} = \lambda_{s-3} = 20$ for the heating zones 1–3, respectively, together with the first-order backward discretization operator and the noise spike filtering strategy introduced in Sect. 3.4 for implementation. Note that only filtered temperature responses are shown in Fig. 4.19 for clarity. It can be seen that the temperatures of zones 1–3 are well maintained within the error band of ±0.5°C, except for the initial load

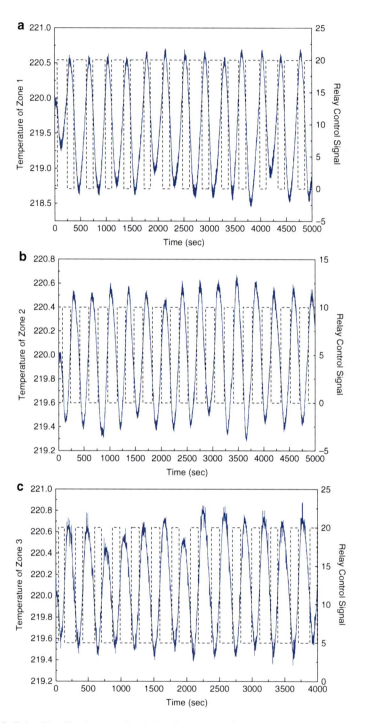

Fig. 4.18 Relay identification tests for the heating zones 1–3

4.5 Application to Barrel Temperature Maintenance in Injection Molding

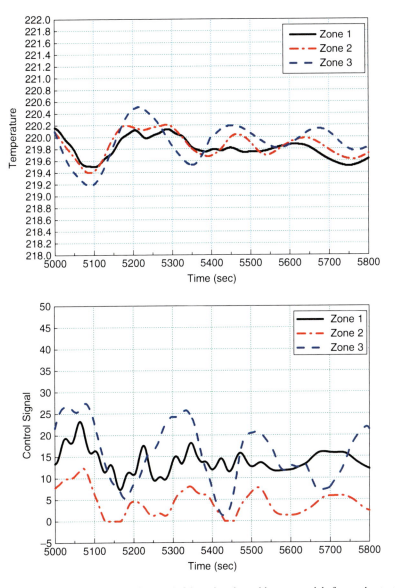

Fig. 4.19 Temperature responses of zones 1–3 by using the stable type models from relay tests

disturbance response arising from feeding the raw materials to start the injection molding (though initial preheating has been considered by taking a nonzero initial moment for the injection molding test, as seen from the time axis in Fig. 4.19).

For comparison, the experimental results based on the heating-up control scheme using the integrating type models presented in Sect. 3.4, are shown in Fig. 4.20. Also, the experimental results based on the Z-N method and the IMC-based PID

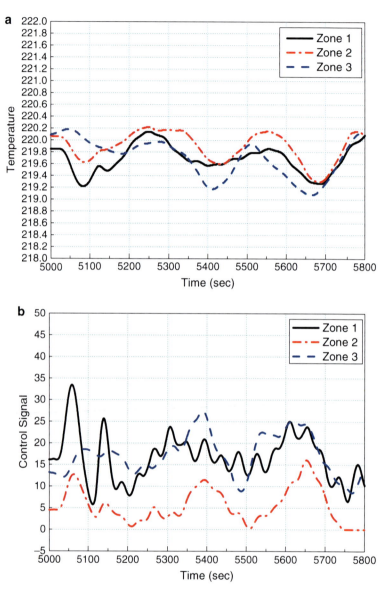

Fig. 4.20 Temperature responses of zones 1–3 by using the heating-up integrating type models

tuning method (Skogestad 2003) introduced in Sect. 3.4, are shown in Fig. 4.21. It can be seen that the obvious improvement in the temperature control performance is obtained by using the above-identified models and the corresponding control scheme. The temperature responses based on the heating-up control scheme using the integrating type models are somewhat oscillatory compared to the use of the

4.5 Application to Barrel Temperature Maintenance in Injection Molding

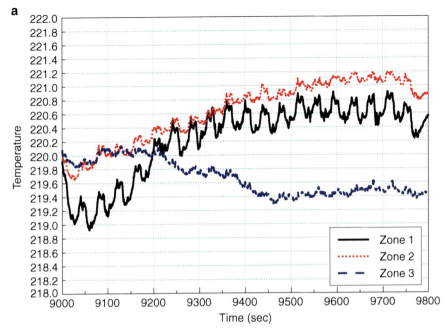

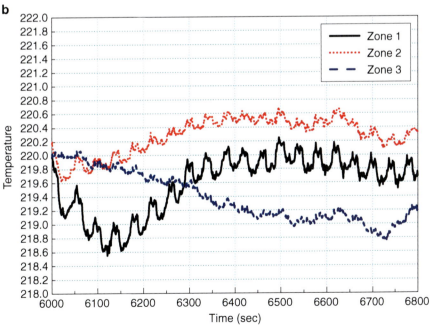

Fig. 4.21 Temperature responses of zones 1–3 by using the Z-N (**a**) and Skogestad (**b**) methods

above models identified from the relay tests, due to a lower accuracy in representing the process response characteristics around the set-point temperature of 220 °C, but are still acceptable from a practical view of system operation.

4.6 Summary

Relay feedback identification has received significantly increasing attention in the process control community, since the pioneering works that occurred in 1980s (Atherton 1982; Tsypkin 1984; Åström and Hägglund 1984; Luyben 1987). An important merit of using a relay test is that the process will not be propelled too far away from the set-point of system operation. This is especially necessary for a piecewise linearized model identification of a highly nonlinear process to facilitate control design. Generally, there are two types of relay feedback test – unbiased (symmetrical) and biased (asymmetrical). The guidelines for implementing such a relay test have been given in Sect. 4.1, along with online denoising strategies to prevent relay chattering. Moreover, offline denoising strategies to consolidate identification robustness against measurement noise have also been presented.

The low-order models of FOPDT and SOPDT have been widely used for model-based control design and online tuning in engineering practices, which can describe typical three types of dynamic response characteristics – overdamped, critically damped, and underdamped – for a wide variety of industrial processes. Reference relay response shapes of these low-order models have, therefore, been presented in Sect. 4.2 for comparison, together with the guidelines for model structure selection.

For assessing the fluctuation range of the process response under a relay test, relay response expressions have been analytically developed in Sect. 4.3 for FOPDT and SOPDT models under a biased or unbiased relay test. Correspondingly, Four alternative identification algorithms (Liu and Gao 2008) – Algorithm-RS-FA1, Algorithm-RS-FA2, Algorithm-RS-FB1, and Algorithm-RS-FB2 – have been presented for obtaining an FOPDT model to meet with different requirements of identification accuracy and computation effort in practical applications. For identifying an SOPDT model, six alternative algorithms (Liu et al. 2008) – Algorithm-RS-SC1 and Algorithm-RS-SC2 for the critically damped type, Algorithm-RS-SO1 and Algorithm-RS-SO2 for the overdamped type, Algorithm-RS-SU1 and Algorithm-RS-SU2 for the underdamped type – have been presented for application under a biased or unbiased relay test. Illustrative examples have demonstrated that all of these identification algorithms can result in good accuracy if the model structure adopted matches the process, together with good identification robustness against measurement noise. Moreover, the improvement on control performance by using the identified model for a higher order process has been well illustrated.

A generalized relay identification method (Liu and Gao 2009) has been presented in Sect. 4.4, for identifying time delay or NMP processes of any order. An important merit is that the process static gain can be separately derived from a unified formula developed here under an unbiased or biased relay test. The relay response expression has been derived based on a general model structure, which can summarize the relay expressions for FOPDT and SOPDT models as given in Sect. 4.3. A frequency response estimation algorithm with good accuracy has been proposed based on the Fourier/Laplace transform analysis of the relay response. By establishing fitting conditions for the process response at the oscillation frequency, three identification algorithms have been transparently developed based on a classification of the model parameters to be identified. To enhance the fitting accuracy over a user specified frequency range for identifying a high-order process, in particular in the low frequency range, a recursive LS algorithm has been given based on estimating multiple frequency response points in the specified frequency range. This algorithm can also be used to identify a very special model with more describing parameters for an industrial process with a known model structure but different from the low-order models as aforementioned. The applications to four examples from existing literature have demonstrated that these algorithms can give good identification accuracy and robustness.

An experimental application to maintain the barrel temperature in injection molding (Liu et al. 2009) has been presented in Sect. 4.5, in comparison with using the heating-up integrating type models introduced in Sect. 3.4, and the Z-N and IMC-based PID tuning methods also introduced in Sect. 3.4. The advantage for using a relay test for online identification around the set-point operation has, therefore, been well demonstrated.

References

Åström KJ, Hägglund T (1984) Automatic tuning of simple regulators with specification on phase angle and amplitude margins. Automatica 20:645–651

Åström KJ, Hägglund T (2005) Advanced PID control. ISA Society of America, Research Triangle Park

Atherton DP (1982) Oscillations in relay systems. Trans Inst Meas Control (London) 3:171–184

Atherton DP (2006) Relay autotuning: an overview and alternative approach. Ind Eng Chem Res 45:4075–4080

Hang CC, Åström KJ, Wang QG (2002) Relay feedback auto-tuning of process controllers—a tutorial review. J Process Control 12:143–163

Huang HP, Jeng JC, Luo KY (2005) Auto-tune system using single-run relay feedback test and model-based controller design. J Process Control 15:713–727

Kaya I, Atherton DP (2001) Parameter estimation from relay autotuning with asymmetric limit cycle data. J Process Control 11:429–439

Li W, Eskinat E, Luyben WL (1991) An improved autotune identification method. Ind Eng Chem Res 30:1530–1541

Liu T, Gao F (2008) Alternative identification algorithms for obtaining a first-order stable/unstable process model from a single relay feedback test. Ind Eng Chem Res 47(4):1140–1149

Liu T, Gao F (2009) A generalized approach for relay identification of time delay and nonminimum phase processes. Automatica 45(4):1072–1079

Liu T, Gao F, Wang YQ (2008) A systematic approach for on-line identification of second-order process model from relay feedback test. AICHE J 54(6):1560–1578

Liu T, Yao K, Gao F (2009) Identification and autotuning of temperature control system with application to injection molding. IEEE Transactions on Control System Technology 17(6):1282–1294

Luyben WL (1987) Derivation of transfer functions for highly nonlinear distillation columns. Ind Eng Chem Res 26:2490–2495

Luyben WL (2001) Getting more information from relay feedback tests. Ind Eng Chem Res 40:4391–4402

Ma M, Zhu X (2006) A simple auto-tuner in frequency domain. Comput Chem Eng 30:581–586

Majhi S (2007a) Relay-based identification of a class of non-minimum phase SISO processes. IEEE Trans Autom Control 52:134–139

Majhi S (2007b) Relay based identification of processes with time delay. J Process Control 17(2):93–101

Morari M, Zafiriou E (1989) Robust process control. Prentice Hall, Englewood Cliff

Ogunnaike BA, Ray WH (1994) Process dynamics, modeling, and control. Oxford University Press, New York

Panda RC (2006) Estimation of parameters of underdamped second order plus dead time processes using relay feedback. Comput Chem Eng 30:832–837

Panda RC, Yu CC (2003) Analytical expressions for relay feed back responses. J Process Control 13:489–501

Panda RC, Yu CC (2005) Shape factor of relay response curves and its use in autotuning. J Process Control 15:893–906

Ramakrishnan V, Chidambaram M (2003) Estimation of a SOPTD transfer function model using a single asymmetrical relay feedback test. Comput Chem Eng 27:1779–1784

Seborg DE, Edgar TF, Mellichamp DA (2004) Process dynamics and control, 2nd edn. Wiley, Hoboken

Shen SH, Wu JS, Yu CC (1996) Use of biased-relay feedback for system identification. AICHE J 42:1174–1180

Shinskey FG (1996) Process control system, 4th edn. McGraw Hill, New York

Skogestad S (2003) Simple analytical rules for model reduction and PID controller tuning. J Process Control 13(4);291–309

Srinivasan K, Chidambaram M (2003) Modified relay feedback method for improved system identification. Comput Chem Eng 27:727–732

Sung SW, Lee I-B (1997) Enhanced relay feedback method. Ind Eng Chem Res 36:5526–5530

Thyagarajan T, Yu CC (2003) Improved autotuning using the shape factor from relay feedback. Ind Eng Chem Res 42:4425–4440

Tsypkin YZ (1984) Relay control system. Cambridge University Press, Oxford

Vivek S, Chidambaram M (2005) Identification using single symmetrical relay feedback test. Comput Chem Eng 29:1625–1630

Wang QG, Hang CC, Zou B (1997) Low-order modelling from relay feedback. Ind Eng Chem Res 36:375–381

Wang QG, Lee TH, Lin C (2003) Relay feedback: analysis, identification and control. Springer, London

Yu CC (2006) Autotuning of PID controllers: a relay feedback approach, 2nd edn. Springer, London

Chapter 5
Relay Feedback Identification of Integrating Processes

For identifying an integrating process under a relay test, a few references presented identification algorithms for obtaining the low-order FOPDT and SOPDT models that are widely used for control system design and online tuning. Early literature (Ho et al. 1996) presented simple relay identification algorithms to facilitate online tuning of integrating processes, based on the standard relay feedback structure as shown in Fig. 4.2. Tsang et al. (2000) suggested the use of a derivative filter in the relay feedback channel to enhance identification accuracy of the process dynamic response characteristics. Using a state-space description of the relay control system, Majhi and Atherton (2000) gave an identification algorithm for obtaining the above low-order models based on the describing function theory. By comparison with Luyben (2003) who used the step-response data for identifying integrating processes with inverse response, Gu et al. (2006) presented an improved relay identification method based on the time domain relay response analysis. Using the so-called A-locus analysis, Kaya (2006) proposed a robust relay identification method against static load disturbance. Few references, however, reported the limiting conditions in forming steady oscillation from relay feedback for integrating processes. To guarantee identification stability, a few relay identification methods using a P- or PI-type controller for closed-loop stabilization have been developed (Kwak et al. 1997; Sung et al. 1998; Sung and Lee 2006).

To address relay feedback stability for integrating processes, the existence of the limit cycle for an integrating process under a relay test is first clarified here, by analytically deriving the relay response expressions of FOPDT and SOPDT integrating type models. Subsequently, identification algorithms for obtaining these low-order models are presented.

5.1 Existence of the Limit Cycle

Generally, the widely used FOPDT and SOPDT models are expressed as

$$G_{I-1} = \frac{k_p e^{-\theta s}}{s} \quad (5.1)$$

$$G_{I-2} = \frac{k_p e^{-\theta s}}{s(\tau_p s + 1)} \quad (5.2)$$

where k_p is the proportional gain, θ the process time delay, and τ_p a time constant reflecting the process inertial characteristics.

For clarity, a biased relay feedback test is used here to derive the relay response expressions for integrating processes. The following proposition is given for a first-order integrating process:

Proposition 5.1. *Under a biased relay feedback test as shown in Fig. 5.1, the output response of a first-order integrating process shown in (5.1) converges to the limit cycle characterized by*

$$y_+(t) = k_p (\Delta\mu + \mu_0) t + k_p (\Delta\mu - \mu_0) t_0, \quad t \in [0, P_+] \quad (5.3)$$

$$y_-(t) = k_p (\Delta\mu - \mu_0) (t + t_0) + k_p (\Delta\mu + \mu_0) P_+, \quad t \in [0, P_-] \quad (5.4)$$

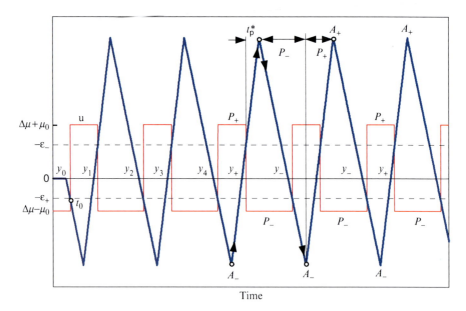

Fig. 5.1 Relay response depiction for an FOPDT integrating process

5.1 Existence of the Limit Cycle

where $y_+(t)$ is the monotonically ascending part for $t \in [0, P_+]$, $y_-(t)$ is the monotonically descending part for $t \in [0, P_-]$ that corresponds to $t \in [P_+, P_u]$ in the limit cycle, and $P_u = P_+ + P_-$ is the oscillation period.

Proof. The initial step response of an FOPDT integrating process arising from the relay output $\Delta \mu - \mu_0$ can be obtained as

$$y_0(t) = k_p (\Delta \mu - \mu_0)(t - \theta) \tag{5.5}$$

When it comes to the first relay switch point, denoted as t_0 in Fig. 5.1, the relay output changes to be $\Delta \mu + \mu_0$, which indicates that a step change of $2\mu_0$ is added to the process input. According to the linear superposition principle, the process output response can be derived as

$$y_1(t) = y_0(t + t_0) + 2k_p \mu_0 (t - \theta) \tag{5.6}$$

By using a time shift of $t_0 + \theta$, (5.6) can be rewritten as

$$y_1(t)|_{\text{shift}} = y_0(t + t_0 + \theta) + 2k_p \mu_0 t \tag{5.7}$$

When it comes to the second relay switch point, the relay output changes to be $\Delta \mu - \mu_0$, indicating that a step change of $-2\mu_0$ is added to the process input. According to the linear superposition principle, the process output response can be derived as

$$y_2(t) = y_1(t + P_+) - 2k_p \mu_0 (t - \theta) \tag{5.8}$$

Using a time shift of $t_0 + \theta + P_+$, (5.8) can be rewritten as

$$y_2(t)|_{\text{shift}} = y_0(t + t_0 + \theta + P_+) + 2k_p \mu_0 t (1 - 1) + 2k_p \mu_0 P_+ \tag{5.9}$$

At the third relay switch point, the relay output changes back to $\Delta \mu + \mu_0$, indicating that a step change of $2\mu_0$ is once again added to the process input. The process output response can be correspondingly derived as

$$y_3(t) = y_2(t + P_-) + 2k_p \mu_0 (t - \theta) \tag{5.10}$$

Using a time shift of $t_0 + \theta + P_u$, (5.10) can be rewritten as

$$y_3(t)|_{\text{shift}} = y_0(t + t_0 + \theta + P_u) + 2k_p \mu_0 t (1 - 1 + 1) + 2k_p \mu_0 P_+ \tag{5.11}$$

The process output response following the fourth relay switch point is the result of four interlaced step changes, respectively, with a magnitude of $2\mu_0$. Using a time shift of $t_0 + \theta + P_u + P_+$, the process output response can be written as

$$y_4(t)|_{\text{shift}} = y_0(t + t_0 + \theta + P_u + P_+) + 2k_p \mu_0 t (1 - 1 + 1 - 1) + 2k_p \mu_0 \cdot 2P_+ \tag{5.12}$$

Hence, the time shifted output response after each relay switch point can be generally expressed as

$$y_{2n+1}(t)|_{\text{shift}} = y_0(t + t_0 + \theta + nP_u) + 2k_p\mu_0 t + 2k_p\mu_0 \cdot nP_+ \quad (5.13)$$

$$y_{2n+2}(t)|_{\text{shift}} = y_0(t + t_0 + \theta + nP_u + P_+) + 2k_p\mu_0 \cdot (n+1)P_+ \quad (5.14)$$

where $n = 0, 1, 2, \ldots$.

Substituting (5.5) into (5.13) and (5.14), respectively, one can obtain

$$\begin{aligned} y_{2n+1}(t)|_{\text{shift}} &= k_p(\Delta\mu + \mu_0)t + k_p(\Delta\mu - \mu_0)t_0 \\ &\quad + k_p n\left[(\Delta\mu - \mu_0)P_u + 2\mu_0 P_+\right] \end{aligned} \quad (5.15)$$

$$\begin{aligned} y_{2n+2}(t)|_{\text{shift}} &= k_p(\Delta\mu - \mu_0)(t + t_0) + k_p(\Delta\mu + \mu_0)P_+ \\ &\quad + k_p n\left[(\Delta\mu - \mu_0)P_u + 2\mu_0 P_+\right] \end{aligned} \quad (5.16)$$

Note that $n \to \infty$ as $t \to \infty$. The condition for the process output response to move into steady oscillation is, therefore, required as

$$(\Delta\mu - \mu_0)P_u + 2\mu_0 P_+ = 0 \quad (5.17)$$

Hence, in the limit cycle the output response is expressed by (5.3) and (5.4). It can be easily seen from (5.3) and (5.4) that $y_+(t)$ increases monotonically for $t \in [0, P_+]$ while $y_-(t)$ decreases monotonically for $t \in (P_+, P_u]$.

Note that there exists an inherent relationship between $y_+(t)$ and $y_-(t)$,

$$y_+(0) = y_-(P_-) \quad (5.18)$$

which can be reformulated using (5.3) and (5.4) as

$$(\Delta\mu - \mu_0)P_- + (\Delta\mu + \mu_0)P_+ = 0 \quad (5.19)$$

It can be easily verified that (5.19) coincides with (5.17). Therefore, it can be concluded that under a relay test the limit cycle can surely be formed for an FOPDT integrating process. In particular, there exist $\Delta\mu = 0$ and $P_u = 2P_+$ when an unbiased relay test is used, so that (5.17) is obviously satisfied for $\forall \mu_0$. This completes the proof. □

If an unbiased relay test is used, there exist $\Delta\mu = 0$ and $P_+ = P_- = P_u/2$. Substituting them into (5.3) and (5.4), the corresponding relay response expression can be obtained as

$$y_+(t) = -y_-(t) = k_p\mu_0(t - t_0) \quad (5.20)$$

5.1 Existence of the Limit Cycle

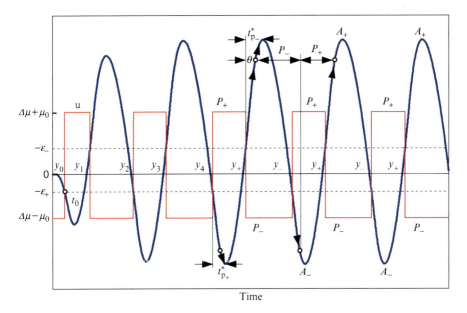

Fig. 5.2 Relay response depiction for an SOPDT integrating process

where $y_+(t)$ is the ascending part for $t \in [0, P_u/2]$, and $y_-(t)$ is the descending part for $t \in (P_u/2, P_u]$ in the limit cycle.

For a second-order integrating process, the following proposition gives the relay response expression for steady oscillation:

Proposition 5.2. *Under a biased relay test as shown in Fig. 5.2, the output response of a second-order integrating process shown in (5.2) converges to the limit cycle characterized by*

$$y_+(t) = k_p \left(\Delta\mu + \mu_0 \right) \left(t - \tau_p \right) + k_p \left(\Delta\mu - \mu_0 \right) t_0 + 2 k_p \mu_0 \tau_p E e^{-\frac{t}{\tau_p}}, \quad t \in [0, P_+] \tag{5.21}$$

$$y_-(t) = k_p \left(\Delta\mu - \mu_0 \right) \left(t + t_0 - \tau_p \right) + k_p \left(\Delta\mu + \mu_0 \right) P_+$$
$$+ 2 k_p \mu_0 \tau_p F e^{-\frac{t}{\tau_p}}, \quad t \in [0, P_-] \tag{5.22}$$

where

$$E = \frac{1 - e^{-\frac{P_-}{\tau_p}}}{1 - e^{-\frac{P_u}{\tau_p}}} \tag{5.23}$$

$$F = -\frac{1 - e^{-\frac{P_+}{\tau_p}}}{1 - e^{-\frac{P_u}{\tau_p}}} \tag{5.24}$$

Proof. The initial step response of an SOPDT integrating process arising from the relay output $\Delta\mu - \mu_0$ can be derived as

$$y_0(t) = k_p (\Delta\mu - \mu_0) \left[t - \theta + \tau_p \left(e^{-\frac{t-\theta}{\tau_p}} - 1 \right) \right] \tag{5.25}$$

Following a similar analysis as in the above proof for Proposition 5.1, the time shifted output response from the initial to the fourth relay switch point can be derived, respectively, as

$$y_1(t)|_{\text{shift}} = y_0 (t + t_0 + \theta) + 2k_p\mu_0 (t - \tau_p) + 2k_p\mu_0\tau_p e^{-\frac{t}{\tau_p}} \tag{5.26}$$

$$y_2(t)|_{\text{shift}} = y_0 (t + t_0 + \theta + P_+) + 2k_p\mu_0 (t - \tau_p)(1-1) + 2k_p\mu_0 P_+$$
$$+ 2k_p\mu_0\tau_p e^{-\frac{t}{\tau_p}} \left(e^{-\frac{P_+}{\tau_p}} - 1 \right) \tag{5.27}$$

$$y_3(t)|_{\text{shift}} = y_0 (t + t_0 + \theta + P_u) + 2k_p\mu_0 (t - \tau_p)(1-1+1) + 2k_p\mu_0 P_+$$
$$+ 2k_p\mu_0\tau_p e^{-\frac{t}{\tau_p}} \left(e^{-\frac{P_u}{\tau_p}} - e^{-\frac{P_-}{\tau_p}} + 1 \right) \tag{5.28}$$

$$y_4(t)|_{\text{shift}} = y_0 (t + t_0 + \theta + P_u + P_+) + 2k_p\mu_0 (t - \tau_p)(1-1+1-1)$$
$$+ 2k_p\mu_0 \cdot 2P_+ + 2k_p\mu_0\tau_p e^{-\frac{t}{\tau_p}} \left(e^{-\frac{P_u+P_+}{\tau_p}} - e^{-\frac{P_u}{\tau_p}} + e^{-\frac{P_+}{\tau_p}} - 1 \right) \tag{5.29}$$

The general relay response can, therefore, be summarized as

$$y_{2n+1}(t)|_{\text{shift}} = y_0 (t + t_0 + \theta + n P_u) + 2k_p\mu_0 (t - \tau_p)$$
$$+ 2k_p\mu_0 \cdot n P_+ + 2k_p\mu_0\tau_p E e^{-\frac{t}{\tau_p}} \tag{5.30}$$

$$y_{2n+2}(t)|_{\text{shift}} = y_0 (t + t_0 + \theta + n P_u + P_+) + 2k_p\mu_0 \cdot (n+1) P_+$$
$$+ 2k_p\mu_0\tau_p F e^{-\frac{t}{\tau_p}} \tag{5.31}$$

where $n = 0, 1, 2, \ldots$, and

$$E = 1 + \sum_{k=1}^{n} \left(e^{-\frac{kP_u}{\tau_p}} - e^{-\frac{(k-1)P_u + P_-}{\tau_p}} \right) \tag{5.32}$$

5.1 Existence of the Limit Cycle

$$F = \sum_{k=0}^{n} \left(e^{-\frac{kP_u+P_+}{\tau_p}} - e^{-\frac{kP_u}{\tau_p}} \right) \quad (5.33)$$

Note that $0 < e^{-P_u/\tau_p} < 1$. It follows for $n \to \infty$ that

$$\sum_{k=0}^{n} e^{-\frac{kP_u}{\tau_p}} = \frac{1}{1-e^{-\frac{P_u}{\tau_p}}} \quad (5.34)$$

Substituting (5.34) into (5.32) and (5.33), respectively, one can obtain

$$E = \frac{1}{1-e^{-\frac{P_u}{\tau_p}}} - \frac{e^{-\frac{P_-}{\tau_p}}}{1-e^{-\frac{P_u}{\tau_p}}} = \frac{1-e^{-\frac{P_-}{\tau_p}}}{1-e^{-\frac{P_u}{\tau_p}}} \quad (5.35)$$

$$F = \frac{e^{-\frac{P_+}{\tau_p}}-1}{1-e^{-\frac{P_u}{\tau_p}}} = -\frac{1-e^{-\frac{P_+}{\tau_p}}}{1-e^{-\frac{P_u}{\tau_p}}} \quad (5.36)$$

Note that

$$\lim_{n \to \infty} e^{-\frac{t+t_0+nP_u}{\tau_p}} = \lim_{n \to \infty} e^{-\frac{t+t_0+nP_u+P_+}{\tau_p}} = 0 \quad (5.37)$$

Substituting (5.25) and (5.37) into (5.30) and (5.31), respectively, one can obtain

$$y_{2n+1}(t)|_{\text{shift}} = k_p (\Delta\mu + \mu_0)(t - \tau_p) + k_p (\Delta\mu - \mu_0) t_0$$
$$+ k_p n \left[(\Delta\mu - \mu_0) P_u + 2\mu_0 P_+ \right] + 2k_p \mu_0 \tau_p E e^{-\frac{t}{\tau_p}} \quad (5.38)$$

$$y_{2n+2}(t)|_{\text{shift}} = k_p (\Delta\mu - \mu_0)(t + t_0 - \tau_p) + k_p (\Delta\mu + \mu_0) P_+$$
$$+ k_p n \left[(\Delta\mu - \mu_0) P_u + 2\mu_0 P_+ \right] + 2k_p \mu_0 \tau_p F e^{-\frac{t}{\tau_p}} \quad (5.39)$$

In view of that $n \to \infty$ as $t \to \infty$, the condition for the output response to move into steady oscillation is, therefore, required as

$$(\Delta\mu - \mu_0) P_u + 2\mu_0 P_+ = 0 \quad (5.40)$$

Note that (5.40) is the same as (5.17) for an FOPDT integrating process.

Hence, in the limit cycle the output response is expressed by (5.21) and (5.22), where $y_+(t)$ denotes the ascending output response in the half period P_+, and $y_-(t)$ is the descending output response in the other half period P_-.

According to the inherent relationship between $y_+(t)$ and $y_-(t)$, there exists

$$y_+(0) = y_-(P_-) \tag{5.41}$$

Substituting (5.21) and (5.22) into (5.41), one can obtain

$$(\Delta\mu - \mu_0) P_- + (\Delta\mu + \mu_0) P_+ = 0 \tag{5.42}$$

It can be easily verified that (5.42) coincides with (5.40), indicating that the limit cycle can surely be formed for an SOPDT integrating process under a biased or unbiased relay test. This completes the proof. □

Following a similar analysis as in the proofs for Propositions 5.1 and 5.2, the conclusion on the existence of the limit cycle can be transparently generalized to a high order integrating process.

5.2 The FOPDT Model

It follows from Proposition 5.1 that the process time delay can intuitively be measured as the time to reach the peak of the process output response from the initial relay switch point in a half period of the relay, which is denoted as t_p^* in Fig. 5.1.

Correspondingly, it follows from (5.3) and (5.4) that

$$y_+(0) = k_p (\Delta\mu - \mu_0) t_0 = A_- \tag{5.43}$$

$$y_-(0) = k_p (\Delta\mu - \mu_0) t_0 + k_p (\Delta\mu + \mu_0) P_+ = A_+ \tag{5.44}$$

Substituting (5.43) into (5.44) yields

$$k_p = \frac{A_+ - A_-}{(\Delta\mu + \mu_0) P_+} \tag{5.45}$$

It should be noted that the process gain cannot be derived from a biased relay test as

$$k_p = G(0) = \frac{\int_0^{P_u} y(t) dt}{\int_0^{P_u} u(t) dt} \tag{5.46}$$

which has been introduced in Chap. 4 for identifying a stable process. The reason lies in $G(0) \to \infty$ for an integrating process, as seen from the low-order process models in (5.1) and (5.2).

5.2 The FOPDT Model

Hence, the above algorithm named Algorithm-RI-F1 for identifying an FOPDT integrating type model can be summarized as follows.

Algorithm-RI-F1

(i) Measure P_+, P_-, A_+, and A_- from the limit cycle;
(ii) Measure the process time delay as t_p^*, which is the time to reach the positive peak (A_+) of the process output response from the initial relay switch point in a negative half period of the relay (P_-);
(iii) Compute the proportional gain, k_p, from (5.45).

Note that the above algorithm is based on the fact that the positive part of the relay response occupies a larger percentage compared to the negative part, as shown in Fig. 5.1. If the opposite case happens in practice, a similar identification algorithm can be developed in terms of the negative part of the limit cycle for model fitting.

Moreover, the process response at the oscillation frequency can be computed in terms of the frequency response estimation formula shown in (4.30), i.e.,

$$G(j\omega_u) = A_u e^{j\varphi_u} = \frac{\int_{t_{os}}^{t_{os}+P_u} y(t)e^{-j\omega_u t} dt}{\int_{t_{os}}^{t_{os}+P_u} u(t)e^{-j\omega_u t} dt} \qquad (5.47)$$

where t_{os} can be taken as any relay switch point in steady oscillation.

For practical applications subject to measurement noise, it is suggested to average the measured values of P_+, P_-, A_+, and A_- in 5 ~ 20 steady oscillation periods as $\bar{P}_+$, $\bar{P}_-$, $\bar{A}_+$, and $\bar{A}_-$, respectively, for applying the above identification algorithm. The oscillation period can, thus, be computed as $\bar{P}_u = \bar{P}_+ + \bar{P}_-$. Accordingly, the process response at the oscillation frequency can be computed as

$$G(j\bar{\omega}_u) = \bar{A}_u e^{j\bar{\varphi}_u} = \frac{\int_{t_{os}}^{t_{os}+N\bar{P}_u} y(t)e^{-j\bar{\omega}_u t} dt}{\int_{t_{os}}^{t_{os}+N\bar{P}_u} u(t)e^{-j\bar{\omega}_u t} dt} \qquad (5.48)$$

where $\bar{\omega}_u = 2\pi/\bar{P}_u$, N is the number of steady oscillation periods used for averaging, which can be practically taken in the range of 5–20. Note that the relay output, $u(t)$, remains as a constant for each half period of the limit cycle, so it can be used for the measurement of the oscillation period.

Note that the above use of the measured parameters is based on the statistical averaging principle for eliminating the random measurement errors, and therefore, can guarantee identification robustness in the presence of a low noise level (e.g., NSR < 10%). For a higher noise level, a low-pass Butterworth filter is suggested for the implementation of a relay test or offline denoising, as introduced in Sect. 4.1.

In the case where a static type load disturbance occurs in a relay identification test, which is denoted as $LD = c$, where $c \in \Re$, it follows that

$$\int_0^{P_u} ce^{-j\omega_u t} dt = c \int_0^{P_u} e^{-j\omega_u t} dt = 0 \qquad (5.49)$$

Thus, one can ensure that

$$G(j\omega_u) = \frac{\int_{t_{os}}^{t_{os}+P_u} y(t)e^{-j\omega_u t}dt}{\int_{t_{os}}^{t_{os}+P_u} u(t)e^{-j\omega_u t}dt} = \frac{\int_{t_{os}}^{t_{os}+P_u} (y(t)+c)e^{-j\omega_u t}dt}{\int_{t_{os}}^{t_{os}+P_u} u(t)e^{-j\omega_u t}dt} \quad (5.50)$$

which indicates that the computation of the process response at the oscillation frequency is not affected by a static type load disturbance.

Note that the influence of nonstatic load disturbance may be intuitively excluded by comparing the uniformity of the sequential steady oscillation periods.

Substituting the FOPDT integrating type model in (5.1) into the left-hand side of (5.47), one can obtain

$$\frac{k_p}{\omega_u} = A_u \quad (5.51)$$

$$-\theta\omega_u - \frac{\pi}{2} = \varphi_u, \quad \varphi_u \in (-\pi, -\pi/2) \quad (5.52)$$

The proportional gain and time delay can, therefore, be derived from (5.51) and (5.52) as

$$k_p = A_u \omega_u \quad (5.53)$$

$$\theta = -\frac{2\varphi_u + \pi}{2\omega_u} \quad (5.54)$$

Hence, an alternative identification algorithm named Algorithm-RI-F2 for obtaining an FOPDT integrating type model can be summarized.

Algorithm-RI-F2

(i) Measure P_u from the limit cycle;
(ii) Compute $G(j\omega_u)$ from (5.47);
(iii) Compute the proportional gain, k_p, from (5.53);
(iv) Compute the process time delay, θ, from (5.54).

Owing to that the process response at the oscillation frequency is exactly represented by the FOPDT model derived from Algorithm-RI-F2, this algorithm can be used to obtain better fitting accuracy for identifying a high-order integrating process compared to Algorithm-RI-F1, but at the cost of more computation effort.

5.3 The SOPDT Model

According to Proposition 5.2, it can be derived from (5.21) and (5.22) that

$$\frac{dy_+(t)}{dt} = k_p(\Delta\mu + \mu_0) - 2k_p\mu_0 E e^{-\frac{t}{\tau_p}} \quad (5.55)$$

5.3 The SOPDT Model

$$\frac{dy_-(t)}{dt} = k_p (\Delta\mu - \mu_0) - 2k_p\mu_0 F e^{-\frac{t}{\tau_p}} \tag{5.56}$$

By letting $\frac{dy_+(t)}{dt} = 0$ and $\frac{dy_-(t)}{dt} = 0$, the time to reach the single extreme value of $y_+(t)$ and $y_-(t)$ can be derived, respectively, as

$$t_{p_+} = \tau_p \ln \frac{2\mu_0 E}{\Delta\mu + \mu_0} \tag{5.57}$$

$$t_{p_-} = \tau_p \ln \frac{2\mu_0 F}{\Delta\mu - \mu_0} \tag{5.58}$$

It can be easily verified that $\frac{d^2 y_+(t)}{dt^2} > 0$ and $\frac{d^2 y_-(t)}{dt^2} < 0$. Therefore, t_{p_+} is the time to reach the minimum of $y_+(t)$, while t_{p_-} is the time to reach the maximum of $y_-(t)$.

Note that the time, $t_{p_-}^*$, to reach the maximum of $y_-(t)$ from the initial relay switch point in a negative half period of the relay can be measured, as shown in Fig. 5.2. It follows that

$$t_{p_-} = t_{p_-}^* - \theta \tag{5.59}$$

Likewise, there exists

$$t_{p_+} = t_{p_+}^* - \theta \tag{5.60}$$

where $t_{p_+}^*$ is the time to reach the minimum of $y_+(t)$ from the initial relay switch point in a positive half period of the relay, as shown in Fig. 5.2.

Subtracting (5.59) from (5.60) yields

$$t_{p_+} - t_{p_-} = t_{p_+}^* - t_{p_-}^* \tag{5.61}$$

Substituting (5.57) and (5.58) into (5.61), one can obtain

$$\ln \frac{\mu_0 - \Delta\mu}{\mu_0 + \Delta\mu} + \ln \frac{1 - e^{-\frac{P_-}{\tau_p}}}{1 - e^{-\frac{P_+}{\tau_p}}} = \frac{t_{p_+}^* - t_{p_-}^*}{\tau_p} \tag{5.62}$$

Define $z = x/P_-$, $a = P_-/P_+$, and

$$f(x) = \frac{1 - e^{-\frac{P_-}{x}}}{1 - e^{-\frac{P_+}{x}}}, \quad x \in (0, \infty) \tag{5.63}$$

It follows that

$$f(z) = \frac{1 - e^{-\frac{1}{z}}}{1 - e^{-\frac{1}{az}}} \tag{5.64}$$

$$\frac{df(z)}{dz} = \frac{e^{-\frac{a+1}{az}} \left(e^{\frac{1}{z}} - ae^{\frac{1}{az}} + a - 1\right)}{az^2 \left(1 - e^{-\frac{1}{az}}\right)^2} \tag{5.65}$$

Let

$$g(z) = e^{\frac{1}{z}} - ae^{\frac{1}{az}} + a - 1 \tag{5.66}$$

It can be verified that

$$\frac{dg(z)}{dz} = \frac{1}{z^2}\left(e^{\frac{1}{az}} - e^{\frac{1}{z}}\right) < 0 \tag{5.67}$$

$$\lim_{z \to \infty} g(z) = 0 \tag{5.68}$$

It can, therefore, be concluded that $g(z) > 0$ for $z \in (0, \infty)$ and

$$\frac{df(z)}{dz} > 0 \tag{5.69}$$

Hence, $f(z)$ increases monotonically with respect to z, and so is $f(x)$ with respect to x.

Consequently, it can be concluded that the left-hand side of (5.62) increases monotonically with respect to τ_p. Meanwhile, it is obvious that the right-hand side of (5.62) decreases monotonically with respect to τ_p. Note that there exists for $\tau_p \in (0, +\infty)$,

$$0 < \ln \frac{1 - e^{-\frac{p_-}{\tau_p}}}{1 - e^{-\frac{p_+}{\tau_p}}} < \ln \frac{p_-}{p_+} \tag{5.70}$$

It can, thus, be ascertained that there exist only finite solutions of τ_p for (5.62), which can be computed using any numerical algorithm such as the Newton-Raphson method.

Substituting the SOPDT model shown in (5.2) into the process response fitting condition at the oscillation frequency as shown in (5.47), one can obtain

$$\frac{k_p}{\omega_u \sqrt{\tau_p^2 \omega_u^2 + 1}} = A_u \tag{5.71}$$

5.3 The SOPDT Model

$$-\theta\omega_u - \frac{\pi}{2} - \arctan(\tau_p\omega_u) = \varphi_u, \quad \varphi_u \in (-\pi, -\pi/2) \tag{5.72}$$

It can be derived from (5.71) and (5.72) that

$$k_p = A_u\omega_u\sqrt{\tau_p^2\omega_u^2 + 1} \tag{5.73}$$

$$\theta = -\frac{1}{\omega_u}\left[\varphi_u + \frac{\pi}{2} + \arctan(\tau_p\omega_u)\right] \tag{5.74}$$

Note that in the case where $\theta/\tau_p > 1$, $y_+(t)$ may decrease monotonically for $t \in (0, P_+)$ while $y_-(t)$ increases monotonically for $t \in (P_+, P_u]$. Correspondingly, there exists

$$t^*_{p_+} = t^*_{p_-} = \theta \tag{5.75}$$

The process time constant can then be derived inversely from (5.74) as

$$\tau_p = \frac{1}{\omega_u}\tan\left(-\theta\omega_u - \frac{\pi}{2} - \varphi_u\right) \tag{5.76}$$

Subsequently, the proportional gain can be derived from (5.73).
However, the process time constant should not be derived from (5.76) if

$$\tan\left(-\theta\omega_u - \frac{\pi}{2} - \varphi_u\right) < 0 \tag{5.77}$$

Therefore, the practical constraint in (5.77) can be used to check whether the process time delay should be derived from (5.75).

Hence, the identification of an SOPDT integrating type model can be summarized in the following algorithm named Algorithm-RI-S1.

Algorithm-RI-S1

(i) Measure P_+, P_-, $t^*_{p_+}$, and $t^*_{p_-}$ from the limit cycle;
(ii) Compute $G(j\omega_u)$ from (5.47);
(iii) Compute the process time delay from (5.75) as $\theta = t^*_{p_+}$, and then check whether (5.77) is satisfied. If yes, go to step (vi);
(iv) Compute the process time constant, τ_p, from (5.76), and then check whether $\theta/\tau_p < 1$ is satisfied. If yes, go to step (vi);
(v) Compute the proportional gain, k_p, from (5.73). Then go to step (ix).
(vi) Compute the process time constant, τ_p, from (5.62) using the Newton-Raphson iteration method. The initial estimation of τ_p for iteration may be taken as $t^*_{p_+}$ (or $t^*_{p_-}$);
(vii) Compute the proportional gain, k_p, from (5.73);
(viii) Compute the process time delay, θ, from (5.74);

(ix) End the algorithm if the fitting condition of relay response, $\sum_{k=1}^{N} \left[y(kT_s + t_{os}) - \hat{y}(kT_s + t_{os}) \right]^2 / N < \varepsilon$, is satisfied, where $y(kT_s + t_{os})$ and $\hat{y}(kT_s + t_{os})$ denote, respectively, the process and model responses in the limit cycle, T_s the sampling period corresponding to $N = P_u/T_s$, and ε is a user-specified fitting threshold that may be practically set no larger than 1%. Otherwise, change the initial estimation of τ_p and then go back to step (vi).

Alternatively, with a known value of the process time delay (θ), which can be initially estimated from the step response in the relay test, the process time constant and the proportional gain can then be derived from (5.76) and (5.73). Correspondingly, a one-dimensional search of θ can be performed in combination with the above fitting condition of relay response to determine the optimal fitting.

Hence, an alternative identification algorithm named Algorithm-RI-S2 for obtaining an SOPDT integrating type model can be summarized.

Algorithm-RI-S2

(i) Measure P_u from the limit cycle;
(ii) Compute $G(j\omega_u)$ from (5.47);
(iii) Compute the process time constant, τ_p, from (5.76) based on a preestimated time delay (θ). A one-dimensional search of θ can be implemented within a possible range as observed from the initial step response in the relay test;
(iv) Compute the proportional gain, k_p, from (5.73);
(v) End the algorithm if the fitting condition of relay response, $\sum_{k=1}^{N} \left[y(kT_s + t_{os}) - \hat{y}(kT_s + t_{os}) \right]^2 / N < \varepsilon$, is satisfied. Otherwise, go back to step (iii) by monotonically varying θ in a one-dimensional search.

5.4 Illustrative Examples

Three examples from existing literature are used to illustrate the effectiveness and merits of the above identification algorithms. Examples 5.1 and 5.2 are given to demonstrate the fitting accuracy of Algorithm-RI-F1, Algorithm-RI-F2, Algorithm-RI-S1, and Algorithm-RI-S2 for identifying a low-order model in terms of the exact model structure. Example 5.3 is given to show the effectiveness of these algorithms for identifying a higher order integrating process. In all the relay tests, the sampling period is taken as $T_s = 0.01$ (s) for computation.

Example 5.1. Consider the FOPDT integrating process studied in the literature (Majhi and Atherton 2000),

$$G_1 = \frac{e^{-5s}}{s}$$

5.4 Illustrative Examples

Based on an unbiased relay test, Majhi and Atherton (2000) derived an FOPDT model with each parameter error within the limit of ±0.1%.

For illustration, an unbiased relay test with $u_+ = -u_- = 0.5$ and $\varepsilon_+ = -\varepsilon_- = 0.2$ is performed here. Correspondingly, Algorithm-RI-F1 is used for model identification, obtaining the exact process model as listed in Table 5.1. Note that Algorithm-RI-F2 can also result in the exact process model, which can be verified using the measured limit cycle data listed in Table 5.1 for computation.

Example 5.2. Consider the SOPDT integrating process studied in the references (Ho et al. 1996; Kaya 2006),

$$G_2 = \frac{e^{-10s}}{s(20s+1)}$$

Based on a biased relay test with $u_+ = 0.7$, $u_- = -0.5$, and $\varepsilon_+ = -\varepsilon_- = 0.1$, Kaya (2006) derived an almost exact SOPDT model, showing quite improved accuracy in comparison with Ho et al. (1996).

For illustration, a biased relay test as used in Kaya (2006) is performed here. Correspondingly, Algorithm-RI-S1 is used for model identification, obtaining the almost exact process model as listed in Table 5.1. Note that Algorithm-RI-S2 can result in the exact process model, based on a one-dimensional search of θ in the range of [5, 15] that is intuitively observed from the initial step response in the relay test.

Example 5.3. Consider the high-order integrating process studied in the literature (Ingimundarson and Hägglund 2001),

$$G_3 = \frac{(-s+1)e^{-5s}}{s(s+1)^5}$$

Based on the step-response data, Ingimundarson and Hägglund (2001) derived an FOPDT model, $G_m = 1.0e^{-11s}/s$.

For illustration, a biased relay test with $u_+ = 0.5$, $u_- = -0.3$, and $\varepsilon_+ = -\varepsilon_- = 0.1$ is performed here. The measured limit cycle data are listed in Table 5.1. Correspondingly, Algorithm-RI-F1 gives an FOPDT model, $G_m = 0.8675e^{-11.25s}/s$, and Algorithm-RI-F2 and Algorithm-RI-S1 result in the FOPDT and SOPDT models listed in Table 5.1. Note that Algorithm-RI-S2 can give a very similar model to that of Algorithm-RI-S1 and, thus, is omitted.

For comparison, the Nyquist plots of these models are shown in Fig. 5.3. It can be seen that the FOPDT model obtained from Algorithm-RI-F2 procures better fitting compared to the other FOPDT model, owing to that the model response coincides with the process at the oscillation frequency, i.e., (−7.3134, −j0.9873). In contrast with these FOPDT models, the SOPDT model obtained using Algorithm-RI-S1 shows apparently improved fitting effect.

Table 5.1 Identified models for three integrating processes under a relay test

Process	Limit cycle						Integrating types model	
	P_+	P_-	A_+	A_-	A_u	φ_u	FOPDT	SOPDT
$\dfrac{e^{-5s}}{s}$	10.8	10.8	2.7	−2.7	3.4377	−3.0267	$\dfrac{1.0000 e^{-5.0000s}}{s}$	
$\dfrac{e^{-10s}}{s(20s+1)}$	42.9	60.06	9.1041	−6.4432	10.3853	−3.0657		$\dfrac{0.9983 e^{-10.027s}}{s(19.9443s+1)}$
$\dfrac{(-s+1)e^{-5s}}{s(s+1)^5}$	17.99	29.99	4.9569	−2.8459	7.3798	−3.0074	$\dfrac{0.9664 e^{-10.9703s}}{s}$	$\dfrac{1.018 e^{-8.5278s}}{s(2.5293s+1)}$

5.5 Experimental Tests for the Barrel Temperature Maintenance

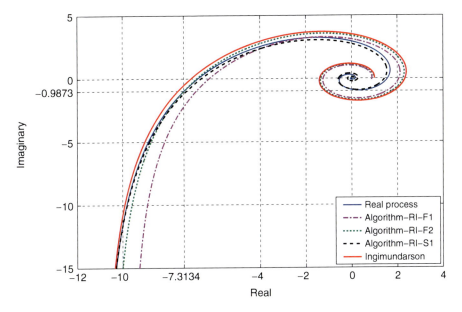

Fig. 5.3 Nyquist fitting of identified models for Example 5.3

5.5 Experimental Tests for the Barrel Temperature Maintenance

Consider the barrel temperature control of an industrial injection molding machine as introduced in Sect. 3.4. To identify the temperature response models of zones 1–3 around the set-point temperature, 220°C, for injection molding, online relay tests are conducted for these three zones, respectively, as presented in Sect. 4.5. Based on the experimental results shown in Fig. 4.18, Algorithm-RI-S2 is applied to obtain the SOPDT integrating type models for these zones, in view of that the time delay of the temperature response of each zone can be intuitively measured from the initial step response for heating or cooling in these relay tests. In contrast with the identification of stable type models around the set-point temperature as presented in Sect. 4.5, the motivation for identifying the integrating type models for these heating zones is that a relatively larger heating power around the set-point temperature may cause the temperature response of each zone to behave in an integrating manner, similar to the heating-up control introduced in Sect. 3.4.

Correspondingly, these identified models are listed in Table 5.2, based on averaging five similar oscillation periods for determining the limit cycle as done in Sect. 4.5.

To demonstrate the effectiveness of these identified models, the control scheme shown in Fig. 3.9 is implemented to maintain the set-point temperature of 220°C in injection molding. According to the desired closed-loop transfer function shown in

Table 5.2 Integrating type temperature response models for zones 1–3

Zone	Limit cycle data P_u	A_u	φ_u	Model for set-point operation
1	367.4	7.9775	−2.9144	$\dfrac{0.2046e^{-29.4s}}{s(65.3475s+1)}$
2	374.4	9.0318	−2.8298	$\dfrac{0.2159e^{-27.8s}}{s(60.4314s+1)}$
3	302.4	4.8005	−2.8246	$\dfrac{0.1386e^{-23.4s}}{s(46.4496s+1)}$

(3.54) for the heating-up control, the corresponding controller can be derived from the nominal relationship, $T_r = G_m C$, as

$$C = \frac{s(\tau_p s + 1)}{k_p (\lambda_s s + 1)^2}$$

where λ_s is an adjustable parameter that corresponds to the closed-loop time constant, which can be monotonically tuned to obtain a desirable closed-loop performance specification.

For comparison with the stable type models identified for these zones as listed in Table 4.9 in Sect. 4.5, the same injection molding tests are performed here. The experimental results for 20 cycles are shown in Fig. 5.4, by taking $\lambda_{s-1} = \lambda_{s-2} = \lambda_{s-3} = 20$ for the heating zones 1–3, respectively, together with the first-order backward discretization operator and the noise spike filtering strategy introduced in Sect. 3.4 for implementation. Note that only filtered temperature responses are shown in Fig. 5.4 for clarity.

It can be seen that the temperatures of zones 1–3 are well maintained within the error band of ±0.5°C, similar to the results obtained by using the stable type models as shown in Fig. 4.19 in Sect. 4.5. Compared to the results shown in Fig. 4.20 that were obtained from the heating-up control scheme with the integrating type models from open-loop step tests, both the integrating and the stable type models identified from the relay tests facilitate evidently better control performance, thus demonstrating the advantage of using an online relay test for model identification.

5.6 Summary

For relay feedback identification of integrating processes, the limit cycle can undoubtedly be formed for an integrating process under a biased or unbiased relay test, which has been clarified by analytically deriving the relay response expressions of the integrating type FOPDT and SOPDT models (Liu and Gao 2008).

Based on the limit cycle information and the relay response expressions derived here, four identification algorithms – Algorithm-RI-F1, Algorithm-RI-F2,

5.6 Summary

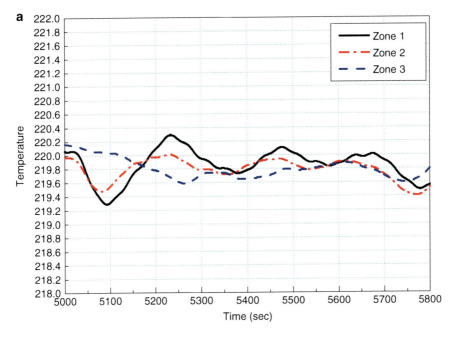

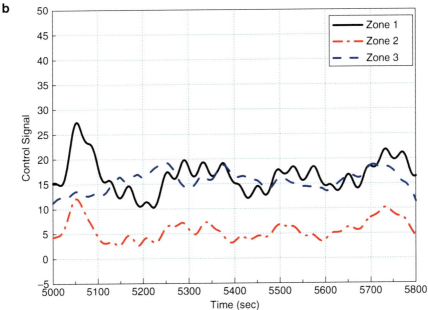

Fig. 5.4 Temperature responses of zones 1–3 using the integrating type models from relay tests

Algorithm-RI-S1, and Algorithm-RI-S2 – have been developed for obtaining the low-order integrating type FOPDT and SOPDT models that are widely used for the control of integrating processes in engineering practice. As illustrated by the three examples from existing references, all of these algorithms can give good accuracy if the model structure matches the process. For identifying higher order integrating processes, these algorithms can be chosen alternatively to meet with different requirements of identification accuracy and computational effort in practice. Compared to an FOPDT model, an SOPDT model identified hereby has been demonstrated to have the capacity of obtaining further improved frequency response fitting for a higher-order integrating process, and therefore, can be effectively used to represent various integrating processes for advanced control system design or controller tuning.

Experimental tests for maintaining the barrel temperature in injection molding (Liu et al. 2009) have been presented in Sect. 5.5, based on the use of integrating type models identified from the presented Algorithm-RI-S2. Compared to the use of heating-up integrating type models identified from step tests as presented in Sect. 3.4, apparent improvement on the control performance has been obtained. The control performance is similar to that obtained from using the stable type models as present in Sect. 4.5, which were also identified from the relay tests. It is, therefore, demonstrated that using a relay test for online identification can effectively facilitate control design and online tuning around the set-point operation.

References

Gu DY, Ou LL, Wang P, Zhang WD (2006) Relay feedback autotuning method for integrating processes with inverse response and time delay. Ind Eng Chem Res 45:3119–3132

Ho WK, Feng EB, Gan OP (1996) A novel relay autotuning technique for processes with integration. Control Eng Pract 4(7):923–928

Ingimundarson A, Hägglund T (2001) Robust tuning procedures of dead-time compensating controllers. Control Eng Pract 9:1195–1208

Kaya I (2006) Parameter estimation for integrating processes using relay feedback control under static load disturbances. Ind Eng Chem Res 45:4726–4731

Kwak HJ, Sung SW, Lee IB (1997) On-line process identification and autotuning for integrating processes. Ind Eng Chem Res 36:5329–5338

Liu T, Gao F (2008) Identification of integrating and unstable processes from relay feedback. Comput Chem Eng 32(12):3038–3056

Liu T, Yao K, Gao F (2009) Identification and autotuning of temperature control system with application to injection molding. IEEE Trans Control Syst Technol 17(6):1282–1294

Luyben WL (2003) Identification and tuning of integrating processes with deadtime and inverse response. Ind Eng Chem Res 42:3030–3035

Majhi M, Atherton DP (2000) Obtaining controller parameters for a new Smith predictor using autotuning. Automatica 36:1651–1658

Sung SW, Lee J (2006) Relay feedback method under large static disturbances. Automatica 42:353–356

Sung SW, Lee IB, Lee BK (1998) On-line process identification and automatic tuning method for PID controllers. Chem Eng Sci 53(10):1847–1859

Tsang KM, Lo WL, Rad AB (2000) Autotuning of phase-lead controller for integrating systems. IEEE Trans Ind Electron 47(1):203–210

Chapter 6
Relay Feedback Identification of Unstable Processes

For identifying an unstable process under a relay test, existing references have been mainly devoted to identification algorithms for obtaining the low-order FOPDT and SOPDT models to facilitate control system design and online tuning. Based on the standard unbiased relay test as shown in Fig. 4.2, Tan et al. (1998) proposed an FOPDT identification algorithm in terms of the gain and phase conditions for steady oscillation; Shiu et al. (1998) used two relay response points with the minimal and maximal gains to estimate the unstable type FOPDT and SOPDT models; Vivek and Chidambaram (2005) adopted the Fourier series approximation of the relay response to derive the FOPDT identification formulas for online PID tuning; Majhi (2007) developed FOPDT and SOPDT identification algorithms based on a state-space description of the relay control system and using the second derivative information of the limit cycle for analysis. Using a biased relay test, Park et al. (1998) presented an FOPDT identification algorithm in terms of the magnitude and phase conditions for steady oscillation; Kaya and Atherton (2001) proposed alternative FOPDT and SOPDT identification algorithms in terms of the so-called A-locus analysis; Ramakrishnan and Chidambaram (2003) presented an SOPDT identification algorithm using the Laplace transform of the relay response to construct model parameter fitting conditions at the oscillation frequency. The use of two different relay tests was suggested by Marchetti et al. (2001) to ensure the identification effectiveness for control-oriented modeling of unstable processes.

It has been recognized that the standard relay feedback structure cannot guarantee steady oscillation for various unstable processes, especially for unstable processes with large time delay. Some limiting conditions for forming steady oscillation under an unbiased relay test have been reported in the literature (Tan et al. 1998; Majhi and Atherton 2000; Thyagarajan and Yu 2003; Lin et al. 2004; Co 2010). To ensure identification stability, closed-loop identification methods with a P-, PI-, or PID-type controller for closed-loop stabilization have been developed (to enumerate a few, Jin et al. 1998; Ananth and Chidambaram 1999; Majhi and Atherton 2000; Forssell and Ljung 2000; Paraskevopoulos et al. 2004; Sree and Chidambaram 2006; Cheres 2006). A modified relay feedback structure with a proportional-derivative (PD) type controller to stabilize an unstable process was presented by Padhy and Majhi (2006) for identifying an FOPDT model based on the describing function analysis.

To address relay feedback stability for unstable processes, a limiting condition to forming steady oscillation for an unstable process under a biased or unbiased relay test (Liu and Gao 2008b) is presented here, along with the relay response expressions of FOPDT and SOPDT unstable type models. Subsequently, identification algorithms for obtaining these low-order models are given for practical application.

6.1 The Limiting Condition for Steady Oscillation

Generally, the widely used FOPDT and SOPDT unstable type models are expressed as

$$G_{U-1} = \frac{k_p e^{-\theta s}}{\tau_p s - 1} \tag{6.1}$$

$$G_{U-2} = \frac{k_p e^{-\theta s}}{(\tau_1 s - 1)(\tau_2 s + 1)} \tag{6.2}$$

where k_p is the proportional gain, θ the process time delay, τ_p (τ_1 and τ_2) positive coefficient(s) reflecting fundamental dynamic response characteristics of the process.

For a first-order unstable process as shown in (6.1), the following proposition gives a limiting condition to forming steady oscillation under a relay test and the corresponding relay response expression:

Proposition 6.1. *Under a biased relay test as shown in Fig. 6.1, a limiting condition to forming steady oscillation for a first-order unstable process shown in (6.1) is*

$$\frac{\theta}{\tau_p} < \min \left\{ \ln \frac{2k_p \mu_0}{k_p (\mu_0 - \Delta \mu) + \varepsilon_+}, \; \ln \frac{2k_p \mu_0}{k_p (\Delta \mu + \mu_0) - \varepsilon_-} \right\} \tag{6.3}$$

and the resulting limit cycle is characterized by

$$y_+(t) = -k_p(\Delta \mu + \mu_0) + \frac{2k_p \mu_0 \left(1 - e^{\frac{P_-}{\tau_p}}\right) e^{\frac{t}{\tau_p}}}{1 - e^{\frac{P_u}{\tau_p}}}, \quad t \in [0, P_+] \tag{6.4}$$

$$y_-(t) = k_p(\mu_0 - \Delta \mu) + \frac{2k_p \mu_0 \left(e^{\frac{P_+}{\tau_p}} - 1\right) e^{\frac{t}{\tau_p}}}{1 - e^{\frac{P_u}{\tau_p}}}, \quad t \in [0, P_-] \tag{6.5}$$

6.1 The Limiting Condition for Steady Oscillation

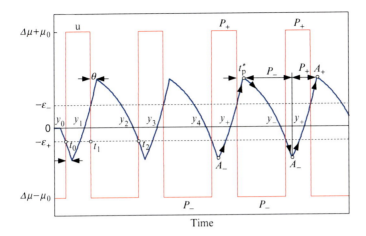

Fig. 6.1 Relay response depiction for an FOPDT unstable process

where $y_+(t)$ increases monotonically for $t \in [0, P_+]$, $y_-(t)$ decreases monotonically for $t \in [0, P_-]$ corresponding to $t \in [P_+, P_u]$ in the limit cycle, and $P_u = P_+ + P_-$ is the oscillation period.

Proof. The unity step response of an FOPDT unstable process shown in (6.1) can be derived as

$$y(t) = k_p \left(e^{\frac{t-\theta}{\tau_p}} - 1 \right) \tag{6.6}$$

It can be seen from Fig. 6.1 that according to the initial relay output $\Delta\mu - \mu_0$, there exist

$$y_0(t) = k_p (\Delta\mu - \mu_0) \left(e^{\frac{t-\theta}{\tau_p}} - 1 \right), \quad t \in [\theta, t_0 + \theta] \tag{6.7}$$

$$y_0(t_0) = -\varepsilon_+ \tag{6.8}$$

Using (6.8) one can obtain

$$y_0(t_0 + \theta) = -k_p(\Delta\mu - \mu_0) + \left[k_p(\Delta\mu - \mu_0) - \varepsilon_+ \right] e^{\frac{\theta}{\tau_p}} \tag{6.9}$$

After the time of $t_0 + \theta$, the process output begins to respond to the switched relay output $\Delta\mu + \mu_0$ in the form of

$$y(t) = y_0(t_0 + \theta) e^{\frac{t-t_0-\theta}{\tau_p}} + k_p (\Delta\mu + \mu_0) \left(e^{\frac{t-t_0-\theta}{\tau_p}} - 1 \right), \quad t \in (t_0 + \theta, t_1 + \theta] \tag{6.10}$$

It follows from (6.10) that

$$\frac{dy(t)}{dt} = \frac{1}{\tau_p}\left[y_0\left(t_0 + \theta\right) + k_p\left(\Delta\mu + \mu_0\right)\right]e^{\frac{t-t_0-\theta}{\tau_p}} \tag{6.11}$$

To form steady oscillation, $y(t)$ should have an ascending tendency in the time interval $(t_0 + \theta, t_1 + \theta]$, therefore requiring that

$$y_0\left(t_0 + \theta\right) + k_p\left(\Delta\mu + \mu_0\right) > 0 \tag{6.12}$$

By substituting (6.9) into (6.12), one can obtain

$$\frac{\theta}{\tau_p} < \ln\frac{2k_p\mu_0}{k_p\left(\mu_0 - \Delta\mu\right) + \varepsilon_+} \tag{6.13}$$

Analogously, it follows from (6.10) and $y(t_1) = -\varepsilon_-$ that

$$y(t_1 + \theta) = -k_p\left(\Delta\mu + \mu_0\right) + \left[k_p\left(\Delta\mu + \mu_0\right) - \varepsilon_-\right]e^{\frac{\theta}{\tau_p}} \tag{6.14}$$

It can be seen from Fig. 6.1 that after the time of $t_1 + \theta$, the process output begins to respond to the switched relay output $\Delta\mu - \mu_0$ in the form of

$$y(t) = y\left(t_1 + \theta\right)e^{\frac{t-t_1-\theta}{\tau_p}} + k_p\left(\Delta\mu - \mu_0\right)\left(e^{\frac{t-t_1-\theta}{\tau_p}} - 1\right), \quad t \in (t_1 + \theta, t_2 + \theta] \tag{6.15}$$

It follows from (6.15) that

$$\frac{dy(t)}{dt} = \frac{1}{\tau_p}\left[y\left(t_1 + \theta\right) + k_p\left(\Delta\mu - \mu_0\right)\right]e^{\frac{t-t_1-\theta}{\tau_p}} \tag{6.16}$$

To form steady oscillation, $y(t)$ should have an descending tendency in the time interval $(t_1 + \theta, t_2 + \theta]$, therefore requiring that

$$y\left(t_1 + \theta\right) + k_p\left(\Delta\mu - \mu_0\right) < 0 \tag{6.17}$$

Substituting (6.14) into (6.17), one can obtain

$$\frac{\theta}{\tau_p} < \ln\frac{2k_p\mu_0}{k_p\left(\mu_0 + \Delta\mu\right) - \varepsilon_-} \tag{6.18}$$

Combining (6.13) with (6.18), one can obtain the limiting condition to forming steady oscillation as shown in Proposition 6.1.

6.1 The Limiting Condition for Steady Oscillation

In the sequel, the time shifted output response from the initial to the fourth relay switch point can be derived, respectively, as

$$y_1(t)|_{\text{shift}} = y_0(t + t_0 + \theta) - 2k_p\mu_0 + 2k_p\mu_0 e^{\frac{t}{\tau_p}} \tag{6.19}$$

$$y_2(t)|_{\text{shift}} = y_0(t + t_0 + \theta + P_+) - 2k_p\mu_0(1 - 1) + 2k_p\mu_0 e^{\frac{t}{\tau_p}}\left(e^{\frac{P_+}{\tau_p}} - 1\right) \tag{6.20}$$

$$y_3(t)|_{\text{shift}} = y_0(t + t_0 + \theta + P_u) - 2k_p\mu_0(1 - 1 + 1) + 2k_p\mu_0 e^{\frac{t}{\tau_p}}\left(e^{\frac{P_u}{\tau_p}} - e^{\frac{P_-}{\tau_p}} + 1\right) \tag{6.21}$$

$$y_4(t)|_{\text{shift}} = y_0(t + t_0 + \theta + P_u + P_+) - 2k_p\mu_0(1 - 1 + 1 - 1)$$
$$+ 2k_p\mu_0 e^{\frac{t}{\tau_p}}\left(e^{\frac{P_u + P_+}{\tau_p}} - e^{\frac{P_u}{\tau_p}} + e^{\frac{P_+}{\tau_p}} - 1\right) \tag{6.22}$$

The general relay response can therefore be summarized as

$$y_{2n+1}(t)|_{\text{shift}} = y_0(t + t_0 + \theta + n P_u) - 2k_p\mu_0 + 2k_p\mu_0 E e^{\frac{t}{\tau_p}} \tag{6.23}$$

$$y_{2n+2}(t)|_{\text{shift}} = y_0(t + t_0 + \theta + n P_u + P_+) + 2k_p\mu_0 F e^{\frac{t}{\tau_p}} \tag{6.24}$$

where $n = 0, 1, 2, \ldots$, and

$$E = 1 + \sum_{k=1}^{n}\left(e^{\frac{k P_u}{\tau_p}} - e^{\frac{(k-1) P_u + P_-}{\tau_p}}\right) = 1 + \left(e^{\frac{P_u}{\tau_p}} - e^{\frac{P_-}{\tau_p}}\right)\frac{1 - e^{\frac{(n-1) P_u}{\tau_p}}}{1 - e^{\frac{P_u}{\tau_p}}} \tag{6.25}$$

$$F = \sum_{k=0}^{n}\left(e^{\frac{k P_u + P_+}{\tau_p}} - e^{\frac{k P_u}{\tau_p}}\right) = \left(e^{\frac{P_+}{\tau_p}} - 1\right)\frac{1 - e^{\frac{n P_u}{\tau_p}}}{1 - e^{\frac{P_u}{\tau_p}}} \tag{6.26}$$

It follows from (6.7) that

$$y_0(t + t_0 + \theta + n P_u) = k_p(\Delta\mu - \mu_0)\left(e^{\frac{t + t_0 + n P_u}{\tau_p}} - 1\right) \tag{6.27}$$

$$y_0(t + t_0 + \theta + n P_u + P_+) = k_p(\Delta\mu - \mu_0)\left(e^{\frac{t + t_0 + n P_u + P_+}{\tau_p}} - 1\right) \tag{6.28}$$

Substituting (6.25)–(6.28) into (6.23) and (6.24), respectively, one can obtain

$$y_{2n+1}|_{\text{shift}}(t) = -k_p (\Delta\mu + \mu_0) + \frac{k_p e^{\frac{L}{\tau_p}}}{1 - e^{\frac{P_u}{\tau_p}}}$$

$$\times \left\{ e^{\frac{nP_u}{\tau_p}} \left[(\Delta\mu - \mu_0) \left(1 - e^{\frac{P_u}{\tau_p}} \right) e^{\frac{t_0}{\tau_p}} - 2\mu_0 \left(1 - e^{-\frac{P_+}{\tau_p}} \right) \right] + 2\mu_0 \left(1 - e^{-\frac{P_-}{\tau_p}} \right) \right\}$$

(6.29)

$$y_{2n+2}|_{\text{shift}}(t) = k_p (\mu_0 - \Delta\mu) + \frac{k_p e^{\frac{L}{\tau_p}}}{1 - e^{\frac{P_u}{\tau_p}}}$$

$$\times \left\{ e^{\frac{nP_u}{\tau_p}} \left[(\Delta\mu - \mu_0) \left(1 - e^{\frac{P_u}{\tau_p}} \right) e^{\frac{t_0 + P_+}{\tau_p}} - 2\mu_0 \left(e^{\frac{P_+}{\tau_p}} - 1 \right) \right] + 2\mu_0 \left(e^{\frac{P_+}{\tau_p}} - 1 \right) \right\}$$

(6.30)

Note that $n \to \infty$ as $t \to \infty$. It can be seen from (6.29) and (6.30) that there exists an inherent constraint for steady oscillation,

$$(\Delta\mu - \mu_0) \left(1 - e^{\frac{P_u}{\tau_p}} \right) e^{\frac{t_0}{\tau_p}} - 2\mu_0 \left(1 - e^{-\frac{P_+}{\tau_p}} \right) = 0 \qquad (6.31)$$

Hence, in the limit cycle the output response is expressed by (6.4) and (6.5). It can easily be seen from (6.4) and (6.5) that $y_+(t)$ increases monotonically for $t \in [0, P_+]$ while $y_-(t)$ decreases monotonically for $t \in (P_+, P_u]$.

It can be verified from (6.4) and (6.5) that the inherent relationship between $y_+(t)$ and $y_-(t)$ is satisfied, that is,

$$y_+(0) = y_-(P_-) \qquad (6.32)$$

$$y_+(P_+) = y_-(0) \qquad (6.33)$$

This completes the proof. □

If an unbiased relay test is used, the limiting condition for steady oscillation can be obtained by substituting $\Delta\mu = 0$ and $\varepsilon_+ = -\varepsilon_- = 0$ into (6.3), that is, $\theta/\tau_p < \ln 2$, which is exactly the limiting condition reported in the references (Tan et al. 1998; Majhi and Atherton 2000; Thyagarajan and Yu 2003). Correspondingly, the relay response expression can be obtained as

$$y_+(t) = -y_-(t) = -k_p\mu_0 + \frac{2k_p\mu_0 e^{\frac{L}{\tau_p}}}{1 + e^{\frac{P_u}{2\tau_p}}} \qquad (6.34)$$

For a second-order unstable process, the following proposition gives the corresponding relay response expression in the limit cycle.

6.1 The Limiting Condition for Steady Oscillation

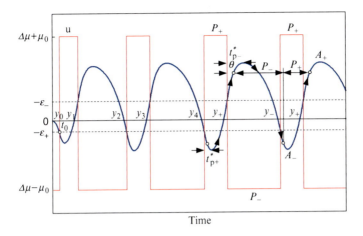

Fig. 6.2 Relay response depiction for an SOPDT unstable process

Proposition 6.2. *Under a biased relay test as shown in Fig. 6.2, the resulting limit cycle for a second-order unstable process shown in (6.2) is characterized by*

$$y_+(t) = -k_p(\Delta\mu + \mu_0) + \frac{2k_p\mu_0\tau_1 E_1}{\tau_1 + \tau_2}e^{\frac{t}{\tau_1}} + \frac{2k_p\mu_0\tau_2 E_2}{\tau_1 + \tau_2}e^{-\frac{t}{\tau_2}}, \quad t \in [0, P_+] \tag{6.35}$$

$$y_-(t) = k_p(\mu_0 - \Delta\mu) + \frac{2k_p\mu_0\tau_1 F_1}{\tau_1 + \tau_2}e^{\frac{t}{\tau_1}} + \frac{2k_p\mu_0\tau_2 F_2}{\tau_1 + \tau_2}e^{-\frac{t}{\tau_2}}, \quad t \in [0, P_-] \tag{6.36}$$

where $0 < E_1 < E_2$, $F_2 < F_1 < 0$, *and*

$$E_1 = \frac{1 - e^{\frac{P_-}{\tau_1}}}{1 - e^{\frac{P_u}{\tau_1}}} \tag{6.37}$$

$$E_2 = \frac{1 - e^{-\frac{P_-}{\tau_2}}}{1 - e^{-\frac{P_u}{\tau_2}}} \tag{6.38}$$

$$F_1 = -\frac{1 - e^{\frac{P_+}{\tau_1}}}{1 - e^{\frac{P_u}{\tau_1}}} \tag{6.39}$$

$$F_2 = -\frac{1 - e^{-\frac{P_+}{\tau_2}}}{1 - e^{-\frac{P_u}{\tau_2}}} \tag{6.40}$$

224   6 Relay Feedback Identification of Unstable Processes

Proof. The initial step response of an SOPDT unstable process arising from the relay output $\Delta\mu - \mu_0$ can be derived as

$$y_0(t) = k_p (\Delta\mu - \mu_0) \left(\frac{\tau_1}{\tau_1 + \tau_2} e^{\frac{t-\theta}{\tau_1}} + \frac{\tau_2}{\tau_1 + \tau_2} e^{-\frac{t-\theta}{\tau_2}} - 1 \right) \quad (6.41)$$

Following a similar analysis as in the above proof for Proposition 6.1, the time shifted process output response from the initial to the fourth relay switch point can be derived, respectively, as

$$y_1(t)|_{\text{shift}} = y_0(t + t_0 + \theta) + 2k_p\mu_0 \left(\frac{\tau_1}{\tau_1 + \tau_2} e^{\frac{t}{\tau_1}} + \frac{\tau_2}{\tau_1 + \tau_2} e^{-\frac{t}{\tau_2}} - 1 \right) \quad (6.42)$$

$$y_2(t)|_{\text{shift}} = y_0(t + t_0 + \theta + P_+) - 2k_p\mu_0(1-1) + \frac{2k_p\mu_0\tau_1}{\tau_1 + \tau_2} e^{\frac{t}{\tau_1}} \left(e^{\frac{P_+}{\tau_1}} - 1 \right)$$

$$+ \frac{2k_p\mu_0\tau_2}{\tau_1 + \tau_2} e^{-\frac{t}{\tau_2}} \left(e^{-\frac{P_+}{\tau_2}} - 1 \right) \quad (6.43)$$

$$y_3(t)|_{\text{shift}} = y_0(t + t_0 + \theta + P_u) - 2k_p\mu_0(1 - 1 + 1) + \frac{2k_p\mu_0\tau_1}{\tau_1 + \tau_2} e^{\frac{t}{\tau_1}}$$

$$\times \left(e^{\frac{P_u}{\tau_1}} - e^{\frac{P_-}{\tau_1}} + 1 \right) + \frac{2k_p\mu_0\tau_2}{\tau_1 + \tau_2} e^{-\frac{t}{\tau_2}} \left(e^{-\frac{P_u}{\tau_2}} - e^{-\frac{P_-}{\tau_2}} + 1 \right) \quad (6.44)$$

$$y_4(t)|_{\text{shift}} = y_0(t + t_0 + \theta + P_u + P_+) - 2k_p\mu_0(1 - 1 + 1 - 1)$$

$$+ \frac{2k_p\mu_0\tau_1}{\tau_1 + \tau_2} e^{\frac{t}{\tau_1}} \left(e^{\frac{P_u+P_+}{\tau_1}} - e^{\frac{P_u}{\tau_1}} + e^{\frac{P_+}{\tau_1}} - 1 \right)$$

$$+ \frac{2k_p\mu_0\tau_2}{\tau_1 + \tau_2} e^{-\frac{t}{\tau_2}} \left(e^{-\frac{P_u+P_+}{\tau_2}} - e^{-\frac{P_u}{\tau_2}} + e^{-\frac{P_+}{\tau_2}} - 1 \right) \quad (6.45)$$

The general relay response can therefore be summarized as

$$y_{2n+1}(t)|_{\text{shift}} = y_0(t + t_0 + \theta + nP_u) - 2k_p\mu_0 + \frac{2k_p\mu_0 E_1^* \tau_1}{\tau_1 + \tau_2} e^{\frac{t}{\tau_1}} + \frac{2k_p\mu_0 E_2 \tau_2}{\tau_1 + \tau_2} e^{-\frac{t}{\tau_2}}$$

$$(6.46)$$

$$y_{2n+2}(t)|_{\text{shift}} = y_0(t + t_0 + \theta + nP_u + P_+) + \frac{2k_p\mu_0 F_1^* \tau_1}{\tau_1 + \tau_2} e^{\frac{t}{\tau_1}} + \frac{2k_p\mu_0 F_2 \tau_2}{\tau_1 + \tau_2} e^{-\frac{t}{\tau_2}}$$

$$(6.47)$$

6.1 The Limiting Condition for Steady Oscillation

where $n = 0, 1, 2, \ldots$, and

$$E_1^* = 1 + \sum_{k=1}^{n}\left(e^{\frac{kP_u}{\tau_1}} - e^{\frac{(k-1)P_u + P_-}{\tau_1}}\right) = 1 + \frac{\left(e^{\frac{P_u}{\tau_1}} - e^{\frac{P_-}{\tau_1}}\right)\left(1 - e^{\frac{(n-1)P_u}{\tau_1}}\right)}{1 - e^{\frac{P_u}{\tau_1}}} \quad (6.48)$$

$$E_2 = 1 + \sum_{k=1}^{n}\left(e^{-\frac{kP_u}{\tau_2}} - e^{-\frac{(k-1)P_u + P_-}{\tau_2}}\right) = \frac{1 - e^{-\frac{P_-}{\tau_2}}}{1 - e^{-\frac{P_u}{\tau_2}}} \quad (6.49)$$

$$F_1^* = \sum_{k=0}^{n}\left(e^{\frac{kP_u + P_+}{\tau_1}} - e^{\frac{kP_u}{\tau_1}}\right) = -\frac{\left(1 - e^{\frac{P_+}{\tau_1}}\right)\left(1 - e^{\frac{nP_u}{\tau_1}}\right)}{1 - e^{\frac{P_u}{\tau_1}}} \quad (6.50)$$

$$F_2 = \sum_{k=0}^{n}\left(e^{-\frac{kP_u + P_+}{\tau_2}} - e^{-\frac{kP_u}{\tau_2}}\right) = -\frac{1 - e^{-\frac{P_+}{\tau_2}}}{1 - e^{-\frac{P_u}{\tau_2}}} \quad (6.51)$$

It follows from (6.41) that

$$y_0(t + t_0 + \theta + nP_u) = k_p(\Delta\mu - \mu_0)\left(\frac{\tau_1}{\tau_1 + \tau_2}e^{\frac{t + t_0 + nP_u}{\tau_1}} + \frac{\tau_2}{\tau_1 + \tau_2}e^{-\frac{t + t_0 + nP_u}{\tau_2}} - 1\right)$$

$$(6.52)$$

$$y_0(t + t_0 + \theta + nP_u + P_+)$$
$$= k_p(\Delta\mu - \mu_0)\left(\frac{\tau_1}{\tau_1 + \tau_2}e^{\frac{t + t_0 + nP_u + P_+}{\tau_1}} + \frac{\tau_2}{\tau_1 + \tau_2}e^{-\frac{t + t_0 + nP_u + P_+}{\tau_2}} - 1\right)$$

$$(6.53)$$

Substituting (6.48)–(6.53) into (6.46) and (6.47), respectively, one can obtain

$$y_{2n+1}(t)|_{\text{shift}} = -k_p(\Delta\mu + \mu_0) + \frac{2k_p\mu_0\tau_1 E_1}{\tau_1 + \tau_2}e^{\frac{t}{\tau_1}} + \frac{2k_p\mu_0\tau_2 E_2}{\tau_1 + \tau_2}e^{-\frac{t}{\tau_2}}$$

$$+ \frac{k_p\tau_1}{\tau_1 + \tau_2}e^{\frac{t+(n-1)P_u}{\tau_1}}\left[(\Delta\mu - \mu_0)e^{\frac{t_0 + P_u}{\tau_1}} - \frac{2\mu_0\left(e^{\frac{P_u}{\tau_1}} - e^{\frac{P_-}{\tau_1}}\right)}{1 - e^{\frac{P_u}{\tau_1}}}\right]$$

$$(6.54)$$

$$y_{2n+2}(t)|_{\text{shift}} = k_p(\mu_0 - \Delta\mu) + \frac{2k_p\mu_0\tau_1 F_1}{\tau_1 + \tau_2} e^{\frac{t}{\tau_1}} + \frac{2k_p\mu_0\tau_2 F_2}{\tau_1 + \tau_2} e^{-\frac{t}{\tau_2}}$$

$$+ \frac{k_p\tau_1}{\tau_1 + \tau_2} e^{\frac{t+nP_u}{\tau_1}} \left[(\Delta\mu - \mu_0) e^{\frac{t_0 + P_+}{\tau_1}} - \frac{2\mu_0 \left(e^{\frac{P_+}{\tau_1}} - 1 \right)}{1 - e^{\frac{P_u}{\tau_1}}} \right]$$

(6.55)

where

$$E_1 = \frac{1 - e^{\frac{P_-}{\tau_1}}}{1 - e^{\frac{P_u}{\tau_1}}}$$
(6.56)

$$F_1 = -\frac{1 - e^{\frac{P_+}{\tau_1}}}{1 - e^{\frac{P_u}{\tau_1}}}$$
(6.57)

In view of that $n \to \infty$ as $t \to \infty$, one can see from (6.54) and (6.55) that there exists an inherent constraint for steady oscillation,

$$(\Delta\mu - \mu_0)\left(1 - e^{\frac{P_u}{\tau_1}}\right) e^{\frac{t_0}{\tau_1}} - 2\mu_0 \left(1 - e^{-\frac{P_+}{\tau_1}}\right) = 0 \quad (6.58)$$

Note that this constraint is essentially the same as that shown in (6.31) for a first-order unstable process.

Hence, in the limit cycle the output response is expressed by (6.35) and (6.36), where $y_+(t)$ denotes the ascending response in the half period P_+, corresponding to a positive step change of the relay output as shown in Fig. 6.2, while $y_-(t)$ denotes the descending response in the other half period P_-, corresponding to a negative step change of the relay output.

It can be verified from (6.35) and (6.36) that the inherent relationship between $y_+(t)$ and $y_-(t)$ is satisfied, i.e.,

$$y_+(0) = y_-(P_-) \quad (6.59)$$

$$y_+(P_+) = y_-(0) \quad (6.60)$$

Define $z = x/\tau_1$, $a = \tau_1/\tau_2$, and

$$f(x) = \frac{1 - e^{\frac{x}{\tau_1}}}{1 - e^{-\frac{x}{\tau_2}}}, \quad x \in (0, \infty) \quad (6.61)$$

6.1 The Limiting Condition for Steady Oscillation

It follows that

$$f(z) = \frac{1-e^z}{1-e^{-az}} \tag{6.62}$$

$$\frac{df(z)}{dz} = \frac{e^z\left[(a+1)e^{-az} - ae^{-(a+1)z} - 1\right]}{(1-e^{-az})^2} \tag{6.63}$$

Let

$$g(z) = (a+1)e^{-az} - ae^{-(a+1)z} - 1 \tag{6.64}$$

It can be derived that

$$\frac{dg(z)}{dz} = a(a+1)e^{-az}(e^{-z} - 1) < 0 \tag{6.65}$$

$$\lim_{z \to 0} g(z) = 0 \tag{6.66}$$

Thus, it can be concluded that $g(z) < 0$ for $z \in (0, \infty)$, and accordingly,

$$\frac{df(z)}{dz} < 0 \tag{6.67}$$

Hence, $f(z)$ decreases monotonically with respect to z, and so is $f(x)$ with respect to x. One can then conclude from $f(p_-) > f(p_u)$ that

$$0 < E_1 < E_2 \tag{6.68}$$

Following a similar analysis, it can be concluded that

$$F_2 < F_1 < 0 \tag{6.69}$$

This completes the proof. □

If an unbiased relay test is used, substituting $\Delta\mu = 0$ and $\varepsilon_+ = -\varepsilon_- = 0$ into (6.35)–(6.40), one can obtain the corresponding relay response expression,

$$y_+(t) = -y_-(t) = -k_p\mu_0 + \frac{2k_p\mu_0\tau_1 E_1}{\tau_1 + \tau_2}e^{\frac{t}{\tau_1}} + \frac{2k_p\mu_0\tau_2 E_2}{\tau_1 + \tau_2}e^{-\frac{t}{\tau_2}} \tag{6.70}$$

where

$$E_1 = \frac{1}{1 + e^{\frac{p_u}{2\tau_1}}} \tag{6.71}$$

$$E_2 = \frac{1}{1 + e^{-\frac{p_u}{2\tau_2}}} \tag{6.72}$$

6.2 The FOPDT Model

It follows from Proposition 6.1 that the process time delay can be measured as the time to reach the peak of the process output response from the initial relay switch point in a half period of the relay, which is denoted as t_p^* in Fig. 6.1.

When a biased relay test is used, the proportional gain, k_p, can be derived as

$$k_p = -G(0) = -\frac{\int_{t_{os}}^{t_{os}+P_u} y(t)dt}{\int_{t_{os}}^{t_{os}+P_u} u(t)dt} \tag{6.73}$$

where t_{os} can be taken as any relay switch point in the steady oscillation.

Correspondingly, it follows from (6.4) and (6.5) that

$$y_+(P_+) = -k_p(\Delta\mu + \mu_0) + \frac{2k_p\mu_0 e^{\frac{P_+}{\tau_p}}\left(1 - e^{\frac{P_-}{\tau_p}}\right)}{1 - e^{\frac{P_u}{\tau_p}}} = A_+ \tag{6.74}$$

$$y_-(P_-) = k_p(\mu_0 - \Delta\mu) + \frac{2k_p\mu_0 e^{\frac{P_-}{\tau_p}}\left(e^{\frac{P_+}{\tau_p}} - 1\right)}{1 - e^{\frac{P_u}{\tau_p}}} = A_- \tag{6.75}$$

$$y_1(P_1 - \theta) = -k_p(\Delta\mu + \mu_0) + \frac{2k_p\mu_0 e^{\frac{P_+-\theta}{\tau_p}}\left(1 - e^{\frac{P_-}{\tau_p}}\right)}{1 - e^{\frac{P_u}{\tau_p}}} = -\varepsilon_- \tag{6.76}$$

$$y_-(P_- - \theta) = k_p(\mu_0 - \Delta\mu) + \frac{2k_p\mu_0 e^{\frac{P_--\theta}{\tau_p}}\left(e^{\frac{P_+}{\tau_p}} - 1\right)}{1 - e^{\frac{P_u}{\tau_p}}} = -\varepsilon_+ \tag{6.77}$$

Substituting (6.74) into (6.76), one can obtain

$$\tau_p = \frac{\theta}{\ln\frac{k_p(\mu_0+\Delta\mu)+A_+}{k_p(\mu_0+\Delta\mu)-\varepsilon_-}} \tag{6.78}$$

Alternatively, substituting (6.75) into (6.77) yields

$$\tau_p = \frac{\theta}{\ln\frac{k_p(\mu_0-\Delta\mu)-A_-}{k_p(\mu_0-\Delta\mu)+\varepsilon_+}} \tag{6.79}$$

The process time constant can then be derived from (6.78) or (6.79). It is preferred in practice to use (6.78) for better fitting accuracy, in view of that the

6.2 The FOPDT Model

positive part of the limit cycle occupies a larger percentage compared to the negative part, as shown in Fig. 6.1. If the opposite case happens, the formula in (6.79) is preferred accordingly.

Hence, the identification of an FOPDT unstable type model under a biased relay test can be summarized in the following algorithm named *Algorithm-RU-FA1*.

Algorithm-RU-FA1

(i) Measure P_+, P_-, A_+, and A_- from the limit cycle.
(ii) Measure the process time delay as t_p^*, which is the time to reach the positive peak (A_+) of the process output response from the initial relay switch point in a negative half period of the relay (P_-).
(iii) Compute the proportional gain, k_p, from (6.73).
(iv) Compute the process time constant, τ_p, from (6.78).

Note that the process response at the oscillation frequency can be computed in terms of the frequency response estimation formula shown in (4.30), i.e.,

$$G(j\omega_u) = A_u e^{j\varphi_u} = \frac{\int_{t_{os}}^{t_{os}+P_u} y(t)e^{-j\omega_u t} dt}{\int_{t_{os}}^{t_{os}+P_u} u(t)e^{-j\omega_u t} dt} \tag{6.80}$$

Substituting the FOPDT unstable type model in (6.1) into the left-hand side of (6.80), one can obtain

$$\tau_p = \frac{1}{\omega_u}\sqrt{\frac{k_p^2}{A_u^2} - 1} \tag{6.81}$$

$$\theta = -\frac{1}{\omega_u}\left[\varphi_u + \pi - \arctan(\tau_p \omega_u)\right], \quad \varphi_u \in (-\pi, -\pi/2) \tag{6.82}$$

Hence, an alternative identification algorithm named *Algorithm-RU-FA2* for obtaining an FOPDT unstable type model under a biased relay test can be summarized.

Algorithm-RU-FA2

(i) Measure P_u from the limit cycle.
(ii) Compute $G(j\omega_u)$ from (6.80).
(iii) Compute the proportional gain, k_p, from (6.73).
(iv) Compute the process time constant, τ_p, from (6.81).
(v) Compute the process time delay, θ, from (6.82).

Note that the process response at the oscillation frequency is exactly represented by the FOPDT model derived from Algorithm-RU-FA2. Therefore, it can be used to obtain better fitting accuracy for identifying a high-order unstable process compared to Algorithm-RU-FA1, but at the cost of computation effort.

230   6 Relay Feedback Identification of Unstable Processes

When an unbiased relay test is used, substituting (6.74) into (6.76) to eliminate k_p, one can obtain

$$\varepsilon_+ \left(1 - e^{-\frac{p_u}{2\tau_p}}\right) = A_+ \left(2e^{-\frac{\theta}{\tau_p}} - e^{-\frac{p_u}{2\tau_p}} - 1\right) \tag{6.83}$$

Note that the left-hand side of (6.83) is monotonically decreasing with respect to τ_p, together with a practical constraint of $0 < \tau_p < P_u$. Therefore, only finite solutions of τ_p exist for (6.83), which can be derived using any numerical algorithm such as the Newton–Raphson method. Correspondingly, the proportional gain can then be derived from (6.74) as

$$k_p = \frac{A_+ \left(e^{\frac{p_u}{2\tau_p}} + 1\right)}{\mu_0 \left(e^{\frac{p_u}{2\tau_p}} - 1\right)} \tag{6.84}$$

Therefore, the identification of an FOPDT unstable type model under an unbiased relay test can be summarized in the following algorithm named *Algorithm-RU-FB1*.

Algorithm-RU-FB1

(i) Measure P_u and A_+ from the limit cycle.
(ii) Measure the process time delay as $\theta = t_p^*$.
(iii) Compute the process time constant, τ_p, from (6.83) by using the Newton–Raphson iteration method. The initial estimation for iteration may be taken as $\hat{\tau}_p = P_u/2$.
(iv) Compute the proportional gain, k_p, from (6.84).
(v) Determine the suitable solution pair of τ_p and k_p by comparing the relay response of the resulting model with that of the process, or checking if $|N(A_+) \hat{G}(j\omega_u)| \to 1$ and $\angle N(A_+) + \angle \hat{G}(j\omega_u) \to -\pi$ are satisfied, where $N(A_+) = 4\mu_0 e^{-j \arcsin(\varepsilon_+/A_+)}/(\pi A_+)$ denotes the describing function of the unbiased relay, and $\hat{G}(j\omega_u)$ is the FOPDT model response at the oscillation frequency.

With the process time delay measured from the limit cycle, the proportional gain and time constant can also be derived from the fitting conditions at the oscillation frequency as shown in (6.81) and (6.82), i.e.,

$$\tau_p = \frac{1}{\omega_u} \tan(\pi + \varphi_u + \theta\omega_u) \tag{6.85}$$

$$k_p = A_u \sqrt{\tau_p^2 \omega_u^2 + 1} \tag{6.86}$$

To procure fitting accuracy for identifying a higher-order unstable process in practice, a one-dimensional search of θ can be implemented in terms of the

6.3 The SOPDT Model

aforementioned fitting condition of the relay response. The optimal fitting can be determined by deriving such a model that yields the smallest value of the fitting error with respect to the limit cycle. Improved fitting accuracy can therefore be obtained for identifying a high-order process compared to Algorithm-RU-FB1.

Hence, an alternative identification named *Algorithm-RU-FB2* for obtaining an FOPDT unstable type model under an unbiased relay test can be summarized.

Algorithm-RU-FB2

(i) Measure P_u and the process time delay ($\theta = t_p^*$) from the limit cycle.
(ii) Compute $G(j\omega_u)$ from (6.80).
(iii) Compute the process time constant, τ_p, from (6.85).
(iv) Compute the proportional gain, k_p, from (6.86).
(v) End the algorithm if the fitting condition of relay response, $\sum_{k=1}^{N}[y(kT_s + t_{os}) - \hat{y}(kT_s + t_{os})]^2/N < \varepsilon$, is satisfied, where $y(kT_s + t_{os})$ and $\hat{y}(kT_s + t_{os})$ denote, respectively, the process and model responses in the limit cycle, T_s is the sampling period corresponding to $N = P_u/T_s$, and ε is a user-specified fitting threshold that may be practically taken no larger than $(0.1-1.0)\%$. Otherwise, go back to step (iii) by monotonically varying θ in a one-dimensional search within a possible range as observed from the initial step response in the relay test.

6.3 The SOPDT Model

According to Proposition 6.2, it can be derived from (6.35) and (6.36) that

$$\frac{dy_+(t)}{dt} = \frac{2k_p\mu_0}{\tau_1 + \tau_2}\left(E_1 e^{\frac{t}{\tau_1}} - E_2 e^{-\frac{t}{\tau_2}}\right) \tag{6.87}$$

$$\frac{dy_-(t)}{dt} = \frac{2k_p\mu_0}{\tau_1 + \tau_2}\left(F_1 e^{\frac{t}{\tau_1}} - F_2 e^{-\frac{t}{\tau_2}}\right) \tag{6.88}$$

By letting $\frac{dy_+(t)}{dt} = 0$ and $\frac{dy_-(t)}{dt} = 0$, the time to reach the single extreme value of $y_+(t)$ and $y_-(t)$ can be derived, respectively, as

$$t_{p+} = \frac{\tau_1 \tau_2}{\tau_1 + \tau_2} \ln \frac{E_2}{E_1} \tag{6.89}$$

$$t_{p-} = \frac{\tau_1 \tau_2}{\tau_1 + \tau_2} \ln \frac{F_2}{F_1} \tag{6.90}$$

It can be easily verified that $\frac{d^2 y_+(t)}{dt^2} > 0$ and $\frac{d^2 y_-(t)}{dt^2} < 0$. Therefore, t_{p+} is the time to reach the minimum of $y_+(t)$, while t_{p-} is the time to reach the maximum of $y_-(t)$.

Note that the time, $t_{p_-}^*$, to reach the maximum of $y_-(t)$ from the initial relay switch point in a negative half period of the relay can be measured, as shown in Fig. 6.2. So is for $t_{p_+}^*$ which is the time to reach the minimum of $y_+(t)$ from the initial relay switch point in a positive half period of the relay. It follows that

$$t_{p_-} = t_{p_-}^* - \theta \qquad (6.91)$$

$$t_{p_+} = t_{p_+}^* - \theta \qquad (6.92)$$

Subtracting (6.91) from (6.92) yields

$$t_{p_+} - t_{p_-} = t_{p_+}^* - t_{p_-}^* \qquad (6.93)$$

Substituting the SOPDT model in (6.2) into the frequency response fitting condition shown in (6.80), one can obtain

$$\frac{k_p}{\sqrt{(\tau_1^2 \omega_u^2 + 1)(\tau_2^2 \omega_u^2 + 1)}} = A_u \qquad (6.94)$$

$$-\theta \omega_u - \pi + \arctan(\tau_1 \omega_u) - \arctan(\tau_2 \omega_u) = \varphi_u, \quad \varphi_u \in (-\pi, -\pi/2) \qquad (6.95)$$

It can be derived from (6.94) and (6.95) that

$$\tau_2 = \frac{1}{\omega_u} \sqrt{\frac{k_p^2}{A_u^2 (\tau_1^2 \omega_u^2 + 1)} - 1} \qquad (6.96)$$

$$\theta = -\frac{1}{\omega_u} [\varphi_u + \pi - \arctan(\tau_1 \omega_u) + \arctan(\tau_2 \omega_u)] \qquad (6.97)$$

Note that the proportional gain, k_p, can be derived from (6.73). Substituting (6.89), (6.90), and (6.96) into (6.93), one can obtain a transcendental equation with respect to τ_1. This equation can be solved numerically using the Newton–Raphson iteration method. The initial estimation of τ_1 for iteration may be taken as

$$\hat{\tau}_1 = \frac{1}{\omega_u} \sqrt{\frac{k_p^2}{A_u^2} - 1} \qquad (6.98)$$

which has been derived in Algorithm-RU-FA2 for computing the single time constant of an FOPDT unstable type model.

Alternatively, substituting (6.96) into (6.95) yields a transcendental equation with respect to τ_1, where the process time delay (θ) is preestimated for computation. Correspondingly, a one-dimensional search of θ is needed in combination with the aforementioned time domain fitting condition of the relay response to determine the optimal fitting.

6.3 The SOPDT Model

Therefore, the above algorithm named *Algorithm-RU-SA* for identifying an SOPDT unstable type model under a biased relay test can be summarized.

Algorithm-RU-SA

(i) Measure P_+, P_-, t^*_{p+}, and t^*_{p-} from the limit cycle.
(ii) Compute $G(j\omega_u)$ from (6.80).
(iii) Compute the proportional gain, k_p, from (6.73).
(iv) Compute the process time constant, τ_1, from the equation resulting from substituting (6.89), (6.90), and (6.96) into (6.93), by using the Newton–Raphson iteration method, or from the equation resulting from substituting (6.96) into (6.95) based on a preestimated time delay (θ). The initial estimation of τ_1 for iteration may be taken from (6.98), or alternatively, a one-dimensional search of θ can be implemented within a possible range as observed from the initial step response in the relay test.
(v) Compute the process time constant, τ_2, from (6.96).
(vi) Compute the process time delay, θ, from (6.97) if τ_1 has been computed from (6.93).
(vii) End the algorithm if the fitting condition of relay response, $\sum_{k=1}^{N}[y(kT_s + t_{os}) - \hat{y}(kT_s + t_{os})]^2/N < \varepsilon$, is satisfied. Otherwise, go back to step (iv) by changing the initial estimation of τ_1 or monotonically varying θ in a one-dimensional search.

When an unbiased relay test is used, it follows from (6.70) that

$$y_+ \left(\frac{P_u}{2} - \theta\right) = -k_p\mu_0 + \frac{2k_p\mu_0\tau_1 E_1}{\tau_1 + \tau_2} e^{-\frac{P_u - 2\theta}{2\tau_1}} + \frac{2k_p\mu_0\tau_2 E_2}{\tau_1 + \tau_2} e^{-\frac{P_u - 2\theta}{2\tau_2}} = \varepsilon_+ \quad (6.99)$$

According to the amplitude fitting condition at the oscillation frequency as shown in (6.94), one can obtain

$$k_p = A_u \sqrt{(\tau_1^2 \omega_u^2 + 1)(\tau_2^2 \omega_u^2 + 1)} \quad (6.100)$$

Substituting (6.100) into (6.99) to eliminate k_p yields an implicit equation with respect to τ_1 and τ_2. By preestimating the process time delay (θ), this implicit equation together with (6.95) of the phase fitting condition at the oscillation frequency can be solved using a nonlinear programming method introduced in Section 4.3.2 (see (4.135)–(4.137)). A one-dimensional search of θ is therefore needed in combination with the aforementioned time domain fitting condition of the relay response to determine the optimal fitting.

Hence, the above algorithm named *Algorithm-RU-SB* for identifying an SOPDT unstable type model under an unbiased relay test can be summarized.

Algorithm-RU-SB

(i) Measure A_+ (or A_-) and P_u from the limit cycle.
(ii) Compute $G(j\omega_u)$ from (6.80).
(iii) Compute τ_1 and τ_2, using a nonlinear programming algorithm as given in (4.135)–(4.137) to solve together (6.95) and the implicit equation resulting from substituting (6.100) into (6.99), based on a preestimated time delay (θ). A one-dimensional search of θ can be implemented within a possible range as observed from the initial step response in the relay test.
(iv) Compute the proportional gain, k_p, from (6.100).
(v) End the algorithm if the fitting condition of relay response, $\sum_{k=1}^{N}[y(kT_s + t_{os}) - \hat{y}(kT_s + t_{os})]^2/N < \varepsilon$, is satisfied. Otherwise, go back to step (iii) by monotonically varying θ in a one-dimensional search.

6.4 Illustrative Examples

Three examples from existing literature are used to illustrate the effectiveness and merits of the above identification algorithms. Examples 6.1 and 6.2 are given to demonstrate the fitting accuracy of these algorithms for identifying a low-order model in terms of the exact model structure, together with measurement noise tests for demonstrating the identification robustness. Example 6.3 is given to show the effectiveness of these algorithms for identifying a higher-order unstable process. In all the relay tests, the sampling period is taken as $T_s = 0.01(s)$ for computation.

Example 6.1. Consider the first-order unstable process studied in the reference (Marchetti et al. 2001),

$$G_1 = \frac{e^{-0.4s}}{s-1}$$

Based on using two different unbiased relay tests, Marchetti et al. (2001) derived the process model, $G_m = 0.928e^{-0.392s}/(0.757s - 1)$, demonstrating enhanced identification accuracy in comparison with previous methods.

For illustration, a biased ($u_+ = 1.2$ and $u_- = -0.9$) and an unbiased ($u_+ = -u_- = 1.0$) relay test, along with $\varepsilon_+ = -\varepsilon_- = 0.1$, are performed for model identification. Correspondingly, Algorithm-RU-FA1, Algorithm-RU-FA2, Algorithm-RU-FB1, and Algorithm-RU-FB2 are used, obtaining the results shown in Table 6.1, together with the measured limit cycle data for computation. It is seen that all of these algorithms can give good accuracy.

Example 6.2. Consider the SOPDT unstable process widely studied in the references (Park et al. 1998; Majhi and Atherton 2000; Ramakrishnan and Chidambaram 2003; Vivek and Chidambaram 2005),

6.4 Illustrative Examples

Table 6.1 Identified models for three unstable processes under a relay test

Process	Relay	\multicolumn{5}{c}{Measured limit cycle data}					\multicolumn{4}{c}{Algorithm-}			
		P_+	P_-	A_+	A_-	A_u	φ_u	RU-FA1	RU-FA2	RU-FB1	RU-FB2
$\frac{e^{-0.4s}}{s-1}$	Biased	1.17	2.25	0.7413	−0.5975	0.4778	−2.8107	$\frac{1.0001e^{-0.4s}}{0.9976s-1}$	$\frac{1.0001e^{-0.4038s}}{1.0009s-1}$		
	Unbiased	1.54	1.54	0.6472	−0.6457	0.4396	−2.8514			$\frac{0.9877e^{-0.4s}}{0.9815s-1}$	$\frac{1.0016e^{-2.0s}}{10.0043s+1}$
								RU-SA		RU-SB	
$\frac{e^{-0.5s}}{(2s-1)(0.5s+1)}$	Biased	4.24	6.35	0.8569	−0.6441	0.6177	−2.8587	$\frac{1.0001e^{-0.5051s}}{(1.9975s-1)(0.4988s+1)}$		$\frac{1.0002e^{-0.5s}}{(1.9997s-1)(0.4998s+1)}$	
$\frac{e^{-0.5s}}{(5s-1)(2s+1)(0.5s+1)}$	Biased	13.76	18.88	0.8698	−0.7199	0.6694	−2.9359	$\frac{1.0001e^{-0.9422s}}{(4.9996s-1)(2.07s+1)}$		$\frac{1.0002e^{-0.94s}}{(4.9997s-1)(2.0696s+1)}$	

$$G_2 = \frac{e^{-0.5s}}{(2s-1)(0.5s+1)}$$

Based on an unbiased relay test, Vivek and Chidambaram (2005) derived an FOPDT model, $G_{m-1} = 0.7534e^{-1.0412s}/(2.1642s - 1)$. Based on a biased relay test, Park et al. (1998) derived an FOPDT model, $G_m = 1.002e^{-1.067s}/(2.347s-1)$, and Ramakrishnan and Chidambaram (2003) gave an SOPDT model, $G_{m-2} = e^{-0.52s}/(1.9999s - 1)(0.4837s + 1)$. Using a P-type controller for closed-loop stabilization to perform an unbiased relay test, Majhi and Atherton (2000) derived an FOPDT model, $G_m = 1.0e^{-1.061s}/(2.875s - 1)$, with a prior knowledge of the proportional gain (k_p).

For illustration, a biased relay test using $u_+ = -u_- = 1.0$, $\varepsilon_+ = 0.1$, and $\varepsilon_- = -0.2$ is performed here. Correspondingly, Algorithm-RU-SA can give almost the exact process model shown in Table 6.1, based on the measured limit cycle data for computation. For comparison, Algorithm-RU-FA1 is used to obtain an FOPDT model, $G_m = 1.0001e^{-1.08s}/(2.9336s - 1)$, and Algorithm-RU-FA2 gives an FOPDT model, $G_m = 1.0001e^{-1.0538s}/(2.1279s - 1)$. The Nyquist plots of the FOPDT models obtained from these two algorithms, Park et al. (1998) and Majhi and Atherton (2000), are shown in Fig. 6.3. It can be seen that apparently improved fitting is captured by Algorithm-RU-FA2 over the low-frequency range, owing to the use of the precise fitting condition of the process response at the oscillation frequency, i.e., $(-0.6449, -j0.1912)$, as shown in Fig. 6.3. With a lower computation effort, the FOPDT model obtained from Algorithm-RU-FA1 gives an inferior Nyquist fitting compared to that of Algorithm-RU-FA2, but is still comparable with that of Majhi and Atherton (2000).

If an unbiased relay test using $u_+ = -u_- = 1.0$ and $\varepsilon_+ = -\varepsilon_- = 0.1$ is performed, Algorithm-RU-SB can give almost the exact process model as shown in Table 6.1. Note that Algorithm-RU-FB1 can be used to obtain an FOPDT model, $G_m = 1.0097e^{-1.36s}/(2.74s - 1)$, while Algorithm-RU-FB2 gives an FOPDT model, $G_m = 0.9936e^{-1.03s}/(2.1703s - 1)$. Compared to the FOPDT model obtained by Vivek and Chidambaram (2005), both Algorithm-RU-FB1 and Algorithm-RU-FB2 give evidently improved identification of the proportional gain (k_p), and it can easily be verified through the Nyquist plot that Algorithm-RU-FB2 gives the best fitting owing to the use of the precise process response information at the oscillation frequency.

Note that Vivek and Chidambaram (2005) gave an FOPDT model, $G_{m-1} = 0.5332e^{-0.8807s}/(1.3524s - 1)$, in the case where a random noise with zero mean and the standard deviation $\sigma_N^2 = 0.01$ is added to the process output measurement, which is then used for relay feedback control. It can be verified that the noise causes NSR = 1.6%. Based on averaging 10 steady oscillation periods to determine the limit cycle data, Algorithm-RU-FA1 and Algorithm-RU-FA2 give the FOPDT models, $G_m = 0.9992e^{-1.806s}/(4.0701s - 1)$ and $G_m = 0.9992e^{-1.047s}/(2.1449s - 1)$, respectively. The identification errors resulting from Algorithm-RU-SA are listed in Table 6.2, which demonstrate good identification robustness.

6.4 Illustrative Examples

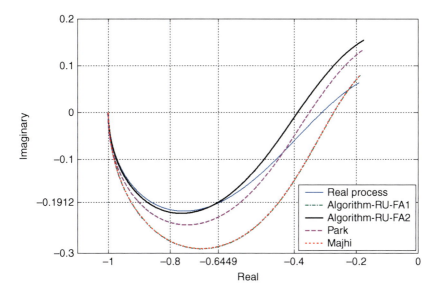

Fig. 6.3 The Nyquist fitting of identified FOPDT models for Example 6.2

To further demonstrate the identification robustness, assume that the noise level is increased to NSR = 10%, it can be seen from Table 6.2 that noticeable errors are caused in measuring the positive and negative magnitudes of the limit cycle, but the computation of the process response at the oscillation frequency is slightly affected owing to the use of all the measured data in the 10 oscillation periods. Accordingly, only the stable time constant (τ_2) of the process is notably underestimated. To enhance the identification robustness, a third-order Butterworth filter with a cutoff frequency, $f_c = 0.5$(HZ), according to the guideline given in (4.3) in Section 4.1, is used to recover the corrupted limit cycle data and also for relay feedback control. The corresponding results listed in Table 6.2 demonstrate that significantly improved accuracy for the measured limit cycle data and model identification can thus be obtained. For illustration, the recovery effect for a noise level of NSR = 20% is shown in Fig. 6.4, using the above filter for off-line denoising in both the forward and reverse directions. It is seen that the corrupted limit cycle has been well-recovered against the severe noise level, corresponding to the small identification errors shown in Table 6.2.

Example 6.3. Consider the high-order unstable process studied in the reference (Majhi 2007),

$$G_3 = \frac{e^{-0.5s}}{(5s-1)(2s+1)(0.5s+1)}$$

Table 6.2 Identification errors for Example 6.3 under different measurement noise levels

NSR	Denoising	% Error in the measured data				% Error in the model parameters			
		A_+	A_-	A_u	φ_u	k_p	τ_1	τ_2	θ
1.6%	Averaging	1.49	−2.6	−0.96	−0.05	−0.08	0.21	−2.44	3.52
10%	Averaging	20.35	−26.19	1.39	0.47	−0.02	−0.82	−11.8	5.94
10%	Filtering	1.96	−0.97	0.58	0.41	−0.01	0.76	−0.78	1.94
20%	Off-line filtering	2.21	−1.86	1.34	0.58	0.02	0.87	−6.04	4.36

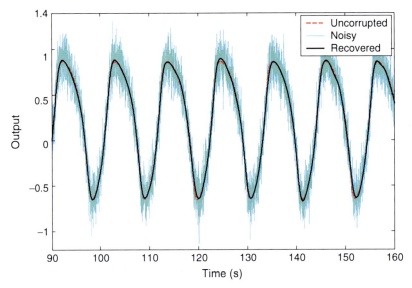

Fig. 6.4 Recovered limit cycle for Example 6.2 subject to NSR = 20%

Based on an unbiased relay test, Majhi (2007) derived an SOPDT model, $G_m = 1.001e^{-0.939s}/(10.354s^2 + 2.932s - 1)$ in terms of a state-space description of the relay response.

For illustration, a biased relay test ($u_+ = -u_- = 1.0$, $\varepsilon_+ = 0.1$, and $\varepsilon_- = -0.15$) and an unbiased relay test ($u_+ = -u_- = 1.0$, $\varepsilon_+ = -\varepsilon_- = 0.1$) are performed here for model identification. Correspondingly, Algorithm-RU-FA1 and Algorithm-RU-FA2 give the FOPDT models $G_m = 1.0001e^{-6.55s}/(13.4749s - 1)$ and $G_m = 1.0001e^{-3.2821s}/(5.7663s - 1)$, respectively. In contrast, the SOPDT models obtained using Algorithm-RU-SA and Algorithm-RU-SB are listed in Table 6.1. It is seen that Algorithm-RU-SA and Algorithm-RU-SB result in the very similar SOPDT models. The Nyquist plots of these models are shown in Fig. 6.5. It is seen that both SOPDT models derived from Algorithm-RU-SA and Majhi (2007) obtain very close fitting in almost all the frequency range. Compared to Algorithm-RU-FA1, Algorithm-RU-FA2 gives better fitting over the low-frequency range, owing to that the model response coincides with the process at the oscillation frequency, i.e., $(-0.6553, -j0.1367)$.

6.5 Summary

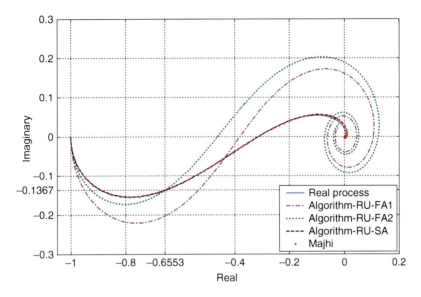

Fig. 6.5 The Nyquist plots of identified models for Example 6.3

6.5 Summary

For relay feedback identification of unstable processes, the standard biased or unbiased relay test may not guarantee steady oscillation for various unstable processes in practice, especially for unstable processes with large time delay. A limiting condition to forming steady oscillation is revealed for reference, by analytically deriving the relay response expressions of FOPDT and SOPDT unstable type models (Liu and Gao 2008a, b).

For the use of a biased relay test, Algorithm-RU-FA1 and Algorithm-RU-FA2 have been developed for obtaining an FOPDT unstable type model, while Algorithm-RU-SA has been given for obtaining an SOPDT unstable type model. For the use of an unbiased relay test, Algorithm-RU-FB1 and Algorithm-RU-FB2 have been developed for obtaining an FOPDT unstable type model, while Algorithm-RU-SB has been given for obtaining an SOPDT unstable type model.

As illustrated by three examples from the existing references, all of these algorithms can give good accuracy if the model structure matches the process. For identifying higher-order unstable processes, these algorithms may be chosen alternatively to meet with different requirements of identification accuracy and computational effort in practice.

The results for identifying a higher-order unstable process subject to model mismatch have demonstrated that an SOPDT model thus obtained evidently outperforms such an FOPDT model for the frequency response fitting, even very close to the process response in the entire low-frequency range of interest to control design. Therefore, advanced control performance can be expected based on using such an SOPDT model for control system design or controller tuning.

References

Ananth I, Chidambaram M (1999) Closed loop identification to transfer function model for unstable systems. J Frankl Inst 336(7):1055–1061
Cheres E (2006) Parameter estimation of an unstable system with a PID controller in a closed loop configuration. J Frankl Inst 343(2):204–209
Co T (2010) Relay stabilization and bifurcations of unstable SISO processes with time delay. IEEE Trans Autom Control 55(5):1131–1141
Forssell U, Ljung L (2000) Identification of unstable systems using output error and Box–Jenkins model structures. IEEE Trans Autom Control 45(1):137–141
Jin HP, Heung IP, Lee I-B (1998) Closed-loop on-line process identification using a proportional controller. Chem Eng Sci 53(9):1713–1724
Kaya I, Atherton DP (2001) Parameter estimation from relay autotuning with asymmetric limit cycle data. J Process Control 11:429–439
Liu T, Gao F (2008a) Alternative identification algorithms for obtaining a first-order stable/unstable process model from a single relay feedback test. Ind Eng Chem Res 47(4):1140–1149
Liu T, Gao F (2008b) Identification of integrating and unstable processes from relay feedback. Comput Chem Eng 32(12):3038–3056
Lin C, Wang QG, Lee TH (2004) Relay feedback: A complete analysis for first-order systems. Ind Eng Chem Res 43:8400–8402
Majhi S (2007) Relay based identification of processes with time delay. J Process Control 17(2):93–101
Majhi S, Atherton DP (2000) Online tuning of controllers for an unstable FOPDT process. IEE Proc Control Theory Appl 147(4):421–427
Marchetti G, Scali C, Lewin DR (2001) Identification and control of open-loop unstable processes by relay methods. Automatica 37(12):2049–2055
Padhy PK, Majhi S (2006) Relay based PI-PD design for stable and unstable FOPDT processes. Comput Chem Eng 30(5):790–796
Paraskevopoulos PN, Pasgianos GD, Arvanitis KG (2004) New tuning and identification methods for unstable first order plus dead-time. IEEE Trans Control Syst Technol 12(3):455–464
Park JH, Sung SW, Lee I-B (1998) An enhanced PID control strategy for unstable processes. Automatica 34(6):751–756
Ramakrishnan V, Chidambaram M (2003) Estimation of a SOPTD transfer function model using a single asymmetrical relay feedback test. Comput Chem Eng 27:1779–1784
Shiu SJ, Hwang SH, Li ML (1998) Automatic tuning of systems with one or two unstable poles. Chem Eng Commun 167:51–72
Sree RP, Chidambaram M (2006) Improved closed loop identification of transfer function model for unstable systems. J Frankl Inst 343:152–160
Tan KK, Wang QG, Lee TH (1998) Finite spectrum assignment control of unstable time delay processes with relay tuning. Ind Eng Chem Res 37:1351–1357
Thyagarajan T, Yu CC (2003) Improved autotuning using the shape factor from feedback. Ind Eng Chem Res 42:4425–4440
Vivek S, Chidambaram M (2005) An improved relay auto tuning of PID controllers for unstable FOPTD systems. Comput Chem Eng 29:2060–2068

Part II
Control System Design

Chapter 7
Control of Single-Input-Single-Output (SISO) Processes

7.1 Control Engineering Specifications

For control system design and engineering practice, a number of performance specifications and robust stability criteria have been developed to evaluate control system performance and stability. In industrial process control, the following widely used error criteria, time domain and frequency domain performance specifications, and robust stability criteria are introduced here for reference.

For clarity, hereafter a capital letter is generally used to denote a variable in frequency domain, while the corresponding lower-case letter denotes the variable in time domain. Exceptions will be definitely stated in the context. For simplicity, the time operator, t, in a time domain variable or function, and the Laplace operator, s, in a frequency domain variable or function, will be omitted in showing the corresponding variable or function when this does not cause any confusion.

7.1.1 Error Criteria

To evaluate the control system performance, a fundamental criterion is that there exists no steady-state output error with respect to the set-point or load disturbance, i.e.,

$$\lim_{t \to \infty} e(t) = \lim_{t \to \infty} \left[y(t) - y_{\text{sp}}(t) \right] = 0 \qquad (7.1)$$

where $y(t)$ and $y_{\text{sp}}(t)$ denotes the process output and the set-point values, respectively. Note that t is limited to a finite time length for industrial batch processes and repetitive control systems.

To further evaluate the dynamic output performance of set-point tracking or load disturbance rejection, the following error criteria are widely used in practical applications,

1. The criterion of integral-of-squared-error (ISE)

$$\text{ISE} = \int_0^\infty e^2(t)dt \quad (7.2)$$

2. The criterion of integral-of-absolute-error (IAE)

$$\text{IAE} = \int_0^\infty |e(t)|dt \quad (7.3)$$

3. The criterion of integral-of-time-weighted-squared-error (ITSE)

$$\text{ITSE} = \int_0^\infty te^2(t)dt \quad (7.4)$$

4. The criterion of integral-of-time-weighted-absolute-error (ITAE)

$$\text{ITAE} = \int_0^\infty t|e(t)|dt \quad (7.5)$$

The above error criteria have been extensively studied in the existing literature for control system design and online controller tuning (Morari and Zafiriou 1989; Seborg et al. 2004; Åström and Hägglund 2005; Skogestad and Postlethwaite 2005). Note that ISE and IAE criteria have been alternatively used for controller design to optimize the set-point tracking or load disturbance rejection. By comparison, the ITSE and ITAE criteria have been primarily used for enhancing the output response speed of set-point tracking or load disturbance rejection, but at the cost of robust stability against process uncertainties.

7.1.2 Time Domain Performance Specifications

Generally speaking, time domain performance specifications have been mainly developed in terms of a step response test to evaluate the dynamic performance of a control system. As shown in Fig. 7.1, the following specifications are widely referenced for control system design:

1. The rise time (t_r), usually defined as the time that the output response first reach 90% of its final steady-state value in response to a step change of the set-point
2. The settling time (t_{set}), usually defined as the time after which the output moves into the error band of $\pm 5\%$ with respect to its final steady-state value in response to a step change of the set-point

7.1 Control Engineering Specifications

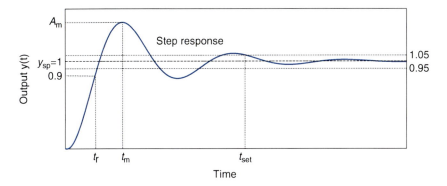

Fig. 7.1 An illustration of time domain performance specifications

Fig. 7.2 The unity feedback control structure

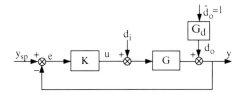

3. The overshoot (A_m), usually defined as the output peak value in response to a step change of the set-point, or alternatively, a ratio of the output peak value to the final steady-state value (A_m/y_{sp})
4. The steady-state offset (e_s), usually defined as $e_s = \lim_{t \to \infty} [y(t) - y_{sp}]$, where y_{sp} denotes the set-point value

7.1.3 Frequency Domain Performance Specifications

For the convenience of introduction, the unity feedback control structure is shown in Fig. 7.2, where G denotes the process and K is the closed-loop controller. $y_{sp}(t)$ denotes the time domain set-point value, corresponding to the Laplace transform of $Y_{sp}(s)$ in frequency domain, $u(t)$ is the controller output, corresponding to $U(s)$ in frequency domain, and $y(t)$ is the process output, corresponding to $Y(s)$ in frequency domain. $d_i(t)$ denotes load disturbance entering into the process at its input side, corresponding to $D_i(s)$ in frequency domain, and $d_o(t)$ denotes load disturbance entering into the process from its output side (through a transfer function, $G_d(s)$, in terms of a normalized input, $\hat{d}_o(t) = 1$), corresponding to $D_o(s)$ in frequency domain.

In Fig. 7.2, it can be seen that the loop transfer function is

$$H = GK \tag{7.6}$$

Fig. 7.3 The Nyquist plot of $H(j\omega)$ for a depiction of GM and PM

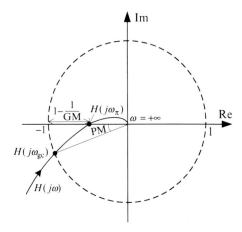

the closed-loop sensitivity function is

$$S = \frac{1}{1+H} \tag{7.7}$$

and the closed-loop complementary sensitivity function is

$$T = \frac{H}{1+H} \tag{7.8}$$

Correspondingly, the following specifications are widely referenced for control system design:

1. The phase crossover frequency (ω_π), as shown in Fig. 7.3, defined as the frequency by which the Nyquist curve of $H(j\omega)$ crosses the negative real axis between these two real points, -1 and 0, i.e.,

$$\angle H(j\omega_\pi) = -\pi \tag{7.9}$$

2. The gain crossover frequency (ω_{gc}), as shown in Fig. 7.3, defined as the frequency by which $|H(j\omega)|$ crosses the unity circle in the complex plane, i.e.,

$$\left|H\left(j\omega_{gc}\right)\right| = 1 \tag{7.10}$$

3. The gain margin (GM), as shown in Fig. 7.3, defined as

$$\text{GM} = 1/\left|H\left(j\omega_\pi\right)\right| \tag{7.11}$$

7.1 Control Engineering Specifications

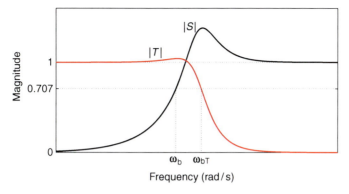

Fig. 7.4 The magnitude plots of $|S(j\omega)|$ and $|T(j\omega)|$ for a depiction of the closed-loop bandwidth

4. The phase margin (PM), as shown in Fig. 7.3, defined as

$$\text{PM} = \angle H\left(j\omega_{\text{gc}}\right) + \pi \tag{7.12}$$

5. The maximal peak of the sensitivity function (M_S), defined as

$$M_S = \max_{\omega} |S(j\omega)| \tag{7.13}$$

6. The maximal peak of the complementary sensitivity function (M_T), defined as

$$M_T = \max_{\omega} |T(j\omega)| \tag{7.14}$$

Note that M_S and M_T differ at most by unity, since $S + T = 1$. A large value of M_S (or M_T) usually indicates poor control performance (or poor robust stability). The upper bound of M_T has been commonly used for control system design in the classical M-circles and Nichols chart methods.

7. The bandwidth, usually referred to the frequency, ω_b, which is shown in Fig. 7.4, by which $|S(j\omega)|$ reaches the value of $1/\sqrt{2} = 0.707$ (≈ -3 dB) from $\omega = 0$. Alternatively, it is referred in some references to the highest frequency, ω_{BT}, by which $|T(j\omega)|$ reaches the value of $1/\sqrt{2} = 0.707$ (≈ -3 dB) from $\omega = 0$.

Note that a larger bandwidth corresponds to a smaller rise time in a step response test, owing to that a wider frequency range of the input is allowed to pass through the control system to excite the output response. However, a higher bandwidth also indicates that the closed-loop system is more sensitive to measurement noise or high-frequency disturbance. On the contrary, a smaller bandwidth will result in a slower output response but in exchange for better robust stability.

7.2 Robust Stability Criteria

In practical applications, there are ubiquitous process uncertainties, e.g., unmodeled dynamic response uncertainty, control valve stiction, and measurement sensor nonlinearity. For the convenience of analysis, different process uncertainties can be represented by the additive or multiplicative form as shown in Fig. 7.5. The process family described by an additive uncertainty can be written as

$$\Pi_A = \left\{ G_A(j\omega) : \left| G(j\omega) - \widehat{G}(j\omega) \right| \leq \bar{\Delta}_A \right\}, \quad \omega \in [0, \infty) \tag{7.15}$$

where $\widehat{G}(s)$ denotes the nominal process or an identified model, and $\bar{\Delta}_A$ indicates the upper bound of $\Delta_A(j\omega)$ for $\omega \in [0, \infty)$, i.e.,

$$|\Delta_A(j\omega)| \leq \bar{\Delta}_A, \quad \omega \in [0, \infty) \tag{7.16}$$

Thus, a member of the above family can be expressed as

$$G_A(j\omega) = \widehat{G}(j\omega) + \Delta_A(j\omega) \tag{7.17}$$

By comparison, the process family described by a multiplicative uncertainty can be written as

$$\Pi_M = \left\{ G_M(j\omega) : \left| \frac{G(j\omega) - \widehat{G}(j\omega)}{\widehat{G}(j\omega)} \right| \leq \bar{\Delta}_M \right\}, \quad \omega \in [0, \infty) \tag{7.18}$$

where $\bar{\Delta}_M$ denotes the upper bound of $\Delta_M(j\omega)$ for $\omega \in [0, \infty)$, i.e.,

$$|\Delta_M(j\omega)| \leq \bar{\Delta}_M, \quad \omega \in [0, \infty) \tag{7.19}$$

Correspondingly, a member of this family can be expressed as

$$G_M(j\omega) = \widehat{G}(j\omega) [1 + \Delta_M(j\omega)] \tag{7.20}$$

A transformation between Δ_A and Δ_M can therefore be defined as

$$\Delta_M(j\omega) = \frac{\Delta_A(j\omega)}{\widehat{G}(j\omega)} \tag{7.21}$$

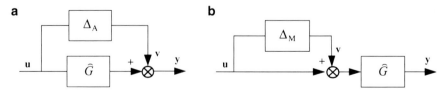

Fig. 7.5 Additive (**a**) and multiplicative (**b**) uncertainty

7.2 Robust Stability Criteria

Fig. 7.6 The Nyquist plot of robust stability conditions in terms of the multiplicative uncertainty

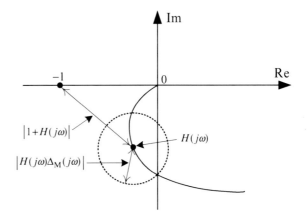

Without loss of generality, unmodeled process dynamics uncertainty together with other sources of system uncertainty can be lumped into a single unified multiplicative uncertainty for robust stability analysis in practical applications (Skogestad and Postlethwaite 2005). Correspondingly, the Nyquist stability condition is given in the following theorem:

Nyquist Stability Theorem. Assuming (by design) stability of the nominal closed-loop system (i.e., $\Delta_M = 0$) as shown in Fig. 7.2, the closed-loop system holds robust stability against the multiplicative uncertainty described by Δ_M as shown in Fig. 7.5b, if and only if the loop transfer function, $H_M = G_M K = \hat{G}K(1 + \Delta_M)$, does not encircle the point $(-1, j0)$ in the complex plane, $\forall G_M \in \Pi_M$.

Figure 7.6 shows a graphical interpretation of the above Nyquist stability condition, where $H(j\omega) = \hat{G}(j\omega)K(j\omega)$ is the nominal loop transfer function. It can be seen from Fig. 7.6 that the stability condition in the above theorem is equivalent to

$$|H(j\omega)\Delta_M(j\omega)| < |1 + H(j\omega)|, \quad \forall \omega \in [0, \infty) \tag{7.22}$$

or

$$|T(j\omega)\Delta_M(j\omega)| < 1, \quad \forall \omega \in [0, \infty) \tag{7.23}$$

where $T(j\omega) = H(j\omega)/[1 + H(j\omega)]$ is the nominal closed-loop transfer function.

Following an algebraic derivation (Skogestad and Postlethwaite 2005), equivalent robust stability conditions can be derived, respectively, as

$$|1 + H_M(j\omega)| \neq 0, \quad \forall G_M \in \Pi_M \text{ and } \omega \in [0, \infty) \tag{7.24}$$

$$|1 + H_M(j\omega)| > 0, \quad \forall G_M \in \Pi_M \text{ and } \omega \in [0, \infty) \tag{7.25}$$

Fig. 7.7 The $M - \Delta$ structure for robust stability analysis

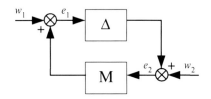

According to the robust control theory (Zhou et al. 1996), the $M - \Delta$ structure can in general be adopted for robust stability analysis, which is shown in Fig. 7.7, where M denotes a stable $p \times q$ transfer function matrix, $\Delta \in RH_\infty$ describes the process uncertainty, e.g., Δ_A or Δ_M shown in Fig. 7.5, w_1 and w_2 denote external signals or disturbances added to the input(s) and output(s) of M. Correspondingly, the following robust stability theorem is given not only for SISO systems but also for MIMO systems:

Small Gain Theorem. Assuming $M \in RH_\infty$ and letting $\gamma > 0$, the interconnected system shown in Fig. 7.7 holds robust stability for all $\Delta \in RH_\infty$ with (a) $\|\Delta\|_\infty \leq 1/\gamma$ if and only if $\|M(s)\|_\infty < \gamma$ (b) $\|\Delta\|_\infty < 1/\gamma$ if and only if $\|M(s)\|_\infty \leq \gamma$.

By letting $\gamma = 1$, it can be seen by comparison with (7.23) that the small gain theorem is essentially equivalent to the above Nyquist stability theorem for SISO systems.

It should be noted that, based on the above robust stability theorems, a number of robust stability conditions have been developed by assuming specific model structures or uncertainty forms in the existing literature. Those not being used in this monograph are omitted.

As far as robust control performance is concerned, the following criterion is commonly used for assessment (Morari and Zafiriou 1989; Zhou et al. 1996; Skogestad and Postlethwaite 2005)

$$|T(j\omega)\Delta_M(j\omega)| + |w_s S(j\omega)| < 1, \quad \forall \omega \in [0, \infty) \tag{7.26}$$

$$\Leftrightarrow \sup_{\omega \in [0,\infty)} [|T(j\omega)\Delta_M(j\omega)| + |w_s S(j\omega)|] < 1 \tag{7.27}$$

where w_s is a weighting function that should be chosen in terms of the set-point type or the frequency range considered for control. For instance, $w_s = 1/s$ may be taken for a step change of the set-point.

7.3 Review of the Internal Model Control (IMC) Design

The IMC design theory (Morari and Zafiriou 1989) has been widely recognized and applied in control system design (Braatz 1995; Seborg et al. 2004; Skogestad

7.3 Review of the Internal Model Control (IMC) Design

Fig. 7.8 The internal model control structure

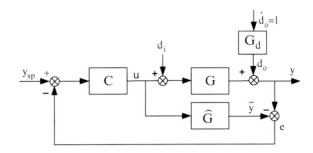

and Postlethwaite 2005). Figure 7.8 shows the standard IMC structure, where G denotes the process, $\widehat{G}$ the process transfer function model in frequency domain, C the controller, and G_d load transfer function corresponding to a normalized input, $\hat{d}_o(t) = 1$.

Comparing Fig. 7.2 with Fig. 7.8, it can be seen that the IMC structure is equivalent to the unity feedback control structure if the controllers satisfy the following relationship,

$$K = \frac{C}{1 - \widehat{G}C} \qquad (7.28)$$

For a stable process, the IMC controller design can be summarized as

1. Factorize the process model ($\widehat{G}$) into an all-pass portion ($\widehat{G}_{ap}$) and a minimum-phase (MP) portion ($\widehat{G}_{mp}$).
2. Choose a low-pass filter (F) such that $F/\widehat{G}_{mp}$ holds bi-properness for implementation. For instance, the type I filter is prescribed for a step-type set-point, i.e.,

$$F = \frac{1}{(\lambda s + 1)^{n_{mp}}} \qquad (7.29)$$

where λ is an adjustable parameter and $n_{mp} = \deg\{\widehat{G}_{mp}\}$. The type II filter is for a ramp-type set-point, i.e.,

$$F = \frac{n_{mp}\lambda s + 1}{(\lambda s + 1)^{n_{mp}}} \qquad (7.30)$$

3. The IMC controller is correspondingly obtained as

$$C = \frac{F}{\widehat{G}_{mp}} \qquad (7.31)$$

which corresponds to the nominal closed-loop transfer function,

$$T = \widehat{G}_{ap} F \qquad (7.32)$$

Note that the above IMC design can achieve the H_2 optimal performance objective (i.e., ISE), $\min \|e\|_2$, for tracking the set-point or rejecting load disturbance from the process output side (denoted as d_o in Fig. 7.8). Another important merit is that there exists only a single adjustable parameter (λ) in the controller, corresponding to the sole time constant of the nominal closed-loop transfer function in (7.32). This parameter can be monotonically tuned to meet a good trade-off between the control performance and robust stability of the closed-loop system.

Example 7.1. Consider a second-order process with time delay,

$$G = \frac{1}{4s^2 + 2s + 1} e^{-3s} \tag{7.33}$$

According to the above IMC method for the nominal case ($G = \hat{G}$), there are

$$\hat{G}_{ap} = e^{-3s} \tag{7.34}$$

$$\hat{G}_{mp} = \frac{1}{4s^2 + 2s + 1} \tag{7.35}$$

Assume that the set-point is of step-type, it follows from (7.29) that

$$F = \frac{1}{(\lambda s + 1)^2} \tag{7.36}$$

Substituting (7.35) and (7.36) into (7.31), the IMC controller can be obtained as

$$C = \frac{4s^2 + 2s + 1}{(\lambda s + 1)^2} \tag{7.37}$$

Accordingly, the nominal closed-loop transfer function can be derived from (7.32) as

$$T = \frac{1}{(\lambda s + 1)^2} e^{-3s} \tag{7.38}$$

The corresponding time domain output response to a step change of the set-point can be analytically derived from (7.38) as

$$y(t) = \begin{cases} 0 & t \leq 3 \\ 1 - \left(1 + \frac{t-3}{\lambda}\right) e^{-\frac{t-3}{\lambda}} & t > 3 \end{cases} \tag{7.39}$$

which indicates that there is no overshoot in the nominal set-point response and its time domain specification can be quantitatively tuned. For instance, the time domain specification, rise time, can be analytically derived from (7.39) as $t_r = 3.8897\lambda + 3$.

7.3 Review of the Internal Model Control (IMC) Design

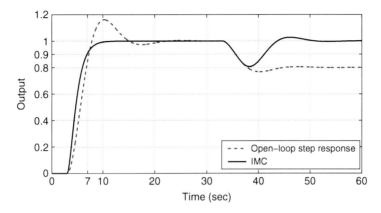

Fig. 7.9 A step response for Example 7.1 using the IMC method

Assume that the rise time is required as $t_r \leq 7(s)$ for implementation. It can be determined from the above formula that the control parameter should be limited to $\lambda \leq 1.0284$. Taking $\lambda = 1$ and adding a unity step change of the set-point and a step load disturbance with a magnitude of 0.2 to the process input at $t = 30(s)$, the output response is shown in Fig. 7.9. It is seen that the IMC method results in smooth output response with no overshoot, while satisfying the required rise time ($y(t = 7) \geq 0.9$). Moreover, there is no steady output offset in the presence of the load disturbance.

It should be noted that the standard IMC structure cannot be directly used for the control of an integrating or unstable process (Morari and Zafiriou 1989). For such a process, a transformation to the unity feedback control structure is required for implementation, subject to the asymptotic constraints associated with controller parameterization for holding the internal stability of the closed-loop system.

The internal stability of a closed-loop system is in theory defined as that the transfer functions relating all possible inputs to the outputs of the system are maintained to be stable (Zhou et al. 1996). In industrial process control, the standard IMC structure shown in Fig. 7.8 holds internal stability in the nominal case ($G = \hat{G}$) if the following transfer function matrix is maintained to be stable,

$$\begin{bmatrix} y \\ u \\ \hat{y} \end{bmatrix} = \begin{bmatrix} GC & G & G(1-GC) & 1-GC \\ C & 0 & -GC & -C \\ GC & G & -G^2C & -GC \end{bmatrix} \begin{bmatrix} y_{sp} \\ \tilde{u} \\ d_i \\ d_o \end{bmatrix} \quad (7.40)$$

where $\tilde{u}$ denotes an external signal that enters into the controller output and then feeds through both the process (G) and its model ($\hat{G}$).

It is seen from (7.40) that the second column elements of the transfer matrix remain uncontrollable, which implies that the standard IMC structure cannot hold internal stability for an integrating or unstable process when there exists an external disturbance like $\tilde{u}$.

In contrast, the unity feedback control structure shown in Fig. 7.2 holds internal stability in the nominal case if the following transfer function matrix is maintained to be stable,

$$\begin{bmatrix} y \\ u \end{bmatrix} = \begin{bmatrix} \dfrac{GK}{1+GK} & \dfrac{G}{1+GK} & \dfrac{1}{1+GK} \\ \dfrac{K}{1+GK} & \dfrac{-GK}{1+GK} & \dfrac{-K}{1+GK} \end{bmatrix} \begin{bmatrix} y_{sp} \\ d_i \\ d_o \end{bmatrix} \quad (7.41)$$

It is seen from (7.41) that the controller is included in all elements of the transfer matrix and, therefore, can hold the internal stability of the closed-loop system if it is designed to maintain the stability of each element in the transfer matrix.

For integrating and unstable processes, the following constraint is required for controller design in the unity feedback control structure to hold the internal stability (Morari and Zafiriou 1989),

$$\lim_{s \to p_i} (1 - T) = 0, \quad i = 1, 2, \ldots, m \quad (7.42)$$

where p_i ($i = 1, 2, \ldots, m$) are closed-RHP poles of such a process.

7.4 Enhanced IMC Design for Load Disturbance Rejection

Load disturbance rejection is one of the most important issues in the context of industrial process control. The standard IMC method can be used for controller tuning, which leads to the H_2 optimal performance, i.e., the minimization of ISE, for tracking the set-point or rejecting a step-type load disturbance entering into the process from its output side (denoted by d_o in Fig. 7.8). In engineering practice, it is often encountered that load disturbance enters into the process from its input side, denoted by d_i in Fig. 7.8. Moreover, in many industrial applications, load disturbance ($\hat{d}_o$) with a transfer function (G_d) can be transformed or lumped into d_i to deal with (Shinskey 1996; Goodwin et al. 2001; Seborg et al. 2004). It can be seen from Fig. 7.8 that the transfer function relating d_i to y is

$$H_{di} = G(1 - T) \quad (7.43)$$

where T is the closed-loop transfer function (i.e., the complementary sensitivity function).

It is seen from (7.43) that the load disturbance response is unavoidably subject to the influence of the process time constant(s), if the standard IMC method is used to only optimize T for tracking the set-point or rejecting the output side disturbance (d_o), which corresponds to $H_{do} = 1 - T$. For a slow process with a large time constant, the recovery trajectory of the disturbance response is subject to "a long tail" (i.e., sluggish load disturbance suppression), as early reported by Horn et al. (1996).

7.4 Enhanced IMC Design for Load Disturbance Rejection

To reduce the influence arising from the process time constant(s) to the load disturbance response, it is of course ideal to eliminate the corresponding pole(s) from the characteristic equation of (7.43). It is thus expected that $1 - T$ (i.e., S), rather than T, has the corresponding zero(s) to cancel the pole(s) of G, such that the characteristic equation is governed only by the time constant of T (i.e., an adjustable parameter in the IMC filter). Using an idea of eliminating the slowest pole of G, Horn et al. (1996) suggested an improved IMC filter design for some delay-free processes and first-order processes with small time delay that can be properly approximated by the first-order Padé expansion. Further extended IMC-based PID tuning methods were presented by Shamsuzzoha and Lee (2007) for improving closed-loop disturbance rejection performance.

For an industrial process modeled with time delay, the numerator in $1 - T$ is unavoidably involved with time delay factor(s), so it cannot be analytically factorized to make exact zero-pole cancelation with the denominator of G. The following asymptotic constraint(s) is, therefore, proposed to realize the above idea,

$$\lim_{s \to -p_i} (1 - T) = 0, \quad i = 1, 2, \ldots, m \tag{7.44}$$

where p_i is the reciprocal of the process time constant(s) and m is the corresponding number.

Based on the widely used low-order process models of FOPDT and SOPDT, a modified IMC design is detailed in the following two subsections to improve load disturbance rejection for stable processes.

7.4.1 For FOPDT Stable Processes

Given an FOPDT process model, $\widehat{G}(s) = k_p e^{-\theta s}/(\tau_p s + 1)$, the conventional IMC filter should be configured as

$$F_{\text{IMC}-1} = \frac{1}{\lambda s + 1} \tag{7.45}$$

where λ is a user-specified time constant, i.e., an adjustable parameter. Correspondingly, the nominal closed-loop complementary sensitivity function is obtained as

$$T_{\text{IMC}-1} = \frac{e^{-\theta s}}{\lambda s + 1} \tag{7.46}$$

To improve the load disturbance response, the conventional IMC filter in (7.45) is here rectified as

$$F_{\text{RIMC}-1} = \frac{\alpha s + 1}{(\lambda_f s + 1)^2} \tag{7.47}$$

where α is an additional parameter used to satisfy the following asymptotic constraint

$$\lim_{s \to -1/\tau_p} (1 - T) = 0 \tag{7.48}$$

Accordingly, it follows that

$$T_{\text{RIMC}-1} = \frac{(\alpha s + 1)e^{-\theta s}}{(\lambda_f s + 1)^2} \tag{7.49}$$

Substituting (7.49) into (7.48) yields

$$\alpha = \tau_p \left[1 - \left(\frac{\lambda_f}{\tau_p} - 1 \right)^2 e^{-\frac{\theta}{\tau_p}} \right] \tag{7.50}$$

It is thus seen that α is a function of λ_f. So, there is essentially a single adjustable parameter, λ_f, in the proposed IMC filter.

Using the nominal closed-loop relationship, $T = \widehat{G}C$, the modified IMC controller is obtained as

$$C_{\text{RIMC}-1} = \frac{(\alpha s + 1)(\tau_p s + 1)}{k_p (\lambda_f s + 1)^2} \tag{7.51}$$

Remark 7.1. Note that $C_{\text{RIMC}-1} = 1/k_p$ when λ_f is tuned as τ_p (or τ_d of G_d), for which $T_{\text{RIMC}-1}$ becomes the same as G (or G_d). When λ_f is tuned larger than τ_p (or τ_d), the load disturbance response will be slower than G (or G_d). Hence, it is suggested to tune $\lambda_f < \tau_p$ for load disturbance rejection, unless it is intentionally violated to obtain sustainable closed-loop stability for accommodating process uncertainties. ◇

By substituting the FOPDT process model and (7.49) into (7.43) and taking an inverse Laplace transform, the time domain output response to a step change of d_i can be derived as

$$y_{d_i}(t) = \begin{cases} 0, & t \leq \theta; \\ k_p(1 - e^{-\frac{t-\theta}{\tau_p}}), & \theta < t \leq 2\theta; \\ k_p \left[1 - e^{-\frac{\theta}{\tau_p}} + \frac{1 + (\frac{\lambda_f}{\tau_p} - 1)e^{-\frac{\theta}{\tau_p}}}{\lambda_f}(t - 2\theta) \right] e^{-\frac{t-2\theta}{\lambda_f}}, & t > 2\theta. \end{cases} \tag{7.52}$$

Note that $y_{d_i}(t)$ increases monotonically in the time interval $t \in (0, 2\theta]$ and $dy_{d_i}(t)/dt \big|_{t=2\theta} \neq 0$. The output peak value should be reached in the time interval $(2\theta, \infty)$. The time to reach the disturbance response peak (DP) can be derived by

7.4 Enhanced IMC Design for Load Disturbance Rejection

solving $dy_{d_i}(t)/dt = 0$ for the final phase as

$$t_{DP} = 2\theta + \frac{\lambda_f^2 e^{-\frac{\theta}{\tau_p}}}{\tau_p + (\lambda_f - \tau_p) e^{-\frac{\theta}{\tau_p}}} \qquad (7.53)$$

Substituting (7.53) into (7.52) yields

$$y_{d_i}(t_{DP}) = k_p \left[1 + \left(\frac{\lambda_f}{\tau_p} - 1\right) e^{-\frac{\theta}{\tau_p}}\right] e^{\frac{-\frac{\lambda_f}{\tau_p} e^{-\frac{\theta}{\tau_p}}}{1+\left(\frac{\lambda_f}{\tau_p}-1\right)e^{-\frac{\theta}{\tau_p}}}} \qquad (7.54)$$

To make clear the tuning relationship between λ_f and DP, the following proposition is given.

Proposition 7.1. *For an FOPDT process, the DP of load disturbance response (y_{di}) increases monotonically with respect to λ_f.*

Proof. Letting $A = 1 + \left(\frac{\lambda_f}{\tau_p} - 1\right) e^{-\frac{\theta}{\tau_p}}$ and $B = 1 - e^{-\frac{\theta}{\tau_p}}$, the first derivative of $y_{d_i}(t_{DP})$ shown in (7.54) can be derived as

$$\frac{dy_{d_i}(t_{DP})}{d\lambda_f} = \frac{k_p}{\tau_p} e^{\frac{B}{A} - \frac{\theta}{\tau_p} - 1} \left(1 - \frac{B}{A}\right)$$

Owing to the fact that $A > B > 0$, there exists $dy_{d_i}(t_{DP})/d\lambda_f > 0$. Hence, the conclusion in Proposition 7.1 follows. □

Note that the above analysis of DP can be utilized to assess the maximal output deviation from the set-point in the presence of load disturbance in practical applications.

Define the recovery time, t_{re}, as the time from the moment that a step change of load disturbance is added to the process to the moment that the load disturbance response recovers into the error band of $\pm 5\%$ that is usually specified for set-point tracking in practice. It follows from (7.52) that

$$1 - e^{-\frac{\theta}{\tau_p}} + \frac{t_{re} - 2\theta}{\lambda_f} \left[1 + \left(\frac{\lambda_f}{\tau_p} - 1\right) e^{-\frac{\theta}{\tau_p}}\right] = 0.05 e^{\frac{t_{re} - 2\theta}{\lambda_f}} \qquad (7.55)$$

Obviously, (7.55) is a transcendental equation, which cannot be solved analytically. Numerical computation based on the Newton–Raphson algorithm is therefore explored to disclose the quantitative tuning relationship between λ_f and t_{re}. By sweeping over the ratio ranges of $\lambda_f/\tau_p \in [0.1, 2.0]$ and $\theta/\tau_p \in [0.1, 2.0]$, numerical results based on the scaled recovery time, t_{re}/k_p, are plotted in Fig. 7.10. Note that $\tau_p = 1$ is assumed to obtain the scaled recovery time shown in Fig. 7.10. Given $\tau_p \neq 1$, the recovery time can be graphically read as $t_{re} = \tau_p t_{re}|_{\tau_p=1}$, in view of that t_{re}/τ_p and $t_{re}|_{\tau_p=1}$ correspond to the identical solution of λ_f/τ_p in (7.55).

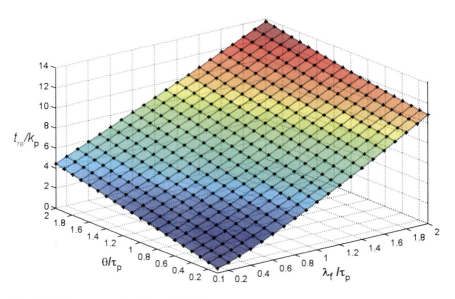

Fig. 7.10 The recovery time for an FOPDT process

Remark 7.2. Though the mathematical Lambert W function, defined as the multi-inverse function for $\omega \mapsto \omega e^{\omega}$, can be employed to solve (7.55) to give an "analytical" formula, $t_{re} = 2\theta + \lambda_f[-W(-e^{-\frac{A}{B}}/B) - A/B]$, where $A = 20(1 - e^{-\frac{\theta}{\tau_p}})$ and $B = 20[1 + (\lambda_f/\tau_p - 1)e^{-\frac{\theta}{\tau_p}}]$, it may lead to an incorrect solution because the principle branch value of the Lambert W function is generally given by commercial symbolic software packages (Hwang and Cheng 2005). Note that in the above formula, $-W(-e^{-\frac{A}{B}}/B) - A/B$ should be positive since $t_{re} > 2\theta$, but this formula may give a negative value because $W(-e^{-\frac{A}{B}}/B) > W(-Ae^{-\frac{A}{B}}/B) = -A/B$ is obtained from the principle branch of the Lambert W function. According to the numerical results shown in Fig. 7.10, the secondary real branch of the Lambert W function should be used to compute t_{re} from the above formula. ◇

According to the standard $M - \Delta$ structure for robust stability analysis (Zhou et al. 1996), the transfer function relating the input to the output of the process multiplicative uncertainty block can be derived in terms of the IMC structure shown in Fig. 7.8, which is exactly equivalent to the closed-loop complementary sensitivity function, T. Hence, it follows from the small gain theorem that the perturbed closed-loop system described with the process multiplicative uncertainty holds robust stability, if and only if

$$\|T\|_{\infty} < \frac{1}{\|\Delta\|_{\infty}} \tag{7.56}$$

where $\Delta = (G - \widehat{G})/\widehat{G}$ denotes the process multiplicative uncertainty.

7.4 Enhanced IMC Design for Load Disturbance Rejection

Note that for a SISO system, there exist $\|T\|_\infty = \sup(|T(j\omega)|)$ and $\|\Delta\|_\infty = \sup(|\Delta(j\omega)|)$, $\forall \omega \in [0, +\infty)$. Denote $|\Delta|_m = \sup(|\Delta(j\omega)|)$ hereafter for the convenience of analysis.

It follows from (7.49) that

$$|T(j\omega)| = \frac{\sqrt{\alpha^2\omega^2 + 1}}{\lambda_f^2\omega^2 + 1} \qquad (7.57)$$

The first derivative can be obtained as

$$\frac{d|T(j\omega)|}{d\omega} = \frac{\omega(\alpha^2 - 2\lambda_f^2 - \alpha^2\lambda_f^2\omega^2)}{(\lambda_f^2\omega^2 + 1)^2 \sqrt{\alpha^2\omega^2 + 1}} \qquad (7.58)$$

It can be verified by solving $d|T(j\omega)|/d\omega = 0$ that $\omega = 0$ is the unique extreme point to reach $\sup(|T(j\omega)|)$ for $\lambda_f \geq \alpha/\sqrt{2}$. That is, $\sup(|T(j\omega)|) = |T(0)| = 1$.

For $0 < \lambda_f < \alpha/\sqrt{2}$, there are two extreme points $\omega_1 = 0$ and $\omega_2 = \sqrt{\alpha^2 - 2\lambda_f^2}/(\alpha\lambda_f)$.

Substituting the latter into (7.57) yields

$$|T(j\omega_2)| = \frac{\alpha^2}{2\lambda_f\sqrt{\alpha^2 - 2\lambda_f^2}}$$

It can be concluded from $\alpha^2 - 2\lambda_f^2 > 0$ that $|T(j\omega_2)| > 1$. Therefore, it follows that

$$\sup(|T(j\omega)|) = \begin{cases} 1, & \lambda_f \geq \alpha/\sqrt{2}; \\ \dfrac{\alpha^2}{2\lambda_f\sqrt{\alpha^2 - 2\lambda_f^2}}, & 0 < \lambda_f < \alpha/\sqrt{2}. \end{cases} \qquad (7.59)$$

To make clear the robust stability constraint to tuning λ_f, the following proposition is given:

Proposition 7.2. *There is only a positive real root,* $\lambda_f = \frac{\tau_p}{2}[2 - \frac{1}{\varepsilon}e^{\frac{\theta}{\tau_p}} + \sqrt{(2 - \frac{1}{\varepsilon}e^{\frac{\theta}{\tau_p}})^2 + 4(e^{\frac{\theta}{\tau_p}} - 1)}]$, *for the equation of* $\lambda_f = \varepsilon\tau_p[1 - (\frac{\lambda_f}{\tau_p} - 1)^2 e^{-\frac{\theta}{\tau_p}}]$, $\forall \varepsilon > 0$, *which increases monotonically with respect to* ε.

Proof. Substituting $\alpha = \tau_p[1 - (\frac{\lambda_f}{\tau_p} - 1)^2 e^{-\frac{\theta}{\tau_p}}]$ into $\lambda_f = \varepsilon\alpha$ yields

$$\left(\frac{\lambda_f}{\tau_p} - 1\right)^2 + \frac{1}{\varepsilon}e^{\frac{\theta}{\tau_p}}\frac{\lambda_f}{\tau_p} - e^{\frac{\theta}{\tau_p}} = 0 \qquad (7.60)$$

Then it can be derived from (7.60) that there exist two real roots

$$\lambda_{f-1,2} = \frac{\tau_p}{2}\left[2 - \frac{1}{\varepsilon}e^{\frac{\theta}{\tau_p}} \pm \sqrt{\left(2 - \frac{1}{\varepsilon}e^{\frac{\theta}{\tau_p}}\right)^2 + 4\left(e^{\frac{\theta}{\tau_p}} - 1\right)}\right] > 0$$

In view of that λ_f is an adjustable parameter of C, the negative real root should be disregarded.

Define a function with respect to ε, $f(\varepsilon) = 2 - \frac{1}{\varepsilon}e^{\frac{\theta}{\tau_p}} + \sqrt{\left(2 - \frac{1}{\varepsilon}e^{\frac{\theta}{\tau_p}}\right)^2 + 4\left(e^{\frac{\theta}{\tau_p}} - 1\right)}$, the first derivative of $f(\varepsilon)$ can be derived as

$$\frac{df(\varepsilon)}{d\varepsilon} = \frac{1}{\varepsilon^2}e^{\frac{\theta}{\tau_p}}\left[1 + \frac{2 - \frac{1}{\varepsilon}e^{\frac{\theta}{\tau_p}}}{\sqrt{\left(2 - \frac{1}{\varepsilon}e^{\frac{\theta}{\tau_p}}\right)^2 + 4\left(e^{\frac{\theta}{\tau_p}} - 1\right)}}\right] > 0$$

Hence, $f(\varepsilon)$ increases monotonically with respect to ε, and so is for λ_f with respect to ε. This completes the proof. □

For $\lambda_f \geq \alpha/\sqrt{2}$, it can be solved using Proposition 7.2 that

$$\lambda_f \geq \frac{\tau}{2}\left[2 - \sqrt{2}e^{\frac{\theta}{\tau}} + \sqrt{\left(2 - \sqrt{2}e^{\frac{\theta}{\tau}}\right)^2 + 4\left(e^{\frac{\theta}{\tau}} - 1\right)}\right] \quad (7.61)$$

Accordingly, it can be seen from (7.59) that once λ_f is increased into this range, the upper bound of $|T(j\omega)|$ is fixed as unity, such that $|\Delta|_m < 1$ must be required for the closed-loop stability, regardless of the tuning of λ_f in this range. In other words, tuning λ_f in this range will not affect the permissible upper bound of $|\Delta|_m$.

For $0 < \lambda_f < \alpha/\sqrt{2}$, it follows from Proposition 7.2 that

$$\lambda_f < \frac{\tau}{2}\left[2 - \sqrt{2}e^{\frac{\theta}{\tau_p}} + \sqrt{\left(2 - \sqrt{2}e^{\frac{\theta}{\tau_p}}\right)^2 + 4\left(e^{\frac{\theta}{\tau_p}} - 1\right)}\right] \quad (7.62)$$

Combining (7.56) with (7.59), the robust stability constraint can be determined as

$$\frac{\alpha^2}{2\lambda_f\sqrt{\alpha^2 - 2\lambda_f^2}} < \frac{1}{|\Delta|_m} \quad (7.63)$$

It follows from solving (7.63) that

$$\sqrt{\frac{1 - \sqrt{1 - |\Delta|_m^2}}{2}}\alpha < \lambda_f < \sqrt{\frac{1 + \sqrt{1 - |\Delta|_m^2}}{2}}\alpha \quad (7.64)$$

7.4 Enhanced IMC Design for Load Disturbance Rejection

Denote

$$\eta = \sqrt{\frac{1 - \sqrt{1 - |\Delta|_m^2}}{2}}, \quad \gamma = \sqrt{\frac{1 + \sqrt{1 - |\Delta|_m^2}}{2}}$$

Using Proposition 7.2 to solve (7.64), it follows that

$$\frac{\tau_p}{2}\left[2 - \frac{1}{\eta}e^{\frac{\theta}{\tau_p}} + \sqrt{\left(2 - \frac{1}{\eta}e^{\frac{\theta}{\tau_p}}\right)^2 + 4\left(e^{\frac{\theta}{\tau_p}} - 1\right)}\right] < \lambda_f$$

$$< \frac{\tau_p}{2}\left[2 - \frac{1}{\gamma}e^{\frac{\theta}{\tau_p}} + \sqrt{\left(2 - \frac{1}{\gamma}e^{\frac{\theta}{\tau_p}}\right)^2 + 4\left(e^{\frac{\theta}{\tau_p}} - 1\right)}\right] \quad (7.65)$$

Note that $\eta < 1/\sqrt{2}$ and $\gamma > 1/\sqrt{2}$. Comparing (7.62) and (7.65), one can obtain the robust tuning constraint,

$$\frac{\tau_p}{2}\left[2 - \frac{1}{\eta}e^{\frac{\theta}{\tau_p}} + \sqrt{\left(2 - \frac{1}{\eta}e^{\frac{\theta}{\tau_p}}\right)^2 + 4\left(e^{\frac{\theta}{\tau_p}} - 1\right)}\right] < \lambda_f$$

$$< \frac{\tau_p}{2}\left[2 - \sqrt{2}e^{\frac{\theta}{\tau_p}} + \sqrt{\left(2 - \sqrt{2}e^{\frac{\theta}{\tau_p}}\right)^2 + 4\left(e^{\frac{\theta}{\tau_p}} - 1\right)}\right] \quad (7.66)$$

Remark 7.3. In the presence of the process gain uncertainty, Δk_p, the robust stability constraint of (7.56) is equivalent to the Nyquist stability criterion, $|T(j\omega)| < 1/|\Delta(j\omega)|$, since $\Delta k_p \in \Re$. However, the above robust tuning constraint of (7.66) may be somewhat conservative for other process uncertainties. For instance, the model uncertainty arising from a variation of the time constant or time delay is far less likely to result in a phase change over $-\pi$ in practice, so $\sup(|T(j\omega)|) < 1/|\Delta|_m$ is not necessary if $|T(j\omega)| < 1/|\Delta(j\omega)|$ is satisfied for $\omega \in [0, \infty)$. ◇

7.4.2 For SOPDT Stable Processes

Consider an SOPDT process described generally in the form of

$$G_2 = \frac{k\omega_n^2}{s^2 + 2\xi\omega_n s + \omega_n^2}e^{-\theta s} \quad (7.67)$$

where ω_n denotes the natural frequency and ξ is the damping ratio.

According to the IMC theory, the conventional IMC filter should be configured as

$$F_{IMC-2} = \frac{1}{(\lambda s + 1)^2} \qquad (7.68)$$

To improve the load disturbance response, the conventional IMC filter shown in (7.68) is rectified as

$$F_{RIMC-2} = \frac{\alpha s^2 + \beta s + 1}{(\lambda_f s + 1)^4} \qquad (7.69)$$

where α and β are utilized to satisfy the following asymptotic constraints

$$\lim_{s \to -p_1} (1 - T) = 0 \qquad (7.70)$$

$$\lim_{s \to -p_2} (1 - T) = 0 \qquad (7.71)$$

where

$$p_1 = \begin{cases} \omega_n \left(\xi - j\sqrt{1-\xi^2}\right), & 0 < \xi < 1; \\ \omega_n \left(\xi - \sqrt{\xi^2-1}\right), & \xi \geq 1. \end{cases}, \quad p_2 = \begin{cases} \omega_n \left(\xi + j\sqrt{1-\xi^2}\right), & 0 < \xi < 1; \\ \omega_n \left(\xi + \sqrt{\xi^2-1}\right), & \xi \geq 1. \end{cases}$$

Note that $-p_1$ and $-p_2$ are the two poles of G_2. When $\xi = 1$, there exists $p_1 = p_2 = \omega_n$, so (7.70) becomes the same as (7.71). Another asymptotic constraint should therefore be imposed to derive α and β, i.e.,

$$\lim_{s \to -p_1} \frac{d}{ds}(1 - T) = 0 \qquad (7.72)$$

Accordingly, the nominal closed-loop complementary sensitivity function can be derived as

$$T_{RIMC-2} = \frac{(\alpha s^2 + \beta s + 1) e^{-\theta s}}{(\lambda_f s + 1)^4} \qquad (7.73)$$

When $\xi \neq 1$, substituting (7.73) into (7.70) and (7.71), respectively, one can obtain

$$\alpha = \frac{p_1 e^{-\theta p_2}(p_2 \lambda_f - 1)^4 - p_2 e^{-\theta p_1}(p_1 \lambda_f - 1)^4 - p_1 + p_2}{p_1 p_2 (p_2 - p_1)} \qquad (7.74)$$

$$\beta = \frac{p_1^2 e^{-\theta p_2}(p_2 \lambda_f - 1)^4 - p_2^2 e^{-\theta p_1}(p_1 \lambda_f - 1)^4 - p_1^2 + p_2^2}{p_1 p_2 (p_2 - p_1)} \qquad (7.75)$$

7.4 Enhanced IMC Design for Load Disturbance Rejection

When $\xi = 1$, substituting (7.73) into (7.70) and (7.72), respectively, one can obtain

$$\alpha = \frac{1}{\omega_n^2}\left[1 + e^{-\omega_n\theta}(\omega_n\lambda_f - 1)^3\left(1 + \omega_n\theta + 3\omega_n\lambda_f - \omega_n^2\theta\lambda_f\right)\right] \qquad (7.76)$$

$$\beta = \frac{1}{\omega_n}\left[2 + e^{-\omega_n\theta}(\omega_n\lambda_f - 1)^3\left(2 + \omega_n\theta + 2\omega_n\lambda_f - \omega_n^2\theta\lambda_f\right)\right] \qquad (7.77)$$

Hence, it is seen that both α and β are functions of λ_f. So, there is still a single adjustable parameter, λ_f, in the proposed IMC filter. Correspondingly, the controller can be derived as

$$C_{\text{RIMC-2}} = \frac{(\alpha s^2 + \beta s + 1)(s^2 + 2\xi\omega_n s + \omega_n^2)}{k\omega_n^2(\lambda_f s + 1)^4} \qquad (7.78)$$

By substituting (7.67) and (7.73) into (7.43) and taking an inverse Laplace transform, the time domain output response to a step change of d_i can be derived from

$$y_{d_i}(t) = L^{-1}\left\{\frac{k\omega_n^2 e^{-\theta s}}{s(s^2 + 2\xi\omega_n s + \omega_n^2)}\left[1 - \frac{\alpha s^2 + \beta s + 1}{(\lambda_f s + 1)^4}e^{-\theta s}\right]\right\} \qquad (7.79)$$

In view of that DP cannot be analytically solved from (7.79), numerical guidelines are explored to disclose the quantitative tuning relationship between DP and λ_f. It can be verified from (7.79) using a scaled complex variable, $\hat{s} = s/(\xi\omega_n)$, that, given the values of $\lambda_f(\xi\omega_n)$ and $\theta(\xi\omega_n)$, DP/k is determined only by ξ, regardless of ω_n. By sweeping over the ranges of $\lambda_f(\xi\omega_n) \in [0.2, 2]$ and $\theta(\xi\omega_n) \in [0.1, 2]$, numerical results based on three cases of $\xi = 0.5$, $\xi = 1.0$, and $\xi = 1.5$ are plotted in Fig. 7.11, respectively. It can be seen that DP/k becomes larger when ξ becomes smaller. Figure 7.11a indicates that the admissible tuning range of λ_f will be severely narrowed when ξ is small. Figure 7.11b, c show that DP/k increases monotonically with respect to $\lambda_f(\xi\omega_n)$ and $\theta(\xi\omega_n)$, respectively. Note that DP for other values of ξ can be quantitatively evaluated using a linear interpolation method.

In the case where the load disturbance $\hat{d}_o$ affects the process output with a first-order transfer function as usually modeled for simplicity in practice,

$$G_d = \frac{k_d}{\tau_d s + 1}e^{-\theta_d s} \qquad (7.80)$$

it can be derived from Fig. 7.8 that

$$\frac{y_{di}}{\hat{d}_o} = G_d(1 - T) \qquad (7.81)$$

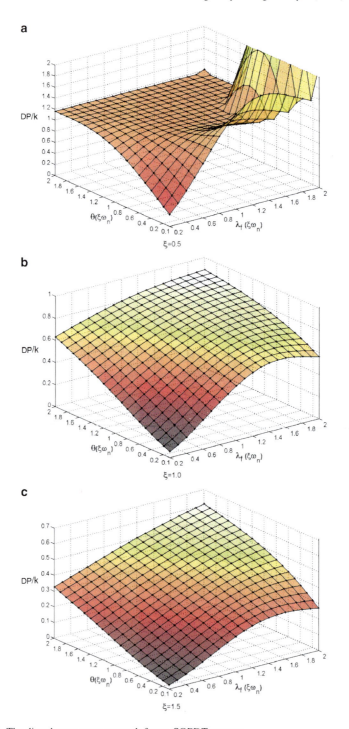

Fig. 7.11 The disturbance response peak for an SOPDT process

7.4 Enhanced IMC Design for Load Disturbance Rejection

To reduce the influence arising from the time constant of G_d to the load disturbance response, the IMC filter is rectified accordingly as

$$F_{\text{RIMC-3}} = \frac{\alpha s + 1}{(\lambda_f s + 1)^3} \tag{7.82}$$

where α is utilized to satisfy the following asymptotic constraint

$$\lim_{s \to -1/\tau_d} (1 - T) = 0 \tag{7.83}$$

Note that the closed-loop complementary sensitivity function is therefore obtained as

$$T_{\text{RIMC-3}} = \frac{\alpha s + 1}{(\lambda_f s + 1)^3} e^{-\theta s} \tag{7.84}$$

Substituting (7.84) into (7.83) yields

$$\alpha = \tau_d \left[1 + \left(\frac{\lambda_f}{\tau_d} - 1 \right)^3 e^{-\frac{\theta}{\tau_d}} \right] \tag{7.85}$$

Accordingly, the IMC controller can be derived as

$$C_{\text{RIMC-3}} = \frac{(\alpha s + 1)\left(s^2 + 2\xi\omega_n s + \omega_n^2\right)}{k\omega_n^2 (\lambda_f s + 1)^3} \tag{7.86}$$

It can be seen from (7.84) and (7.85) that DP/k_d is related to λ_f/τ_d and θ/τ_d, regardless of the time delay, θ_d. By sweeping over the ranges of $\lambda_f/\tau_d \in [0.2, 2]$ and $\theta/\tau_d \in [0.1, 2]$, the numerical tuning relationship is plotted in Fig. 7.12.

In the controller design for rejecting a load disturbance entering into the process from its input side, to ascertain the robust tuning constraints of λ_f, it follows from (7.73) that

$$|T(j\omega)| = \frac{\sqrt{(\alpha^2\omega^2 - 1)^2 + \beta^2\omega^2}}{\left(\lambda_f^2 \omega^2 + 1\right)^2} \tag{7.87}$$

The first derivative can be obtained as

$$\frac{d|T(j\omega)|}{d\omega} = \frac{\omega}{\left(\lambda_f^2 \omega^2 + 1\right)^3} \left[\frac{2\alpha\left(\alpha\omega^2 - 1\right) + \beta^2}{\sqrt{(\alpha\omega^2 - 1)^2 + \beta^2\omega^2}} \left(\lambda_f^2 \omega^2 + 1\right) \right.$$
$$\left. - 4\lambda_f^2 \sqrt{(\alpha\omega^2 - 1)^2 + \beta^2\omega^2} \right] \tag{7.88}$$

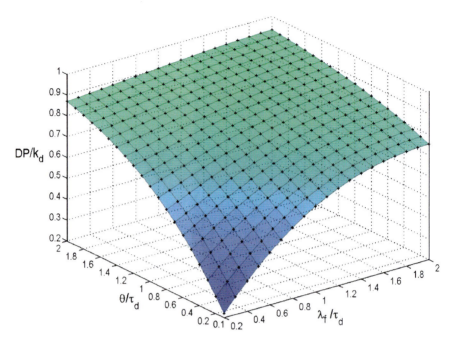

Fig. 7.12 The disturbance response peak for a first-order load disturbance transfer function

Obviously, $\omega = 0$ is an extreme point corresponding to $|T(0)| = 1$. For $\omega \neq 0$, it follows from $d|T(j\omega)|/d\omega = 0$ that

$$2\alpha^2 \lambda_f^2 x^2 + \left[3\lambda_f^2(\beta^2 - 2\alpha) - 2\alpha^2\right] x + 4\lambda_f^2 + 2\alpha - \beta^2 = 0 \quad (7.89)$$

where $x = \omega^2$. Denote $A_0 = 2\alpha^2 \lambda_f^2$, $B_0 = 3\lambda_f^2(\beta^2 - 2\alpha) - 2\alpha^2$, and $C_0 = 4\lambda_f^2 + 2\alpha - \beta^2$, the quadratic discriminant of (7.89) can be expressed as $\delta_0 = B_0^2 - 4A_0C_0$. Correspondingly, a robust stability constraint to the tuning of λ_f is given in the following proposition.

Proposition 7.3. *For an SOPDT process described in (7.67), the closed-loop system shown in Fig.7.8 holds robust stability if and only if* $\sup(|T(j\omega)|) < 1/|\Delta|_m$, *where*

$$\sup(|T(j\omega)|) = \begin{cases} \max\{1, |T(j\omega_1)|\} & \delta_0 > 0, C_0 \leq 0; \\ \max\{1, |T(j\omega_1)|, |T(j\omega_2)|\} & \delta_0 > 0, B_0 < 0, C_0 > 0; \\ \max\left\{1, \left|T(j\sqrt{\frac{-B_0}{2A_0}})\right|\right\} & \delta_0 = 0, B_0 < 0; \\ 1 & else. \end{cases}$$

7.4 Enhanced IMC Design for Load Disturbance Rejection

$$\omega_1 = \sqrt{\frac{-B_0 + \sqrt{B_0^2 - 4A_0C_0}}{2A_0}} \quad \text{and} \quad \omega_2 = \sqrt{\frac{-B_0 - \sqrt{B_0^2 - 4A_0C_0}}{2A_0}}.$$

Proof. There is a classification of six cases for deriving sup $(|T(j\omega)|)$. In the above expression, $\delta_0 > 0$ is in cases (i) and (ii). Case (iii) is "$\delta_0 = 0, B_0 < 0$." Case (iv), "else," includes three subcases, i.e., "$\delta_0 < 0$," "$\delta_0 > 0, B_0 > 0, C_0 > 0$," and "$\delta_0 = 0, B_0 \geq 0$," which lead to no positive real solution of (7.89).

For $\delta_0 > 0$, there are two real roots for (7.89), i.e.,

$$x_{1,2} = \frac{-B_0 \pm \sqrt{B_0^2 - 4A_0C_0}}{2A_0}$$

If $C_0 \leq 0$, i.e., case (i), it can be seen that $\sqrt{\delta_0} \geq |B_0|$, corresponding to $x_1 \geq 0$ and $x_2 < 0$. Accordingly, there is only one positive real root of $x = \omega^2$, i.e., $\omega_1 = \sqrt{x_1}$. So, sup $(|T(j\omega)|)$ can be reached at $\omega = 0$ or ω_1. In case (ii), it can be seen that $\sqrt{\delta_0} < |B_0|$, corresponding to $x_1 > 0$ and $x_2 > 0$. Accordingly, there exists two positive real roots of $x = \omega^2$, i.e., $\omega_1 = \sqrt{x_1}$ and $\omega_2 = \sqrt{x_2}$. So, sup $(|T(j\omega)|)$ can be reached at $\omega = 0, \omega_1$ or ω_2.

In case (iii), it can be seen that (7.89) has a dual positive root at $x = -B_0/(2A_0)$. Correspondingly, sup $(|T(j\omega)|)$ can be reached at $\omega = 0$ or $\sqrt{-B_0/(2A_0)}$.

In the case where $\delta_0 < 0$, there is no real roots for (7.89) according to the solvability of a linear quadratic equation. Thus, sup $(|T(j\omega)|)$ can only be reached at $\omega = 0$. This conclusion can be similarly drawn for the other two subcases of case (iv).

Hence, using the small gain theorem, the conclusion in Proposition 7.3 follows. □

In the controller design for rejecting $\hat{d}_o$ from the process output side with a first-order transfer function, it follows from (7.84) that

$$|T(j\omega)| = \frac{\sqrt{\alpha^2\omega^2 + 1}}{\left(\lambda_f^2\omega^2 + 1\right)^{\frac{3}{2}}} \tag{7.90}$$

The first derivative can be derived as

$$\frac{d|T(j\omega)|}{d\omega} = \frac{\omega}{\left(\lambda_f^2\omega^2 + 1\right)^{\frac{5}{2}}}\left[\frac{\alpha^2\left(\lambda_f^2\omega^2 + 1\right)}{\sqrt{\alpha^2\omega^2 + 1}} - 3\lambda_f^2\sqrt{\alpha^2\omega^2 + 1}\right]$$

Note that $\omega = 0$ is an obvious extreme point corresponding to $|T(0)| = 1$. A robust stability constraint to tuning λ_f is given in the following proposition.

Proposition 7.4. *For an SOPDT process described in (7.67), the closed-loop system shown in Fig.7.8 for rejecting load disturbance from the process output side with a first-order transfer function holds robust stability if and only if* $|\Delta|_m < 1$.

Proof. It follows from $d\,|T(j\omega)|/d\omega = 0$ that

$$\alpha^2 \left(\lambda_f^2 \omega^2 + 1\right) - 3\lambda_f^2 \left(\alpha^2 \omega^2 + 1\right) = 0 \qquad (7.91)$$

The solution of (7.91) can be derived as

$$\omega^2 = \frac{\alpha^2 - 3\lambda_f^2}{2\alpha^2 \lambda_f^2} \qquad (7.92)$$

To obtain a positive solution of ω, it requires that

$$\lambda_f < \frac{\alpha}{\sqrt{3}} \qquad (7.93)$$

The extreme point can thus be derived as

$$\omega = \frac{\sqrt{\alpha^2 - 3\lambda_f^2}}{\sqrt{2}\alpha \lambda_f} \qquad (7.94)$$

Substituting (7.94) into (7.90) yields

$$|T(j\omega)| = \frac{2\alpha^3}{3\sqrt{3}\lambda_f \left(\alpha^2 - \lambda_f^2\right)} \qquad (7.95)$$

Let $g(\lambda_f)$ be a function with respect to λ_f,

$$g(\lambda_f) = 3\sqrt{3}\lambda_f \left(\alpha^2 - \lambda_f^2\right) \qquad (7.96)$$

The first derivative of $g(\lambda_f)$ can be derived as

$$g'(\lambda_f) = 3\sqrt{3} \left(\alpha^2 - 3\lambda_f^2\right) \qquad (7.97)$$

Combining (7.93) with (7.97), it can be seen that $g'(\lambda_f) > 0$. Thus, $\lambda_f = \alpha/\sqrt{3}$ is the unique extreme point to reach the maximum of $g(\lambda_f)$, i.e.,

$$\max\{g(\lambda_f)\} = 3\sqrt{3} \cdot \frac{\alpha}{\sqrt{3}} \left[\alpha^2 - \left(\frac{\alpha}{\sqrt{3}}\right)^2\right] = 2\alpha^3 \qquad (7.98)$$

7.5 Proportional-Integral-Derivative (PID) Tuning

Substituting (7.98) into (7.95) yields

$$|T(j\omega)| < 1, \quad \forall \omega > 0.$$

Hence, sup $(|T(j\omega)|)$ is reached only at $\omega = 0$, i.e., sup $(|T(j\omega)|) = 1$.

Using the small gain theorem, one can conclude that sup $|\Delta(j\omega)| < 1$ is required to hold the closed-loop stability. □

Remark 7.4. It is implied by Proposition 7.4 that tuning λ_f does not affect the closed-loop robust stability since sup $(|T(j\omega)|) = 1$ for $\lambda_f \in (0, +\infty)$. Hence, tuning λ_f can be purely focused on the closed-loop performance for rejecting such load disturbance. ◇

7.5 Proportional-Integral-Derivative (PID) Tuning

In the unity feedback control structure shown in Fig. 7.2, a PI- or PID-type controller is widely used in practical applications. Here, an IMC-based PID tuning method is given accordingly.

From the equivalent relationship between the IMC structure and the unity feedback control structure, as shown in (7.28), it can be easily verified from the IMC controller design, e.g., (7.31), (7.51), or (7.78), that

$$\lim_{s \to 0} K(s) = \infty \tag{7.99}$$

This indicates that an IMC-based controller in the unity feedback control structure has a property of integral, which can thus eliminate the output error with respect to the set-point.

Therefore, one can approximate the above IMC-based closed-loop controller into a PID form for implementation. The analytical approximation method based on the mathematical Maclaurin series, as developed in the reference (Lee et al. 1998; Zhang et al. 2002; Liu et al. 2005), is presented here for simplicity. Let $K(s) = M/s$, it follows that

$$K(s) = \frac{1}{s}\left[M(0) + M'(0)s + \frac{M''(0)}{2!}s^2 + \cdots\right] \tag{7.100}$$

Obviously, the first two terms in the above Maclaurin expansion can constitute a PI controller, and the first three terms can constitute a PID controller.

Generally, a PID controller is in the form of

$$K = k_C + \frac{1}{\tau_I s} + \frac{\tau_D s}{\tau_F s + 1} \tag{7.101}$$

where k_C denotes the controller gain, τ_I the integral constant, τ_D the derivative constant, and τ_F a filter constant that is usually taken as $\tau_F = (0.01 \sim 0.1)\tau_D$ for practical implementation.

Comparing (7.100) with (7.101), the PI or PID parameters can be derived as

$$\begin{cases} k_C = M'(0) \\ \tau_I = 1/M(0) \\ \tau_D = M''(0)/2 \end{cases} \qquad (7.102)$$

Remark 7.5. Based on a low-order process model of FOPDT or SOPDT, an IMC-based PI or PID controller can also be analytically derived by simply approximating the time delay with a first-order Taylor or Padé expansion. Such an exercise may save computation effort but at the cost of an inferior approximation accuracy, compared to the PID formula given in (7.102). ◇

According to the small gain theorem (Zhou et al. 1996), the closed-loop system shown in Fig. 7.2 holds robust stability in the presence of the process multiplicative uncertainty (Δ), if and only if

$$\left| \frac{\widehat{G}(j\omega)K(j\omega)}{1 + G(j\omega)K(j\omega)} \right| < \frac{1}{|\Delta(j\omega)|}, \quad \forall \omega \in [0, \infty) \qquad (7.103)$$

Hence, given a specified norm bound of $\Delta(j\omega)$ in practice, an admissible tuning range of the essentially single adjustable parameter, λ (or λ_f), in the above PI or PID controller can be determined by observing whether the magnitude plot of the left-hand side of (7.103) is below that of the right-hand side for $\omega \in [0, \infty)$. For practical application, it is generally suggested to initially take $\lambda = \theta$ (or $\lambda_f = \theta$). A desirable trade-off between the closed-loop disturbance rejection performance and its robust stability can be conveniently reached through monotonically tuning λ online.

Alternatively, one can use the robust stability constraint in (7.103) to evaluate an admissible upper bound of $|\Delta(j\omega)|$ to a fixed tuning of λ (or λ_f), based on the above controller design.

7.6 Illustrative Examples

Examples 7.2 and 7.3 are given to demonstrate the advantage of the modified IMC design for load disturbance rejection for first- and second-order processes with slow dynamics. Example 7.4 is given to illustrate the effectiveness of the presented PID tuning method for disturbance rejection for a high-order process.

Example 7.2. Consider a slow industrial process modeled in an FOPDT form of

$$G = \frac{e^{-30s}}{100s + 1}$$

7.6 Illustrative Examples

A conventional IMC filter should determine the controller shown in Fig. 7.8 for load disturbance rejection as

$$C_{IMC-1} = \frac{100s + 1}{\lambda s + 1}$$

Using the proposed IMC design formula in (7.51), it follows that

$$C_{RIMC-1} = \frac{(\alpha s + 1)(100s + 1)}{(\lambda_f s + 1)^2}$$

where $\alpha = 100\left[1 - (0.01\lambda_f - 1)^2 e^{-0.3}\right]$.

For comparison, adding a unit step change for d_i as shown in Fig. 7.8 to the process and taking $\lambda = \lambda_f = 40$, the corresponding output response is shown in Fig. 7.13. It is seen that obviously improved load disturbance rejection is obtained by the proposed IMC filter. The conventional IMC filter has led to a long "tail" in the load disturbance response, due to the influence of the slow time constant of the process. To obtain the same DP with the proposed IMC filter, $\lambda = 20$ is required in the conventional IMC filter, as shown in Fig. 7.13, but the recovery time is still about 50% longer.

Assume that there exists 30% error in the process modeling. The worst case is that the process time constant is actually 30% smaller and the time delay 30% larger. The corresponding output response is shown in Fig. 7.14, which indicates that the proposed IMC filter performs better against the severe process uncertainties.

Example 7.3. Consider the second-order process studied by Skogestad (2003),

$$G = \frac{e^{-s}}{(20s + 1)(2s + 1)}$$

Skogestad (2003) derived an IMC-based PID controller, $C = 10(0.125/s + 1)(2s + 1)$, to optimize the system performance against load disturbance, of which the adjustable parameter, $\tau_c = 1.0$, corresponds to $\lambda = 1.0$ in the conventional IMC filter shown in (7.68). The IMC controller in Fig. 7.8 should be configured as

$$C_{IMC-2} = \frac{(20s + 1)(2s + 1)}{(\lambda s + 1)^2}$$

Using the proposed IMC filter shown in (7.69), it follows from (7.74), (7.75), and (7.78) that

$$C_{RIMC-2} = \frac{(\alpha s^2 + \beta s + 1)(20s + 1)(2s + 1)}{(\lambda_f s + 1)^4}$$

where $\alpha = 2.6957 (0.5\lambda_f - 1)^4 - 42.2769 (0.05\lambda_f - 1)^4 + 40$, $\beta = 0.1348 (0.5\lambda_f - 1)^4 - 21.1384 (0.05\lambda_f - 1)^4 + 22$.

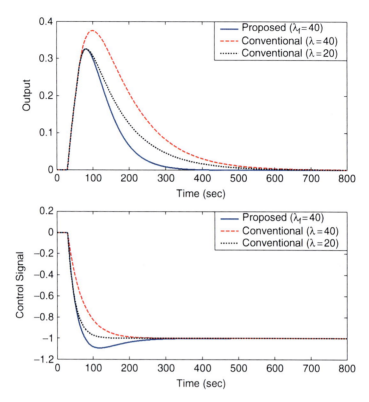

Fig. 7.13 Nominal output responses of Example 7.2

For comparison, adding a unit step change for d_i as shown in Fig. 7.8 to the process and taking $\lambda = \lambda_f = 1.0$, the output response is shown in Fig. 7.15. It is again seen that apparently improved load disturbance response is obtained using the proposed IMC filter. To obtain the same DP with the proposed IMC filter, $\lambda = 0.09$ and $\tau_c = 0.15$ are required in the conventional IMC filter and the IMC-based controller of Skogestad (2003). It is seen that the recovery time cannot be efficiently reduced by using the conventional IMC filter, and in contrast, the controller of Skogestad (2003) turns out a slightly oscillatory response. Moreover, it can be verified that, if the process time delay is actually 20% larger and the slower time constant ($\tau_1 = 20$) is 20% smaller, the proposed IMC filter can maintain well the closed-loop stability, whereas the conventional IMC filter with $\lambda = 0.09$ cannot hold the closed-loop stability any longer, and the IMC-based controller of Skogestad (2003) with $\tau_c = 0.15$ will give very oscillatory response.

To demonstrate the achievable performance for rejecting load disturbance from the process output side (denoted as $\hat{d}_o$ in Fig. 7.8) with a slow dynamics, e.g., $G_d = e^{-2s}/(10s + 1)$, the proposed IMC filter shown in (7.82) should be used to design the controller as

7.6 Illustrative Examples

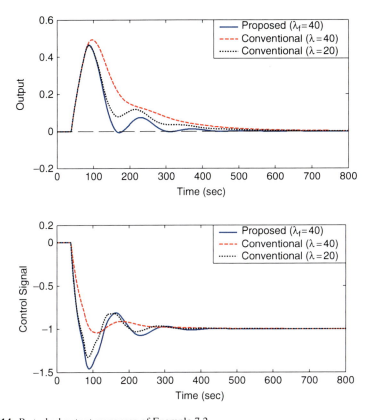

Fig. 7.14 Perturbed output responses of Example 7.2

$$C_{\text{RIMC-3}} = \frac{(\alpha s + 1)(20s + 1)(2s + 1)}{(\lambda_f s + 1)^3}$$

where $\alpha = 10\left[1 + (0.1\lambda_f - 1)^3 e^{-0.1}\right]$. Assume that there exists 50% error in estimating the time constant of the load disturbance transfer function. The worst case is that the time constant is actually 50% smaller. Note that the time delay, $\theta_d = 2$, in the load disturbance transfer function does not affect the disturbance response performance as discussed in the filter design procedure in Sect. 7.4.2. By adding a unit step change for $\hat{d}_o$, and taking $\lambda = \lambda_f = 1.0$, $\lambda = 0.45$, and $\tau_c = 0.6$ for comparison, the output response is shown in Fig. 7.16. It is seen that the proposed IMC filter results in apparently improved disturbance rejection, given the severe modeling error of the time constant in the load disturbance transfer function.

Example 7.4 Consider a high-order process studied by Huang et al. (2005),

$$G(s) = \frac{1}{(4s^2 + 2.8s + 1)(s + 1)^2} e^{-2.2s}$$

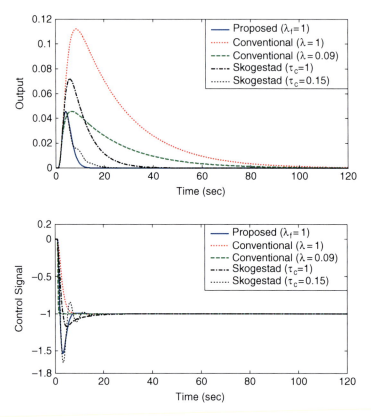

Fig. 7.15 Nominal output responses of Example 7.3

Based on a closed-loop relay feedback test for the frequency response estimation, Huang et al. (2005) gave a PID controller, $C = 0.314(1 + 1/2.59s + 2.103s)/(0.1s + 1)$, which has shown evident improvement for load disturbance rejection, compared to other model-based PID tuning methods including the standard IMC method.

Based on an SOPDT model, $G_{m-2} = 0.9934e^{-3.54s}/(5.5069s^2 + 3.4095s + 1)$, which is obtained using the identification algorithm presented in Sect. 2.5.2 in terms of a closed-loop step test with the above PID controller (see Example 2.12), the controller formulas of (7.74), (7.75), (7.78), (7.100), and (7.102) give the PID controller parameters, $k_C = 0.3717$, $\tau_I = 9.2698$, and $\tau_D = 0.6079$, by taking $\lambda_f = 2.25$ for comparison with Huang et al. (2005). By adding a unit step change for the load disturbance to the process input, the load disturbance response is shown in Fig. 7.17. It is seen that the recovery time is reduced almost by 25% in terms of the same disturbance response peak.

7.6 Illustrative Examples

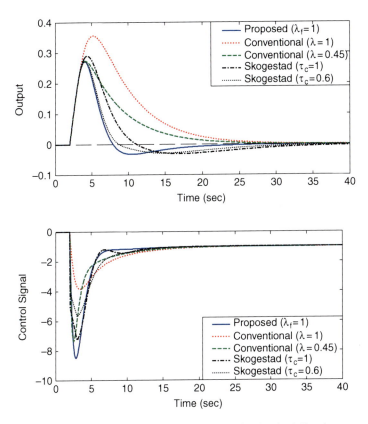

Fig. 7.16 Output responses of Example 7.3 in the presence of a slow load disturbance

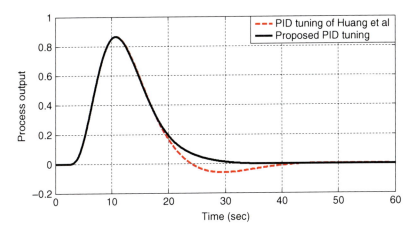

Fig. 7.17 Comparison of load disturbance response for Example 7.4

7.7 Summary

For the control of SISO processes, the widely used output error criteria, control performance specifications in the time and frequency domains, and robust stability conditions have been introduced for reference. Subsequently, a brief overview of the IMC theory (Morari and Zafiriou 1989) has been given for control system design and controller tuning, along with an illustrative example for application.

For load disturbance rejection, which is a common concern in industrial process operation, a modified IMC design (Liu and Gao 2010) has been presented for improving the disturbance rejection performance for slow processes with large time constant(s). The key lies with countering the influence from the slow time constant(s) of the process or load disturbance by establishing asymptotic canceling constraints through the closed-loop transfer function. Based on the widely used low-order process models of FOPDT and SOPDT, the corresponding controller formulas have been analytically developed, together with the quantitative tuning guidelines and robust stability constraints. In view of the fact that a PI- or PID-type controller is widely applied in engineering practice, an IMC-based PID tuning method has been presented for practical application.

Three examples from existing references have been used to demonstrate the effectiveness and merits of the enhanced IMC design and PID tuning method, based on a low-order model of the process or the inherent load disturbance.

References

Åström KJ, Hägglund T (2005) Advanced PID control. ISA Society of America, Research Triangle Park
Braatz RD (1995) Internal model control: The control handbook, FL: CRC Press, Boca Raton
Goodwin GC, Graebe L, Salgado ME (2001) Control system design, Prentice Hall, Upper Saddle River, NJ
Horn IG, Arulandu JR, Braatz RD (1996) Improved filter design in internal model control. Ind Eng Chem Res 35(10):3437–3441
Huang HP, Jeng JC, Luo KY (2005) Auto-tune system using single-run relay feedback test and model-based controller design. J Process Control 15:713–727
Hwang C, Cheng YC (2005) A note on the use of the Lambert W function in the stability analysis of time-delay systems. Automatica 41(11):1979–1985
Lee Y, Park S, Lee M, Brosilow C (1998) PID controller tuning for desired closed-loop responses for SI/SO systems. AICHE J 44(1):106–115
Liu T, Gao F (2010) New insight into internal model control filter design for load disturbance rejection. IET Control Theory Appl 4(3):448–460
Liu T, Zhang WD, Gu DY (2005) Analytical two-degree-of-freedom tuning design for open-loop unstable processes with time delay. J Process Control 15:559–572
Morari M, Zafiriou E (1989) Robust process control. Prentice Hall, Englewood Cliff
Seborg DE, Edgar TF, Mellichamp DA (2004) Process dynamics and control, 2nd edn. Wiley, Hoboken
Shamsuzzoha M, Lee M (2007) IMC-PID controller design for improved disturbance rejection of time-delayed processes. Ind Eng Chem Res 46(7):2077–2091

Shinskey FG (1996) Process control system, 4th edn. McGraw Hill, New York

Skogestad S (2003) Simple analytical rules for model reduction and PID controller tuning. J Process Control 13(4):291–309

Skogestad S, Postlethwaite I (2005) Multivariable feedback control: analysis and design, 2nd edn. Wiley, Chichester

Zhang WD, Xi YG, Yang GK, Xu XM (2002) Design PID controllers for desired time-domain or frequency-domain response. ISA Trans 41(4):511–520

Zhou KM, Doyle JC, Glover K (1996) Robust and optimal control. Prentice Hall, Englewood Cliff

Chapter 8
Two-Degrees-of-Freedom (2DOF) Control of SISO Processes

8.1 The Advantage of a 2DOF Control Scheme

A two-degrees-of-freedom (2DOF) IMC structure is shown in Fig. 8.1, where G denotes the process, $\hat{G}$ is the process transfer function model in frequency domain, C_s is a feedforward controller, C_f is a feedback controller, and G_d is the load disturbance transfer function corresponding to a normalized input, $\hat{d}_o(t) = 1$. In the nominal case ($G = \hat{G}$), it follows that

$$\frac{y}{r} = GC_s \tag{8.1}$$

$$\frac{y}{d_i} = G(1 - GC_f) \tag{8.2}$$

$$\frac{y}{\hat{d}_o} = G_d(1 - GC_f). \tag{8.3}$$

It is seen that C_s is responsible for the set-point tracking, and C_f is for the load disturbance rejection, such that both the set-point tracking performance and load disturbance rejection performance can be separately tuned or optimized. Compared to the standard IMC structure shown in Fig. 7.8, where the IMC controller is responsible for both set-point tracking and load disturbance rejection, the advantage of a 2DOF control structure is evident, in particular, for rejecting various load disturbances that are different from the set-point in the signal type (Morari and Zafiriou 1989; Skogestad and Postlethwaite 2005).

Note that even for the same type of set-point and load disturbance, the waterbed effect between the set-point response and the load disturbance response will become very severe for an integrating or unstable process in the unity feedback control structure (Zhou et al. 1996). Since the standard IMC structure cannot hold internal stability for an integrating or unstable process, as discussed in Sect. 7.3,

Fig. 8.1 Two-degrees-of-freedom IMC structure

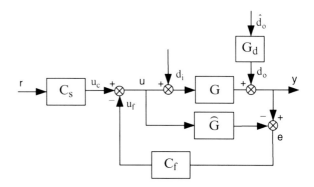

the standard 2DOF IMC structure shown in Fig. 8.1 can only be used for stable processes to guarantee the internal stability for system operation. A number of 2DOF control schemes have therefore been developed for integrating and unstable processes (Huang and Chen 1997; Tian and Gao 1999; Kwak et al. 2001; Chien et al. 2002; Yang et al. 2002; Zhong and Mirkin 2002; Zhong and Normey-Rico 2002; Normey-Rico and Camacho 2002, 2009; Hang et al. 2003; Tan et al. 2003; Kaya 2004; Zhang et al. 2004, 2008; Liu et al. 2005a, b; Lu et al. 2005; Garcia et al. 2006; Garcia and Albertos 2008; Rao and Chidambaram 2007).

For clarity, a 2DOF IMC design based on a classification on different cases of load disturbance entering into a stable process will be first presented. Subsequently, a unified 2DOF control scheme based on the 2DOF control methods developed by Zhang et al. (2004) and Liu et al. (2005a, b) will be presented for integrating and unstable processes.

8.2 The 2DOF IMC Design for Optimizing the Set-Point Tracking and Load Disturbance Rejection

In the standard 2DOF IMC structure shown in Fig. 8.1, the controller C_s for set-point tracking can be designed the same as that in the standard IMC structure (see Sect. 7.3); by the virtue of that, the standard IMC design can achieve the H_2 optimal performance objective (i.e., ISE), min $\|e\|_2$, for set-point tracking (Morari and Zafiriou 1989).

For the design of C_f to optimize the load disturbance rejection performance, it depends on individual cases of load disturbance entering into the process in practice. Four cases that are mostly encountered in practical applications are considered here:

Case (i): There exists only load disturbance that affects the process from its output side, denoted as d_o in Fig. 8.1, which is the same type with the set-point. Correspondingly, C_f can be designed the same as C_s, owing to the above merit for performance optimization.

Case (ii): There exists only load disturbance, denoted as $\hat{d}_o$ in Fig. 8.1, with obviously different dynamics compared to the set-point, corresponding to $d_o = G_d \hat{d}_o$. If d_o is a ramp-type signal while the set-point is a step-type signal, the type II IMC filter, see (7.30), should be used to design C_f, while the type I IMC filter, see (7.29), should be used to design C_s, and vice versa. If G_d is an evidently slower transfer function compared to $\widehat{G}$, the enhanced IMC design in Sect. 7.4 can be used to design C_f for optimizing the disturbance rejection performance based on an identified low-order disturbance model of FOPDT or SOPDT.

Case (iii): There exists only load disturbance denoted as d_i in Fig. 8.1. If the process dynamic characteristics do not give sluggish load disturbance response, i.e., a long "tail," the above controller design for Case 1 can be used. Otherwise, the enhanced IMC design in Sect. 7.4 can be used to design C_f for improving the disturbance rejection performance based on an identified low-order process model of FOPDT or SOPDT.

Case (iv): There exists two different types of load disturbance, denoted as d_i and d_o in Fig. 8.1, e.g., step and ramp types. The worst case, i.e., the corresponding transfer function has more poles at the origin ($s = 0$), should be considered to design C_f for optimizing the disturbance rejection performance, which can therefore guarantee no steady-state output error in the presence of both d_i and d_o.

In the case where load disturbance is unexpected or unknown, it is generally suggested to use the controller design for Case 1 for implementation. Based on knowledge or experience of the process operation, the controller designs for other cases can be adopted accordingly to deal with a deterministic-type load disturbance.

8.3 A 2DOF Control Scheme for Integrating Processes

For an integrating or unstable process, the 2DOF control methods developed by Zhang et al. (2004) and Liu et al. (2005a, b) have shown evident superiority in comparison with many existing methods. The control structures in these authors' methods are shown in Fig. 8.2a, b, respectively, where G denotes the process, C_s the feedforward controller for set-point tracking, and C_f the feedback controller for load disturbance rejection. T_r in Fig. 8.2a is a desired transfer function for set-point tracking, and G_c in Fig. 8.2b is an auxiliary controller of P or PD type used for stabilizing the set-point response and the delay-free part (G_{mo}) of the process model ($G_m = G_{mo} e^{-\theta s}$). r is the set-point, y the process output, y_r the desired (reference) output response, u the process control input, u_c the output of C_s, u_f the output of C_f, d_i the load disturbance entering into the process from its input side, and $\hat{d}_o$ the load disturbance entering into the process from its output side with a transfer function, G_d.

Note that by comparing the expressions derived in Zhang et al. (2004) and Liu et al. (2005a) for the set-point response of a first-order unstable process as

Fig. 8.2 The 2DOF control structures: (**a**) Zhang et al. (2004) and (**b**) Liu et al. (2005b)

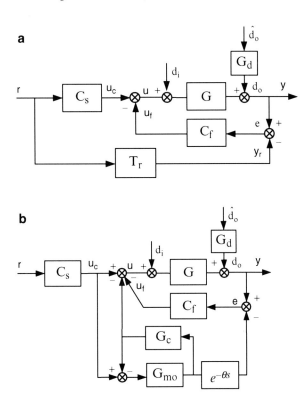

commonly studied, and comparing the expressions derived in Zhang et al. (2004) and Liu et al. (2005a) for the load disturbance response, it can be seen that the achievable system performance in both set-point tracking and load disturbance rejection is identical in these two control structures. Therefore, a unified 2DOF control scheme will be given here.

In engineering practice, first- and second-order integrating process models are widely used for control system design and controller tuning, which are respectively in the form of

$$G_{\text{I-1}} = \frac{k_p}{s} e^{-\theta s} \tag{8.4}$$

$$G_{\text{I-2}} = \frac{k_p}{s(\tau_p s + 1)} e^{-\theta s}, \tag{8.5}$$

where k_p denotes the proportional gain, θ the process time delay, and τ_p a time constant reflecting the inertial characteristics. Note that the above SOPDT model can be effectively used to describe the dynamic response characteristics for a wide variety of higher-order integrating processes (Liu and Gao 2008, 2010a).

8.3 A 2DOF Control Scheme for Integrating Processes

Based on the above models, the following controller design formulae are developed in terms of the 2DOF control structure shown in Fig. 8.2a, which can be similarly derived in terms of the equivalent 2DOF control structure shown in Fig. 8.2b.

8.3.1 Controller Design

According to the IMC controller design presented in Sect. 7.3, the feedforward controller for set-point tracking can be derived for an FOPDT integrating process model in (8.4) as

$$C_{s\text{-}I\text{-}1} = \frac{s}{k_p(\lambda_c s + 1)} \tag{8.6}$$

which corresponds to a desired set-point response transfer function,

$$T_{r\text{-}I\text{-}1} = \frac{1}{\lambda_c s + 1} e^{-\theta s}, \tag{8.7}$$

where λ_c is an adjustable parameter for tuning the set-point tracking speed.

Similarly, the feedforward controller for an SOPDT integrating process model in (8.5) can be derived as

$$C_{s\text{-}I\text{-}2} = \frac{s(\tau_p s + 1)}{k_p(\lambda_c s + 1)^2} \tag{8.8}$$

which corresponds to a desired set-point response transfer function,

$$T_{r\text{-}I\text{-}2} = \frac{1}{(\lambda_c s + 1)^2} e^{-\theta s}. \tag{8.9}$$

By performing the inverse Laplace transform for $T_{r\text{-}I\text{-}1}$ and $T_{r\text{-}I\text{-}2}$, one can obtain the corresponding time domain set-point responses to a step change,

$$y_{r\text{-}1}(t) = \begin{cases} 0 & t \leq \theta \\ 1 - e^{-(t-\theta)/\lambda_c} & t > \theta \end{cases} \tag{8.10}$$

$$y_{r\text{-}2}(t) = \begin{cases} 0 & t \leq \theta \\ 1 - (1 + \frac{t-\theta}{\lambda_c})e^{-(t-\theta)/\lambda_c} & t > \theta \end{cases}. \tag{8.11}$$

It can be seen from (8.10) or (8.11) that there is no overshoot in the nominal set-point response, and the time domain performance specification can be quantitatively tuned through the single adjustable parameter λ_c of $C_{s\text{-}I\text{-}1}$ or $C_{s\text{-}I\text{-}2}$.

For instance, a tuning formula for the rise time of the set-point response can be numerically determined from (8.10) as $t_r = 2.3026\lambda_c + \theta$ for a first-order integrating process as shown in (8.4), and from (8.11) as $t_r = 3.8897\lambda_c + \theta$ for a second-order integrating process as shown in (8.5).

For load disturbance rejection, the transfer functions relating d_i and $\hat{d}_o$ to y for the nominal case can be derived from Fig. 8.2 as

$$\frac{y}{d_i} = \frac{G}{1+GC_f} \quad (8.12)$$

$$\frac{y}{\hat{d}_o} = \frac{G_d}{1+GC_f}. \quad (8.13)$$

Note that the complementary sensitivity function of the closed-loop structure set between the process input and output can be derived as

$$T_d = \frac{GC_f}{1+GC_f} \quad (8.14)$$

which is exactly equivalent to the transfer function relating d_i to the controller output, u_f.

Hence, (8.12) and (8.13) can be rewritten as

$$\frac{y}{d_i} = G(1-T_d) \quad (8.15)$$

$$\frac{y}{\hat{d}_o} = G_d(1-T_d). \quad (8.16)$$

In the ideal case, a desired complementary sensitivity function of the closed-loop structure should be $T_d(s) = e^{-\theta s}$. That is to say, when a load disturbance, d_i, enters into the process, the feedback controller C_f should detect the corresponding output error immediately after the process time delay and then compute an equivalent signal u_f to counteract it. There, however, exist the following asymptotic tracking constraints in practice:

$$\lim_{s\to 0}(1-T_d) = 0 \quad (8.17)$$

$$\lim_{s\to p_i}(1-T_d) = 0, \quad i=1,2,\ldots,m. \quad (8.18)$$

where p_i ($i = 1, 2, \ldots, m$) is the RHP pole of the process. The constraints in (8.17) and (8.18) must be satisfied in order to hold internal stability of the closed-loop structure for load disturbance rejection.

For an integrating process, the constraints in (8.17) and (8.18) are identical in essence. Therefore, another asymptotic constraint should be imposed,

8.3 A 2DOF Control Scheme for Integrating Processes

$$\lim_{s \to 0} \frac{d}{ds}(1 - T_d) = 0. \quad (8.19)$$

If p_i is a double RHP pole of the process, the following asymptotic constraint should be imposed accordingly:

$$\lim_{s \to p_i} \frac{d}{ds}(1 - T_d) = 0 \quad (8.20)$$

If p_i is a multiple RHP pole of the process, additional asymptotic constraints can be imposed by taking higher order derivatives to $1 - T_d$, as in (8.20).

Based on the H_2 optimal performance objective of the IMC theory, a desired closed-loop complementary sensitivity function is proposed as

$$T_d = \frac{\sum_{i=1}^{m} \alpha_i s^i + 1}{(\lambda_f s + 1)^{l+m}} e^{-\theta s}, \quad (8.21)$$

where λ_f is an adjustable parameter of the closed-loop transfer function, l the relative degree of the process model, m the number of RHP poles of the process (a multiple RHP pole should be counted multiple times), and α_i can be determined from the asymptotic constraints in (8.17)–(8.20).

Accordingly, the feedback controller can be inversely derived from (8.14) as

$$C_f = \frac{1}{G} \cdot \frac{T_d}{1 - T_d}. \quad (8.22)$$

For a first-order integrating process in (8.4), there are $l = 1$ and $m = 1$. It follows from (8.21) that

$$T_{d\text{-I-1}} = \frac{\alpha_1 s + 1}{(\lambda_f s + 1)^2} e^{-\theta s}. \quad (8.23)$$

Obviously, it satisfies the asymptotic constraint in (8.17). Then, it follows from (8.19) that

$$\lim_{s \to 0} \frac{d}{ds} \left[1 - \frac{\alpha_1 s + 1}{(\lambda_f s + 1)^2} e^{-\theta s} \right] = 0. \quad (8.24)$$

Solving (8.24) obtains $\alpha_1 = 2\lambda_f + \theta$.

Substituting (8.4) and (8.23) into (8.22) yields

$$C_{f\text{-I-1}} = \frac{s(\alpha_1 s + 1)}{k_p (\lambda_f s + 1)^2} \cdot \frac{1}{1 - \frac{\alpha_1 s + 1}{(\lambda_f s + 1)^2} e^{-\theta s}}. \quad (8.25)$$

Fig. 8.3 Closed-loop unit for implementation

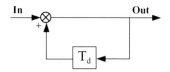

Note that the second multiplier in (8.25) satisfies

$$\lim_{s \to \infty} \frac{1}{1 - \frac{\alpha_1 s + 1}{(\lambda_f s + 1)^2} e^{-\theta s}} = 1 \quad (8.26)$$

$$\lim_{s \to 0} \frac{s}{1 - \frac{\alpha_1 s + 1}{(\lambda_f s + 1)^2} e^{-\theta s}} = \infty \quad (8.27)$$

$$\lim_{s \to 0} \frac{s^2}{1 - \frac{\alpha_1 s + 1}{(\lambda_f s + 1)^2} e^{-\theta s}} = 2\lambda_f^2 + 4\lambda_f \theta + \theta^2. \quad (8.28)$$

Hence, this multiplier can be viewed as a special bi-proper integrator with a double zero at $s = 0$, thus capable of eliminating the output error caused by a step change of d_i or d_o. In fact, it can be practically implemented using the closed-loop unit shown in Fig. 8.3.

For a second-order integrating process in (8.5), there are $l = 2$ and $m = 1$. It follows from (8.21) that

$$T_{\text{d-I-2}} = \frac{\alpha_1 s + 1}{(\lambda_f s + 1)^3} e^{-\theta s}. \quad (8.29)$$

Similarly, it follows from (8.19) that

$$\lim_{s \to 0} \frac{d}{ds} \left[1 - \frac{\alpha_1 s + 1}{(\lambda_f s + 1)^3} e^{-\theta s} \right] = 0. \quad (8.30)$$

Solving (8.30) obtains $\alpha_1 = 3\lambda_f + \theta$.
Substituting (8.5) and (8.29) into (8.22) yields

$$C_{\text{f-I-2}} = \frac{s(\tau_p s + 1)[(3\lambda_f + \theta)s + 1]}{k_p(\lambda_f s + 1)^3} \cdot \frac{1}{1 - \frac{(3\lambda_f + \theta)s + 1}{(\lambda_f s + 1)^3} e^{-\theta s}}, \quad (8.31)$$

where the second multiplier can also be implemented using the closed-loop unit shown in Fig. 8.3.

For an integrating process with slow dynamics, corresponding to a large time constant (τ_p) in the SOPDT model shown in (8.5), in order to reduce the influence

8.3 A 2DOF Control Scheme for Integrating Processes

of the slow dynamics to the load disturbance response, the following asymptotic constraint is imposed to the controller design:

$$\lim_{s \to -1/\tau} (1 - T_d) = 0, \tag{8.32}$$

where $\tau = \tau_p$.

Correspondingly, the closed-loop transfer function for rejecting a step-type load disturbance occurring at the process input side is proposed as

$$T_{\text{d-IS-2}} = \frac{\eta_2 s^2 + \eta_1 s + 1}{(\lambda_f s + 1)^4} e^{-\theta s}, \tag{8.33}$$

where η_1 and η_2 are taken to satisfy the asymptotic constraints in (8.19) and (8.32). Substituting (8.33) into (8.19) and (8.32), one can obtain

$$\begin{cases} \eta_1 = 4\lambda_f + \theta \\ \eta_2 = \tau_p \eta_1 + \tau_p^2 \left[\left(\frac{\lambda_f}{\tau_p} - 1 \right)^4 e^{-\frac{\theta}{\tau_p}} - 1 \right] \end{cases}. \tag{8.34}$$

Hence, substituting (8.5) and (8.33) into (8.22) yields the corresponding feedback controller,

$$C_{\text{f-IS-2}} = \frac{s(\tau_p s + 1)(\eta_2 s^2 + \eta_1 s + 1)}{k_p (\lambda_f s + 1)^4} \cdot \frac{1}{1 - \frac{\eta_2 s^2 + \eta_1 s + 1}{(\lambda_f s + 1)^4} e^{-\theta s}}. \tag{8.35}$$

To reject a step-type load disturbance occurring at the process output side with a slow transfer function of $G_d = k_d/(\tau_d s + 1)$, the closed-loop transfer function is correspondingly proposed as

$$T_{\text{d-IO-2}} = \frac{\eta_1 s + 1}{(\lambda_f s + 1)^3} e^{-\theta s}. \tag{8.36}$$

Substituting (8.36) into (8.32) with $\tau = \tau_d$, one can obtain

$$\eta_1 = \tau_d \left[\left(\frac{\lambda_f}{\tau_d} - 1 \right)^3 e^{-\frac{\theta}{\tau_d}} + 1 \right]. \tag{8.37}$$

Accordingly, substituting (8.5) and (8.37) into (8.22) yields the corresponding feedback controller,

$$C_{\text{f-IO-2}} = \frac{s(\tau_p s + 1)(\eta_1 s + 1)}{k_p (\lambda_f s + 1)^3} \cdot \frac{1}{1 - \frac{\eta_1 s + 1}{(\lambda_f s + 1)^3} e^{-\theta s}}. \tag{8.38}$$

Note that if a PID-type controller is preferred in practice, the Maclaurin approximation formulae in (7.100)–(7.102) can be used to approximate the above controllers for implementation.

8.3.2 Robust Stability Analysis

Based on a low-order process model shown in (8.4) or (8.5), together with the process uncertainties described in a multiplicative form, one can use the standard $M - \Delta$ structure for robust stability analysis (Zhou et al. 1996). It can be seen from Fig. 8.2a or b that, as far as the closed-loop structure between the process input and output is concerned, the transfer function connecting the input and output of the process multiplicative uncertainty is exactly equivalent to the closed-loop complementary sensitivity function, T_d. Hence, it follows from the small gain theorem that the perturbed closed-loop system with the process multiplicative uncertainty holds robust stability if and only if

$$\|T_d\|_\infty < \frac{1}{\|\Delta\|_\infty}, \tag{8.39}$$

where $\Delta = (G - \hat{G})/\hat{G}$ denotes the process multiplicative uncertainty.

For a first-order integrating process shown in (8.4), substituting (8.23) into (8.39), one obtains the closed-loop robust stability constraint for tuning λ_f in $C_{f\text{-I-1}}$,

$$\left\| \frac{(2\lambda_f + \theta)s + 1}{(\lambda_f s + 1)^2} \right\|_\infty < \frac{1}{\|\Delta(s)\|_\infty}. \tag{8.40}$$

For instance, given the process gain uncertainty, $\Delta = \Delta k_p/k_p$, the robust stability constraint for tuning λ_f can be determined as

$$\frac{\lambda_f^2 \omega^2 + 1}{\sqrt{(2\lambda_f + \theta)^2 \omega^2 + 1}} > \frac{|\Delta k_p|}{k_p}, \quad \forall \omega \in [0, +\infty). \tag{8.41}$$

For the time delay uncertainty, $\Delta \theta$, which can be converted to a multiplicative uncertainty form, $\Delta(s) = e^{-\Delta \theta s} - 1$, the robust stability constraint for tuning λ_f can be determined as

$$\frac{\sqrt{(2\lambda_f + \theta)^2 \omega^2 + 1}}{\lambda_f^2 \omega^2 + 1} < \frac{1}{|e^{-j\Delta \theta \omega} - 1|}, \quad \forall \omega \in [0, +\infty). \tag{8.42}$$

8.3 A 2DOF Control Scheme for Integrating Processes

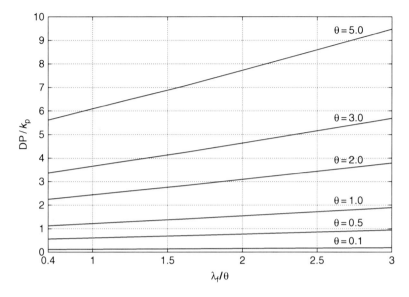

Fig. 8.4 The tuning relationship between DP and λ_f for a first-order integrating process

For the process uncertainty of both gain and time delay, which can be converted to a multiplicative uncertainty form, $\Delta(s) = (1 + \frac{\Delta k_p}{k_p})e^{-\Delta\theta s} - 1$, the robust stability constraint for tuning λ_f can be similarly derived and thus is omitted.

For the corresponding PID tuning in practical applications, simulation tests are motivated to explore the quantitative tuning guideline. Define the disturbance response peak (DP) as the closed-loop output response peak to a unit step change for d_i shown in Fig. 8.2a. Also define the recovery time as the time interval from the moment that d_i is added to the moment that the closed-loop output response is damped down to no larger than 5% of DP, subject to 10% error in estimating the process time delay. Simulation results based on using the PID form in (7.100) for approximating (8.25) are shown in Figs. 8.4 and 8.5 for reference.

For a second-order integrating process shown in (8.5), substituting (8.29) into (8.39), one obtains the closed-loop robust stability constraint for tuning λ_f in $C_{f\text{-}I\text{-}2}$,

$$\left\| \frac{(3\lambda_f + \theta)s + 1}{(\lambda_f s + 1)^3} \right\|_\infty < \frac{1}{\|\Delta(s)\|_\infty}. \tag{8.43}$$

Following a similar analysis, the robust stability constraint for tuning λ_f in $C_{f\text{-}IS\text{-}2}$ (or $C_{f\text{-}IO\text{-}2}$) for an integrating process with slow dynamics can be derived and thus is omitted.

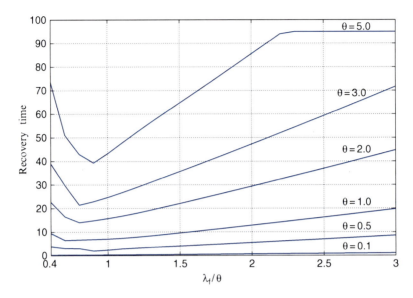

Fig. 8.5 The tuning relationship between the recovery time and λ_f in the presence of 10% error in estimating θ

It should be noted that, in order to compromise the control performance with the robust stability of the closed-loop structure for load disturbance rejection, the following robust performance constraint should be satisfied (Morari and Zafiriou 1989):

$$|\Delta(j\omega)T_d(j\omega)| + |W(j\omega)[1 - T_d(j\omega)]| < 1, \quad \forall \omega \in [0, +\infty). \quad (8.44)$$

where $W(j\omega)$ is a weighting function of the closed-loop sensitivity function, $S(j\omega) = 1 - T_d(j\omega)$. For instance, $W(j\omega)$ can be taken as $1/s$ for a step-type load disturbance.

Hence, in practical applications, tuning λ_f in C_f (e.g., $C_{f\text{-}I\text{-}1}$ or $C_{f\text{-}I\text{-}2}$) aims at a good trade-off between the nominal disturbance rejection performance of the closed-loop structure and its robust stability. Generally speaking, decreasing λ_f can speed up the disturbance response of the closed-loop structure, but degrade its robust stability in the presence of the process uncertainties. In the opposite direction, increasing λ_f can strengthen the closed-loop robust stability, but in exchange for a degradation in the disturbance rejection performance.

8.4 A 2DOF Control Scheme for Unstable Processes

Since the dynamic response characteristics of unstable processes may be different from each other to a large extent, a variety of model structures have been studied to

8.4 A 2DOF Control Scheme for Unstable Processes

describe various unstable processes in industry. Here the following low-order model structures that are widely used for control system design and controller tuning in practical applications are considered,

$$G_{\text{U-1}}(s) = \frac{k_p e^{-\theta s}}{\tau_p s - 1} \tag{8.45}$$

$$G_{\text{U-2}}(s) = \frac{k_p e^{-\theta s}}{(\tau_1 s - 1)(\tau_2 s + 1)} \tag{8.46}$$

$$G_{\text{U-3}}(s) = \frac{k_p e^{-\theta s}}{s(\tau s - 1)} \tag{8.47}$$

$$G_{\text{U-4}}(s) = \frac{k_p e^{-\theta s}}{(\tau_1 s - 1)(\tau_2 s - 1)}, \tag{8.48}$$

where k_p denotes the proportional gain, θ the process time delay, and τ_p, τ, τ_1, and τ_2 are positive time constants reflecting the process dynamic response characteristics.

The corresponding controller design formulae are developed in terms of the 2DOF control structure shown in Fig. 8.2a, which can also be similarly derived in terms of the control structure shown in Fig. 8.2b.

8.4.1 Controller Design

According to the IMC design presented in Sect. 7.3, the feedforward controller for set-point tracking can be determined for an FOPDT unstable process model in (8.45) as

$$C_{\text{s-U-1}} = \frac{\tau_p s - 1}{k_p (\lambda_c s + 1)} \tag{8.49}$$

which corresponds to a desired set-point response transfer function,

$$T_{\text{r-U-1}} = \frac{1}{\lambda_c s + 1} e^{-\theta s}, \tag{8.50}$$

where λ_c is an adjustable parameter for set-point tracking.

Similarly, the feedforward controller for an SOPDT unstable process model in (8.46) can be derived as

$$C_{\text{s-U-2}} = \frac{(\tau_1 s - 1)(\tau_2 s + 1)}{k_p (\lambda_c s + 1)^2} \tag{8.51}$$

which corresponds to a desired set-point response transfer function,

$$T_{r\text{-U-2}} = \frac{1}{(\lambda_c s + 1)^2} e^{-\theta s}. \tag{8.52}$$

For a second-order model with an integrator shown in (8.47), the feedforward controller can be derived accordingly as

$$C_{s\text{-U-3}} = \frac{s(\tau s - 1)}{k_p (\lambda_c s + 1)^2} \tag{8.53}$$

which corresponds to a desired set-point response transfer function,

$$T_{r\text{-U-3}} = \frac{1}{(\lambda_c s + 1)^2} e^{-\theta s}. \tag{8.54}$$

For another second-order model with two RHP poles shown in (8.48), the feedforward controller can be derived accordingly as

$$C_{s\text{-U-4}} = \frac{(\tau_1 s - 1)(\tau_2 s - 1)}{k_p (\lambda_c s + 1)^2} \tag{8.55}$$

which corresponds to a desired set-point response transfer function,

$$T_{r\text{-U-4}} = \frac{1}{(\lambda_c s + 1)^2} e^{-\theta s}. \tag{8.56}$$

Following a similar analysis as in Sect. 8.3.1, the time domain performance specification can be quantitatively tuned through the single adjustable parameter λ_c in each of the above controllers, such as the rise time for a step change of the set-point.

For load disturbance rejection, the desired closed-loop complementary sensitivity function in (8.21) can be used to derive the feedback controller, C_f, in Fig. 8.2a.

For a first-order unstable process shown in (8.45), there are $l = 1$ and $m = 1$. It follows from (8.21) that

$$T_{d\text{-U-1}} = \frac{\alpha_1 s + 1}{(\lambda_f s + 1)^2} e^{-\theta s}. \tag{8.57}$$

Obviously it satisfies the asymptotic constraint in (8.17). Then, it follows from (8.18) that

$$\lim_{s \to 1/\tau_p} \left[1 - \frac{\alpha_1 s + 1}{(\lambda_f s + 1)^2} e^{-\theta s} \right] = 0. \tag{8.58}$$

8.4 A 2DOF Control Scheme for Unstable Processes

Solving (8.58) obtains

$$\alpha_1 = \tau_p \left[\left(\frac{\lambda_f}{\tau_p} + 1 \right)^2 e^{\frac{\theta}{\tau_p}} - 1 \right]. \tag{8.59}$$

Substituting (8.45) and (8.57) into (8.22) yields

$$C_{\text{f-U-1}} = \frac{\alpha_1 s + 1}{k_p (\lambda_f s + 1)^2} \cdot \frac{\tau_p s - 1}{1 - \frac{\alpha_1 s + 1}{(\lambda_f s + 1)^2} e^{-\theta s}}. \tag{8.60}$$

Note that there exists an implicit RHP zero-pole canceling at $s = 1/\tau_p$ in (8.60), which may cause the controller to work unstably. Therefore, the second multiplier at the right-hand side of (8.60) cannot be directly implemented using the closed-loop unit shown in Fig. 8.3. A rational approximation is therefore needed for implementation. The analytical approximation based on the Maclaurin expansion series, as shown in (7.100), can be used to yield a PID form by letting $C_{\text{f-U-1}} = M(s)/s$, since the second multiplier has an integral property to eliminate the system output deviation from the set-point.

For better approximation, the following analytical approximation formula based on the linear fractional Padé expansion (Liu et al. 2005b) can be used,

$$D_{N/N}(s) = \frac{\sum_{j=0}^{N} d_j s^j}{s \sum_{i=0}^{N-1} c_i s^i}, \tag{8.61}$$

where N is a user-specified order to achieve the desirable performance specification for load disturbance rejection, and c_i and d_j are determined by the following two linear matrix equations:

$$\begin{bmatrix} d_0 \\ d_1 \\ \vdots \\ d_N \end{bmatrix} = \begin{bmatrix} b_0 & 0 & 0 & \cdots & 0 \\ b_1 & b_0 & 0 & \cdots & 0 \\ \vdots & \vdots & \ddots & \cdots & \vdots \\ b_N & b_{N-1} & b_{N-2} & \cdots & b_1 \end{bmatrix} \begin{bmatrix} c_0 \\ c_1 \\ \vdots \\ c_{N-1} \end{bmatrix} \tag{8.62}$$

$$\begin{bmatrix} b_N & b_{N-1} & \cdots & b_2 \\ b_{N+1} & b_N & \cdots & b_3 \\ \vdots & \vdots & \ddots & \vdots \\ b_{2N-2} & b_{2N-3} & \cdots & b_N \end{bmatrix} \begin{bmatrix} c_1 \\ c_2 \\ \vdots \\ c_{N-1} \end{bmatrix} = - \begin{bmatrix} b_{N+1} \\ b_{N+2} \\ \vdots \\ b_{2N-1} \end{bmatrix}, \tag{8.63}$$

where $b_i = M^{(i)}(0)/i!$ and $i = 0, 1, \ldots, 2N - 1$ are the Maclaurin coefficients of (7.100) and c_0 should be taken as

$$c_0 = \begin{cases} 1, & c_i \geq 0; \\ -1, & c_i < 0. \end{cases} \quad (8.64)$$

Note that there exists an integrator in (8.61), which guarantees such an approximation essentially consistent with the ideally desired controller in (8.60).

For instance, letting $N = 2$ in (8.61) obtains an approximation formula in the form of PID,

$$F_{2/2}(s) = \left(k_C + \frac{1}{\tau_I s} + \tau_D s\right) \frac{1}{\tau_F s + 1}, \quad (8.65)$$

where

$$k_C = \frac{d_1}{c_0}, \quad \tau_I = \frac{1}{b_0}, \quad \tau_D = \frac{d_2}{c_0}, \quad \tau_F = \frac{c_1}{c_0},$$

$$d_0 = b_0 c_0, \quad d_1 = b_1 c_0 + b_0 c_1, \quad d_2 = b_2 c_0 + b_1 c_1, \quad c_1 = -\frac{b_3}{b_2}.$$

Letting $N = 3$ gives a third-order approximation formula,

$$F_{3/3}(s) = \frac{d_3 s^3 + d_2 s^2 + d_1 s + d_0}{s(c_2 s^2 + c_1 s + c_0)}, \quad (8.66)$$

where

$$c_1 = \frac{b_2 b_5 - b_3 b_4}{b_3^2 - b_2 b_4}, \quad c_2 = \frac{b_4^2 - b_3 b_5}{b_3^2 - b_2 b_4},$$

$$d_0 = b_0 c_0, \quad d_1 = b_1 c_0 + b_0 c_1, \quad d_2 = b_2 c_0 + b_1 c_1 + b_0 c_2, \quad d_3 = b_3 c_0 + b_2 c_1 + b_1 c_2.$$

In fact, the above third-order approximation can be implemented by three low-order controllers, i.e.,

$$F_{3/3}(s) = \frac{d_3 s^2 + d_2 s + d_1}{c_2 s^2 + c_1 s + c_0} + \frac{d_0}{s(c_2 s^2 + c_1 s + c_0)}, \quad (8.67)$$

where the first part is exactly a second-order lead-lag controller, and the second part is an integrator in tandem with a second-order lag controller.

Remark 8.1. Note that each of the above approximation formulae for implementing the feedback controller is actually tuned by the single adjustable parameter λ_f shown in (8.60). The choice of c_0 is to keep all of c_i ($i = 0, 1, \ldots, N - 1$) the same sign in order to prevent any RHP zero from being enclosed in the denominator of such an

8.4 A 2DOF Control Scheme for Unstable Processes

approximation. According to the Routh-Hurwitz stability criterion, it is obvious that a low-order approximation in terms of $N \leq 3$ can ensure the closed-loop internal stability. For $N \geq 4$, the same sign of c_i ($i = 0, 1, \ldots, N - 1$) still cannot guarantee no presence of RHP poles, and therefore, the Routh-Hurwitz criterion should be used for verification in combination with tuning λ_f, before such an approximation is adopted for implementation. It is obvious that a higher-order approximation facilitates better closed-loop performance for load disturbance rejection. Hence, it depends on the user choice of a trade-off between the achievable disturbance rejection performance and the approximation complexity for implementation. ◇

For a second-order unstable process in (8.46), there are $l = 2$ and $m = 1$. It follows from (8.21) that

$$T_{\text{d-U-2}} = \frac{\alpha_1 s + 1}{(\lambda_f s + 1)^3} e^{-\theta s}. \tag{8.68}$$

Similarly, it follows from (8.18) that

$$\lim_{s \to 1/\tau_1} \left[1 - \frac{\alpha_1 s + 1}{(\lambda_f s + 1)^3} e^{-\theta s} \right] = 0. \tag{8.69}$$

Solving (8.69) obtains

$$\alpha_1 = \tau_1 \left[\left(\frac{\lambda_f}{\tau_1} + 1 \right)^3 e^{\frac{\theta}{\tau_1}} - 1 \right]. \tag{8.70}$$

Substituting (8.46) and (8.68) into (8.22) yields

$$C_{\text{f-U-2}} = \frac{(\tau_2 s + 1)(\alpha_1 s + 1)}{k_p (\lambda_f s + 1)^3} \cdot \frac{\tau_1 s - 1}{1 - \frac{\alpha_1 s + 1}{(\lambda_f s + 1)^3} e^{-\theta s}}. \tag{8.71}$$

For a second-order model with an integrator shown in (8.47), there are $l = 2$ and $m = 2$. It follows from (8.21) that

$$T_{\text{d-U-3}} = \frac{\alpha_2 s^2 + \alpha_1 s + 1}{(\lambda_f s + 1)^4} e^{-\theta s}. \tag{8.72}$$

Using the asymptotic constraints in (8.18) and (8.19), one can obtain

$$\lim_{s \to 1/\tau} \left[1 - \frac{\alpha_2 s^2 + \alpha_1 s + 1}{(\lambda_f s + 1)^4} e^{-\theta s} \right] = 0 \tag{8.73}$$

$$\lim_{s \to 0} \frac{d}{ds} \left[1 - \frac{\alpha_2 s^2 + \alpha_1 s + 1}{(\lambda_f s + 1)^4} e^{-\theta s} \right] = 0. \tag{8.74}$$

Solving (8.73) and (8.74) yields

$$\begin{cases} \alpha_1 = 4\lambda_f + \theta; \\ \alpha_2 = \tau^2 \left[\left(\dfrac{\lambda_f}{\tau} + 1 \right)^4 e^{\frac{\theta}{\tau}} - \dfrac{\alpha_1}{\tau} - 1 \right]. \end{cases} \quad (8.75)$$

Substituting (8.47) and (8.72) into (8.22) yields

$$C_{\text{f-U-3}} = \frac{\alpha_2 s^2 + \alpha_1 s + 1}{k_p (\lambda_f s + 1)^4} \cdot \frac{s(\tau s - 1)}{1 - \frac{\alpha_2 s^2 + \alpha_1 s + 1}{(\lambda_f s + 1)^4} e^{-\theta s}}. \quad (8.76)$$

For another second-order model with two RHP poles shown in (8.48), there are $l = 2$ and $m = 2$. It follows from (8.21) that

$$T_{\text{d-U-4}} = \frac{\alpha_2 s^2 + \alpha_1 s + 1}{(\lambda_f s + 1)^4} e^{-\theta s}. \quad (8.77)$$

Similarly, it follows from (8.18) that

$$\lim_{s \to 1/\tau_1} \left[1 - \frac{\alpha_2 s^2 + \alpha_1 s + 1}{(\lambda_f s + 1)^4} e^{-\theta s} \right] = 0 \quad (8.78)$$

$$\lim_{s \to 1/\tau_2} \left[1 - \frac{\alpha_2 s^2 + \alpha_1 s + 1}{(\lambda_f s + 1)^4} e^{-\theta s} \right] = 0. \quad (8.79)$$

Solving (8.78) and (8.79) yields

$$\begin{cases} \alpha_1 = \dfrac{1}{\tau_2 - \tau_1} \left\{ \tau_2^2 \left[\left(\dfrac{\lambda_f}{\tau_2} + 1 \right)^4 e^{\frac{\theta}{\tau_2}} - 1 \right] - \tau_1^2 \left[\left(\dfrac{\lambda_f}{\tau_1} + 1 \right)^4 e^{\frac{\theta}{\tau_1}} - 1 \right] \right\} \\ \alpha_2 = \dfrac{\tau_1^2 \tau_2}{\tau_2 - \tau_1} \left[\left(\dfrac{\lambda_f}{\tau_1} + 1 \right)^4 e^{\frac{\theta}{\tau_1}} - 1 \right] - \dfrac{\tau_1 \tau_2^2}{\tau_2 - \tau_1} \left[\left(\dfrac{\lambda_f}{\tau_2} + 1 \right)^4 e^{\frac{\theta}{\tau_2}} - 1 \right] \end{cases}. \quad (8.80)$$

Substituting (8.48) and (8.77) into (8.22) yields

$$C_{\text{f-U-4}} = \frac{\alpha_2 s^2 + \alpha_1 s + 1}{k_p (\lambda_f s + 1)^4} \cdot \frac{(\tau_1 s - 1)(\tau_2 s - 1)}{1 - \frac{\alpha_2 s^2 + \alpha_1 s + 1}{(\lambda_f s + 1)^4} e^{-\theta s}}. \quad (8.81)$$

Due to the fact that there exists RHP zero-pole canceling in the controller formulae of (8.71), (8.76), and (8.81) for second-order unstable processes, none of them can be directly implemented in practice. The analytical approximation

8.4 A 2DOF Control Scheme for Unstable Processes

formulae in (7.100)–(7.102) should be used to give a corresponding PID form for implementation. Alternatively, higher-order approximation formulae in (8.61)–(8.64) can be used to improve the control performance.

For an unstable process with slow dynamics, corresponding to a large time constant (τ_2) in the SOPDT model shown in (8.46), in order to reduce the influence of the slow dynamics to the load disturbance response, the following asymptotic constraint is imposed to the controller design:

$$\lim_{s \to -1/\tau_2} (1 - T_d) = 0. \tag{8.82}$$

Correspondingly, the closed-loop transfer function for rejecting a step-type load disturbance occurring at the process input side is proposed as

$$T_{\text{d-US-2}} = \frac{\eta_2 s^2 + \eta_1 s + 1}{(\lambda_f s + 1)^4} e^{-\theta s}, \tag{8.83}$$

where η_1 and η_2 are taken to satisfy the asymptotic constraints in (8.18) and (8.82). Substituting (8.83) into (8.18) and (8.82), one can obtain

$$\begin{cases} \eta_1 = \dfrac{1}{\tau_1 + \tau_2} \left[\tau_1^2 \left(\dfrac{\lambda_f}{\tau_1} + 1 \right)^4 e^{\frac{\theta}{\tau_1}} - \tau_2^2 \left(\dfrac{\lambda_f}{\tau_2} - 1 \right)^4 e^{-\frac{\theta}{\tau_2}} \right] + \tau_2 - \tau_1 \\[2mm] \eta_2 = \tau_1^2 \left[\left(\dfrac{\lambda_f}{\tau_1} + 1 \right)^4 e^{\frac{\theta}{\tau_1}} - 1 \right] - \tau_1 \eta_1 \end{cases} \tag{8.84}$$

Hence, substituting (8.46) and (8.83) into (8.22) yields the corresponding feedback controller,

$$C_{\text{f-US-2}} = \frac{(\tau_2 s + 1)(\eta_2 s^2 + \eta_1 s + 1)}{k_p (\lambda_f s + 1)^4} \cdot D(s), \tag{8.85}$$

where

$$D(s) = \frac{\tau_1 s - 1}{1 - \frac{\eta_2 s^2 + \eta_1 s + 1}{(\lambda_f s + 1)^4} e^{-\theta s}}. \tag{8.86}$$

It is seen that there exists RHP zero-pole canceling in (8.86). A rational approximation is therefore needed for implementation, which can be obtained using the above formulae in (8.61)–(8.64).

To reject a step-type load disturbance occurring at the process output side with a slow transfer function of $G_d = k_d/(\tau_d s + 1)$, the closed-loop transfer function is correspondingly proposed as

$$T_{\text{d-UO-2}} = \frac{\eta_1 s + 1}{(\lambda_f s + 1)^3} e^{-\theta s}. \tag{8.87}$$

Substituting (8.87) into (8.32) with $\tau = \tau_d$, one can obtain

$$\eta_1 = \tau_d \left[\left(\frac{\lambda_f}{\tau_d} - 1 \right)^3 e^{-\frac{\theta}{\tau_d}} + 1 \right]. \tag{8.88}$$

Accordingly, substituting (8.46) and (8.88) into (8.22) yields the corresponding feedback controller,

$$C_{\text{f-UO-2}} = \frac{(\tau_1 s - 1)(\tau_2 s + 1)(\eta_1 s + 1)}{k_p (\lambda_f s + 1)^3} \cdot \frac{1}{1 - \frac{\eta_1 s + 1}{(\lambda_f s + 1)^3} e^{-\theta s}}, \tag{8.89}$$

where the second multiplier can be practically implemented using the closed-loop unit shown in Fig. 8.3.

8.4.2 Robust Stability Analysis

Since the same control structure shown in Fig. 8.2 is used for both integrating and unstable processes, the sufficient and necessary condition for maintaining robust stability of the closed-loop structure can be similarly developed. Specifically, when the process multiplicative uncertainty is considered for the convenience of analysis, robust stability constraints can be derived similar to those given in Sect. 8.3.2, according to the small gain theorem (Zhou et al. 1996).

For a first-order unstable process shown in (8.45), substituting (8.57) and (8.59) into (8.39), one can obtain the closed-loop robust stability constraint for tuning λ_f in $C_{\text{f-U-1}}$,

$$\left\| \frac{\tau_p [(\lambda_f / \tau_p + 1)^2 e^{\theta/\tau_p} - 1]s + 1}{(\lambda_f s + 1)^2} \right\|_\infty < \frac{1}{\|\Delta(s)\|_\infty}. \tag{8.90}$$

For a second-order unstable process shown in (8.46), substituting (8.68) and (8.70) into (8.39), one can obtain the closed-loop robust stability constraint for tuning λ_f in $C_{\text{f-U-2}}$,

$$\left\| \frac{\tau_1 [(\frac{\lambda_f}{\tau_1} + 1)^3 e^{\theta/\tau_1} - 1]s + 1}{(\lambda_f s + 1)^3} \right\|_\infty < \frac{1}{\|\Delta(s)\|_\infty}. \tag{8.91}$$

For a second-order model with an integrator as shown in (8.47), substituting (8.72) and (8.75) into (8.39), one can obtain the closed-loop robust stability constraint for tuning λ_f in $C_{\text{f-U-3}}$,

$$\left\| \frac{\tau^2[(\frac{\lambda_f}{\tau}+1)^4 e^{\frac{\theta}{\tau}} - \frac{4\lambda_f+\theta}{\tau} - 1]s^2 + (4\lambda_f+\theta)s + 1}{(\lambda_f s + 1)^4} \right\|_\infty < \frac{1}{\|\Delta(s)\|_\infty}. \qquad (8.92)$$

For another second-order model with two RHP poles, as shown in (8.48), substituting (8.77) and (8.80) into (8.39), one can obtain the closed-loop robust stability constraint for tuning λ_f in $C_{\text{f-U-4}}$,

$$\left\| \frac{\alpha_2 s^2 + \alpha_1 s + 1}{(\lambda_f s + 1)^4} \right\|_\infty < \frac{1}{\|\Delta(s)\|_\infty}. \qquad (8.93)$$

For a second-order model shown in (8.46) with a large time constant (τ_2), substituting (8.83) and (8.84) into (8.39), one can obtain the closed-loop robust stability constraint for tuning λ_f in $C_{\text{f-US-2}}$ to reject a step-type load disturbance occurring at the process input side,

$$\left\| \frac{\eta_2 s^2 + \eta_1 s + 1}{(\lambda_f s + 1)^4} \right\|_\infty < \frac{1}{\|\Delta(s)\|_\infty}. \qquad (8.94)$$

Similarly, the robust constraint for tuning λ_f in $C_{\text{f-UO-2}}$ to reject a step-type load disturbance occurring at the process output side can be obtained by substituting (8.87) and (8.88) into (8.39).

8.5 Illustrative Examples

Six examples from existing references are used to demonstrate the effectiveness and merits of the presented 2DOF control scheme and controller designs, respectively, for integrating and unstable processes.

Example 8.1. Consider a first-order integrating process studied in the references (Majhi and Atherton 2000; Mataušek and Micic 1999),

$$G_1 = \frac{e^{-5s}}{s}.$$

In the modified Smith predictor (SP) control scheme for 2DOF tuning (Majhi and Atherton 2000), the parameters of three controllers were taken as $k_p = 0.5$, $T_i = 1$, $K_f = 1$, and $K_d = 0.105$. In the modified SP scheme of Mataušek and Micic (1999), the parameters for tuning three controllers were taken as $k_p = 1$, $\tau = 5$, $T_r = 1/0.6$, $\Phi_{pm} = 64°$, and $\alpha = 0.4$.

Using the controller formulae of (8.6) and (8.25), it follows for the 2DOF control scheme shown in Fig. 8.2a that

$$C_{s\text{-}I\text{-}1} = \frac{s}{\lambda_c s + 1}$$

$$C_{f\text{-}I\text{-}1} = \frac{s(\alpha_1 s + 1)}{(\lambda_f s + 1)^2} \cdot \frac{1}{1 - \frac{\alpha_1 s + 1}{(\lambda_f s + 1)^2} e^{-5s}},$$

where $\alpha_1 = 2\lambda_f + 5$.

For comparison, take $\lambda_c = 2.0$ in order to obtain a similar rising speed of the set-point response with those of Majhi's and Matausek's methods, and take $\lambda_f = 3.0$ for load disturbance rejection. Using the PID tuning formula in (7.100) to approximate $C_{f\text{-}I\text{-}1}$ for implementation, it follows that

$$C_{f\text{-PID}} = 0.2576 + \frac{1}{51.5s} + \frac{0.5069s}{0.05s + 1}.$$

By adding a unit step change to the set-point and a negative step load disturbance with a magnitude of 0.1 to the process input at $t = 50(s)$, the control results are shown in Fig. 8.6. It is seen that the proposed control design gives obviously improved load disturbance response, and the IMC-based PID tuning shows a comparable performance.

Assume that there exists 10% error in estimating the process time delay, e.g., it is actually 10% larger. The perturbed system responses are shown in Fig. 8.7. It is seen that the proposed control design holds robust stability well.

For illustrating the 2DOF performance optimization, simulation tests for the above perturbed process are made for three cases of tuning the controller parameters of λ_c and λ_f, which are listed in Table 8.1 together with the corresponding controller forms. The control results are shown in Fig. 8.8. It is seen that there exists relative independence in tuning λ_c and λ_f for separate optimization of the set-point tracking and load disturbance rejection. Monotonically increasing λ_f decreases the oscillation in the load disturbance response. On the other hand, gradually increasing λ_c decreases the set-point response oscillation. Hence, it is convenient to tune λ_f monotonically to reach a good trade-off between the disturbance rejection performance and robust stability of the closed-loop structure in the 2DOF control scheme shown in Fig. 8.2. So is for tuning λ_c to meet a good trade-off between the tracking performance and robustness of the set-point response.

Example 8.2. Consider a second-order integrating process studied in the reference (Normey-Rico and Camacho 2009),

$$G_2 = \frac{0.1e^{-5s}}{s(5s + 1)}.$$

8.5 Illustrative Examples

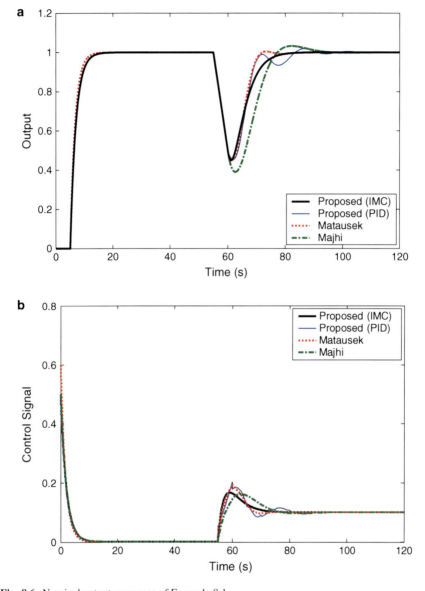

Fig. 8.6 Nominal output responses of Example 8.1

A 2DOF SP control scheme was given by Normey-Rico and Camacho (2009), where the two controllers for the set-point tracking and load disturbance rejection were designed as $C = 1.667(5s + 1)/(0.5s + 1)$ and $F_r = (17s + 1)/(6s + 1)$.

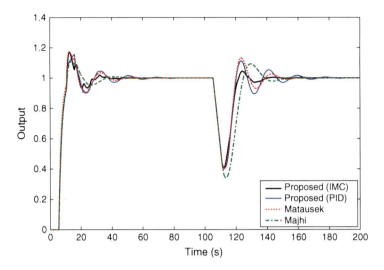

Fig. 8.7 Perturbed system responses of Example 8.1

Table 8.1 Tuning controller parameters for the perturbed case of Example 8.1

Tuning parameters	$C_{\text{s-I-1}}$	$C_{\text{f-I-1}}$ (PID form)		
$\lambda_c = 2.0$, $\lambda_f = 3.0$	$\dfrac{s}{2s+1}$	$k_C = 0.2576$,	$\tau_I = 51.5$,	$\tau_D = 0.5069$
$\lambda_c = 2.0$, $\lambda_f = 5.0$	$\dfrac{s}{2s+1}$	$k_C = 0.1932$,	$\tau_I = 87.5$,	$\tau_D = 0.3306$
$\lambda_c = 6.0$, $\lambda_f = 5.0$	$\dfrac{s}{6s+1}$	$k_C = 0.1932$,	$\tau_I = 87.5$,	$\tau_D = 0.3306$

Using the controller formulae of (8.8) and (8.31), it follows that

$$C_{\text{s-I-2}} = \frac{s(5s+1)}{0.1(\lambda_c s + 1)^2}$$

$$C_{\text{f-I-2}} = \frac{s(5s+1)[(3\lambda_f + 5)s + 1]}{0.1(\lambda_f s + 1)^3} \cdot \frac{1}{1 - \frac{(3\lambda_f + 5)s + 1}{(\lambda_f s + 1)^3} e^{-5s}}$$

Note that the second multiplier in $C_{\text{f-I-2}}$ can be implemented using the closed-loop unit shown in Fig. 8.3.

To further enhance disturbance rejection design, the controller formulae of (8.34) and (8.35) for rejecting a step-type load disturbance occurring at the process input side give

8.5 Illustrative Examples

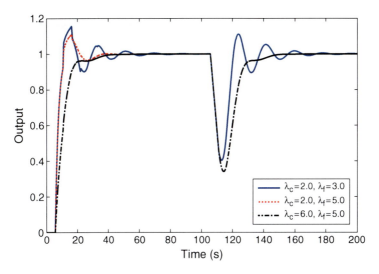

Fig. 8.8 Perturbed system responses of Example 8.1 in terms of tuning the controller parameters

$$C_{\text{f-IS-2}} = \frac{s(5s+1)(\eta_2 s^2 + \eta_1 s + 1)}{0.1(\lambda_f s + 1)^4} \cdot \frac{1}{1 - \frac{\eta_2 s^2 + \eta_1 s + 1}{(\lambda_f s + 1)^4} e^{-5s}}$$

$$\begin{cases} \eta_1 = 4\lambda_f + 5 \\ \eta_2 = 5\eta_1 + 25(0.2\lambda_f - 1)^4 e^{-1} - 25 \end{cases}$$

For comparison, taking $\lambda_c = 3$ and $\lambda_f = 3.6$, the control results are shown in Fig. 8.9 for adding a unit step change to the set-point and then to the process input at $t = 50(s)$. It is seen that the load disturbance response is recovered faster by the enhanced IMC design of (8.35), in terms of the similar set-point tracking speed and the same magnitude of disturbance response peak. Note that the IMC-based disturbance rejection design of (8.31) results in a very close response with the enhanced IMC design of (8.35) when using the above controller parameters. Given the same set-point tracking speed (by taking $\lambda_c = 3$), if $\lambda_f = 2$ is tuned to improve the load disturbance response, it is seen from Fig. 8.9 that $\lambda_f = 1.5$ is required by the IMC-based design of (8.31) to obtain the same disturbance response peak, but still with a longer recovery time.

Note that, if the process time constant is larger, e.g., $\tau_p = 20$, the advantage of the enhanced IMC design of (8.35) for load disturbance rejection will become more obvious, as shown in Fig. 8.10, where the recovery time for load disturbance response is reduced almost by 30% in terms of the same disturbance response peak.

Then assume that the process proportional gain (k_p) and time constant (τ_p) are actually 20% larger. The perturbed output response is shown in Fig. 8.11, indicating good robust stability of the 2DOF control scheme shown in Fig. 8.2.

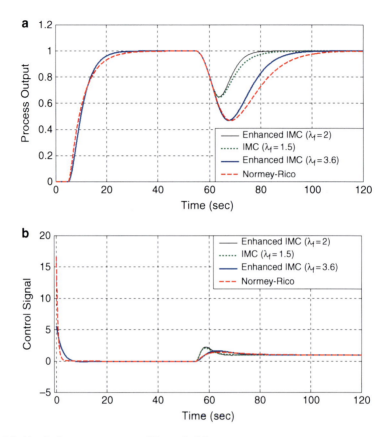

Fig. 8.9 Nominal output responses of Example 8.2

To demonstrate the control performance for rejecting a load disturbance from the process output side with a slow dynamics, assume that the load disturbance transfer function is $G_d = 1/(10s + 1)$, but is actually estimated with 20% error for the control design, i.e., $\widehat{G}_d = 1/(8s + 1)$. The corresponding controller formulae of (8.37) and (8.38) give

$$C_{f\text{-}IO\text{-}2} = \frac{s(5s+1)(\eta_1 s + 1)}{0.1(\lambda_f s + 1)^3} \cdot \frac{1}{1 - \frac{\eta_1 s + 1}{(\lambda_f s + 1)^3} e^{-5s}}$$

$$\eta_1 = 8(0.125\lambda_f - 1)^3 e^{-0.625} + 8.$$

Taking $\lambda_f = 2$ for comparison, the control results are shown in Fig. 8.12 in the presence of a unit step change of the load disturbance. It is seen that apparently improved disturbance response is obtained by the enhanced IMC design of (8.38),

8.5 Illustrative Examples

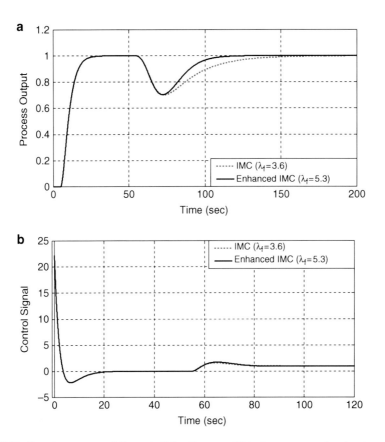

Fig. 8.10 Output responses of Example 8.2 with $\tau_p = 20$ in the presence of a step-type load disturbance from the process input side

which demonstrates that based on estimating the deterministic load disturbance dynamics, further enhanced disturbance rejection performance can thus be obtained.

Example 8.3. Consider a first-order unstable process widely studied in existing literature,

$$G_3 = \frac{e^{-0.4s}}{s-1}.$$

In the modified 2DOF IMC scheme (Tan et al. 2003), the controller parameters were taken as $k_0 = 2$, $\lambda = 0.4$, $K_c = 2.079$, and $T_c = 0.156$. In the modified SP scheme (Majhi and Atherton 2000), the control parameters are taken as $k_p = 1$, $T_i = 0.4$, $T_f = -0.3$, $K_f = 2$, $K_d = 1.5811$, according to the tuning formulae.

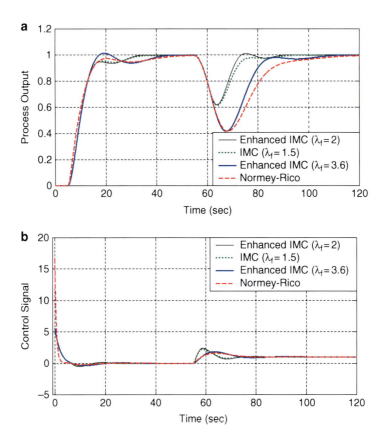

Fig. 8.11 Perturbed system responses of Example 8.2

Using the controller formulae of (8.49) and (8.60), it follows that

$$C_{\text{s-U-1}} = \frac{s-1}{\lambda_c s + 1}$$

$$C_{\text{f-U-1}} = \frac{(s-1)(\alpha_1 s + 1)}{(\lambda_f s + 1)^2} \cdot \frac{1}{1 - \frac{\alpha_1 s + 1}{(\lambda_f s + 1)^2} e^{-0.4s}},$$

where $\alpha_1 = (\lambda_f + 1)^2 e^{0.4} - 1$.

For comparison, take $\lambda_c = 0.4$ in order to obtain the same rise speed of the set-point response with those of Tan's and Majhi's methods. For load disturbance rejection, taking $\lambda_f = 0.4$ and using the PID tuning formula in (7.100) to approximate $C_{\text{f-U-1}}$ for implementation, it follows that

$$C_{\text{f-PID}} = 2.8972 + \frac{1}{0.724s} + \frac{0.469s}{0.04s + 1}.$$

8.5 Illustrative Examples

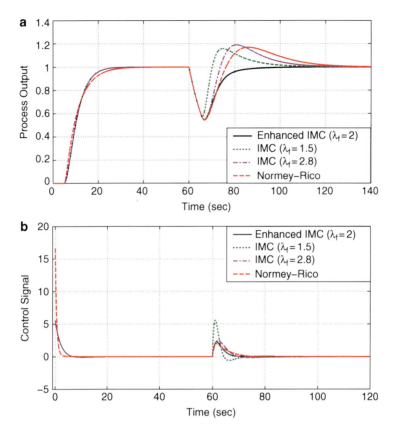

Fig. 8.12 Output responses of Example 8.2 in the presence of a slow load disturbance from the output side

Table 8.2 Comparison of ISE for load disturbance rejection in Example 3

ISE	Proposed	Tan	Majhi
Attenuation	0.3098	0.3429	0.966

By adding a unit step change to the set-point and an inverse step change of load disturbance to the process input at $t = 5(s)$, the control results are shown in Fig. 8.13. The ISE specifications of load disturbance response resulting from these methods are listed in Table 8.2, which indicates that improved disturbance rejection performance is obtained by the proposed PID tuning.

Now suppose that there exists 20% error in estimating the process time delay and the unstable time constant, e.g., both of them are actually 20% larger. The perturbed system responses are shown in Fig. 8.14, demonstrating that the proposed PID tuning method holds well the control system robust stability compared to the other two methods. Note that by monotonically increasing the single adjustable parameter, λ_f, in $C_{f\text{-PID}}$, the control system robust stability can be further enhanced, but at

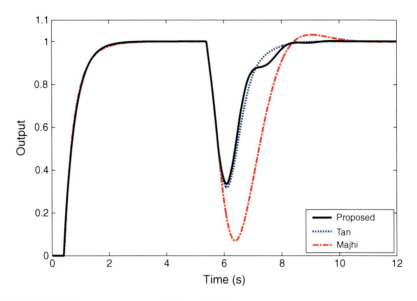

Fig. 8.13 Nominal output responses of Example 8.3

Table 8.3 Controller settings for Example 8.3

Tuning parameters	$C_{\text{s-U-1}}$	$C_{\text{f-U-1}}$ (PID form)
$\lambda_c = 0.4$, $\lambda_f = 0.5$	$\dfrac{s-1}{0.4s+1}$	$k_C = 2.634$, $\tau_I = 0.9566$, $\tau_D = 0.4058$
$\lambda_c = 0.8$, $\lambda_f = 0.6$	$\dfrac{s-1}{0.8s+1}$	$k_C = 2.4394$, $\tau_I = 1.2191$, $\tau_D = 0.3596$

the cost of a degradation in the disturbance rejection performance. For illustration, two groups of simulation test are made for the above perturbed unstable process, as listed in Table 8.3. The control results are shown in Fig. 8.15. It is seen that increasing λ_f has decreased the oscillation in the load disturbance response. On the other hand, increasing λ_c has decreased the set-point response oscillation. Hence, it is convenient to monotonically tune λ_c and λ_f for the optimization of the set-point tracking and load disturbance rejection, respectively, especially in the presence of process uncertainties as usually encountered in practice.

Example 8.4. Consider the third-order unstable process studied in the reference (Tan et al. 2003),

$$G_4 = \frac{1}{(5s-1)(2s+1)(0.5s+1)}e^{-0.5s}.$$

An IMC-based 2DOF control scheme was presented by Tan et al. (2003), where three low-order controllers of PD or lead-lag type were configured for the set-point tracking and load disturbance rejection.

8.5 Illustrative Examples

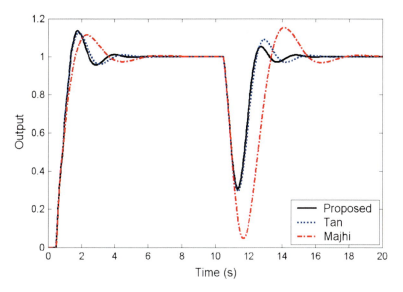

Fig. 8.14 Perturbed system responses of Example 8.3

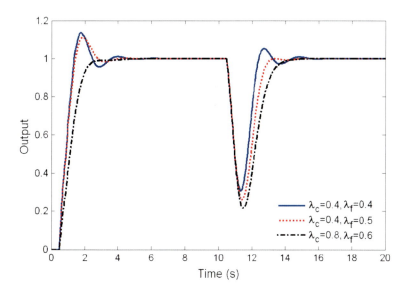

Fig. 8.15 Perturbed system responses of Example 8.3 in terms of increasing the controller tuning parameters

Based on an SOPDT model identified in example 6.3 of Chap. 6, $G_m = 1.0e^{-0.94s}/(5s-1)(2.07s+1)$, the IMC design in (8.51) and (8.71) gives the 2DOF controllers,

$$C_{\text{s-U-2}} = \frac{(5s-1)(2.07s+1)}{(\lambda_c s+1)^2}$$

$C_{\text{f-U-2}}$

$$= \frac{(2.07s+1)[5s(0.2\lambda_f+1)^3 e^{0.188} - 5s+1]}{(\lambda_f s+1)^3} \cdot \frac{5s-1}{1 - \frac{5[(0.2\lambda_f+1)^3 e^{0.188}-1]s+1}{(\lambda_f s+1)^3} e^{-0.94s}}.$$

Note that the second multiplier in $C_{\text{f-U-2}}$ includes a RHP zero-pole canceling and therefore needs to be approximated by using (8.61)–(8.64) for implementation.

For a unit-step type load disturbance entering into the process input as assumed by Tan et al. (2003), taking $\lambda_f = 0.85$ to obtain the same disturbance response peak with that of Tan et al. (2003) for comparison, it follows from (8.61) to (8.64) for the second multiplier in $C_{\text{f-U-2}}$ that

$$D_{\text{IMC}-4/2} = \frac{0.1611s^4 + 1.0761s^3 + 2.4857s^2 + 2.4195s + 0.8515}{s(0.0489s^2 + 0.0886s + 1)}.$$

In contrast, using the enhance IMC design in (8.84)–(8.86) for load disturbance rejection gives the controller,

$$C_{\text{f-US-2}} = \frac{(2.07s+1)(\eta_2 s^2 + \eta_1 s + 1)}{(\lambda_f s+1)^4} \cdot \frac{5s-1}{1 - \frac{\eta_2 s^2 + \eta_1 s + 1}{(\lambda_f s+1)^4} e^{-0.94s}}$$

$$\begin{cases} \eta_1 = 3.5361(0.2\lambda_f+1)^4 e^{0.188} - 0.6061(0.4831\lambda_f - 1)^4 e^{-0.4541} - 2.93 \\ \eta_2 = 25[(0.2\lambda_f+1)^4 e^{0.188} - 1] - 5\eta_1 \end{cases}.$$

By taking $\lambda_f = 1.05$ for comparison, the second multiplier in $C_{\text{f-US-2}}$ is approximated using (8.61)–(8.64) as

$$D_{4/2} = \frac{1.953s^4 + 5.7424s^3 + 7.1852s^2 + 4.1925s + 0.9479}{s(0.1414s^2 + 2.0973s + 1)}.$$

The corresponding output responses are shown in Fig. 8.16. It is seen that the load disturbance response is recovered faster by the enhanced IMC design in terms of the similar set-point tracking speed and the same magnitude of disturbance response peak. Note that, compared to the IMC design in (8.71), more obvious improvement can be observed by tuning a smaller value of λ_f. For instance, take $\lambda_f = 0.8$ in the

8.5 Illustrative Examples

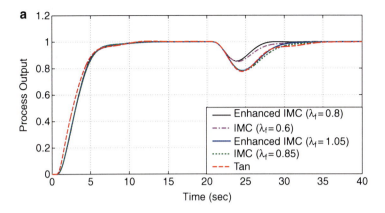

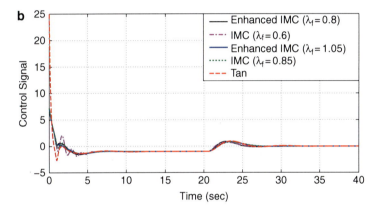

Fig. 8.16 Output responses of Example 8.4 in the presence of a step-type load disturbance from the input side

enhanced IMC design in (8.84)–(8.86) and $\lambda_f = 0.6$ in the IMC design in (8.71), corresponding to

$$D_{4/2} = \frac{1.4729s^4 + 5.1566s^3 + 7.9381s^2 + 5.8091s + 1.6604}{s(0.0347s^2 + 2.0795s + 1)}$$

$$D_{\text{IMC}-4/2} = \frac{0.1045s^4 + 0.7826s^3 + 2.2491s^2 + 2.8652s + 1.3558}{s(0.0379s^2 + 0.016s + 1)}.$$

The resulting output responses are also plotted in Fig. 8.16 for comparison. Note that the control signal of the IMC design in (8.71) becomes somewhat oscillatory, implying a marginal stability to allow for process uncertainties.

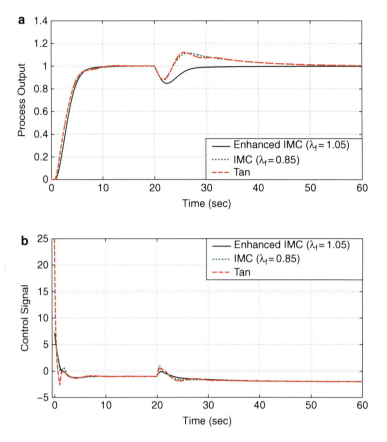

Fig. 8.17 Output responses of Example 8.4 in the presence of a slow load disturbance from the output side

To demonstrate the control performance for rejecting a load disturbance from the process output side with a slow dynamics, the case used in example 8.2 is also performed here. The enhance IMC design in (8.88) and (8.89) for such case gives

$$C_{\text{f-UO-2}} = \frac{(5s-1)(2.07s+1)(\eta_1 s+1)}{(\lambda_f s+1)^3} \cdot \frac{1}{1 - \frac{\eta_1 s+1}{(\lambda_f s+1)^3} e^{-0.94s}}$$

$$\eta_1 = 8(0.125\lambda_f - 1)^3 e^{-0.1175} + 8.$$

Note that the second multiplier in $C_{\text{f-UO-2}}$ can be implemented using the closed-loop unit shown in Fig. 8.3.

With the above controller parameter settings, the output responses are shown in Fig. 8.17. It is seen that apparently improved disturbance response without a long "tail" is obtained by the enhance IMC design in (8.88) and (8.89), which

8.5 Illustrative Examples

once again demonstrates that using an estimation on the deterministic-type load disturbance dynamics for the controller design can evidently improve disturbance rejection performance.

Example 8.5. Consider the integrating and unstable process studied by Lee et al. (2000),

$$G_5 = \frac{e^{-0.2s}}{s(s-1)}.$$

In Lee's method, the tuning parameters of the closed-loop PID controller were taken as $K_C = 0.8412, \tau_I = 3.3066$, and $\tau_D = 2.8113$, and the set-point filter was chosen as $f_R = 1/(8.4593s^2 + 3.3607s + 1)$.

Using the controller formulae of (8.53) and (8.76), it follows that

$$C_{\text{s-U-3}} = \frac{s(s-1)}{(\lambda_c s + 1)^2}$$

$$C_{\text{f-U-3}} = \frac{s(s-1)(\alpha_2 s^2 + \alpha_1 s + 1)}{(\lambda_f s + 1)^4} \cdot \frac{1}{1 - \frac{\alpha_2 s^2 + \alpha_1 s + 1}{(\lambda_f s + 1)^4} e^{-0.2s}},$$

where $\alpha_1 = 4\lambda_f + 0.2$ and $\alpha_2 = (\lambda_f + 1)^4 e^{0.2} - \alpha_1 - 1$. The controller parameters are taken as $\lambda_c = \lambda_f = 3\theta = 0.6$ to test the control performance and robustness. Using the PID tuning formula in (7.100) to approximate $C_{\text{f-U-3}}$ for implementation, it follows that

$$C_{\text{f-PID}} = 1.4738 + \frac{1}{1.7446s} + \frac{2.4804s}{0.05s + 1}.$$

By adding a unit step change to the set-point and an inverse step change of load disturbance to the process input at $t = 25(s)$, the control results are shown in Fig. 8.18. It is seen that the proposed control scheme results in significantly improved system performance in both the set-point tracking and load disturbance rejection.

Now suppose that there exists 20% error in estimating the process time delay and the unstable time constant ($\tau = 1.0$). The worst case is that the process time delay is actually 20% larger and the unstable time constant is 20% smaller. The corresponding output responses are shown in Fig. 8.19, indicating that the proposed control scheme holds good robust stability in the presence of the severe process uncertainty.

Example 8.6. Consider the unstable process with two RHP poles studied by Tan et al. (2003),

$$G_6 = \frac{2e^{-0.3s}}{(3s-1)(s-1)}.$$

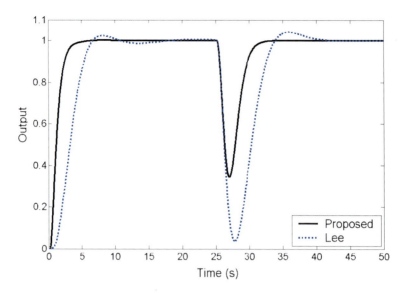

Fig. 8.18 Nominal output responses of Example 8.5

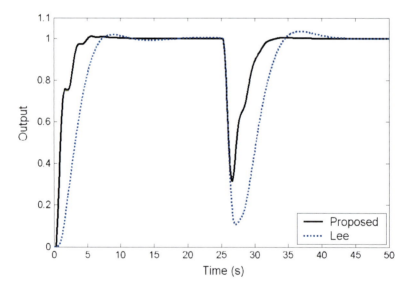

Fig. 8.19 Perturbed system responses of Example 8.5

In Tan's method, the controllers for the set-point response were taken as $k_0 = 4s$, $k_1 = (3s+1)(s+1)/[2(0.5s+1)^2]$, and the controller, k_2, for load disturbance rejection, had two choices, a PD form of $k_2 = 1 + 3.7s$ and a high-order form of

8.5 Illustrative Examples

$$k_2 = \frac{66.8(s + 0.27)(s + 6.667)}{s^2 + 14s + 121.31}.$$

Using the controller formulae of (8.55) and (8.81), it follows that

$$C_{\text{s-U-4}} = \frac{(3s - 1)(s - 1)}{2(\lambda_c s + 1)^2}$$

$$C_{\text{f-U-4}} = \frac{(3s - 1)(s - 1)(\alpha_2 s^2 + \alpha_1 s + 1)}{2(\lambda_f s + 1)^4} \cdot \frac{1}{1 - \frac{\alpha_2 s^2 + \alpha_1 s + 1}{(\lambda_f s + 1)^4} e^{-0.3s}},$$

where $\alpha_1 = 4.5(\lambda_f/3 + 1)^4 e^{0.1} - 0.5(\lambda_f + 1)^4 e^{0.3} - 4$ and $\alpha_2 = 1.5(\lambda_f + 1)^4 e^{0.3} - 4.5(\lambda_f/3 + 1)^4 e^{0.1} + 3$.

For comparison, take $\lambda_c = 1.7\theta = 0.51$ for obtaining the same rising speed of the set-point response with Tan's method. For load disturbance rejection, taking $\lambda_f = 1.7\theta = 0.51$ and using the PID tuning formula in (7.100) to approximate $C_{\text{f-U-4}}$ for implementation, it follows that

$$C_{\text{f-PID}} = 1.7638 + \frac{1}{1.059s} + \frac{4.0642s}{0.05s + 1}.$$

In contrast, taking $N = 3$ and $\lambda_f = 1.5\theta = 0.45$ to apply the analytical approximation formula in (8.67) for implementation, it follows that

$$D_{3/3}(s) = \frac{32.82s^2 + 439.41s + 232.64}{0.56s^2 + 0.8s + 100} + \frac{129.79}{s(0.56s^2 + 0.8s + 100)}.$$

By adding a unit step change to the set-point and an inverse step change of load disturbance to the process input at $t = 15(s)$, the control results are shown in Fig. 8.20. It is seen that the proposed third-order controller results in the best disturbance rejection performance. The IMC-based PID controller is similar to the PD controller of Tan's method for load disturbance rejection, but both are inferior to the high-order controllers.

Now suppose that there exists 10% error in estimating the process time delay and the two unstable time constants, e.g., all of these parameters are actually 10% larger. The perturbed system responses in terms of these two high-order controllers are shown in Fig. 8.21. It is seen that the proposed third-order controller maintains well the load disturbance response robustness against the severe process uncertainty. Note that both the PID form of the proposed feedback controller ($C_{\text{f-U-4}}$) and the PD controller of Tan's method cannot hold the control system stability any longer, which, therefore, demonstrates that the conventional PID controllers are indeed incapable of rejecting load disturbances with robustness for unstable processes with multiple RHP poles, as studied in some references, e.g., Yang et al. (2002).

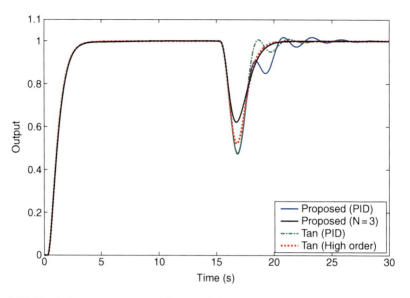

Fig. 8.20 Nominal output responses of Example 8.6

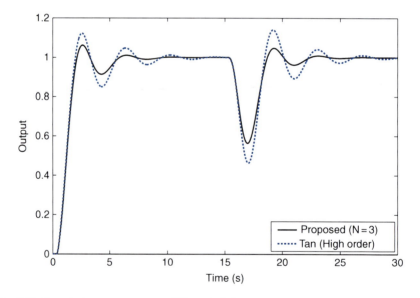

Fig. 8.21 Perturbed system responses of Example 8.6

8.6 Summary

The advantage of a 2DOF control structure has been elucidated in comparison with the conventional unity feedback control structure or the standard IMC structure. An important merit is that both the set-point tracking and load disturbance rejection can

8.6 Summary

be separately tuned or optimized in terms of a 2DOF control scheme. This can be more important for the control of an integrating or unstable process for which the water-bed effect between the set-point response and the load disturbance response may become severe or even inadmissible if using the conventional unity feedback control structure.

Based on a classification on different cases of load disturbance entering into a stable process, the corresponding controller designs in terms of a 2DOF IMC structure have been itemized. Moreover, an enhanced IMC design (Liu and Gao 2010b) has been suggested to further improve disturbance rejection performance for slow processes with large time constant(s).

Since the standard 2DOF IMC structure cannot be used for the control of integrating and unstable processes with internal stability (Morari and Zafiriou 1989), a united 2DOF control scheme based on the 2DOF control methods developed by Zhang et al. (2004) and Liu et al. (2005a, b) has been presented for such a process. Correspondingly, analytical controller formulae have been presented in terms of a few low-order model structures that are widely used for describing integrating and unstable processes in practical applications. Moreover, to overcome sluggish load disturbance rejection involved in integrating and unstable processes with slow dynamics, further enhanced IMC design formulae (Liu and Gao 2011) have been presented based on a classification on the ways through which a deterministic load disturbance enters into the process as detected from a preliminary knowledge or experience of the process operation. In the presence of unknown load disturbance, the proposed controller design for the case of a step-type load disturbance entering into the process from its input side can in general be used to improve disturbance rejection performance.

For all the proposed controller designs, the corresponding PID tuning method has also been presented to facilitate practical applications based on the Maclaurin approximation approach introduced in Sect. 7.5. Due to the fact that there exists implicit RHP zero-pole canceling in the desired feedback controller for an unstable process, an analytical approximation method based on the linear fractional Padé expansion (Liu et al. 2005b) has been presented for implementation or improving the approximation accuracy in contrast with the Maclaurin approximation approach. At the same time, the robust stability constraints for controller tuning have been given in terms of the small gain theorem and the multiplicative description of process uncertainties.

Six examples from existing references have been used to illustrate the proposed 2DOF control scheme and controller designs, respectively, for integrating and unstable processes. By comparison with the control results given in these references, the effectiveness and merits of the proposed 2DOF control scheme and controller tuning methods have been well demonstrated.

References

Chien I-L, Peng SC, Liu JH (2002) Simple control method for integrating processes with long deadtime. J Process Control 12(3):391–404

Garcia P, Albertos P (2008) A new dead-time compensator to control stable and integrating processes with long dead-time. Automatica 44:1062–1071

Garcia P, Albertos P, Hägglund T (2006) Control of unstable non-minimum-phase delayed systems. J Process Control 16:1099–1111

Hang CC, Wang QG, Yang XP (2003) A modified Smith predictor for a process with an integrator and long dead time. Ind Eng Chem Res 42:484–489

Huang H-P, Chen C-C (1997) Control-system synthesis for open-loop unstable process with time delay. IEE Proc Control Theory Appl 144(4):334–346

Kaya I (2004) Two-degree-of-freedom IMC structure and controller design for integrating processes based on gain and phase-margin specifications. IEE Proc Control Theory Appl 151(4):401–407

Kwak H, Whan S, Lee IB (2001) Modified Smith predictor for integrating processes: comparisons and proposition. Ind Eng Chem Res 40:1500–1506

Lee Y, Lee J, Park S (2000) PID controllers tuning for integrating and unstable processes with time delay. Chem Eng Sci 55:3481–3493

Liu T, Gao F (2008) Identification of integrating and unstable processes from relay feedback. Comput Chem Eng 32(12):3038–3056

Liu T, Gao F (2010a) Closed-loop step response identification of integrating and unstable processes. Chem Eng Sci 65(10):2884–2895

Liu T, Gao F (2010b) New insight into internal model control filter design for load disturbance rejection. IET Control Theory Appl 4(3):448–460

Liu T, Gao F (2011) Enhanced IMC design of load disturbance rejection for integrating and unstable processes with slow dynamics. ISA Trans 50(2):239–248

Liu T, Cai YZ, Gu DY, Zhang WD (2005a) New modified Smith predictor scheme for controlling integrating and unstable processes. IEE Proc Control Theory Appl 152(2):238–246

Liu T, Zhang WD, Gu DY (2005b) Analytical two-degree-of-freedom tuning design for open-loop unstable processes with time delay. J Process Control 15:559–572

Lu X, Yang YS, Wang QG, Zheng WX (2005) A double two-degree-of-freedom control scheme for improved control of unstable delay processes. J Process Control 15:605–614

Majhi S, Atherton DP (2000) Obtaining controller parameters for a new Smith predictor using autotuning. Automatica 36:1651–1658

Matausek MR, Micic AD (1999) On the modified Smith predictor for controlling a process with an integrator and long dead-time. IEEE Trans Autom Control 44(8):1603–1606

Morari M, Zafiriou E (1989) Robust process control. Prentice Hall, Englewood Cliff

Normey-Rico JE, Camacho EF (2002) A unified approach to design dead-time compensators for stable and integrative processes with dead-time. IEEE Trans Autom Control 47(2):299–305

Normey-Rico JE, Camacho EF (2009) Unified approach for robust dead-time compensator design. J Process Control 19:38–47

Rao AS, Chidambaram M (2007) Simple analytical design of modified Smith predictor with improved performance for unstable first-order plus time delay (FOPTD) processes. Ind Eng Chem Res 46:4561–4571

Skogestad S, Postlethwaite I (2005) Multivariable feedback control: analysis and design, 2nd edn. Wiley, Chichester

Tan W, Marquez HJ, Chen T (2003) IMC design for unstable processes with time delays. J Process Control 13:203–213

Tian YC, Gao F (1999) Control of integrator processes with dominant time delay. Ind Eng Chem Res 39:2979–2983

Yang XP, Wang QG, Hang CC, Lin C (2002) IMC-Based control system design for unstable processes. Ind Eng Chem Res 41(17):4288–4294

References

Zhang WD, Gu DY, Wang W, Xu X (2004) Quantitative performance design of a modified Smith predictor for unstable processes with time delay. Ind Eng Chem Res 43:56–62

Zhang WD, Gu DY, Rieber JM (2008) Optimal dead-time compensator design for stable and integrating processes with time delay. J Process Control 18:449–457

Zhong Q-C, Mirkin L (2002) Control of integral processes with dead time. Part II: quantitative analysis. IEE Proc Control Theory Appl 149(4):291–296

Zhong Q-C, Normey-Rico JE (2002) Control of integral processes with dead time. Part I: disturbance observer-based 2DOF control scheme. IEE Proc Control Theory Appl 149(4):285–290

Zhou KM, Doyle JC, Glover K (1996) Robust and optimal control. Prentice Hall, Englewood Cliff

Chapter 9
Cascade Control System

9.1 The Advantage of Cascade Control and Implementation Requirements

It is well known that the conventional unity feedback control system can only give a corrective action to load disturbance after the system output has deviated from the set-point, i.e., output error entails feedback control. If a measurement on a secondary (intermediate) process output is available to detect the final output error in advance, a secondary feedback controller or structure can be correspondingly constructed, such that the load disturbance rejection performance can be significantly improved, in particular for slow processes with large time constant(s) or time delay, e.g., industrial furnaces for heating raw materials (Ogunnaike and Ray 1994; Seborg et al. 2004), and chemical distillation columns with large time delays (Morari and Zafiriou 1989; Shinskey 1996). This control strategy, called cascade control, has been widely applied in industrial and chemical processes relating to temperature, flow, and pressure control issues (Shinskey 1996; Seborg et al. 2004).

Generally, a cascade control structure is composed of two control loops, i.e., a secondary inner loop nested in the primary outer loop for closed-loop system operation, based on a precondition that the intermediate process can be measured conveniently or cost-effectively in practice. Load disturbance that enters into the inner loop is expected to be counteracted before it extends to the primary outer loop. Therefore, it is crucial that the inner loop gives a faster dynamical response compared to the outer loop, in the presence of such load disturbance.

Rule-of-thumb tuning rules for operating cascade control systems have been introduced in the control bibliographies (Luyben 1990; Åström and Hägglund 1995; Seborg et al. 2004). Improved PID tuning methods can be found in the references (Hang et al. 1994; Huang et al. 1998; Lee et al. 1998, 2002; Tan et al. 2000; Song et al. 2003). The Smith predictor was also introduced into the conventional cascade control structure to improve disturbance rejection performance for time delay processes (Kaya 2001).

Note that cascade control tuning procedures in existing methods are mostly implemented in a sequential manner. That is, the primary loop controller is first put on manual and the secondary loop controller is tuned. After the secondary loop controller is commissioned, the primary controller is tuned to accomplish the tuning procedure. If the resulting control system performance is not satisfactory, the whole tuning procedure has to be repeated over again. This can be very time consuming and cumbersome, especially for slow processes with large time constant(s) or time delay. Moreover, the load disturbance response of the inner loop is heavily coupled with the set-point response supervised by the outer loop.

To overcome the above deficiencies, two 2DOF cascade control schemes and a 3DOF cascade control scheme are presented here for application, based on a classification of open-loop stable and unstable cascade systems. The corresponding control designs are detailed in the following two sections.

9.2 Two 2DOF Control Schemes for Open-Loop Stable Cascade Processes

For an open-loop stable process, if the intermediate process can be measured for cascade control design, two 2DOF cascade control structures are proposed, as shown in Fig. 9.1a, b, where P_1 denotes the secondary process and P_2 the final process, while $\widehat{P}_1$ and $\widehat{P}_2$ denote their models, and $\widehat{P}$ indicates the overall process model identified for control design. C_s is the primary outer loop controller for set-point tracking and load disturbance rejection of P_2, and C_f is the secondary controller in the inner loop for rejecting load disturbance that enters into the secondary process P_1, which is therefore called as load disturbance estimator. y_1 is the secondary process output and y_2 is the final process output.

In the nominal case, i.e., $P_1 = \widehat{P}_1$ and $P_2 = \widehat{P}_2$ (or $\widehat{P} = \widehat{P}_1 \widehat{P}_2$), there is an "open-loop" control from the primary set-point r to the final output y_2 in the cascade control structures shown in Fig. 9.1a, b. This corresponds to

$$H_r(s) = \frac{y_2}{r} = C_s P_1 P_2. \tag{9.1}$$

Hence, the nominal set-point response is decoupled from the inner loop load disturbance response. That is, both responses can be separately tuned by the set-point tracking controller C_s and the load disturbance estimator C_f.

Note that the cascade control structure shown in Fig. 9.1a is based on using $\widehat{P}_1$ and $\widehat{P}_2$ obtained from process modeling such as the process energy equations or equilibrium relations, or from process identification test(s). Usually the secondary process model, $\widehat{P}_1$, can easily be derived from such a modeling or identification method. The final process model, $\widehat{P}_2$, however, may not be able to carry out in a similar way. Indirect modeling methods are therefore required for cascade control design (Shinskey 1996). For instance, by identifying the overall process model, $\widehat{P}$, as shown in Fig. 9.1b, the final process model may be derived from $\widehat{P}_2 = \widehat{P}/\widehat{P}_1$.

9.2 Two 2DOF Control Schemes for Open-Loop Stable Cascade Processes

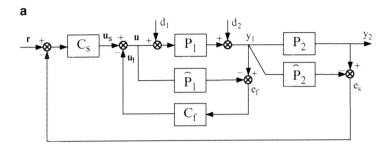

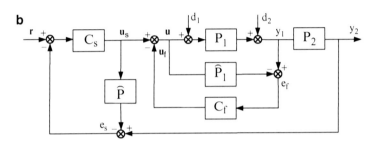

Fig. 9.1 2DOF cascade control structures using the final process model (**a**) and the overall process model (**b**)

In fact, such an indirect modeling method may increase model error or mislead the description of the final process response characteristics, causing undesired control performance degradation. In such a case, the cascade control structure shown in Fig. 9.1b is preferred for practical application. However, when a load disturbance, denoted as d_1 or d_2, enters into the secondary process (P_2), not only the load disturbance estimator (C_f) but also the set-point tracking controller (C_s) in the cascade control structure shown in Fig. 9.1b will give a counteraction, which will cause oscillation of the final output (y_2) as encountered in the conventional cascade control structure. Therefore, the cascade control structure shown in Fig. 9.1b will unavoidably be subject to certain performance degradation for load disturbance rejection when compared to the cascade control structure shown in Fig. 9.1a.

For general application, the following model structures are considered for controller design:

$$P_1(s) = k_1 \frac{B_{1+}(s)B_{1-}(s)}{A_1(s)} e^{-\theta_1 s} \tag{9.2}$$

$$P_2(s) = k_2 \frac{B_{2+}(s)B_{2-}(s)}{A_2(s)} e^{-\theta_2 s}, \tag{9.3}$$

where $A_1(0) = B_{1+}(0) = B_{1-}(0) = 1$, $A_2(0) = B_{2+}(0) = B_{2-}(0) = 1$, and all zeros of $A_1(s)$, $A_2(s)$, $B_{1-}(s)$, and $B_{2-}(s)$ are located in LHP, while all zeros of

$B_{1+}(s)$ and $B_{2+}(s)$ are located in RHP. Denote $\deg\{A_1(s)\} = n_1$, $\deg\{B_{1-}(s)\} = m_{11}$, $\deg\{B_{1+}(s)\} = m_{12}$, $\deg\{A_2(s)\} = n_2$, $\deg\{B_{2-}(s)\} = m_{21}$, and $\deg\{B_{2+}(s)\} = m_{22}$. Practically, it follows that $m_{11} + m_{12} < n_1$ and $m_{21} + m_{22} < n_2$, which indicate that a cascade process to be controlled is physically proper.

Based on the above description of a cascade process, the controller design of C_s and C_f can be identical for both cascade control structures shown in Fig. 9.1a, b for practical implementation, which is therefore uniformly presented in the following section.

9.2.1 Controller Design

For the set-point tracking, the H_2 optimal performance objective, $\min \| e \|_2^2$, is adopted to design C_s. That is, this controller should be designed to achieve the output performance specification, $\min \| W(1 - H_r(s)) \|_2^2$, where $H_r(s)$ is the nominal set-point transfer function shown in (9.1), and W is a weighting function of the set-point. Since a step change of r is typical in practice, W is correspondingly taken as $1/s$.

Using a v/v order all-pass Padé approximation for the time delay, it follows that

$$P_1(s) = k_1 \frac{B_{1+}(s)B_{1-}(s)}{A_1(s)} \cdot \frac{Q_{vv}(-\theta_1 s)}{Q_{vv}(\theta_1 s)} \tag{9.4}$$

$$P_2(s) = k_2 \frac{B_{2+}(s)B_{2-}(s)}{A_2(s)} \cdot \frac{Q_{vv}(-\theta_2 s)}{Q_{vv}(\theta_2 s)}, \tag{9.5}$$

where

$$Q_{vv}(\theta_i s) = \sum_{j=0}^{v} \frac{(2v-j)!v!}{(2v)!j!(v-j)!}(\theta_i s)^j, \quad i = 1, 2$$

and v can be chosen large enough to guarantee that the introduced approximation error is neglectable in comparison with the process uncertainties.

Substituting (9.4) and (9.5) into the above performance specification, one can obtain

$$\| W(1 - H_r(s)) \|_2^2 =$$

$$\left\| \frac{1}{s} \left(1 - C_s(s) \frac{k_1 k_2 B_{1+}(s) B_{2+}(s) B_{1-}(s) B_{2-}(s) Q_{vv}(-\theta_1 s) Q_{vv}(-\theta_2 s)}{A_1(s) A_2(s) Q_{vv}(\theta_1 s) Q_{vv}(\theta_2 s)}\right)\right\|_2^2$$

$$= \left\| \frac{Q_{vv}(\theta_1 s) Q_{vv}(\theta_2 s) B_{1+}^*(s) B_{2+}^*(s)}{s Q_{vv}(-\theta_1 s) Q_{vv}(-\theta_2 s) B_{1+}(s) B_{2+}(s)} \right.$$

$$\left. - C_s(s) \frac{k_1 k_2 B_{1+}^*(s) B_{2+}^*(s) B_{1-}(s) B_{2-}(s)}{s A_1(s) A_2(s)} \right\|_2^2,$$

9.2 Two 2DOF Control Schemes for Open-Loop Stable Cascade Processes

where $B_{1+}^*(s)$ and $B_{2+}^*(s)$ are complex conjugates of $B_{1+}(s)$ and $B_{2+}(s)$, respectively. It is obvious that $B_{1+}(s)/B_{1+}^*(s)$ and $B_{2+}(s)/B_{2+}^*(s)$ are all-pass filters.

Note that $Q_{vv}(0) = B_{1+}^*(0) = B_{2+}^*(0) = 1$ and all zeros of $Q_{vv}(-\theta_1 s)$, $Q_{vv}(-\theta_2 s)$, $B_{1+}(s)$, and $B_{2+}(s)$ are in RHP. It follows from an orthogonal property of the H_2 norm that

$$\|W(1-H_r(s))\|_2^2 = \left\| \frac{A_1(s)A_2(s) - k_1 k_2 C_s(s) B_{1+}^*(s) B_{2+}^*(s) B_{1-}(s) B_{2-}(s)}{s A_1(s) A_2(s)} \right\|_2^2$$

$$+ \left\| \frac{Q_{vv}(\theta_1 s) Q_{vv}(\theta_2 s) B_{1+}^*(s) B_{2+}^*(s) - Q_{vv}(-\theta_1 s) Q_{vv}(-\theta_2 s) B_{1+}(s) B_{2+}(s)}{s Q_{vv}(-\theta_1 s) Q_{vv}(-\theta_2 s) B_{1+}(s) B_{2+}(s)} \right\|_2^2.$$

Minimizing the right-hand side, i.e., letting its first term equal zero, one can obtain the ideally optimal controller,

$$C_{s-\text{ideal}}(s) = \frac{A_1(s)A_2(s)}{k_1 k_2 B_{1+}^*(s) B_{2+}^*(s) B_{1-}(s) B_{2-}(s)}. \tag{9.6}$$

It can be easily verified from the process models in (9.2) and (9.3) that $C_{s-\text{ideal}}(s)$ is not physically proper and thus cannot be realized in practice. A low-pass filter is therefore introduced to allow it to be implemental,

$$F(s) = \frac{1}{(\lambda_c s + 1)^{l_s}}, \tag{9.7}$$

where $l_s = \sum_{i=1}^{2} \left(n_i - \sum_{j=1}^{2} m_{ij} \right)$.

Hence, a practically suboptimal controller is obtained as

$$C_s(s) = \frac{A_1(s)A_2(s)}{k_1 k_2 B_{1+}^*(s) B_{2+}^*(s) B_{1-}(s) B_{2-}(s)(\lambda_c s + 1)^{l_s}}, \tag{9.8}$$

where λ_c is an adjustable parameter. When λ_c is tuned to zero, the controller recovers the optimality.

Substituting (9.8) into (9.1) yields the nominal set-point transfer function,

$$H_r(s) = \frac{1}{(\lambda_c s + 1)^{l_s}} \frac{B_{1+}(s) B_{2+}(s)}{B_{1+}^*(s) B_{2+}^*(s)} e^{-(\theta_1 + \theta_2)s}. \tag{9.9}$$

It is seen that the nominal set-point response can be quantitatively tuned through the adjustable controller parameter, λ_c. For example, if P_1 and P_2 have no RHP zero, it follows that

$$H_r(s) = \frac{1}{(\lambda_c s + 1)^{l_s}} e^{-(\theta_1 + \theta_2)s}. \tag{9.10}$$

By performing an inverse Laplace transform with respect to a step change of the set-point, one can obtain

$$y_2(t) = \begin{cases} 0 & t \leq \theta_1 + \theta_2 \\ 1 - \left(1 + \dfrac{t}{\lambda_c} + \dfrac{t^2}{\lambda_c^2} + \cdots + \dfrac{t^{l_s-1}}{(l_s-1)!\lambda_c^{l_s-1}}\right) e^{-(t-\theta_1-\theta_2)/\lambda_c} & t > \theta_1 + \theta_2 \end{cases}.$$
(9.11)

It shows that there is no overshoot in the nominal set-point response, and the time domain system response specification can be quantitatively achieved by tuning λ_c. For instance, a tuning formula for the rise time can be numerically determined as $t_r = 2.3026\lambda_c + \theta_1 + \theta_2$ for $l_s = 1$, or $t_r = 3.8897\lambda_c + \theta_1 + \theta_2$ for $l_s = 2$, and so on.

For rejecting load disturbance that enters into the inner loop, it can be derived from Fig. 9.1a or b that the nominal load disturbance transfer functions of the inner loop are

$$H_{d_1}(s) = \frac{y_1}{d_1} = P_1(1 - T_{d_1})$$
(9.12)

$$H_{d_2}(s) = \frac{y_1}{d_2} = 1 - T_{d_1},$$
(9.13)

where T_{d_1} is the complementary sensitivity function of the inner loop,

$$T_{d_1} = \frac{u_f}{d_1} = P_1 C_f.$$
(9.14)

Following a similar analysis as in Sect. 8.3.1, a desired closed-loop complementary sensitivity function is proposed as

$$T_{d_1}(s) = \frac{1}{(\lambda_f s + 1)^{l_f}} \frac{B_{1+}(s)}{B_{1+}^*(s)} e^{-\theta_1 s},$$
(9.15)

where $l_f = n_1 - m_{11} - m_{12}$, and λ_f is an adjustable parameter for tuning load disturbance response of the inner loop.

Substituting (9.2) and (9.15) into (9.14) yields the load disturbance estimator,

$$C_f(s) = \frac{A_1(s)}{k_1 B_{1+}^*(s) B_{1-}(s)(\lambda_f s + 1)^{l_f}}.$$
(9.16)

Remark 9.1 Note that the controller in (9.16) was derived in the reference (Liu et al. 2005a) for application in the inner loop shown in Fig. 9.2. In effect it can be verified that the corresponding disturbance rejection performance is the same with

9.2 Two 2DOF Control Schemes for Open-Loop Stable Cascade Processes

Fig. 9.2 An equivalent inner loop

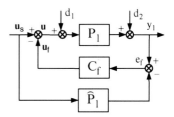

the inner loop shown in Fig. 9.1a or b because there is the same desired closed-loop complementary sensitivity function, as shown in (9.15). Hence, the inner loop shown in Fig. 9.1a or b can be viewed as a simplified version of that given in the reference (Liu et al. 2005a). ◊

9.2.2 Robust Stability Analysis

It is obvious that the primary outer loop in either of the cascade control structures shown in Fig. 9.1a, b is the standard IMC structure. According to the IMC theory (Morari and Zafiriou 1989), the outer loop holds robust stability if and only if

$$\|H_r(s)\|_\infty < \frac{1}{\|\Delta\|_\infty}, \qquad (9.17)$$

where $\Delta = (P_2 - \hat{P}_2)/\hat{P}_2$ for Fig. 9.1a (or $\Delta = (P - \hat{P})/\hat{P}$ for Fig. 9.1b) denotes the final (or overall) process multiplicative uncertainty. Note that the secondary process uncertainty in the inner loop may be lumped into the final (or overall) process multiplicative uncertainty for the convenience of analysis in practice.

Substituting (9.9) into (9.17), a robust stability constraint for tuning λ_c can be obtained as

$$\left(\lambda_c^2 \omega^2 + 1\right)^{\frac{l_s}{2}} > |\Delta(j\omega)|, \quad \forall \omega \in [0, +\infty). \qquad (9.18)$$

For the inner loop shown in Fig. 9.1a, b, it is similar to the closed-loop structure set between the process input and output in a 2DOF IMC structure shown in Fig. 8.1. Following the robust stability analysis given in Sect. 8.3.2, one can conclude that the inner loop holds robust stability if and only if

$$\|\Delta_1 T_{d_1}\|_\infty < 1, \qquad (9.19)$$

where Δ_1 defines the multiplicative uncertainty of the secondary process P_1. That is, it describes the process family, $\Pi_1 = \{P_1(s) : P_1(s) = (1 + \Delta_1)\hat{P}_1(s)\}$.

Substituting (9.15) into (9.19), a robust stability constraint for tuning λ_f can be obtained as

$$\left(\lambda_f^2 \omega^2 + 1\right)^{\frac{l_f}{2}} > |\Delta_1(j\omega)|, \quad \forall \omega \in [0, +\infty). \tag{9.20}$$

It is seen that the above robust stability constraints respectively for tuning λ_c and λ_f are similar to each other, which therefore facilitates the stability analysis in practical applications.

According to the closed-loop performance analysis in the IMC theory (Morari and Zafiriou 1989), the following constraints between the control performance and robust stability of the output and inner loops need to be satisfied for tuning λ_c and λ_f:

$$|\Delta(j\omega)H_r(j\omega)| + |W_1(j\omega)[1 - H_r(j\omega)]| < 1, \quad \forall \omega \in [0, +\infty) \tag{9.21}$$

$$|\Delta_1(j\omega)T_{d_1}(j\omega)| + |W_2(j\omega)[1 - T_{d_1}(j\omega)]| < 1, \quad \forall \omega \in [0, +\infty), \tag{9.22}$$

where W_1 and W_2 are weighting functions of the corresponding closed-loop sensitivity functions. For a step-type load disturbance entering into the outer or inner loop, both can be chosen as $1/s$ for assessment.

It can be seen from (9.9) and (9.15) that tuning λ_c (or λ_f) to a small value can speed up the set-point response (or the disturbance response of the inner loop), but will degrade the robust stability in the presence of process uncertainties. In the opposite, increasing λ_c (or λ_f) can strengthen the outer (or inner) loop robust stability, but in exchange for a degradation in its control performance. In general, it is suggested to initially tune λ_c (or λ_f) around the overall (or secondary) process time delay. By monotonically increasing or decreasing λ_c (or λ_f) on line, a good trade-off between the closed-loop system performance and robust stability can be obtained in a transparent manner.

9.3 A 3DOF Control Scheme for Open-Loop Unstable Cascade Processes

In some unstable industrial processes, e.g., continuous-stirred-tank-reactors (CSTRs) and thermal distillation columns, a measurement on the process intermediate variable such as temperature or flow rate can be obtained for cascade control design to improve disturbance rejection performance (Ogunnaike and Ray 1994; Shinskey 1996; Seborg et al. 2004). Due to that the water-bed effect between the set-point response and the load disturbance response will become very severe for an unstable process in the unity feedback control structure, the conventional cascade control structure is inevitably subject to this deficiency because of the double-loop structure. By adding two prefilters or weighting functions respectively

9.3 A 3DOF Control Scheme for Open-Loop Unstable Cascade Processes

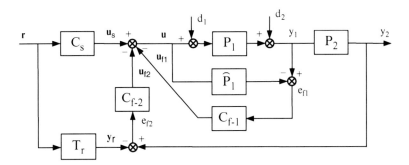

Fig. 9.3 A 3DOF cascade control structure for open-loop unstable cascade processes

to the set-points of the inner and outer loops in the conventional cascade control structure, a few cascade control methods have been developed to alleviate rather than overcome the above deficiency (Lee et al. 2002; Nagrath et al. 2002; Saraf et al. 2003).

An open-loop unstable process is usually divided into a stable secondary process and an unstable final process, for the convenience of cascade control design in practical applications. Correspondingly, the following low-order model structures are mostly adopted for the cascade control system design in the existing literature:

$$P_1(s) = \frac{k_1 e^{-\theta_1 s}}{\tau_1 s + 1} \tag{9.23}$$

$$P_2(s) = \frac{k_2 e^{-\theta_2 s}}{\tau_2 s - 1}, \tag{9.24}$$

where k_1 and k_2 denote the proportional gains of the secondary and final processes, respectively, θ_1 and θ_2 denote the corresponding time delays, τ_1 and τ_2 are positive time constants reflecting the dynamic response characteristics.

Given the above description of an open-loop unstable process in practice, a 3DOF cascade control scheme (Liu et al. 2005b) is presented here. The control structure is shown in Fig. 9.3, where P_1 denotes the secondary process, P_2 is the final process, and $\widehat{P}_1$ is a model identified for P_1. T_r is a desired transfer function for set-point tracking, C_s is a feedforward controller for set-point tracking, C_{f-1} is the inner loop feedback controller set for rejecting load disturbance that enters into P_1, which is therefore called as load disturbance estimator, and C_{f-2} is the outer loop feedback controller for load disturbance rejection of P_2. y_1 is the secondary process output, y_2 is the final process output, and y_r is a desired (referential) output response.

9.3.1 Controller Design

For the set-point tracking, there is an "open-loop" control for the nominal case ($P_1 = \hat{P}_1$ and $P_2 = \hat{P}_2$), corresponding to the transfer function,

$$T_r(s) = H_r(s) = \frac{y_2}{r} = C_s P_1 P_2. \tag{9.25}$$

Hence, following a similar procedure as in Sect. 9.2.1, one can obtain

$$C_s(s) = \frac{(\tau_1 s + 1)(\tau_2 s - 1)}{k_1 k_2 (\lambda_c s + 1)^2} \tag{9.26}$$

$$T_r(s) = \frac{1}{(\lambda_c s + 1)^2} e^{-(\theta_1 + \theta_2)s}, \tag{9.27}$$

where λ_c is an adjustable parameter for tuning the set-point response. It follows from the time domain response shown in (9.11) that the time domain specifications, e.g., the rise time, can be conveniently satisfied by monotonically tuning λ_c.

For rejecting load disturbance that enters into the secondary process P_1, it can be seen that the inner loop structure is exactly the same with that in Fig. 9.1a or b. Therefore, the corresponding controller design method in Sect. 9.2.1 can be used. Substituting (9.23) into (9.16) yields

$$C_{f-1}(s) = \frac{\tau_1 s + 1}{k_1(\lambda_{f-1} s + 1)}, \tag{9.28}$$

where λ_{f-1} is an adjustable parameter for tuning the load disturbance response of the inner loop. The corresponding closed-loop complementary sensitivity function is

$$T_{d-1}(s) = \frac{1}{\lambda_{f-1} s + 1} e^{-\theta_1 s}. \tag{9.29}$$

For rejecting load disturbance that enters into the final process P_2, it can be seen that the outer loop structure is exactly the same with the 2DOF control structure shown in Fig. 8.2a for an unstable process. Therefore, the corresponding controller design method in Sect. 8.4.1 can be used. Substituting (9.23) and (9.24) into (8.22) along with (8.71) yields

$$C_{f-2}(s) = \frac{(\tau_1 s + 1)(\tau_2 s - 1)(\alpha_1 s + 1)}{k_1 k_2 (\lambda_{f-2} s + 1)^3} \cdot \frac{1}{1 - \frac{\alpha_1 s + 1}{(\lambda_{f-2} s + 1)^3} e^{-(\theta_1 + \theta_2)s}} \tag{9.30}$$

$$\alpha_1 = \tau_2 \left[\left(\frac{\lambda_{f-2}}{\tau_2} + 1 \right)^3 e^{\frac{\theta_1 + \theta_2}{\tau_2}} - 1 \right], \tag{9.31}$$

9.3 A 3DOF Control Scheme for Open-Loop Unstable Cascade Processes

where λ_{f-2} is an adjustable parameter for tuning the load disturbance response of the outer loop. The corresponding closed-loop complementary sensitivity function is

$$T_{d-2}(s) = \frac{\alpha_1 s + 1}{(\lambda_{f-2} s + 1)^3} e^{-(\theta_1 + \theta_2)s}. \tag{9.32}$$

Note that there exists a RHP zero-pole canceling at $s = 1/\tau_2$ in (9.30), which may cause the controller to work unstably. A rational approximation is therefore needed for implementation. The analytical approximation based on the Maclaurin expansion series, as shown in (7.100), can be used to give a PID controller for implementation, by letting $C_{f-2} = M(s)/s$, since it has an integral property to eliminate the final output deviation from the set-point. For better approximation, the higher-order analytical approximation formulae given in (8.61)–(8.64) can be used to obtain further enhanced control performance.

Remark 9.2 Note that the 3DOF control structure shown in Fig. 9.3 is essentially equivalent to that in the reference (Liu et al. 2005b), which may be verified through the same nominal set-point response transfer function and the complementary sensitivity functions of the inner and outer loops. Hence, the 3DOF control structure shown in Fig 9.3 can be viewed as a simplified version of that given in the reference (Liu et al. 2005b). ◊

9.3.2 Robust Stability Analysis

Owing to an open-loop control manner for the set-point tracking in the presented 3DOF cascade control structure, robust stability analysis may be focused on the inner and outer loops.

For the convenience of analysis, process uncertainties may be lumped into the multiplicative forms with regard to the secondary process and the final process, respectively. Accordingly, the small gain theorem (Zhou et al. 1996) can be used to derive the robust stability constraints.

Since the inner loop is the same with that in Fig. 9.1a or b, a robust stability constraint can be derived similar to that given in Sect. 9.2.2 as

$$\sqrt{\lambda_{f-1}^2 \omega^2 + 1} > |\Delta_1(j\omega)|, \quad \forall \omega \in [0, +\infty), \tag{9.33}$$

where Δ_1 defines the multiplicative uncertainty of the secondary process P_1. That is, it describes the process family, $\Pi_1 = \{P_1(s) : P_1(s) = (1 + \Delta_1)\widehat{P}_1(s)\}$.

For instance, given the process time delay uncertainty $\Delta\theta_1$, which may be converted to the multiplicative uncertainty, $\Delta_1(s) = e^{-\Delta\theta_1 s} - 1$, a robust stability constraint for tuning λ_{f-1} can be derived as

$$\sqrt{\lambda_{f-1}^2 \omega^2 + 1} > \left| e^{-j\Delta\theta_1 \omega} - 1 \right|, \quad \forall \omega \in [0, +\infty). \tag{9.34}$$

In the presence of the control actuator uncertainty of the secondary process, e.g., $\Delta_1(s) = (s + 0.2)/(s + 1)$, which may be loosely interpreted as the secondary process input fed by the corresponding actuator increases by up to 100% uncertainty at high frequencies and by almost 20% uncertainty in the low frequency range, the corresponding robust stability constraint for tuning λ_{f-1} can be derived as

$$\sqrt{\lambda_{f-1}^2 \omega^2 + 1} > \sqrt{\frac{\omega^2 + 0.04}{\omega^2 + 1}}, \quad \forall \omega \in [0, +\infty). \tag{9.35}$$

In the case where the secondary process output measurement has an uncertainty, e.g., $\Delta_1(s) = -(s + 0.3)/(2s + 1)$, which may be physically viewed as the secondary process output measurement offered by the corresponding sensor decreases by up to 50% uncertainty at high frequencies and by almost 30% uncertainty in the low frequency range, the corresponding robust stability constraint for tuning λ_{f-1} can be derived as

$$\sqrt{\lambda_{f-1}^2 \omega^2 + 1} > \sqrt{\frac{\omega^2 + 0.09}{4\omega^2 + 1}}, \quad \forall \omega \in [0, +\infty). \tag{9.36}$$

Regarding the outer loop in the presented 3DOF cascade control structure, substituting (9.32) into the small gain condition yields the robust stability constraint for tuning λ_{f-2},

$$\left\| \frac{\tau_2 \left[\left(\frac{\lambda_{f-2}}{\tau_2} + 1 \right)^3 e^{\frac{\theta_1 + \theta_2}{\tau_2}} - 1 \right] s + 1}{(\lambda_{f-2} s + 1)^3} \right\|_\infty < \frac{1}{\|\Delta(s)\|_\infty}, \tag{9.37}$$

where $\Delta = (P_1 P_2 - \widehat{P}_1 \widehat{P}_2)/(\widehat{P}_1 \widehat{P}_2)$ defines the overall multiplicative uncertainty of the cascade process.

Meanwhile, according to the closed-loop performance analysis in the IMC theory (Morari and Zafiriou 1989), the following constraints between the control performance and robust stability of the inner and output loops need to be satisfied for tuning λ_{f-1} and λ_{f-2}:

$$|\Delta_1(j\omega) T_{d-1}(j\omega)| + |W_1(j\omega)[1 - T_{d-1}(j\omega)]| < 1, \quad \forall \omega \in [0, +\infty) \tag{9.38}$$

$$|\Delta(j\omega) T_{d-2}(j\omega)| + |W_2(j\omega)[1 - T_{d-2}(j\omega)]| < 1, \quad \forall \omega \in [0, +\infty), \tag{9.39}$$

9.4 Illustrative Examples and Real-Time Tests

where W_1 and W_2 are weighting functions of the corresponding closed-loop sensitivity functions. For a step-type load disturbance that enters into the inner or outer loop, both can be chosen as $1/s$ for assessment.

It can be seen from (9.29) and (9.32) that tuning λ_{f-1} (or λ_{f-2}) to a small value can speed up the disturbance response of the inner (or outer) loop, but will degrade the robust stability in the presence of process uncertainties. In the opposite direction, increasing λ_{f-1} (or λ_{f-2}) can strengthen the inner (or outer) loop robust stability, but in exchange for a degradation in its control performance. In general, it is suggested to initially tune λ_{f-1} (or λ_{f-2}) around the secondary (or overall) process time delay. By monotonically increasing or decreasing λ_{f-1} (or λ_{f-2}) on line, a good trade-off between the disturbance rejection performance and robust stability can be obtained in a transparent manner.

9.4 Illustrative Examples and Real-Time Tests

Five examples from the existing literature are used to demonstrate the effectiveness and merits of the presented 2DOF and 3DOF control schemes and controller designs. Examples 9.1 and 9.3 are given for open-loop stable processes with time delay or RHP zeros. Real-time tests are performed for Example 9.2, based on the KI 101 type process simulator of KentRidge Instruments in Singapore. Examples 9.4 and 9.5 are given for open-loop unstable processes with time delay.

Example 9.1 Consider a stable cascade process studied in the references (Tan et al. 2000; Song et al. 2003),

$$P_1(s) = \frac{e^{-0.1s}}{0.1s + 1}$$

$$P_2(s) = \frac{e^{-s}}{(s+1)^2}.$$

Both Tan's and Song's methods were based on the conventional cascade control structure. In Tan's method, the primary and secondary controllers were respectively designed as a PI-type controller, i.e., $G_{c1} = 0.39(1 + 1/1.44s)$ and $G_{c2} = 0.53(1 + 1/0.17s)$. In Song's method, the controllers were tuned as $G_{c1} = 0.6592 + 0.3536/s + 0.2886s/(1.4392s + 1)$ and $G_{c2} = 0.603 + 2.277/s$. In the proposed 2DOF cascade control schemes shown in Fig. 9.1a, b, using the controller formulae given in (9.8) and (9.16), and taking $\lambda_c = 1.0$ and $\lambda_f = 0.1$ for comparison, it follows that

$$C_s(s) = \frac{0.1s + 1}{s + 1}$$

$$C_f(s) = 1.$$

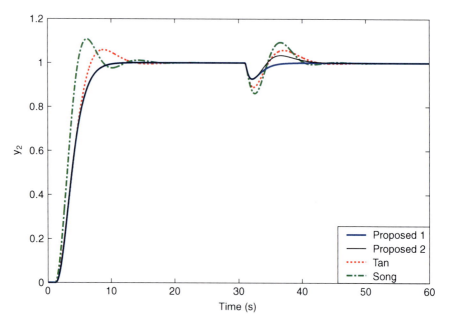

Fig. 9.4 Nominal system responses of Example 9.1

By adding a unit step change to the set-point and an inverse unit step change of load disturbance to the secondary process output at $t = 30(s)$, the control results are shown in Fig. 9.4. It is seen that there is no overshoot in the set-point response by using either of the proposed cascade control structures. Moreover, apparently improved load disturbance response is obtained using either of the proposed cascade control structures. The cascade control structure shown in Fig. 9.1a (thick solid line) outperforms the other cascade control structure shown in Fig. 9.1b (thin solid line) in that no oscillation occurs in the load disturbance response.

Then, assume that both time delays of the secondary and final processes are actually 20% larger and the time constant of the secondary process is 20% smaller. The perturbed system responses are shown in Fig. 9.5. It is seen that the proposed cascade control structures hold good robust stability.

Example 9.2 Consider a real-time cascade process studied by Tan et al. (2000),

$$P_1(s) = \frac{e^{-0.1s}}{s+1}$$

$$P_2(s) = \frac{e^{-s}}{(s+1)^2}$$

9.4 Illustrative Examples and Real-Time Tests

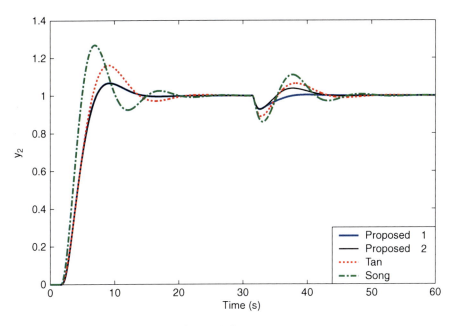

Fig. 9.5 Perturbed system responses of Example 9.1

which is reconstructed here through the KI 101 type process simulator of KentRidge Instruments in Singapore for real-time tests, as shown in Fig. 9.6. The simulator is connected to an industrial-type computer (PCI-1730) of ADVANTECH in Taiwan, via two 8-bit data acquisition cards (PCL-727 and PCL-818 L) respectively for analog-to-digital and digital-to-analog conversions. A window-based software platform, LabView 6.0, is used for system monitoring. The sampling period is taken as $T_s = 10$ (ms). Correspondingly, the one-step backward discretization operator, $\dot{e}(kT_s) = [e(kT_s) - e((k-1)T_s)]/T_s$, is used for differential computation of the control output. Real-time measurements of the secondary and final process outputs are corrupted by a random noise with the amplitude of 0.05.

Note that in the conventional cascade control structure used by Tan et al. (2000), the primary and secondary controllers were respectively tuned as a PI-type controller, $G_{c1} = 0.34(1 + 1/1.88s)$ and $G_{c2} = 0.21(1 + 1/0.583s)$.

Using the proposed 2DOF cascade control schemes shown in Fig. 9.1a, b, it follows from the controller formulae given in (9.8) and (9.16) that

$$C_s(s) = \frac{(s+1)^3}{(\lambda_c s + 1)^3}$$

$$C_f(s) = \frac{s+1}{\lambda_f s + 1}.$$

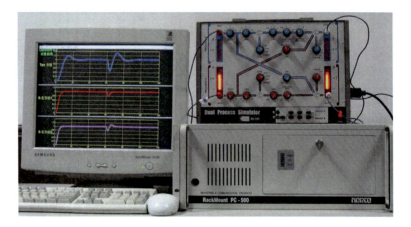

Fig. 9.6 The experimental setup for real-time tests of Example 9.2

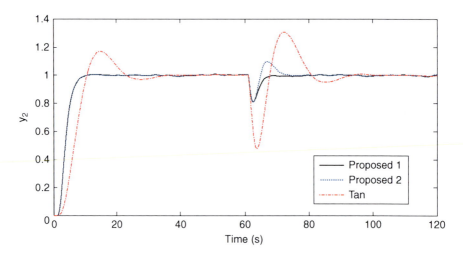

Fig. 9.7 Nominal system responses of Example 9.2

For comparison, $\lambda_c = 1.0$ and $\lambda_f = 0.5$ are taken for real-time tests. By adding a unit step change to the set-point and an inverse unit step change of load disturbance to the secondary process output at $t = 60(s)$, the control results are shown in Fig. 9.7. It is seen that quick set-point response without overshoot is obtained by either of the proposed cascade control structures, and obviously enhanced load disturbance response is obtained in comparison with that of Tan et al. (2000). For load disturbance rejection, the cascade control structure shown in Fig. 9.1a (solid line) outperforms the other cascade control structure shown in Fig. 9.1b (dotted line), owing to the use of the final process model identified.

9.4 Illustrative Examples and Real-Time Tests

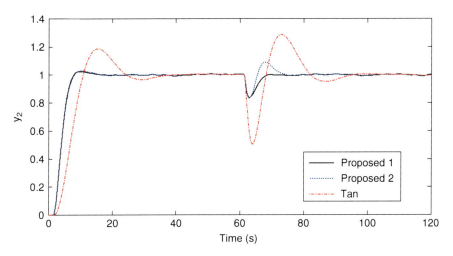

Fig. 9.8 Perturbed system responses of Example 9.2

Then, the cascade process is perturbed for test. The time constant of the secondary process is modified 30% smaller in the simulator, while the time delay of the final process is enlarged by 20%. The perturbed system responses are shown in Fig. 9.8, which again demonstrates that the proposed cascade control structures hold good robust stability.

Example 9.3 Consider a NMP stable cascade process studied by Lee et al. (2002),

$$P_1(s) = \frac{3e^{-3s}}{13.3s+1}$$

$$P_2(s) = \frac{10(-5s+1)e^{-5s}}{(30s+1)^3(10s+1)^2}.$$

The conventional cascade control structure was used in Lee's method, where the primary and secondary controllers were respectively configured as a PID controller, i.e., $G_{cp} = 0.12(1 + 1/87.88s + 21.21s)$, $G_{cs} = 1.62(1 + 1/7.44s + 1.37s)$. Moreover, there were two prefilters in front of the outer and inner loops, respectively chosen as $q_{f1} = 1/(59.34s+1)$ and $q_{f2} = 1/(5.78s+1)$. In the proposed 2DOF cascade control schemes shown in Fig. 9.1a, b, using the controller formulae given in (9.8) and (9.16), and taking $\lambda_c = 10$ and $\lambda_f = 1.5$ for comparison, it follows that

$$C_s(s) = \frac{(13.3s+1)(30s+1)^3}{30(5s+1)(10s+1)^3}$$

$$C_f(s) = \frac{13.3s+1}{3(1.5s+1)}.$$

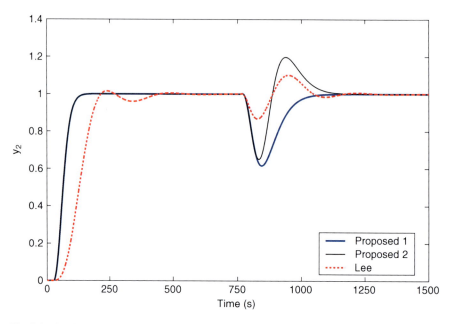

Fig. 9.9 Nominal system responses of Example 9.3

By adding a unit step change to the set-point and an inverse unit step change of load disturbance to the secondary process output at $t = 750(s)$, the control results are shown in Fig. 9.9. It is seen that a smooth set-point response without overshoot is obtained using either of the proposed 2DOF cascade control structures. Lee's method results in sluggish set-point response due to the use of a prefilter with a relatively large time constant to the primary set-point. It is true that better load disturbance response is obtained by Lee's method in terms of the same recovery time, but this method seems to be more sensitive to the process uncertainties. For instance, assume that the secondary process is actually perturbed to $P_1 = 3e^{-3.9s}/(9.31s + 1)$. It can be verified that Lee's method cannot hold the system stability any longer, but the proposed cascade control structures maintain robust stability well, as demonstrated by the results shown in Fig. 9.10.

For further illustration, assume that both time delays of the secondary and final processes are actually 50% larger and the time constant of the secondary process is 50% smaller. Neither of the proposed cascade control structures in terms of the above controller parameters can maintain stability any longer due to the severe process uncertainties. Nevertheless, gradually increasing the single adjustable parameter λ_f in C_f to 3.0 can recover the stability easily, as indicated by the results shown in Fig. 9.11.

Example 9.4 Consider a chemical CSTR studied by Lee et al. (2002), of which the secondary stable process and the final unstable process are respectively identified as

9.4 Illustrative Examples and Real-Time Tests

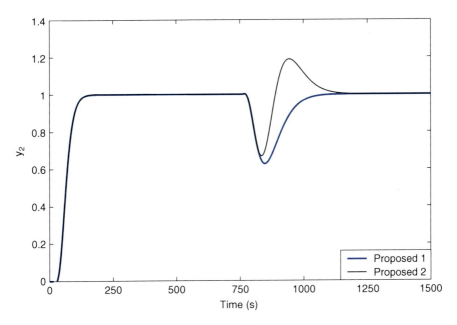

Fig. 9.10 Perturbed system responses of Example 9.3

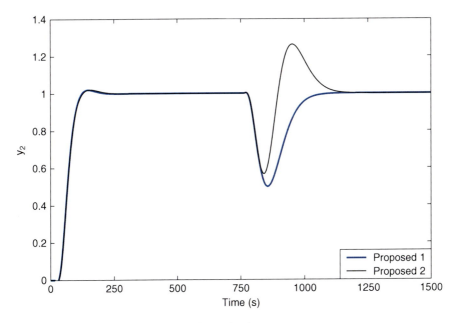

Fig. 9.11 Perturbed system responses by increasing λ_f

$$P_1(s) = \frac{2e^{-2s}}{20s + 1}$$

$$P_2(s) = \frac{e^{-4s}}{20s - 1}.$$

In Lee's method, the primary and secondary controllers in the conventional cascade control structure were respectively tuned as $G_{cp} = 3.31(1 + 1/36.22s + 3.08s)$ and $G_{cs} = 6.92(1 + 1/4.6s + 0.79s)$, together with two prefilters in front of the outer and inner loops which were chosen as $q_{f1} = 1/(32.91s + 1)$ and $q_{f2} = 1/(3.66s + 1)$. In the proposed 3DOF cascade control structure shown in Fig. 9.3, it follows from the controller design formulae given in (9.26)–(9.31) that

$$C_s(s) = \frac{(20s + 1)(20s - 1)}{2(\lambda_c s + 1)^2}$$

$$T_r(s) = \frac{1}{(\lambda_c s + 1)^2} e^{-6s}$$

$$C_{f-1}(s) = \frac{20s + 1}{2(\lambda_{f-1} s + 1)}$$

$$C_{f-2}(s) = \frac{(20s + 1)(20s - 1)(\alpha_1 s + 1)}{2(\lambda_{f-2} s + 1)^3} \cdot \frac{1}{1 - \frac{\alpha_1 s + 1}{(\lambda_{f-2} s + 1)^3} e^{-6s}},$$

where $\alpha_1 = 20\left[(0.05\lambda_{f-2} + 1)^3 e^{0.3} - 1\right]$.

For comparison, take $\lambda_c = 6$, $\lambda_{f-1} = 0.5$, and $\lambda_{f-2} = 6$. Using the analytical approximation formula in (7.100), a PID form to approximate C_{f-2} for implementation can be obtained as

$$C_{f-2-\text{PID}} = 1.9785 + \frac{1}{30.6256s} + \frac{28.0736s}{0.2s + 1}.$$

Also taking $N = 3$ in (8.61) for a higher-order approximation of C_{f-2}, it follows that

$$C_{f-2-3/3}(s) = \frac{40.4166s^2 + 28.7214s + 1.9881}{2.1204s^2 + 0.2925s + 1} + \frac{0.0327}{s(2.1204s^2 + 0.2925s + 1)}.$$

By adding a unit step change to the set-point and an inverse unit step change of load disturbance to the secondary process output at $t = 100(s)$ and an inverse step change of load disturbance with a magnitude of 0.2 to the final process output at $t = 200(s)$, the control results are shown in Fig. 9.12. It is seen that the proposed cascade control structure gives obviously improved set-point response without overshoot for the nominal system. Note that the rise time of the set-point

9.4 Illustrative Examples and Real-Time Tests

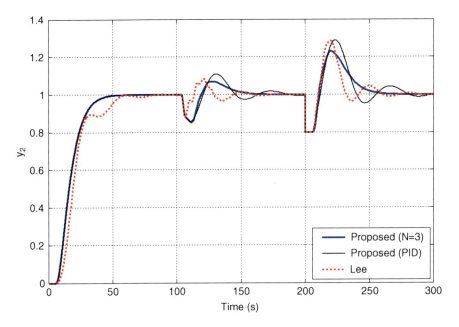

Fig. 9.12 Nominal system responses of Example 9.4

response, which can be numerically computed from (9.11) as $t_r = 29.3382(s)$, is identical with the result shown in Fig. 9.12. Besides, it is seen that compared with the PID form of C_{f-2}, the third order approximation leads to further improved load disturbance response, owing to a better approximation for implementation.

Assume that a random noise with the amplitude of 0.02 is added to both measurements of the secondary and final process outputs, which are then used for feedback control. The corresponding results are shown in Fig. 9.13, demonstrating good robustness of using the proposed cascade control method.

Then, assume that there exists 10% error in estimating the process parameters of time delay and time constant. The worst case is that both θ_1 and θ_2 are actually 10% larger while both τ_1 and τ_2 are 10% smaller. The perturbed system responses are shown in Fig. 9.14. It is seen that the proposed cascade control method maintains the closed-loop system robust stability. Note that Lee's method cannot hold the system stability any longer and thus is omitted. In fact, further enhanced robust stability of the proposed control system can be conveniently obtained on line by monotonically increasing the single adjustable parameter of C_s, C_{f-1}, and C_{f-2}, respectively. For illustration, letting $\lambda_c = 8$ and $\lambda_{f-2} = 8$ gives

$$C_{f-2-\text{PID}} = 1.5527 + \frac{1}{48.1605s} + \frac{23.4708s}{0.2s + 1}$$

$$C_{f-2-3/3}(s) = \frac{35.1847s^2 + 24.8678s + 1.5707}{2.56s^2 + 0.8655s + 1} + \frac{0.0208}{s(2.56s^2 + 0.8655s + 1)}.$$

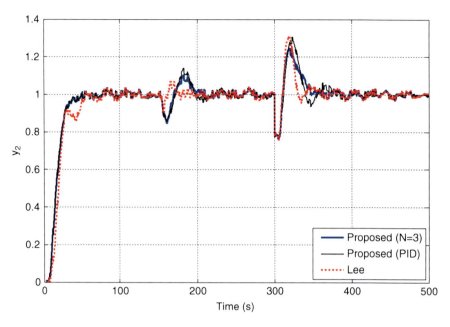

Fig. 9.13 Output responses of Example 9.4 in the presence of measurement noise

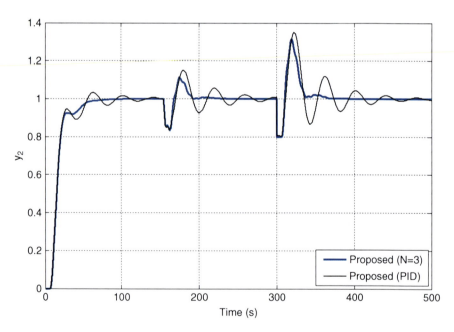

Fig. 9.14 Perturbed system responses of Example 9.4 in the presence of the process uncertainties

9.4 Illustrative Examples and Real-Time Tests

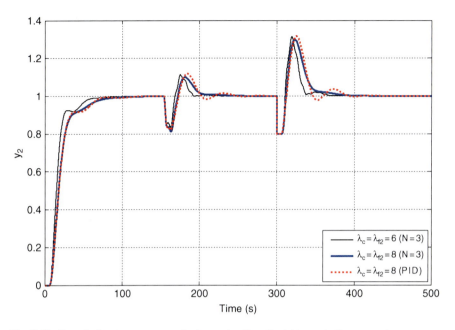

Fig. 9.15 Perturbed system responses by increasing the adjustable controller parameters

The corresponding control results are shown in Fig. 9.15, which demonstrate that increasing these adjustable parameters can effectively improve control robustness for set-point tracking and load disturbance rejection. Hence, it is convenient in practice to monotonically vary λ_{f-1} and λ_{f-2} on line to achieve the best trade-off between the nominal performance and robust stability of the inner and outer loops, respectively, while λ_c can be monotonically tuned to optimize the set-point response.

To demonstrate the robust stability of the proposed cascade control system in the presence of the control valve uncertainties, assume that there exists a multiplicative input uncertainty, $\Delta_1(s) = (s+0.2)/(s+1)$. The perturbed system responses are shown in Fig. 9.16. It is once again demonstrated that the proposed cascade control method facilitates good robust stability. Note that Lee's method still cannot hold the system stability and thus is omitted.

Example 9.5 Consider an unstable process studied by Tan et al. (2003),

$$P(s) = \frac{e^{-0.939s}}{(2.07s+1)(5s-1)}.$$

Here it is used to demonstrate that the proposed 3DOF cascade control structure is superior to a 2DOF control scheme for an unstable process, if the secondary process can be measured for control design. The process is assumed to be composed

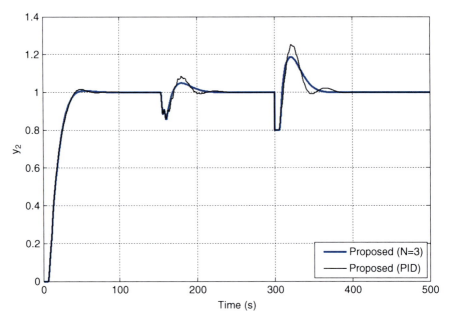

Fig. 9.16 Perturbed system responses of Example 9.4 in the presence of the control valve uncertainties

of the secondary process, $P_1(s) = e^{-0.6s}/(2.07s + 1)$, and the final process, $P_2(s) = e^{-0.339s}/(5s - 1)$.

In Tan's method, the controllers were taken as $K_0(s) = 2(2.07s + 1)$, $K_1(s) = (5s + 1)/(0.2s + 1)$, and $K_2(s) = 3.584(2.4s + 1)$. In the proposed 3DOF cascade control structure shown in Fig. 9.3, it follows from the controller design formulae given in (9.26)–(9.31) that

$$C_s(s) = \frac{(2.07s + 1)(5s - 1)}{(\lambda_c s + 1)^2}$$

$$T_r(s) = \frac{1}{(\lambda_c s + 1)^2} e^{-0.939s}$$

$$C_{f-1}(s) = \frac{2.07s + 1}{\lambda_{f-1} s + 1}$$

$$C_{f-2}(s) = \frac{(2.07s + 1)(5s - 1)(\alpha_1 s + 1)}{(\lambda_{f-2} s + 1)^3} \cdot \frac{1}{1 - \frac{\alpha_1 s + 1}{(\lambda_{f-2} s + 1)^3} e^{-0.939s}},$$

where $\alpha_1 = 5[(0.2\lambda_{f-2} + 1)^3 e^{0.1878} - 1]$.

For comparison, take $\lambda_c = 1$, $\lambda_{f-1} = 0.6$, and $\lambda_{f-2} = 1.2$. Using the analytical approximation formula in (7.100), a PID form to approximate C_{f-2} for implementation may be obtained as

9.5 Summary

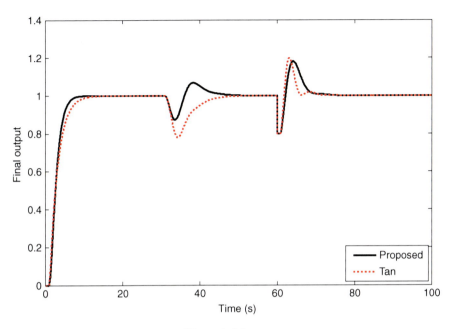

Fig. 9.17 Nominal system responses of Example 9.5

$$C_{f-2-\text{PID}} = 4.4091 + \frac{1}{1.9636s} + \frac{7.1874s}{0.07s + 1}.$$

By adding a unit step change to the set-point and an inverse unit step change of load disturbance to the secondary process input at $t = 30(s)$ and an inverse step load disturbance with a magnitude of 0.2 to the final process output at $t = 60(s)$, the control results are shown in Fig. 9.17. It is seen that the proposed 3DOF cascade control structure results in obviously improved load disturbance response, based on the secondary process output measurement for feedback control.

Then, assume that there exists 10% error in estimating the process parameters of time delay and time constant. The worst case is that both θ_1 and θ_2 are actually 10% larger while both τ_1 and τ_2 are 10% smaller. The perturbed system responses are shown in Fig. 9.18. It is again seen that the proposed cascade control structure holds good robust stability in the presence of the severe process uncertainty.

9.5 Summary

Cascade control strategies have been explored to improve load disturbance rejection performance for long time delay or slow processes in industry, based on measuring the secondary (intermediate) process output for feedback control. To overcome

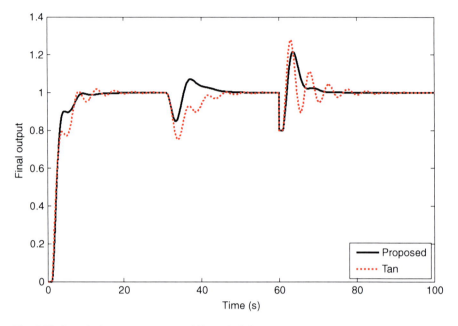

Fig. 9.18 Perturbed system responses of Example 9.5

the deficiencies of the conventional cascade control structure, two 2DOF cascade control structures have been presented for open-loop stable processes, respectively for whether the final process model can be obtained for control design or not. These two cascade control structures are simplified versions of those developed by Liu et al. (2005a). For the convenience of application, the controller designs are unified for these two cascade control structures. An important merit is that there exists only a single adjustable parameter in each of these two controllers, which can be monotonically tuned with relative independence to optimize the outer loop performance of set-point tracking and the inner loop performance of load disturbance rejection. Robust stability constraints for tuning the two controllers have been given for practical applications.

To alleviate the water-bed effect between the set-point response and the load disturbance response that is commonly encountered in the unity feedback control structure for unstable processes, a 3DOF cascade control structure has been presented for open-loop unstable processes. Accordingly, the set-point tracking and load disturbance rejection can be separately optimized. This cascade control structure is a simplified version of that developed by Liu et al. (2005b). There are three controllers which are responsible for tuning the set-point response, the inner loop response, and the outer loop response, respectively. Each controller has a single adjustable parameter which can be monotonically tuned for performance optimization. Note that the proposed 3DOF cascade control scheme can be transparently extended to the case where the secondary process is unstable and the final process is stable. If higher-order models are identified for control design, the model-based controller designs given in Chaps. 7 and 8 can be used for the corresponding controller formulation.

Five examples from the existing literature have been used to illustrate the effectiveness and merits of the presented 2DOF and 3DOF cascade control schemes, respectively for open-loop stable and unstable processes. Real-time tests have also been performed using the KI 101 type process simulator of KentRidge Instruments in Singapore to demonstrate the effectiveness and robustness of the proposed cascade control methods against measurement noise and process uncertainties.

References

Åström KJ, Hägglund T (1995) PID controller: theory, design, and tuning, 2nd edn. ISA Society of America, Research Triangle Park
Hang CC, Loh AP, Vasnani VU (1994) Relay feedback auto-tuning of cascade controllers. IEEE Trans Control Syst Technol 2(1):42–45
Huang HP, Chien I-L, Lee YC (1998) Simple method for tuning cascade control systems. Chem Eng Commun 165:89–121
Kaya I (2001) Improving performance using cascade control and a Smith predictor. ISA Trans 40:223–234
Lee YH, Park SW, Lee MY (1998) PID controller tuning to obtain desired closed loop responses for cascade control systems. Ind Eng Chem Res 37(5):1859–1865
Lee YH, Oh SG, Park SW (2002) Enhanced control with a general cascade control structure. Ind Eng Chem Res 41(11):2679–2688
Liu T, Gu DY, Zhang WD (2005a) Decoupled two-degree-of-freedom control strategy for cascade control systems. J Process Control 15(2):159–167
Liu T, Zhang WD, Gu DY (2005b) New IMC-based control strategy for open-loop unstable cascade processes. Ind Eng Chem Res 44(4):900–909
Luyben WL (1990) Process modeling, simulation, and control for chemical engineers. McGraw Hill, New York
Morari M, Zafiriou E (1989) Robust process control. Prentice Hall, Englewood Cliff
Nagrath D, Prasad V, Bequette BW (2002) A model predictive formulation for control of open-loop unstable cascade systems. Chem Eng Sci 57:365–378
Ogunnaike BA, Ray WH (1994) Process dynamics, modeling, and control. Oxford University Press, New York
Saraf V, Zhao FT, Bequette BW (2003) Relay autotuning of cascade-controlled open-loop unstable reactors. Ind Eng Chem Res 42(20):4488–4494
Seborg DE, Edgar TF, Mellichamp DA (2004) Process dynamics and control, 2nd edn. Wiley, Hoboken
Shinskey FG (1996) Process control system, 4th edn. McGraw Hill, New York
Song SH, Cai WJ, Wang YG (2003) Auto-tuning of cascade control systems. ISA Trans 42(1):63–72
Tan KK, Lee TH, Ferdous R (2000) Simultaneous online automatic tuning of cascade control for open loop stable processes. ISA Trans 39:233–242
Tan W, Marquez HJ, Chen T (2003) IMC design for unstable processes with time delays. J Process Control 13:203–213
Zhou KM, Doyle JC, Glover K (1996) Robust and optimal control. Prentice Hall, Englewood Cliff

Chapter 10
Multiloop Control of Multivariable Processes

10.1 Selection of the Input–Output Pairing

For a multivariable process, a change in a manipulated variable, say u_1, will affect not only the corresponding output variable (y_1) but also other output variables (y_2, y_3, ...), which is generally defined as interaction in process operation. Suitable input–output pairing is therefore a prerequisite for multivariable control system design. For instance, consider a multivariable process with n manipulated variables and n controlled variables, there exist $n!$ possible choices for the pairing selection! Moreover, an incorrect pairing will severely hinder the closed-loop control performance or jeopardize the system stability (Shinskey 1996). A number of criteria have been explored for analyzing cross interaction in multivariable systems (see Bristol (1966), McAvoy (1983), Jensen et al. (1986), Huang et al. (1994, 2003), Shinskey (1996), Lee and Edgar (2004), Salgado and Conley (2004), Skogestad and Postlethwaite (2005), and He et al. (2009)). Among these criteria, relative gain array (RGA) and singular value decomposition (SVD) have been widely recognized in practice, which are briefly introduced as follows.

10.1.1 Relative Gain Array (RGA)

The concept of relative gain was early proposed by Bristol (1966). For a multivariable process with n manipulated variables and n controlled variables, the relative gain, λ_{ij}, between a controlled variable y_i and a manipulated variable u_j is defined as a dimensionless ratio of two steady-state gains:

$$\lambda_{ij} \triangleq \frac{\left(\frac{\partial y_i}{\partial u_j}\right)_{u_k=0, k \neq j}}{\left(\frac{\partial y_i}{\partial u_j}\right)_{y_k=0, k \neq i}}, \quad i, j = 1, 2, \ldots, n. \tag{10.1}$$

where $(\partial y_i/\partial u_j)_{u_k=0, k\neq j}$ is usually called open-loop gain, which denotes a partial derivative that is evaluated with all the manipulated variables except u_j being held constant. It therefore corresponds to the process static gain matrix element, $g(0)_{ij}$. By comparison, $(\partial y_i/\partial u_j)_{y_k=0, k\neq i}$ is called closed-loop gain, which is evaluated with all the controlled variables except y_i maintained to the corresponding set-point values (or as closely as possible by using a multiloop control).

Correspondingly, the RGA is denoted by

$$\Lambda = \begin{bmatrix} \lambda_{11} & \lambda_{11} & \cdots & \lambda_{1n} \\ \lambda_{21} & \lambda_{22} & \cdots & \lambda_{2n} \\ \vdots & \vdots & \cdots & \vdots \\ \lambda_{n1} & \lambda_{n2} & \cdots & \lambda_{nn} \end{bmatrix} \tag{10.2}$$

and this has several important properties for assessing the interaction:

1. It is dimensionless and thus independent of input and output scaling.
2. Its rows (or columns) sum to 1. A positive λ_{ij} indicates positive interaction between y_i and u_j. In contrast, a negative value indicates inverse interaction.
3. It is the identity matrix if the process transfer matrix is upper or lower triangular.
4. Its diagonal dominance means weak interaction between y_i and u_j ($i \neq j$), which facilitates multiloop control of individual output variables with relative independence.
5. A large λ_{ij} implies strong interaction between y_i and u_j ($i \neq j$), corresponding to an ill-conditioned transfer matrix.

Based on the evaluation of RGA, the following pairing rules suggested by Bristol (1966) and McAvoy (1983) have been widely used in industrial applications:

Pairing rule 1. Select input–output pairings along the diagonal that results in RGA close to the identity matrix as much as possible.

Pairing rule 2. Avoid pairing on negative RGA elements.

Note that only static gains of the process transfer matrix are considered for the interaction measure in the above pairing rules. To further consider the process dynamic response characteristics, a modified RGA for dynamic interaction measure was proposed by Skogestad and Postlethwaite (2005):

$$\text{RGA}(G) = \Lambda(G) \triangleq G \oplus \left(G^{-1}\right)^{\text{T}} \tag{10.3}$$

where G denotes the process transfer function. Accordingly, an enhanced pairing rule 1 was given:

Enhanced pairing rule 1. Prefer pairings such that the rearranged system, with the selected pairings along the diagonal, has an RGA matrix close to the identity matrix at frequencies around the closed-loop bandwidth.

10.1 Selection of the Input–Output Pairing

For applying the above enhanced rule 1, one can compute the RGA-number for the diagonal pairings:

$$\text{RGA} - \text{number} \triangleq \|\Lambda(G) - I\|_{\text{sum}} \qquad (10.4)$$

where the norm is defined as $\|A\|_{\text{sum}} = \sum_{i,j} |a_{ij}|$. Obviously, a small RGA-number is preferred for pairing selection. The RGA-number for other pairings may be computed by subtracting 1 for the selected pairings. For example, the off-diagonal pairing for a two-input-two-output (TITO) process may be selected in terms of computing the RGA-number:

$$\text{RGA} - \text{number} = \left\| \Lambda(G) - \begin{bmatrix} 0 & 1 \\ 1 & 0 \end{bmatrix} \right\|_{\text{sum}}. \qquad (10.5)$$

10.1.2 Singular Value Decomposition (SVD)

In contrast with the above RGA approach for pairing selection, an alternative approach has been developed in terms of the singular value decomposition (SVD) analysis (Skogestad and Postlethwaite 2005).

Consider a multivariable process described by

$$Y(s) = G(s)U(s) \qquad (10.6)$$

where $G(s) = [g_{ij}]_{m \times n}$. Correspondingly, the static gain matrix is written as $G(0) = [g_{ij}(0)]_{m \times n}$.

It is obvious by checking if the determinant of $G(0)$ is zero to verify the linear independence between the controlled variables.

For a fixed frequency, $\omega \in [0, \infty)$, the transfer function matrix can be expressed by SVD as:

$$G(j\omega) = U \Sigma V^T \qquad (10.7)$$

where $\Sigma = [\Sigma_1 \, 0]$ if $m \leq n$, or $\Sigma = [\Sigma_1 \, 0]^T$ if $m \geq n$, $\Sigma_1 = \text{diag}[\sigma_i]_{l \times l}$, $l = \min\{m, n\}$,

$$\sigma_i = \sqrt{\lambda_i(GG^*)}, \quad i = 1, 2, \ldots, l \qquad (10.8)$$

which are listed in a descending order along the diagonal of Σ to indicate the maximal singular value ($\bar{\sigma} = \sigma_1$) and the minimal singular value ($\underline{\sigma} = \sigma_l$). $U \in \mathbb{C}^{m \times m}$ and $V \in \mathbb{C}^{n \times n}$ are unitary matrices satisfying

$$UU^T = I \qquad (10.9)$$

$$VV^T = I. \qquad (10.10)$$

Note that all columns in U, denoted by u_i ($i = 1, 2, \ldots, l$) satisfying $\|u_i\|_2 = 1$ and orthogonal to each other, indicate the output directions of the transfer function matrix. All columns in V, denoted by v_i ($i = 1, 2, \ldots, l$) satisfying $\|v_i\|_2 = 1$ and orthogonal to each other, indicate the input directions of the transfer function matrix. Correspondingly, the singular value σ_i indicates the proportional gain of the transfer function matrix in the ith direction, that is,

$$\sigma_i = \frac{\|G v_i\|_2}{\|v_i\|_2} = \|G v_i\|_2, \quad i = 1.2, \ldots, l. \tag{10.11}$$

Hence, two important merits of SVD for analyzing the gains and directions of a multivariable process transfer function matrix are that

1. The singular values indicate the process gains in the corresponding directions.
2. The directions of the process transfer function matrix thus obtained are orthogonal to each other.

The condition number of G can be used for the interaction measure, which is evaluated as a dimensionless ratio of the maximal to the minimal singular value:

$$\gamma(G) = \frac{\bar{\sigma}(G)}{\underline{\sigma}(G)}. \tag{10.12}$$

A large condition number usually refers to an ill-conditioned transfer matrix. For a non-singular square process, it follows from $\underline{\sigma}(G) = 1/\bar{\sigma}\left(G^{-1}\right)$ that

$$\gamma(G) = \bar{\sigma}(G)\bar{\sigma}\left(G^{-1}\right) \tag{10.13}$$

which implies that the condition number tends to be large if G and G^{-1} have large elements.

10.2 Multiloop Structure Controllability

For the control of a multivariable process, a widely used control structure is multiloop (or named decentralized control). This control structure has a few merits for implementation, e.g., economic configuration, tuning simplicity, loop failure tolerance, etc. (Shinskey 1996; Seborg et al. 2004). Due to the interactions between individual loops, closed-loop tuning methods developed for SISO processes cannot be simply extended to multiloop systems (Morari and Zafiriou 1989; Skogestad and Postlethwaite 2005). To avoid complexity, TITO processes are usually considered for multiloop control in practical applications. Many industrial processes with a dimension beyond two are preferred to be divided as several TITO subsystems for the convenience of control design and system operation (Luyben 1990; Ogunnaike and Ray 1994; Seborg et al. 2004).

10.2 Multiloop Structure Controllability

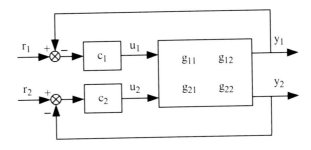

Fig. 10.1 Multiloop control structure for a TITO process

Consider a TITO process generally described in the form of:

$$G(s) = \begin{bmatrix} g_{11}(s) & g_{12}(s) \\ g_{21}(s) & g_{22}(s) \end{bmatrix} \quad (10.14)$$

where $g_{ij}(s) = g_{0,ij}(s)e^{-\theta_{ij}s}$, $i, j = 1, 2$, and $g_{0,ij}(s)$ are physically proper and stable transfer functions.

A multiloop control structure for a TITO process described in (10.14) is shown in Fig. 10.1, where c_1 and c_2 are the multiloop controllers and u_1 and u_2 denote the corresponding controller outputs.

The closed-loop system transfer function matrix can be derived accordingly as:

$$H = GC(I + GC)^{-1} \quad (10.15)$$

where C is a diagonal controller matrix, i.e., $C = diag\{c_1, c_2\}_{2\times 2}$.

It implies by (10.15) that absolute decoupling control of the binary system is impractical in the framework of a multiloop control structure. This can be clarified by a proof of contradiction as given below.

Assume that absolute decoupling control could be realized, that is, the nominal closed-loop system transfer matrix could be led to the diagonal form, $H = \bar{H} = diag\{h_1, h_2\}_{2\times 2}$. Taking the matrix inverse for both sides of (10.15) gives $H^{-1} = (GC)^{-1} + I$. Note that H^{-1} is also a diagonal matrix in such a case. Hence, $(GC)^{-1}$ must be a diagonal matrix, thus requiring the process transfer matrix G to be a diagonal matrix. The requirement contradicts the process description in (10.14). This completes the proof.

For analysis, the multiloop control structure shown in Fig. 10.1 is rearranged into the block diagonal closed-loop structure shown in Fig. 10.2, where $\bar{G}$ is composed of the diagonal elements in G, i.e., $\bar{G} = diag\{g_{11}, g_{22}\}_{2\times 2}$, which connects the desired pairings of the binary system inputs and outputs. Meanwhile, $G - \bar{G}$ is viewed as an additive uncertainty of $\bar{G}$. According to the small gain theorem (Zhou et al. 1996), the larger is the H infinity norm of $G - \bar{G}$, the worse is the closed-loop system stability, and correspondingly, the lower is the multiloop structure controllability. Hence, it is necessary to configure the process transfer matrix in terms of diagonal dominance. Specifically, it is desirable to have the

Fig. 10.2 A block diagram of cross interactions as the additive uncertainty

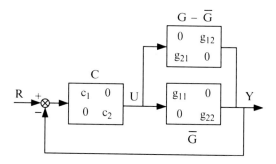

column diagonal dominance of a TITO transfer matrix, i.e., $|g_{11}(j\omega)| > |g_{21}(j\omega)|$ and $|g_{22}(j\omega)| > |g_{12}(j\omega)|$ for $\omega \in [0, +\infty)$, which can be verified by comparing the corresponding magnitude plots of frequency response. An intuitive but not strict judgment on the diagonal dominance is to compare these static gains, i.e., check if $|g_{11}(0)| > |g_{21}(0)|$ and $|g_{22}(0)| > |g_{12}(0)|$.

In the case where the process transfer function matrix cannot be configured diagonal dominance as constrained by the process operation, some existing methods, e.g., Chen and Seborg (2002, 2003), suggested to add a static decoupler, i.e., $D(0) = G^{-1}(0)$, in front of the process inputs, and then to design the multiloop controllers for the augmented plant, in expectation of further enhanced control performance. In fact, such a static decoupler may be incorporated into the multiloop controller design and therefore may not be regarded as an advantage. This will be illustrated by Example 10.2 later.

For additional discussions on multiloop structure controllability, the readers are referred to the existing literature, e.g., Campo and Morari (1994), Lee and Edgar (2000), Cui and Jacobsen (2002), and Skogestad and Postlethwaite (2005).

10.3 Multiloop Control Design

Different approaches have been developed for multiloop control design. To enumerate a few, the Gershgorin band criterion in combination with the frequency domain specifications of gain and phase margins have been intensively studied for tuning multiloop PI/PID controllers (Ho et al. 1997; Lee et al. 1998; Wang et al. 1998; Chen and Seborg 2002, 2003). Using the linear fractional transformation (LFT) in terms of a $M - \Delta$ structure, Hovd and Skogestad (1993) and Gündes and Özgüler (2002) developed independent tuning methods for multiloop controllers. Zhang et al. (2002) further extended the dominant pole placement approach for SISO systems (Åström and Hägglund 1995) to TITO processes, which can lead to evidently improved system performance compared to the well-known biggest log modulus tuning (BLT) method developed by Luyben (1986). To overcome the deficiency of the conventional PID-type multiloop controllers causing excessive

10.3 Multiloop Control Design

oscillation in the set-point responses compared to the PI-type controllers, Chien et al. (1999) proposed a modified implementation of such PID controllers in a multiloop system. Desbiens et al. (1996) developed the Smith predictor (SP)-based multiloop control scheme for TITO processes with time delays. Using the sequential relay feedback identification to realize autotuning of multiloop PI controllers has been studied by Chiu and Arkun (1992), Loh et al. (1993), Shen and Yu (1994), etc. In contrast, Palmor et al. (1995), Halevi et al. (1997), Wang et al. (1997), etc. developed simultaneous relay identification methods to obtain the ultimate frequency response information for tuning multiloop PI/PID controllers. Following the IMC theory (Morari and Zafiriou 1989), Jung et al. (1999) suggested a one-parameter tuning method for multiloop control systems, and Cha et al. (2002) developed a two-step IMC tuning of multiloop PID controllers. Both the IMC-based tuning methods, however, were based on using numerical frequency response fitting algorithms, requiring relatively high computation effort for implementation.

Here an analytical multiloop PI/PID controller design method (Liu et al. 2005) is presented for application, by proposing a practically desired closed-loop diagonal transfer matrix and introducing a dynamic detuning matrix to reduce the interactions between individual loops.

10.3.1 Desired Diagonal Transfer Matrix

From Fig. 10.2, it can be seen that the nominal diagonal transfer matrix of the closed-loop system without the additive uncertainty is

$$\bar{H} = \bar{G}C(I + \bar{G}C)^{-1}. \tag{10.16}$$

Following some linear algebra, the diagonal controller matrix can be inversely derived as

$$C = \bar{G}^{-1}(\bar{H}^{-1} - I)^{-1}. \tag{10.17}$$

Correspondingly, the multiloop controllers can be obtained from (10.17) as

$$c_i = \frac{1}{g_{ii}} \cdot \frac{h_i}{1 - h_i}, \quad i = 1, 2. \tag{10.18}$$

Note that g_{ii} contains time delay θ_{ii}. It can be seen from (10.18) that if the desired transfer function h_i relating the system input r_i to the output y_i were not to include θ_{ii}, the corresponding controller c_i would behave in a predictive manner. Moreover, if g_{ii} has any RHP zeros, h_i is required to include them such that the resulting controller c_i will not include them as unstable poles.

Based on the above observation, using the H$_2$ optimal performance objective of the IMC theory (Morari and Zafiriou 1989), the desired closed-loop diagonal transfer functions for implementation are proposed as

$$h_i = \frac{e^{-\theta_{ii} s}}{(\lambda_i s + 1)^{p_i}} \prod_{k=1}^{q_i} \left(\frac{-s + z_k}{s + z_k^*} \right), \quad i = 1, 2 \quad (10.19)$$

where λ_i is an adjustable parameter to tune the i-th loop response, $\deg(g_{0,ii}) = p_i$, $s = z_k$ with $k = 1, 2, \ldots, q_i$ are RHP zeros of $g_{0,ii}$, and z_k^* is the complex conjugate of z_k.

Substituting (10.17) into (10.15) yields the real multiloop control system transfer matrix, i.e., the transfer matrix of the perturbed diagonal closed-loop system with the additive uncertainty $G - \bar{G}$ shown in Fig. 10.2:

$$H = G\bar{G}^{-1}(\bar{H}^{-1} - I)^{-1} \left(I + G\bar{G}^{-1}(\bar{H}^{-1} - I)^{-1} \right)^{-1}$$

$$= G \left((\bar{H}^{-1} - I)\bar{G} + G \right)^{-1}$$

$$= G \left(\bar{H}^{-1} \left((I - \bar{H})\bar{G} + \bar{H}G \right) \right)^{-1}$$

$$= G \left(\bar{G} + \bar{H}(G - \bar{G}) \right)^{-1} \bar{H} \quad (10.20)$$

Usually there exists $G \neq \bar{G}$ in practice. Therefore, the diagonal transfer functions relating the system inputs to the outputs will not be in the form of (10.19) if the multiloop controllers are derived directly from (10.18). This can be practically interpreted as that the additive uncertainty, $G - \bar{G}$, will inevitably result in the interactions between individual loops. To realize the desired closed-loop diagonal transfer functions shown in (10.19) for the selected pairings of the system inputs and outputs, a diagonal dynamic detuning matrix, $D = \text{diag}\{d_1, d_2\}_{2 \times 2}$, is therefore proposed to modify the diagonal system transfer matrix shown in (10.16), that is,

$$D\bar{H} = \bar{G}C(I + \bar{G}C)^{-1}. \quad (10.21)$$

Accordingly, the multiloop controller matrix can be inversely derived as

$$C = \bar{G}^{-1}(\bar{H}^{-1}D^{-1} - I)^{-1}. \quad (10.22)$$

Substituting (10.22) into (10.15) obtains the corresponding multiloop system transfer matrix

$$H = G \left(D^{-1}\bar{G} + \bar{H}(G - \bar{G}) \right)^{-1} \bar{H} \quad (10.23)$$

10.3 Multiloop Control Design

Letting

$$\text{diag}\left\{G\left(D^{-1}\bar{G} + \bar{H}\left(G - \bar{G}\right)\right)^{-1}\right\} = I, \tag{10.24}$$

one can obtain the diagonal transfer functions of the resulting multiloop transfer matrix as shown in (10.19).

Substituting (10.14) and (10.19) into (10.24), the dynamic detuning factors can be derived as

$$d_1 = \frac{2g_{11}g_{22}}{(h_1 - h_2)g_{12}g_{21} + g_{11}g_{22} + (-1)^m \sqrt{[(h_1-h_2)g_{12}g_{21} - g_{11}g_{22}]^2 - 4g_{11}g_{22}g_{12}g_{21}(1-h_1)h_2}} \tag{10.25}$$

$$d_2 = \frac{2g_{11}g_{22}}{(h_2-h_1)g_{12}g_{21} + g_{11}g_{22} + (-1)^m \sqrt{[(h_1 - h_2)g_{12}g_{21} - g_{11}g_{22}]^2 - 4g_{11}g_{22}g_{12}g_{21}(1-h_1)h_2}} \tag{10.26}$$

where

$$m = \begin{cases} 0, & g_{11}(0)g_{22}(0) > 0; \\ 1, & g_{11}(0)g_{22}(0) < 0. \end{cases} \tag{10.27}$$

Note that the choice of m in (10.27) is to ensure $d_1(0) = d_2(0) = 1$, such that the multiloop transfer matrix shown in (10.23) will be led to the identity matrix in the final steady state, i.e., $H(0) = I$. That is, the diagonal dynamic detuning matrix is reduced to the identity matrix in the steady-state system transfer matrix, introducing no steady offset in the system outputs. This is also the reason for discarding the other solution of (10.24). Concerning $d_1(0) = d_2(0) = 1$, it can be easily verified using $h_1(0) = h_2(0) = 1$ from (10.19).

Even in the case where $G = \bar{G}$, i.e., $g_{12} = g_{21} = 0$, it can be verified from (10.25) and (10.26) that $d_1 = d_2 = 1$. This indicates that the proposed diagonal dynamic detuning matrix is reduced to the identity matrix for a diagonal process transfer matrix and therefore is generally applicable to different TITO processes.

Hence, to realize the desired closed-loop diagonal transfer functions shown in (10.19), the modified diagonal transfer matrix for deriving the multiloop controllers can be determined as

$$\hat{H} = D\bar{H} = \text{diag}\left\{\frac{d_i e^{-\theta_{ii}s}}{(\lambda_i s + 1)^{p_i}} \prod_{k=1}^{q_i} \frac{(-z_k s + 1)}{(z_k s + 1)}\right\}, \quad i = 1, 2. \tag{10.28}$$

10.3.2 Multiloop PI/PID Controller Design

Substituting (10.28) into (10.22), one can obtain the ideally desired multiloop controllers in the form of

$$c_{\text{ideal}-i} = \frac{1}{g_{ii}} \cdot \frac{d_i h_i}{1 - d_i h_i}, \quad i = 1, 2. \tag{10.29}$$

By substituting (10.19), (10.25), and (10.26) into (10.29), it can be seen that both the numerator and denominator of (10.29) are involved with time delays in a complex manner and therefore are difficult for practical implementation. Moreover, if g_{ii} has any RHP zeros, there will exist RHP zero-pole canceling in (10.29), causing the controllers to work in an unstable manner. Therefore, a rational approximation is needed for practical implementation of the ideally desired controller form shown in (10.29).

Since PI- and PID-type controllers have been widely used in industry to facilitate economic operation of multiloop control systems, the analytical approximation formula based on the Maclaurin series expansion shown in (7.100) is used here for multiloop PI/PID controller design. Using (10.19), (10.25), and (10.26), it can be verified that

$$\lim_{s \to 0} (1 - d_i h_i) = 0, \quad i = 1, 2 \tag{10.30}$$

which implies that the ideally desired multiloop controllers shown in (10.29) have an integral property that can eliminate the output deviations from the corresponding set-points. Therefore, letting

$$M_i(s) = s c_{\text{ideal}-i}(s), \quad i = 1, 2, \tag{10.31}$$

one can obtain the corresponding PID-type controllers in the form of (7.102), where the first two terms may compose a PI-type controller.

Obviously, the above PID-type controller is capable of better closed-loop performance than such a PI-type controller. Further enhanced control performance can be obtained using the higher-order approximation formulae given in (8.61)–(8.64).

Note that each of the proposed multiloop PI/PID controllers is essentially tuned by a single adjustable parameter λ_i ($i = 1, 2$), which is used to tune the ith output response as shown in (10.19).

10.4 Robust Stability Analysis

To evaluate robust stability of a multiloop control system in the presence of process uncertainties, the following stability theorem developed from the generalized Nyquist stability theorem is first presented:

10.4 Robust Stability Analysis

Theorem 10.1. (Skogestad and Postlethwaite 2005) *Assume that the nominal system $M(s)$ and the perturbations $\Delta(s)$ are stable in the $M - \Delta$ structure shown in Fig. 7.7. Consider the convex set of perturbations Δ, such that if Δ' is an allowed perturbation then so is $\varepsilon\Delta'$ where ε is any real scalar such that $|\varepsilon| \leq 1$. Then the $M - \Delta$ system is stable for all allowed perturbations, if and only if any of the following equivalent conditions is satisfied:*

1. *The Nyquist plot of $\det(I - M\Delta(s))$ does not encircle the origin, $\forall \Delta$, i.e., $\det(I - M\Delta(j\omega)) \neq 0, \forall \omega, \forall \Delta$*
2. $\lambda_i(M\Delta(j\omega)) \neq 1, \forall i, \forall \omega, \forall \Delta$
3. $\rho(M\Delta(j\omega)) < 1, \forall \omega, \forall \Delta$
4. $\max_\Delta \rho(M\Delta(j\omega)) < 1, \forall \omega$

Based on the above stability lemma, the following corollary is evolved for a TITO multiloop control system shown in Fig. 10.1.

Corollary 10.1. *A TITO multiloop control system holds the nominal stability if and only if*

1. $c_1/(1 + g_{11}c_1)$ *and* $c_2/(1 + g_{22}c_2)$ *are stable*

2. $\rho\left(\begin{bmatrix} 0 & \frac{g_{12}c_1}{1+g_{11}c_1} \\ \frac{g_{21}c_2}{1+g_{22}c_2} & 0 \end{bmatrix}\right) < 1, \forall \omega.$

Proof. By transforming the multiloop control structure shown in Fig. 10.2 into the standard $M - \Delta$ structure shown in Fig. 7.7, one can see that the transfer matrix relating the input to the output of the additive uncertainty $\Delta = G - \bar{G}$ is

$$M = -C(I + \bar{G}C)^{-1} \tag{10.32}$$

which is obviously a diagonal transfer matrix owing to $\bar{G} = \text{diag}\{g_{11}, g_{22}\}_{2\times 2}$ and $C = \text{diag}\{c_1, c_2\}_{2\times 2}$ and consists of the two transfer functions shown in the first condition of Corollary 10.1.

Since the multiloop controllers have been devised in the form of PI/PID for implementation, the first condition is surely satisfied for a stable transfer matrix G.

Then by substituting $\Delta = G - \bar{G}$ and (10.32) into the third equivalent stability condition in Theorem 10.1, the second condition of Corollary 10.1 follows accordingly. □

Note that the second condition in Corollary 10.1 can be practically checked by observing if the magnitude plot of the spectral radius falls below the unity for $\omega \in [0, +\infty)$. An admissible tuning range of the adjustable parameters $\lambda_i (i = 1, 2)$ can thus be determined. Moreover, it can be seen from the desired closed-loop diagonal transfer functions shown in (10.19) that each of the output responses is essentially regulated by λ_1 and λ_2, respectively. When $\lambda_i (i = 1, 2)$ is tuned to a smaller value, the corresponding output response will become faster, but the output energy of c_i and the corresponding actuator action needs to be larger, and more

Fig. 10.3 Multiplicative input (**a**) and output (**b**) uncertainties

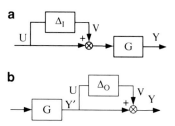

aggressive dynamic behavior of the corresponding output response will occur in the presence of process uncertainties. Alternatively, tuning λ_i to a larger value can slow down the corresponding output response, but the output energy of c_i and the corresponding actuator action is required smaller, and accordingly, less aggressive dynamic behavior of the corresponding output response will occur in the presence of process uncertainties. Hence, tuning the adjustable parameters $\lambda_i (i = 1, 2)$ aims at a good trade-off between individual loop performance, robustness, and the output capacities of $c_i (i = 1, 2)$ and the corresponding actuators.

In the presence of process uncertainties, the multiloop system transfer matrix may become very complex, and therefore, the closed-loop system may lose stability in an intangible manner. A practical way to evaluate the closed-loop system robust stability in the presence of process uncertainties such as the process parameters perturbations, actuator uncertainties, and output measurement uncertainties is to lump multiple sources of uncertainty into a multiplicative form (Skogestad and Postlethwaite 2005). In view of the multiplication sequence of transfer matrices, two cases consisting of the process multiplicative input and output uncertainties are analyzed here, which are shown in Fig. 10.3. The process multiplicative input uncertainty shown in Fig. 10.3a describes the process family $\Pi_I = \{\widehat{G}_I(s) : \widehat{G}_I(s) = G(s)(I + \Delta_I)\}$, where Δ_I is assumed to be stable. The process multiplicative output uncertainty shown in Fig. 10.3b describes the process family $\Pi_O = \{\widehat{G}_O(s) : \widehat{G}_O(s) = (I + \Delta_O) G(s)\}$, where Δ_O is assumed to be stable.

According to the standard $M - \Delta$ structure shown in Fig. 7.7 for robust stability analysis, the transfer matrices relating the outputs to the inputs of Δ_I and Δ_O can be derived, respectively, as

$$M_I = -C(I + GC)^{-1}G \qquad (10.33)$$

$$M_O = -GC(I + GC)^{-1} \qquad (10.34)$$

Note that the nominal system stability has been ensured by tuning λ_1 and λ_2. Accordingly, the closed-loop transfer matrix shown in (10.15) is stable. That is, $C(I + GC)^{-1}$ is maintained stable. Hence, M_I and M_O in (10.33) and (10.34) are also maintained stable for a stable transfer matrix G.

10.5 Illustrative Examples

Substituting (10.33) and (10.34) into the third equivalent stability condition in Theorem 10.1, respectively, one can obtain

$$\rho\left(C(I+GC)^{-1}G\Delta_I\right) < 1, \forall \omega \qquad (10.35)$$

$$\rho\left(GC(I+GC)^{-1}\Delta_O\right) < 1, \forall \omega. \qquad (10.36)$$

Hence, given a specified bound of Δ_I or Δ_O in practice, one can use (10.35) or (10.36) to evaluate the multiloop system robust stability, that is, observe if the magnitude plot of the left-hand sides of (10.35) or (10.36) falls below unity for $\omega \in [0, +\infty)$.

Generally, it is suggested to first tune each of the adjustable parameters λ_i ($i = 1, 2$) around the time delay θ_{ii} ($i = 1, 2$) of the diagonal transfer functions in the process transfer matrix, respectively. Then by monotonically varying these adjustable parameters on line, desirable output response performance of individual loops can be obtained in a transparent manner. To cope with process uncertainties in practice, it is suggested to monotonically increase λ_1 and λ_2 on line, so that the output responses will be slowed down to obtain enhanced closed-loop robust stability. If by doing so, the control system performance and robust stability still cannot be acceptable from a practical view, process re-identification can be performed to obtain a more precise process model for deriving the mutiloop controllers, such that the process unmodeled dynamics can be effectively reduced to facilitate improving system performance and robust stability.

10.5 Illustrative Examples

Two examples from the existing literature are used to demonstrate the effectiveness and merits of the presented multiloop control method. Example 10.1 is the widely studied Wood-Berry process with diagonal dominance. Example 10.2 is a TITO process with off-diagonal dominance, which is used not only to demonstrate the effectiveness of the proposed multiloop control method but also to compare control results with or without a static decoupler. For simulation tests, the simulation solver option is chosen as ode5 (Dormand-prince) and the simulation step size is fixed as 0.02(s).

Example 10.1. Consider the widely studied Wood-Berry binary distillation column process (Wood and Berry 1973):

$$G = \begin{bmatrix} \dfrac{12.8e^{-s}}{16.7s+1} & \dfrac{-18.9e^{-3s}}{21s+1} \\ \dfrac{6.6e^{-7s}}{10.9s+1} & \dfrac{-19.4e^{-3s}}{14.4s+1} \end{bmatrix}.$$

Table 10.1 PI controller parameters for Example 10.1

PI parameters	k_{C1}	τ_{I1}	k_{C2}	τ_{I2}
Chen	0.436	25.2294	−0.0945	−164.0212
Jung	0.19	44.7895	−0.099	−86.6667
Proposed	0.2448	22.2954	−0.0723	−86.8922

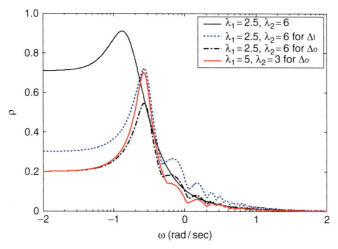

Fig. 10.4 The magnitude plots of spectral radius for Example 10.1

Two multiloop PI/PID control methods given by Chen and Seborg (2003) and Jung et al. (1999) are used here for comparison. The corresponding PI controller parameters are listed in Table 10.1.

In the proposed method, take $\lambda_1 = 2.5$ and $\lambda_2 = 6$ for obtaining the similar rise speed of the set-point response for comparison with the above two methods. Using the controller design formula in (10.29), together with (10.19), (10.25)–(10.27), and the analytical approximation formula in (7.100), the resulting PI controller parameters are also listed in Table 10.1.

Figure 10.4 shows the magnitude plot of the spectral radius condition in Corollary 10.1 to verify the nominal system stability in terms of the proposed method (thin solid line). It is seen that the maximum magnitude of the spectral radius for the nominal system is smaller than unity, indicating good system stability.

By adding a unity step change to the binary set-points at $t = 0(s)$ and $t = 100(s)$, respectively, and an inverse step change of load disturbance with a magnitude of 0.1 to both of the binary process inputs at $t = 200(s)$, the control results are shown in Fig. 10.5. Note that the time scale is altered from the minute unit of the original process to the second unit of the simulation model for the convenience of test and illustration. It is seen that the proposed multiloop PI controllers result in comparable binary output responses in comparison with the above two methods. For further comparison, the binary output responses obtained using the PID-type

10.5 Illustrative Examples

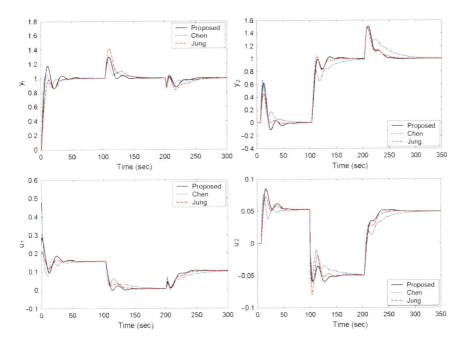

Fig. 10.5 Nominal system responses of Example 10.1 using PI controllers

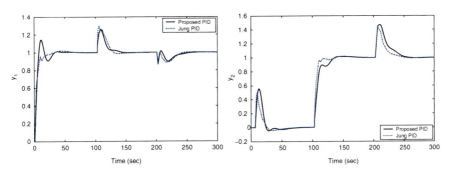

Fig. 10.6 Nominal system responses of Example 10.1 using PID controllers

controllers in terms of the above adjustable parameter settings (i.e., adding the derivative terms, $\tau_{D1} = 0.0624$ and $\tau_{D2} = -0.078$, to the PI controllers listed in Table 10.1) are shown in Fig. 10.6 in comparison with the PID controllers of Jung et al. (1999), $c_1 = 0.27 \left(1 + 1/6.91\text{s} + 3.935\text{s}\right)/(1.81\text{s} + 1)$ and $c_2 = -0.103 \left(1 + 1/5.9\text{s} + 1.88\text{s}\right)/(0.175\text{s} + 1)$. It is seen that both PID tunings lead to improved output response compared to the results in Fig. 10.5. Note that the PID tuning in Jung et al. (1999) was obtained by using numerical calculation with a considerable computation effort rather than simply adding a derivative term to the PI tuning that was also based on numerical calculation.

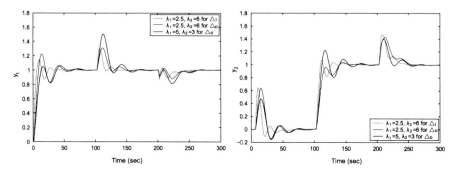

Fig. 10.7 Perturbed system responses of Example 10.1

To demonstrate the multiloop system stability by using the proposed method, assume that there exists the process multiplicative input uncertainty, $\Delta_I = diag\{(s+0.3)/(s+1), (s+0.3)/(s+1)\}_{2\times 2}$, which may be practically interpreted as the binary process inputs fed by the corresponding actuators increasing by 100% at high frequencies and by almost 30% in the low-frequency range. In the other case, assume that there exists the process multiplicative output uncertainty, $\Delta_O = diag\{-(s+0.2)/(2s+1), -(s+0.2)/(2s+1)\}_{2\times 2}$, which may be practically viewed as the binary process measurements offered by the output sensors decreasing by 50% at high frequencies and by almost 20% uncertainty in the low-frequency range. Figure 10.4 shows the magnitude plots of the spectral radius conditions in (10.35) and (10.36) for the above Δ_I and Δ_O, both of which indicate that the proposed method can maintain good robust stability. The corresponding output responses are shown in Fig. 10.7. It is seen that the proposed multiloop control system retains robust stability well. Moreover, in the proposed method the set-point response oscillation of the output y_1 can be alleviated by gradually increasing λ_1 on line at the cost of a degradation in the disturbance rejection performance. On the other hand, a faster response of the output y_2 can be conveniently obtained by gradually decreasing λ_2 on line at the cost of a degradation in the loop stability. For illustration, take $\lambda_1 = 5$ and $\lambda_2 = 3$, corresponding to the PI controllers of $k_{C1} = 0.1807$, $\tau_{I1} = 38.2206$, $k_{C2} = -0.091$, and $\tau_{I2} = -57.9281$. The control result is also shown in Fig. 10.7 for comparison, which demonstrates the relative independence for tuning individual output responses.

Example 10.2. Consider the industrial polymerization reactor studied by Chen and Seborg (2002) and Chien et al. (1999):

$$\begin{bmatrix} y_1(s) \\ y_2(s) \end{bmatrix} = \begin{bmatrix} \dfrac{22.89e^{-0.2s}}{4.572s+1} & \dfrac{-11.64e^{-0.4s}}{1.807s+1} \\ \dfrac{4.689e^{-0.2s}}{2.174s+1} & \dfrac{5.8e^{-0.4s}}{1.801s+1} \end{bmatrix} \begin{bmatrix} u_1(s) \\ u_2(s) \end{bmatrix} + \begin{bmatrix} \dfrac{-4.243e^{-0.4s}}{3.445s+1} \\ \dfrac{-0.601e^{-0.4s}}{1.982s+1} \end{bmatrix} d(s).$$

10.5 Illustrative Examples

Table 10.2 PI controller parameters for Example 10.2

PI parameters	k_{C1}	τ_{I1}	k_{C2}	τ_{I2}
Chen	6.67	0.1559	1.67	0.9401
Chien	0.263	5.3992	0.163	10.8589
Proposed without decoupler	0.2908	16.1502	0.0869	15.5504
Proposed with decoupler	7.7294	0.5	1.2136	1.7

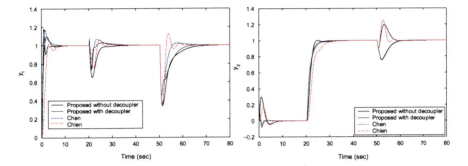

Fig. 10.8 System output responses of Example 10.2

It is obvious that the second column of the process transfer matrix has off-diagonal dominance. Chen and Seborg (2002) suggested to use a static decoupler $D(0) = G^{-1}(0)$ in front of the binary process inputs and then to design multiloop PI controllers for the augmented system. Chien et al. (1999) gave a multiloop PI controller tuning method for the original process. For comparison, two cases are performed by using the proposed method: One case is for the original process, and the other case is for the augmented system with the above static decoupler. By taking $\lambda_1 = 0.3$ and $\lambda_2 = 1.5$ to obtain the similar rising speeds of the set-point responses with those of Chen and Seborg (2002), the corresponding PI controller parameters are listed in Table 10.2, along with those of the above two methods.

By adding a unit step change to the binary set-points at $t = 0(s)$ and $t = 20(s)$, respectively, and then a unity step change of the load disturbance $d(s)$ at $t = 50(s)$, the control results are shown in Fig. 10.8. It is seen that improved system performance for both the set-point tracking and load disturbance rejection is obtained using the proposed method. Moreover, it is demonstrated that using a static decoupler does not help to obtain further improved system performance. Note that simulation tests under process uncertainties as assumed in example 10.1 demonstrate that the multiloop system robust stability is maintained well using the proposed method, regardless of whether the static decoupler is used or not, and therefore are omitted.

10.6 Summary

For multivariable processes, input–output pairing is the first step for control system design. Two widely used approaches developed respectively based on the RGA and SVD analysis for the pairing selection have been briefly introduced.

The multiloop control structure has been widely applied in various industrial and chemical contexts, owing to the simplicity and economic configuration for operation. Based on a TITO process description as mostly adopted in practical applications to facilitate the design and implementation of a multiloop control system, the multiloop structure controllability has been analyzed, along with a conclusion that absolute decoupling regulation is generally impractical to be realized by a multiloop structure.

An analytical multiloop PI/PID controller design method (Liu et al. 2005) has been presented for TITO processes with time delays. By proposing the desired closed-loop diagonal transfer functions relating the selected pairings of the process inputs and outputs and introducing a dynamic detuning matrix to realize these diagonal transfer functions in the multiloop system, the ideally desired multiloop controllers are inversely derived. For practical application, the Maclaurin approximation approach introduced in Sect. 7.5 can be used to derive the corresponding PI or PID controllers for implementation. An important merit is that there is essentially a single adjustable parameter in the proposed PI or PID multiloop controllers, respectively, which can be monotonically tuned to optimize individual loop responses, therefore facilitating the multiloop system operation in practice. Sufficient and necessary constraints for tuning each of the adjustable parameters to maintain the nominal system stability and its robust stability have been derived with respect to the process multiplicative input and output uncertainties. These stability constraints can be graphically checked through the magnitude plots of the corresponding spectral radius conditions.

Two examples from existing literature have been used to illustrate the proposed multiloop control method, one with diagonal dominance and the other with off-diagonal dominance. It has also been demonstrated that a static decoupler obtained as the inverse of the process static gain matrix has little advantage for the multiloop control design.

References

Åström KJ, Hägglund T (1995) PID controller: theory, design, and tuning, 2nd edn. ISA Society of America, Research Triangle Park

Bristol EH (1966) On a new measure of interaction for multivariable process control. IEEE Trans Autom Control 11(1):133–134

Campo PJ, Morari M (1994) Achievable closed-loop properties of systems under decentralized control: involving the steady-state gain. IEEE Trans Autom Control 39(3):932–943

References

Cha S, Chun D, Lee J (2002) Two-step IMC-PID method for multiloop control system design. Ind Eng Chem Res 41(12):3037–3041

Chen D, Seborg DE (2002) Multiloop PI/PID controller design based on Gershgorin bands. IEE Process Control Theory Appl 149(1):68–73

Chen D, Seborg DE (2003) Design of decentralized PI control systems based on Nyquist stability analysis. J Process Control 13(1):27–39

Chien IL, Huang HP, Yang JC (1999) A simple multiloop tuning method for PID controllers with no proportional kick. Ind Eng Chem Res 38(4):1456–1468

Chiu MS, Arkun Y (1992) A methodology for sequential design of robust decentralized control systems. Automatica 28(5):997–1002

Cui H, Jacobsen EW (2002) Performance limitations in decentralized control. J Process Control 12:485–494

Desbiens A, Pomerleau A, Hodouin D (1996) Frequency based tuning of SISO controllers for two-by-two processes. IEE Proccess Control Theory Appl 143(1):49–56

Gündes AN, Özgüler AB (2002) Two-channel decentralized integral-action controller design. IEEE Trans Autom Control 47(12):2084–2088

Halevi Y, Palmor ZJ, Efrati T (1997) Automatic tuning of decentralized PID controllers for MIMO processes. J Process Control 7(2):119–128

He MJ, Cai WJ, Ni W, Xie L-H (2009) RNGA based control system configuration for multivariable processes. J Process Control 19:1036–1042

Ho WK, Lee TH, Gan OP (1997) Tuning of multiloop proportional-integral-derivative controllers based on gain and phase margin specification. Ind Eng Chem Res 36:2231–2238

Hovd M, Skogestad S (1993) Improved independent design of robust decentralized controllers. J Process Control 3:43–51

Huang HP, Ohshima M, Hashimoto L (1994) Dynamic interaction and multiloop control system design. J Process Control 4(1):15–22

Huang HP, Jeng JC, Chiang CH, Pan W (2003) A direct method for multi-loop PI/PID controller design. J Process Control 13(8):769–786

Jensen N, Fisher DG, Shah SL (1986) Interaction analysis in multivariable control systems. AICHE J 32(6):959–970

Jung J, Choi JY, Lee J (1999) One-parameter method for a multiloop control system design. Ind Eng Chem Res 38:1580–1588

Lee J, Cho W, Edgar TF (1998) Multiloop PI controller tuning for interacting multivariable processes. Comput Chem Eng 22(11):1711–1723

Lee J, Edgar TF (2000) Phase conditions for stability of multi-loop control systems. Comput Chem Eng 23:1623–1630

Lee J, Edgar TF (2004) Dynamic interaction measures for decentralized control of multivariable processes. Ind Eng Chem Res 43(2):283–287

Liu T, Zhang W, Gu DY (2005) Analytical multiloop PI/PID controller design for two-by-two processes with time delays. Ind Eng Chem Res 44(6):1832–1841

Loh AP, Hang CC, Quek CK, Vasnani VU (1993) Autotuning of multiloop proportional-integral controllers using relay feedback. Ind Eng Chem Res 32(6):1102–1107

Luyben WL (1986) Simple method for tuning SISO controllers in multivariable systems. Ind Eng Chem Process Des Dev 25:654–660

Luyben WL (1990) Process modeling, simulation, and control for chemical engineers. McGraw Hill, New York

McAvoy TJ (1983) Interaction analysis. ISA Society of America, Research Triangle Park

Morari M, Zafiriou E (1989) Robust process control. Prentice Hall, Englewood Cliff

Ogunnaike BA, Ray WH (1994) Process dynamics, modeling, and control. Oxford University Press, New York

Palmor ZJ, Halevi Y, Krasney N (1995) Automatic tuning of decentralized PID controllers for TITO processes. Automatica 31(7):1001–1010

Salgado ME, Conley A (2004) MIMO interaction measure and controller structure selection. Int J Control 77(4):367–383

Seborg DE, Edgar TF, Mellichamp DA (2004) Process dynamics and control, 2nd edn. Wiley, Hoboken

Shen SH, Yu CC (1994) Use of relay-feedback test for automatic tuning of multivariable systems. AICHE J 40(4):627–646

Shinskey FG (1996) Process control system, 4th edn. McGraw Hill, New York

Skogestad S, Postlethwaite I (2005) Multivariable feedback control: analysis and design, 2nd edn. Wiley, Chichester

Wang QG, Lee TH, Zhang Y (1998) Mutiloop version of the modified Ziegler-Nichols method for two input two output processes. Ind Eng Chem Res 37:4725–4733

Wang QG, Zou B, Lee TH (1997) Auto-tuning of multivariable PID controllers from decentralized relay feedback. Automatica 33(3):319–330

Wood RK, Berry MW (1973) Terminal composition control of binary distillation column. Chem Eng Sci 28(10):1707–1717

Zhang Y, Wang QG, Åström KJ (2002) Dominant pole placement for multi-loop control systems. Automatica 38(7):1213–1220

Zhou KM, Doyle JC, Glover K (1996) Robust and optimal control. Prentice Hall, Englewood Cliff

Chapter 11
Decoupling Control of Multivariable Processes

11.1 Decoupling Control Design for Two-Input-Two-Output (TITO) Processes

For multivariable process control, two-input-two-output (TITO) processes are mostly established for the convenience of control system design and system operation. Many industrial processes with higher dimensions are practically divided as several TITO subsystems for operation (Luyben 1990; Ogunnaike and Ray 1994; Seborg et al. 2004).

As time delay is usually associated with process operation in engineering practice, its presence in an individual loop in a multivariable process can severely prevent a high gain of the closed-loop controller from being used, causing sluggish output performance (Holt and Morari 1985). Moreover, multiloop control performance may be degraded severely for TITO processes without diagonal dominance, causing inadmissible interactions between individual loops, which can be seen from the examples given in Chap. 10. To improve decoupling regulation of the binary system outputs, different control strategies have been developed in the literature. Earlier references (Alevisakis and Seborg 1973; Ogunnaike and Ray 1979; Watanabe et al. 1983; Jerome and Ray 1986, 1992; Desbiens et al. 1996; Wang et al. 2000b) applied the Smith predictor (SP) structure that is specially for time-delay SISO systems to TITO processes with time delays for obtaining a delay-free characteristic equation of such a system transfer matrix, and then extended decoupling control methods well developed for linear multivariable systems free of time delay to enhance decoupling regulation performance. Based on frequency response data estimated from relay feedback identification tests, online sequential tuning methods for decoupling control have been developed by Shiu and Hwang (1998), Toh and Rangaiah (2002), and Gilbert et al. (2003). Using a decoupler in front of the process inputs to obtain diagonal dominance of the augmented process transfer matrix, a number of decoupling control methods have been devoted to tuning the multiloop controllers, such as the references (Åström et al. 2002; Chen and Seborg 2003; Lee et al. 2005) based on a static decoupler, i.e., the inverse of the process static gain

Fig. 11.1 Multivariable internal model control structure

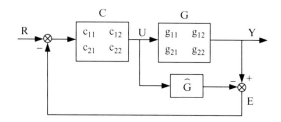

transfer matrix, and other references based on a dynamic decoupler (Perng and Ju 1994; Wang et al. 2000a, b; Pomerleau and Pomerleau 2001; Waller et al. 2003). It should be noted that a static decoupler has little impact on the dynamic system output responses, and in contrast, a dynamic decoupler is difficult to be configured precisely in practice, especially for TITO processes with long time delays, due to the requirements of physical properness and causality for implementation (Waller et al. 2003).

Here, an IMC-based decoupling control method (Liu et al. 2006) is presented for application. A multivariable IMC structure is shown in Fig. 11.1, where $G = [g_{ij}]_{2\times 2}$ denotes the process, $\widehat{G}$ is the process model, and $C = [c_{ij}]_{2\times 2}$ is a controller matrix for decoupling control. R is the set-point vector, U is the process control input vector, and Y is the process output vector.

A TITO process with time delays is usually described as

$$G(s) = \begin{bmatrix} \dfrac{k_{11}e^{-\theta_{11}s}}{\tau_{11}s+1} & \dfrac{k_{12}e^{-\theta_{12}s}}{\tau_{12}s+1} \\ \dfrac{k_{21}e^{-\theta_{21}s}}{\tau_{21}s+1} & \dfrac{k_{22}e^{-\theta_{22}s}}{\tau_{22}s+1} \end{bmatrix} \qquad (11.1)$$

where k_{ij}, τ_{ij}, and θ_{ij} denote the static gain, time constant, and time delay in the transfer functions relating the corresponding input and output variables, respectively.

Before proceeding with decoupling control design, it is preferred to have an analysis of the preconditions for decoupling control of TITO processes as follows.

11.1.1 Decoupling Control Preconditions

From a multivariable IMC structure shown in Fig. 11.1, it can be seen that for the nominal case, i.e., $G = \widehat{G}$, there is an "open-loop" control for the set-point responses, so the system transfer matrix can be simplified as

$$H = GC = \begin{bmatrix} g_{11} & g_{12} \\ g_{21} & g_{22} \end{bmatrix} \begin{bmatrix} c_{11} & c_{12} \\ c_{21} & c_{22} \end{bmatrix} \qquad (11.2)$$

11.1 Decoupling Control Design for Two-Input-Two-Output (TITO) Processes

Obviously, the nominal system can maintain stability if the controller matrix C is designed to be stable. The decoupled output responses correspond to a diagonal system transfer matrix,

$$H = \begin{bmatrix} h_1 & 0 \\ 0 & h_2 \end{bmatrix} \quad (11.3)$$

where the diagonal elements h_1 and h_2 are stable and proper transfer functions. Therefore, two fundamental decoupling control preconditions can be ascertained as below:

(a) Both G and C should be nonsingular at $s = 0$, i.e., $\det[G(0)] \neq 0$ and $\det[C(0)] \neq 0$.
(b) There exists no cross-coupling in tuning the controllers in each column of C.

The first condition (a) is a sufficient and necessary condition for the decoupling control of a TITO process, which can be verified from (11.2) and (11.3). Moreover, it can be easily seen from (11.2) that the first condition implies $k_{11}k_{22} \neq k_{12}k_{21}$. Note that although some industrial processes can be modeled in a stable form, like (11.1), the output responses in a closed-loop control scheme appear unexpectedly to be sensitive to model mismatch or process perturbations. The reason lies in essence with $\det[G(0)] \to 0$. Such process modeling should be avoided for the decoupling control design.

The second condition (b) is an operational requirement to facilitate the decoupling regulation in practice. The reason can be intuitively seen from the postmultiplication relationship between G and C in (11.2) for obtaining a desired diagonal system transfer matrix shown in (11.3).

11.1.2 Desired System Transfer Matrix

It can be seen from (11.2) that if the desired diagonal system transfer matrix in the nominal case can be determined in the first place, the decoupling controller matrix can then be inversely derived as

$$C = G^{-1}H = \frac{adj(G)}{\det(G)} H \quad (11.4)$$

where $adj(G) = \left[G^{ij}\right]^T_{2 \times 2}$ is the adjoint of G and G^{ij} denotes the complement minor corresponding to g_{ij} in G, $i, j = 1, 2$.

Note that the process transfer matrix determinant can be formulated as

$$\det(G) = \begin{cases} G^{11}G^{22}\left(1 - G^{\circ}e^{-\Delta\theta s}\right), & \theta_{11} + \theta_{22} \leq \theta_{12} + \theta_{21} \\ -G^{12}G^{21}\left(1 - \frac{e^{-\Delta\theta s}}{G^{\circ}}\right), & \theta_{11} + \theta_{22} > \theta_{12} + \theta_{21} \end{cases} \quad (11.5)$$

where

$$\Delta\theta = |\theta_{11} + \theta_{22} - \theta_{12} - \theta_{21}|$$

$$G^\circ = \frac{k_{12}k_{21}}{k_{11}k_{22}} \cdot \frac{(\tau_{11}s + 1)(\tau_{22}s + 1)}{(\tau_{12}s + 1)(\tau_{21}s + 1)}$$

Without loss of generality, consider the case where $\theta_{11} + \theta_{22} \leq \theta_{12} + \theta_{21}$ in some TITO processes. It can be seen from (11.2) and (11.3) that each column controllers in C are related to the same diagonal element in H. For instance, the first column controllers are

$$c_{11} = \frac{G^{11}}{\det(G)} h_1 = \frac{1}{G^{22}\left(1 - G^\circ e^{-\Delta\theta s}\right)} h_1 = \frac{(\tau_{11}s + 1)e^{\theta_{11}s}}{k_{11}\left(1 - G^\circ e^{-\Delta\theta s}\right)} h_1 \quad (11.6)$$

$$c_{21} = \frac{G^{12}}{\det(G)} h_1 = \frac{G^{12}}{G^{11}G^{22}\left(1 - G^\circ e^{-\Delta\theta s}\right)} h_1 = -\frac{k_{21}(\tau_{11}s + 1)(\tau_{22}s + 1)e^{(\theta_{11}+\theta_{22}-\theta_{21})s}}{k_{11}k_{22}(\tau_{21}s + 1)\left(1 - G^\circ e^{-\Delta\theta s}\right)} h_1 \quad (11.7)$$

Obviously, it can be seen from (11.6) and (11.7) that if the desired diagonal transfer function h_1 were not to include an equivalent time delay to balance θ_{11}, the controller c_{11} will behave in a predictive manner, and so does c_{21} if $\theta_{11} + \theta_{22} > \theta_{21}$. The predictive manner violates the causal law in nature and therefore cannot be implemented in practice. This can be perceived through the fact that either of the binary process outputs can only begin to track the corresponding set-point after certain time delay of the process. Moreover, if the polynomial $1 - G^\circ e^{-\Delta\theta s}$ in the denominators of (11.6) and (11.7) contains RHP zeros, h_1 is required to include these RHP zeros so that c_{11} and c_{21} will not contain them as unstable poles.

Based on the above analysis and using the robust H_2 optimal performance objective for set-point tracking (Morari and Zafiriou 1989), the desired diagonal transfer function h_1 is proposed as

$$h_1 = \frac{e^{-\theta_1 s}}{\lambda_1 s + 1} \prod_{i=1}^{q} \left(\frac{-s + z_i}{s + z_i^*}\right) \quad (11.8)$$

where λ_1 is an adjustable parameter used to tune the first process output y_1, $\theta_1 = \max\{\theta_{11}, \theta_{11} + \theta_{22} - \theta_{21}\}$, $s = z_i$ ($i = 1, 2, \ldots, q$) are RHP zeros in $1 - G^\circ e^{-\Delta\theta s}$, and z_i^* is the complex conjugate of z_i.

Thus, one of the first column controllers in C can be implemented in a proper and rational form, while the other controller in the first column can be practically implemented in tandem with a specified dead-time compensator such that independent regulation of the first process output y_1 can be realized.

11.1 Decoupling Control Design for Two-Input-Two-Output (TITO) Processes

Following a similar analysis, a desired diagonal transfer function h_2 is proposed as

$$h_2 = \frac{e^{-\theta_2 s}}{\lambda_2 s + 1} \prod_{i=1}^{q} \left(\frac{-s + z_i}{s + z_i^*} \right) \tag{11.9}$$

where λ_2 is an adjustable parameter for tuning the second process output y_2 and $\theta_2 = \max\{\theta_{22}, \theta_{11} + \theta_{22} - \theta_{12}\}$.

In the case where $\theta_{11} + \theta_{22} > \theta_{12} + \theta_{21}$ in other TITO processes, the desired diagonal transfer functions h_1 and h_2 can be proposed almost the same as above. The only difference is that $\theta_1 = \max\{\theta_{12}, \theta_{12} + \theta_{21} - \theta_{22}\}$, $\theta_2 = \max\{\theta_{21}, \theta_{12} + \theta_{21} - \theta_{11}\}$, and $s = z_i$ ($i = 1, 2, \ldots, q$) are RHP zeros in $1 - e^{-\Delta\theta s}/G^\circ$.

Note that by proposing a desired diagonal system transfer matrix as above, the time-domain performance specifications of the system output responses can be quantitatively tuned through the adjustable parameters λ_1 and λ_2. For instance, for TITO processes with no RHP zeros in $\det(G)$, i.e., $1 - G^\circ e^{-\Delta\theta s}$ (if $\theta_{11} + \theta_{22} \leq \theta_{12} + \theta_{21}$) or $1 - e^{-\Delta\theta s}/G^\circ$ (if $\theta_{11} + \theta_{22} > \theta_{12} + \theta_{21}$) has no RHP zeros, the desired diagonal transfer functions h_1 and h_2 can be simplified as

$$h_1 = \frac{1}{\lambda_1 s + 1} e^{-\theta_1 s} \tag{11.10}$$

$$h_2 = \frac{1}{\lambda_2 s + 1} e^{-\theta_2 s} \tag{11.11}$$

By performing an inverse Laplace transform with respect to unit step changes in the binary set-points, the time-domain binary output responses can be obtained as

$$y_1(t) = \begin{cases} 0 & t \leq \theta_1 \\ 1 - e^{-(t-\theta_1)/\lambda_1} & t > \theta_1 \end{cases} \tag{11.12}$$

$$y_2(t) = \begin{cases} 0 & t \leq \theta_2 \\ 1 - e^{-(t-\theta_2)/\lambda_2} & t > \theta_2 \end{cases} \tag{11.13}$$

It is therefore demonstrated that there is no overshoot in either of the nominal output responses, and the time-domain response specifications can be quantitatively tuned through the adjustable parameters λ_1 and λ_2, as analyzed in (9.11) in Sect. 9.2.1.

Hence, it is convenient to tune λ_1 and λ_2 to obtain the desirable process output responses, which in fact are executed through each column controllers in C.

Fig. 11.2 Positive feedback control unit

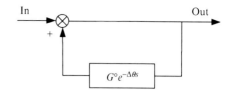

11.1.3 Decoupling Controller Matrix Design

According to the desired diagonal system transfer functions h_1 and h_2 shown in (11.8) and (11.9), the ideally optimal decoupling controller matrix C can be inversely derived from (11.4). However, there exist some constraints for implementing G^{-1} in practice. For instance, when there exist RHP zeros in $\det(G)$, it can be seen from (11.4) together with (11.8) and (11.9) that there will be RHP zero-pole canceling in all controllers in the ideally optimal decoupling controller matrix, causing the controller matrix to be unstable for implementation. To determine a practical form for implementing the above ideal decoupling controller matrix, a design procedure is proposed in terms of two cases: Case 1 is that there exist no RHP zeros in $\det(G)$, and case 2 is the opposite.

In case 1, it can be seen from (11.5) that no RHP zeros in $\det(G)$ indicates that $1 - G°e^{-\Delta\theta s}$ (if $\theta_{11} + \theta_{22} \leq \theta_{12} + \theta_{21}$) or $1 - e^{-\Delta\theta s}/G°$ (if $\theta_{11} + \theta_{22} > \theta_{12} + \theta_{21}$) has no RHP zeros. Correspondingly, $1/(1 - G°e^{-\Delta\theta s})$ or $1/(1 - e^{-\Delta\theta s}/G°)$ is a stable transfer function.

First, consider the case where $\theta_{11} + \theta_{22} \leq \theta_{12} + \theta_{21}$. Substituting (11.8) into (11.6) yields

$$c_{11} = \frac{\tau_{11}s + 1}{k_{11}\left(1 - G°e^{-\Delta\theta s}\right)} \cdot \frac{e^{-(\theta_1 - \theta_{11})s}}{\lambda_1 s + 1} \tag{11.14}$$

which can be rearranged in the form of

$$c_{11} = \frac{(\tau_{11}s + 1)\,e^{-(\theta_1 - \theta_{11})s}}{k_{11}(\lambda_1 s + 1)} \cdot F \tag{11.15}$$

where $F = 1/(1 - G°e^{-\Delta\theta s})$. Obviously, the first part in c_{11} can be practically implemented using a first-order lead-lag controller in tandem with a dead-time compensator, and the second part F can be implemented using a positive feedback unit, as shown in Fig. 11.2. Note that this control unit maintains internal stability because its transfer function has no RHP poles and $G°$ is bi-proper and stable for implementation.

Similarly, the remaining controllers in C can be derived as

$$c_{21} = -\frac{k_{21}}{k_{11}k_{22}} \cdot \frac{(\tau_{11}s + 1)(\tau_{22}s + 1)\,e^{-(\theta_1 + \theta_{21} - \theta_{11} - \theta_{22})s}}{(\tau_{21}s + 1)(\lambda_1 s + 1)} \cdot F \tag{11.16}$$

11.1 Decoupling Control Design for Two-Input-Two-Output (TITO) Processes

$$c_{12} = -\frac{k_{12}}{k_{11}k_{22}} \cdot \frac{(\tau_{11}s + 1)(\tau_{22}s + 1)e^{-(\theta_2+\theta_{12}-\theta_{11}-\theta_{22})s}}{(\tau_{12}s + 1)(\lambda_2 s + 1)} \cdot F \quad (11.17)$$

$$c_{22} = \frac{(\tau_{22}s + 1)e^{-(\theta_2-\theta_{22})s}}{k_{22}(\lambda_2 s + 1)} \cdot F \quad (11.18)$$

Then, in the case where $\theta_{11} + \theta_{22} > \theta_{12} + \theta_{21}$, each column controllers in C can be similarly derived as

$$c_{11} = -\frac{k_{22}}{k_{12}k_{21}} \cdot \frac{(\tau_{12}s + 1)(\tau_{21}s + 1)e^{-(\theta_1+\theta_{22}-\theta_{12}-\theta_{21})s}}{(\tau_{22}s + 1)(\lambda_1 s + 1)} \cdot F \quad (11.19)$$

$$c_{21} = \frac{(\tau_{12}s + 1)e^{-(\theta_1-\theta_{12})s}}{k_{12}(\lambda_1 s + 1)} \cdot F \quad (11.20)$$

$$c_{12} = \frac{(\tau_{21}s + 1)e^{-(\theta_2-\theta_{21})s}}{k_{21}(\lambda_2 s + 1)} \cdot F \quad (11.21)$$

$$c_{22} = -\frac{k_{11}}{k_{12}k_{21}} \cdot \frac{(\tau_{12}s + 1)(\tau_{21}s + 1)e^{-(\theta_2+\theta_{11}-\theta_{12}-\theta_{21})s}}{(\tau_{11}s + 1)(\lambda_2 s + 1)} \cdot F \quad (11.22)$$

where $F = 1/(1 - e^{-\Delta\theta s}/G^\circ)$. Note that G° is bi-proper and stable and so is $1/G^\circ$. Therefore, F can be implemented using a control unit similar to that shown in Fig. 11.2.

It is seen from the above controller formulas that each column controllers in C are exactly tuned in common by a single adjustable parameter (λ_i). Consequently, there is no cross-coupling in tuning each column controllers in C. Moreover, C is maintained nonsingular at $s = 0$.

Combining the desired diagonal transfer functions shown in (11.8) and (11.9), it can be seen that the adjustable parameters λ_1 and λ_2 can be monotonically tuned to obtain desirable output responses. That is, when λ_1 and λ_2 are tuned to smaller values, the output responses y_1 and y_2 will become faster, but the output energy of C and its corresponding actuators will be required to be larger, and vice versa.

Based on a large amount of simulation analysis, it is suggested to first tune λ_1 and λ_2 in the ranges of $(2-10)\theta_1$ and $(2-10)\theta_2$, respectively. If the resulting output responses are not satisfactory, then by monotonically varying λ_1 and λ_2 online, a good trade-off between the output performance and the capacities of C and the corresponding actuators can be conveniently reached.

In case 2, it can be seen from (11.5) that all RHP zeros are located in $1 - G^\circ e^{-\Delta\theta s}$ (if $\theta_{11} + \theta_{22} \leq \theta_{12} + \theta_{21}$) or $1 - e^{-\Delta\theta s}/G^\circ$ (if $\theta_{11} + \theta_{22} > \theta_{12} + \theta_{21}$). The number of these RHP zeros can be ascertained by observing the Nyquist curve of $-G^\circ e^{-\Delta\theta s}$ (or $-e^{-\Delta\theta s}/G^\circ$). Note that the times of encircling the point $(-1, j0)$ in the complex plane by the Nyquist curve is equal to the RHP zero number of det(G) since it has no RHP pole. Alternatively, all RHP zeros in det(G) can be numerically computed by solving $1 - G°e^{-\Delta\theta s} = 0$ (or $1 - e^{-\Delta\theta s}/G° = 0$) from a mathematical software package like the MATLAB toolbox.

When $\theta_{11} + \theta_{22} \leq \theta_{12} + \theta_{21}$, substituting the desired diagonal system transfer functions shown in (11.8) and (11.9) into (11.4) yields

$$c_{11} = \frac{(\tau_{11}s + 1)\, e^{-(\theta_1 - \theta_{11})s}}{k_{11}(\lambda_1 s + 1)\prod_{i=1}^{n}(s + z_i^*)} \cdot D \qquad (11.23)$$

$$c_{21} = -\frac{k_{21}}{k_{11}k_{22}} \cdot \frac{(\tau_{11}s + 1)(\tau_{22}s + 1)\, e^{-(\theta_1 + \theta_{21} - \theta_{11} - \theta_{22})s}}{(\tau_{21}s + 1)(\lambda_1 s + 1)\prod_{i=1}^{n}(s + z_i^*)} \cdot D \qquad (11.24)$$

$$c_{12} = -\frac{k_{12}}{k_{11}k_{22}} \cdot \frac{(\tau_{11}s + 1)(\tau_{22}s + 1)\, e^{-(\theta_2 + \theta_{12} - \theta_{11} - \theta_{22})s}}{(\tau_{12}s + 1)(\lambda_2 s + 1)\prod_{i=1}^{n}(s + z_i^*)} \cdot D \qquad (11.25)$$

$$c_{22} = \frac{(\tau_{22}s + 1)\, e^{-(\theta_2 - \theta_{22})s}}{k_{22}(\lambda_2 s + 1)\prod_{i=1}^{n}(s + z_i^*)} \cdot D \qquad (11.26)$$

where

$$D = \frac{\prod_{i=1}^{n}(-s + z_i)}{1 - G°e^{-\Delta\theta s}} \qquad (11.27)$$

Obviously, the first part in all controllers shown in (11.23)–(11.26) can be implemented using a lead-lag controller, together with a dead-time compensator or not, but the second part, D, cannot be directly implemented due to the occurrence of RHP zero-pole canceling. A rational approximation is therefore needed for practical implementation.

Following the analytical approximation method introduced in Sect. 8.4.1, the mathematical Padé expansion is used here to construct a linear fractional approximation formula,

$$D_{U/V} = \frac{\sum_{i=0}^{U} a_i s^i}{\sum_{j=0}^{V} b_j s^j} \qquad (11.28)$$

11.1 Decoupling Control Design for Two-Input-Two-Output (TITO) Processes

where U and V are user-specified orders for approximation and the constant coefficients a_i ($i = 1, 2, \ldots, U$) and b_j ($j = 1, 2, \ldots, V$) are determined by the following two matrix equations:

$$\begin{bmatrix} a_0 \\ a_1 \\ \vdots \\ a_U \end{bmatrix} = \begin{bmatrix} d_0 & 0 & 0 & \cdots & 0 \\ d_1 & d_0 & 0 & \cdots & 0 \\ \vdots & \vdots & \ddots & \cdots & \vdots \\ d_U & d_{U-1} & d_{U-2} & \cdots & d_{U-V} \end{bmatrix} \begin{bmatrix} b_0 \\ b_1 \\ \vdots \\ b_V \end{bmatrix} \quad (11.29)$$

$$\begin{bmatrix} d_U & d_{U-1} & \cdots & d_{U-V+1} \\ d_{U+1} & d_U & \cdots & d_{U-V+2} \\ \vdots & \vdots & \ddots & \vdots \\ d_{U+V-1} & d_{U+V-2} & \cdots & d_U \end{bmatrix} \begin{bmatrix} b_1 \\ b_2 \\ \vdots \\ b_V \end{bmatrix} = - \begin{bmatrix} d_{U+1} \\ d_{U+2} \\ \vdots \\ d_{U+V} \end{bmatrix} \quad (11.30)$$

where d_k ($k = 0, 1, \ldots, U + V$) are constant coefficients in the Maclaurin series expansion of D shown in (11.27),

$$d_k = \frac{1}{k!} \lim_{s \to 0} \frac{d^k D}{d s^k}, \quad k = 0, 1, \ldots, U + V. \quad (11.31)$$

$$b_0 = \begin{cases} 1, & b_j \geq 0; \\ -1, & b_j < 0. \end{cases} \quad (11.32)$$

Note that (11.29) and (11.30) can be transparently derived by substituting (11.28) into the Maclaurin series expansion of D and then comparing the constant coefficients of each term with the same index exponent of the complex variable (s) at both sides.

For instance, letting $U = V = 1$ gives a first-order approximation formula,

$$D_{1/1} = \frac{a_1 s + a_0}{b_1 s + b_0} \quad (11.33)$$

where

$$b_1 = -\frac{d_2}{d_1}, \quad a_1 = d_1 b_0 + d_0 b_1, \quad a_0 = d_0 b_0.$$

It can be seen from (11.27) that $G°e^{-\Delta \theta s}$ goes to the origin much faster when $s \to \infty$, compared to the rational numerator polynomial. Hence, it is definite that a rational linear fractional approximation for D can give good accuracy. For the convenience of implementation, a simple approximation can be obtained as

$$D = \frac{\prod\limits_{i=1}^{n}(-s + z_i)}{1 - G°(0)} = \frac{\prod\limits_{i=1}^{n}(-s + z_i)}{1 - \frac{k_{12}k_{21}}{k_{11}k_{22}}} \quad (11.34)$$

Note that whether det(G) has infinite many RHP zeros or not can be verified by checking if $[k_{12}k_{21}\tau_{11}\tau_{22}/(k_{11}k_{22}\tau_{12}\tau_{21})] > 1$ (when $\theta_{11} + \theta_{22} \leq \theta_{12} + \theta_{21}$) or $[k_{11}k_{22}\tau_{12}\tau_{21}/(k_{12}k_{21}\tau_{11}\tau_{22})] > 1$ (when $\theta_{11} + \theta_{22} > \theta_{12} + \theta_{21}$). The reason is that the Nyquist curve of $-G^\circ e^{-\Delta\theta s}$ (or $-e^{-\Delta\theta s}/G^\circ$) will encircle the origin for infinite many times with a radius of $k_{12}k_{21}\tau_{11}\tau_{22}/(k_{11}k_{22}\tau_{12}\tau_{21})$ (or $k_{11}k_{22}\tau_{12}\tau_{21}/(k_{12}k_{21}\tau_{11}\tau_{22})$) when $\theta_{11} + \theta_{22} > \theta_{12} + \theta_{21}$) for $\omega \in [0, +\infty)$, if there exist infinite many RHP zeros in det(G). Since off-dominant RHP zeros in a system transfer function have little impact on the achievable system performance (Åström and Hägglund 1995; Skogestad and Postlethwaite 2005), it is suggested to use only dominant RHP zeros in det(G) to formulate the desired diagonal transfer functions, as shown in (11.8) and (11.9), such that the decoupling controller matrix can be analytically derived in a simple way except for a slight degradation in the output performance.

It should be noted that the choice of b_0 shown in (11.32) is to keep all b_j ($j = 1, 2, \ldots, V$) the same sign in order to exclude any RHP zeros from being enclosed in the denominator of (11.28). Note that such a high-order approximation (i.e., $V \geq 3$) may be still involved with RHP poles, which can be further verified by the Routh–Hurwitz stability criterion. Therefore, it is suggested to use the Routh–Hurwitz stability criterion to verify the stability of such a high-order approximation before using it in practice to obtain further enhanced output performance. Nevertheless, such a linear fractional approximation in terms of $V \leq 2$ can be reliably implemented without such verification, and thus is preferred for implementation simplicity.

When $\theta_{11} + \theta_{22} > \theta_{12} + \theta_{21}$, following a similar design procedure as above, each column controllers in C can be derived as

$$c_{11} = -\frac{k_{22}}{k_{12}k_{21}} \cdot \frac{(\tau_{12}s + 1)(\tau_{21}s + 1) e^{-(\theta_1 + \theta_{22} - \theta_{12} - \theta_{21})s}}{(\tau_{22}s + 1)(\lambda_1 s + 1) \prod_{i=1}^{n}(s + z_i^*)} \cdot D \quad (11.35)$$

$$c_{21} = \frac{(\tau_{12}s + 1) e^{-(\theta_1 - \theta_{12})s}}{k_{12}(\lambda_1 s + 1) \prod_{i=1}^{n}(s + z_i^*)} \cdot D \quad (11.36)$$

$$c_{12} = \frac{(\tau_{21}s + 1) e^{-(\theta_2 - \theta_{21})s}}{k_{21}(\lambda_2 s + 1) \prod_{i=1}^{n}(s + z_i^*)} \cdot D \quad (11.37)$$

$$c_{22} = -\frac{k_{11}}{k_{12}k_{21}} \cdot \frac{(\tau_{12}s + 1)(\tau_{21}s + 1) e^{-(\theta_2 + \theta_{11} - \theta_{12} - \theta_{21})s}}{(\tau_{11}s + 1)(\lambda_2 s + 1) \prod_{i=1}^{n}(s + z_i^*)} \cdot D \quad (11.38)$$

11.1 Decoupling Control Design for Two-Input-Two-Output (TITO) Processes

where

$$D = \frac{\prod_{i=1}^{n}(-s + z_i)}{1 - \frac{e^{-\Delta\theta s}}{G^\circ}} \qquad (11.39)$$

Note that D can be implemented as well using the analytical approximation formula shown in (11.28).

11.1.4 Robust Stability Analysis

Given a stable decoupling controller matrix as designed in the above section, a multivariable IMC structure shown in Fig. 11.1 obviously maintains stability for the nominal system response of a stable TITO process. When there exist process uncertainties, the system transfer matrix becomes

$$H = GC\left[I + \left(G - \widehat{G}\right)C\right]^{-1} \qquad (11.40)$$

which may be very complex in the presence of various process uncertainties and tends to lose stability in an intangible manner. The controller matrix C designed above in terms of the nominal process model may not guarantee the control system robust stability anymore. An evaluation of the control system robust stability is therefore necessary for practical application.

There are three types of process uncertainty commonly encountered in practice: additive, multiplicative input and output uncertainties. Note that many other types of unstructured or structured process uncertainties can be treated by being lumped into these uncertainties (Skogestad and Postlethwaite 2005).

First, consider the process additive uncertainty shown in Fig. 11.3, which describes the process family, $\Pi_A = \{\widehat{G}_A(s) : \widehat{G}_A(s) = G(s) + \Delta_A\}$, where Δ_A is assumed to be stable. According to the standard $M - \Delta$ structure for robust stability analysis, it can be derived from Fig. 11.3 that

$$U = CE \qquad (11.41)$$

$$E = R - (Y - \widehat{G}U) \qquad (11.42)$$

$$Y = GU + V \qquad (11.43)$$

Solving (11.41)–(11.43) yields

$$U = C\left[I + \left(G - \widehat{G}\right)C\right]^{-1} R - C\left[I + \left(G - \widehat{G}\right)C\right]^{-1} V \qquad (11.44)$$

Fig. 11.3 Perturbed control system with additive uncertainty

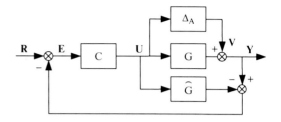

Thus, the transfer matrix from V to U is obtained as

$$T_A = -C\left[I + \left(G - \hat{G}\right)C\right]^{-1} \quad (11.45)$$

Given $G = \hat{G}$ for the nominal system, one can further simplify (11.45) as

$$T_A = -C \quad (11.46)$$

Owing to that C has been designed stable for the nominal system, T_A is surely maintained stable.

According to the small gain theorem, the perturbed system with additive uncertainty maintains robust stability if and only if

$$\|C\|_\infty < \frac{1}{\|\Delta_A\|_\infty} \quad (11.47)$$

In view of the equivalent relationship between the small gain theorem and the spectral radius stability criterion (Skogestad and Postlethwaite 2005), the above robust stability constraint can be equivalently transformed as

$$\rho(C\Delta_A) < 1, \forall \omega \quad (11.48)$$

Hence, it is practical to evaluate robust stability of the multivariable IMC structure by checking if the magnitude plot of the spectral radius in (11.48) falls below unity for $\omega \in [0, +\infty)$.

Then, consider the process multiplicative input and output uncertainties that describe the process families, $\Pi_I = \{\hat{G}_I(s) : \hat{G}_I(s) = G(s)(I + \Delta_I)\}$ and $\Pi_O = \{\hat{G}_O(s) : \hat{G}_O(s) = (I + \Delta_O)G(s)\}$, respectively, where Δ_I and Δ_O are assumed to be stable; the transfer matrices relating the corresponding inputs and outputs can be derived, respectively, as

$$T_I = -C\left[I + (G - \hat{G})C\right]^{-1}G \quad (11.49)$$

$$T_O = -GC\left[I + (G - \hat{G})C\right]^{-1} \quad (11.50)$$

11.1 Decoupling Control Design for Two-Input-Two-Output (TITO) Processes

Similarly, the equivalent spectral radius robust stability constraints can be obtained as

$$\rho(CG\Delta_I) < 1, \forall \omega \tag{11.51}$$

$$\rho(GC\Delta_O) < 1, \forall \omega \tag{11.52}$$

Hence, given a specified bound of Δ_I or Δ_O in practice, the control system robust stability can be intuitively evaluated by checking if the magnitude plot of the spectral radius in (11.51) (or (11.52)) falls below unity for $\omega \in [0, +\infty)$. In this way, admissible tuning ranges of the adjustable parameters λ_1 and λ_2 in the decoupling controller matrix can be numerically ascertained. To accommodate unknown process uncertainties in engineering practice, it is suggested to monotonically increase the adjustable parameters λ_1 and λ_2 in the decoupling controller matrix online so that the nominal system response will be slowed down in exchange for further enhanced robust stability.

11.1.5 Illustrative Examples

Two examples from the existing literature are used to demonstrate the effectiveness and merits of the presented TITO decoupling control method, one with no RHP zero in $\det(G)$ and the other with infinite many RHP zeros in $\det(G)$.

Example 11.1. Consider the widely studied Wood–Berry distillation column process (Wood and Berry 1973),

$$G = \begin{bmatrix} \dfrac{12.8e^{-s}}{16.7s+1} & \dfrac{-18.9e^{-3s}}{21s+1} \\ \dfrac{6.6e^{-7s}}{10.9s+1} & \dfrac{-19.4e^{-3s}}{14.4s+1} \end{bmatrix}$$

According to the classification of a TITO process transfer matrix determinant in (11.5), there is $\theta_{11} + \theta_{22} = 4 < \theta_{12} + \theta_{21} = 10$. It can be easily verified using the Nyquist curve criterion that there is no RHP zero in $\det(G)$. Hence, using the analytical design formulas given in (11.15)–(11.18) obtains the decoupling controller matrix,

$$C = F \cdot \begin{bmatrix} \dfrac{16.7s+1}{12.8(\lambda_1 s+1)} & \dfrac{-0.0761(16.7s+1)(14.4s+1)e^{-2s}}{(21s+1)(\lambda_2 s+1)} \\ \dfrac{0.0266(16.7s+1)(14.4s+1)e^{-4s}}{(10.9s+1)(\lambda_1 s+1)} & \dfrac{-(14.4s+1)}{19.4(\lambda_2 s+1)} \end{bmatrix}$$

where

$$F = \cfrac{1}{1 - \cfrac{0.5023(16.7s+1)(14.4s+1)}{(21s+1)(10.9s+1)}e^{-6s}}$$

which can be implemented using the control unit shown in Fig. 11.2.

For illustration, two groups of simulation test are performed by taking $\lambda_1 = 2$ and $\lambda_2 = 4$, and $\lambda_1 = 4$ and $\lambda_2 = 6$, respectively. By adding a unit step change to the binary set-point inputs at $t = 0$ (min) and $t = 100$ (min), respectively, the output responses are shown in Fig. 11.4. It is seen that the binary output responses have been absolutely decoupled from each other, and there is no overshoot in the set-point responses. Moreover, according to the time-domain system response analysis given in (11.12) and (11.13), the rise time of set-point response can be conveniently obtained as $t_{r1} = 2.3026\lambda_1 + 1$ for the first output y_1, and $t_{r2} = 2.3026\lambda_2 + 3$ for the second output y_2. Hence, it is convenient to tune the adjustable parameters λ_1 and λ_2 for obtaining desirable binary output responses, respectively. Note that tuning λ_1 and λ_2 aims at a good trade-off between the output response performance and the output capacities of the decoupling controller matrix and the corresponding actuators. Figure 11.4 has also illustrated the compromise effect, i.e., the binary control outputs in terms of $\lambda_1 = 2$ and $\lambda_2 = 4$ are required to be more aggressive than those of $\lambda_1 = 4$ and $\lambda_2 = 6$.

Besides, it can be seen from Fig. 11.4 that the binary control outputs u_1 and u_2 are somewhat oscillatory in the transient response, which may wear out the corresponding actuators in a quick fashion. The reason lies with the implementation of F in the decoupling controller matrix C, of which the denominator is involved with a time-delay factor unfavorably entailing the decoupling controller matrix to create oscillatory output signals. This phenomenon is not allowed for operating many industrial and chemical processes in practice. It is therefore suggested to use the analytical approximation formulas given in (11.28)–(11.32) to approximate F for implementation. For instance, a second-order approximation formula can be correspondingly obtained as

$$F_{2/2} = \cfrac{73.648s^2 + 51.077s + 2.01}{150.662s^2 + 32.283s + 1}$$

Substituting it into the above decoupling controller matrix C and also taking $\lambda_1 = 4$ and $\lambda_2 = 6$ for comparison, the control results are shown in Fig. 11.5. It is seen that the binary process output responses are still completely decoupled from each other, while the binary control outputs u_1 and u_2 become evidently smoothed. In contrast, the degradation in the output performance appears to be negligible from a practical view.

To demonstrate the control system robust stability, assume that the static gains of the first column elements in the process transfer matrix are actually 20% larger and those of the second column elements are 30% larger, while all time delays and

11.1 Decoupling Control Design for Two-Input-Two-Output (TITO) Processes

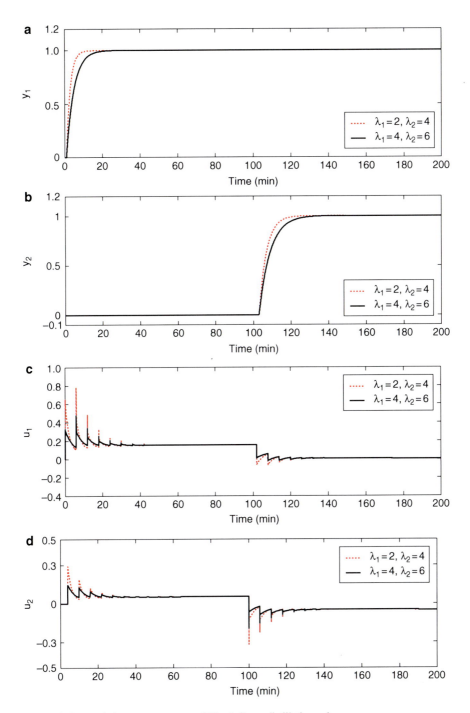

Fig. 11.4 Decoupled output responses of Wood–Berry distillation column

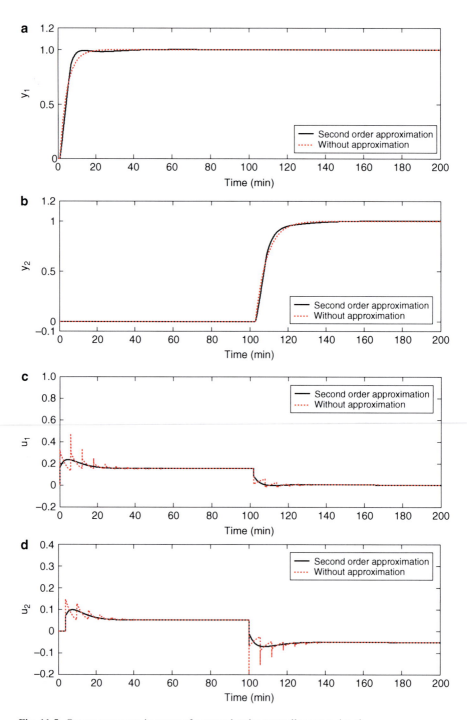

Fig. 11.5 Output responses in terms of a second-order controller approximation

11.1 Decoupling Control Design for Two-Input-Two-Output (TITO) Processes

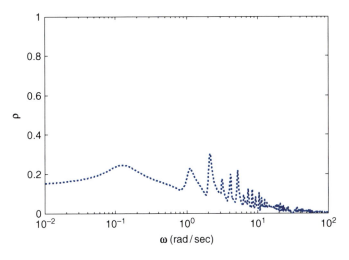

Fig. 11.6 The magnitude plot of spectral radius for the perturbed system

time constants in the process transfer matrix are actually 20% larger. According to the robust stability analysis in Sect. 11.1.4, the magnitude plot of the spectral radius in (11.48) for verifying the control system robust stability is shown in Fig. 11.6. It is seen that the peak value is far smaller than unity, indicating that the proposed control system maintains good robust stability. The corresponding output responses are similar to those shown in Fig. 11.5 and thus are omitted.

Example 11.2. Consider a TITO process with time delays studied by Wang et al. (2000a),

$$G = \begin{bmatrix} \dfrac{-0.51e^{-7.5s}}{(32s+1)^2(2s+1)} & \dfrac{1.68e^{-2s}}{(28s+1)^2(2s+1)} \\ \dfrac{-1.25e^{-2.8s}}{(43.6s+1)(9s+1)} & \dfrac{4.78e^{-1.15s}}{(48s+1)(5s+1)} \end{bmatrix}$$

For decoupling control design, a first-order transfer matrix model was identified by Wang et al. (2000a) as

$$G_m = \begin{bmatrix} \dfrac{-0.5332e^{-19.5838s}}{67.7099s+1} & \dfrac{1.7171e^{-14.8791s}}{48.3651s+1} \\ \dfrac{-1.2585e^{-8.4505s}}{48.7805s+1} & \dfrac{4.7861e^{-4.9768s}}{49.7512s+1} \end{bmatrix}$$

Here, it is also used to derive the decoupling controller matrix in the presented IMC-based decoupling control method. The Nyquist curve of $\det(G_m)$ is plotted in Fig. 11.7, which indicates that there are infinite many RHP zeros in $\det(G_m)$. Using

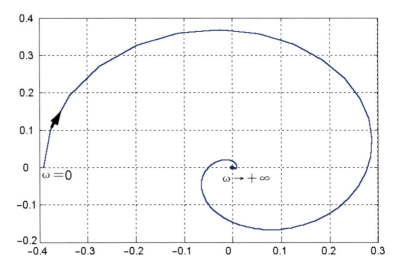

Fig. 11.7 The Nyquist curve of the transfer matrix determinant of the TITO process model

the MATLAB toolbox to numerically solve $\det(G_m)$, it can be verified that there is only one dominant RHP zero at $s = z_1 = 0.0129$, which is therefore used to determine the desired system transfer matrix. Note that $\theta_{11} + \theta_{22} = 24.5606 > \theta_{12} + \theta_{21} = 23.3296$. Using the corresponding design formulas in (11.35)–(11.39) obtains the decoupling controller matrix,

$$C = D \begin{bmatrix} \dfrac{2.2148\,(48.3651s+1)\,(48.7805s+1)}{(49.7512s+1)\,(77.5194s+1)\,(\lambda_1 s+1)} & \dfrac{-(48.7805s+1)}{1.2585\,(77.5194s+1)\,(\lambda_2 s+1)} \\[2mm] \dfrac{(48.3651s+1)\,e^{-3.4737s}}{1.7171\,(77.5194s+1)\,(\lambda_1 s+1)} & \dfrac{-0.2467\,(48.3651s+1)\,(48.7805s+1)\,e^{-4.7047s}}{(67.7099s+1)\,(77.5194s+1)\,(\lambda_2 s+1)} \end{bmatrix}$$

where

$$D = \dfrac{1 - 77.5194s}{1 - \dfrac{1.1809\,(48.3651s+1)\,(48.7805s+1)}{(67.7099s+1)\,(49.7512s+1)} e^{-1.231s}}$$

For simplicity, the first-order approximation formula shown in (11.33) is adopted to approximate D which is involved with a RHP zero-pole canceling, obtaining $D_{1/1} = -(373.2751s + 5.5271) / (4.4191s + 1)$ for practical implementation.

For comparison, take the adjustable parameters, $\lambda_1 = 40$ and $\lambda_2 = 100$, in order to obtain similar rising speeds of set-point tracking with those of Wang et al. (2000a). By adding a unit step change to the binary set-point inputs at $t = 0$ (s) and $t = 1,500$ (s), respectively, and then adding an inverse step change of load disturbance with a magnitude of 0.1 to both the process inputs at $t = 3,000$ (s), the results are shown in Fig. 11.8. It is seen that there is no overshoot in the binary

11.1 Decoupling Control Design for Two-Input-Two-Output (TITO) Processes

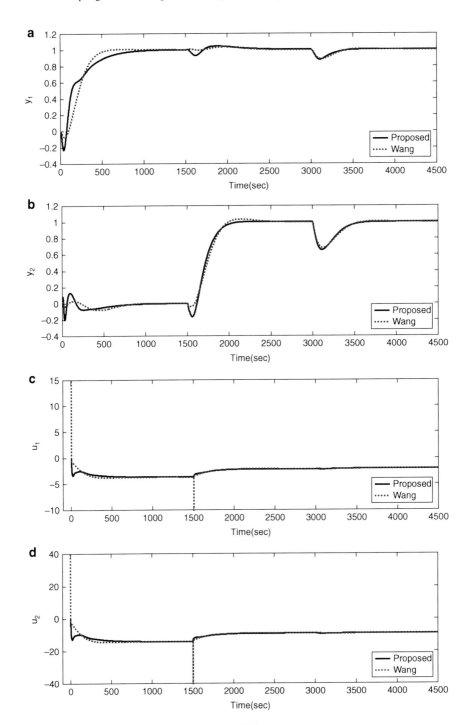

Fig. 11.8 Nominal output responses of Example 11.2

output responses using the presented IMC-based decoupling control method, while both the output responses are almost decoupled from each other. It should be noted that further enhanced output performance can be obtained in the proposed method by using a higher-order approximation formula for D or simply decreasing the adjustable parameters.

To compare the control system robust stability, assume that the static gains of the first column elements in the process transfer matrix are actually 10% larger and those of the second column elements are 20% larger, while all time delays are assumed to be 10% larger. The perturbed system responses are shown in Fig. 11.9, which demonstrates that the presented decoupling control method maintains good robust stability.

11.2 Decoupling Control Design for Multiple-Input-Multiple-Output (MIMO) Processes

For a multivariable process with higher input-output dimensions ($n > 2$), the inverse transfer matrix of the process (G^{-1}) is much more difficult to be configured for practical implementation compared to a TITO process, especially in the presence of multiple time delays in individual channels. Based on classifying different cases of the RHP zero distribution in det(G), a decoupling control design is presented here in terms of the classical unity feedback control structure that is widely used in engineering practice. The control diagram is shown in Fig. 11.10, where G denotes the multivariable process required for decoupling regulation, C is the decoupling controller matrix, R denotes the set-point vector, Y is the output vector, U is the controller output vector, D_I and D_O indicate load disturbances entering into the process from the input and output sides, respectively, and N_O denotes output measurement noise vector.

Before proceeding with decoupling control design, it is preferred to have an analysis on the preconditions for decoupling control of MIMO processes as follows.

11.2.1 Decoupling Control Preconditions

Consider a general transfer matrix for a MIMO process with time delays,

$$G = \begin{bmatrix} g_{11} & \cdots & g_{1m} \\ \vdots & \vdots & \vdots \\ g_{m1} & \cdots & g_{mm} \end{bmatrix} \quad (11.53)$$

where $g_{ij} = g_{0,ij} e^{-\theta_{ij} s}$, $i, j = 1, 2, \ldots, m$, and $g_{0,ij}$ is a rational and stable transfer function.

11.2 Decoupling Control Design for Multiple-Input-Multiple-Output (MIMO) Processes

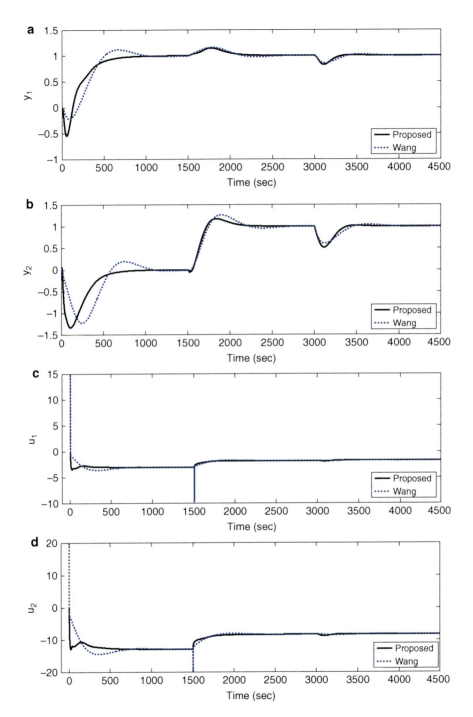

Fig. 11.9 Perturbed output responses of Example 11.2

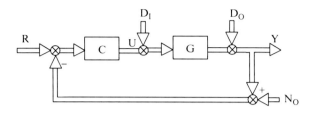

Fig. 11.10 Multivariable unity feedback control structure

From Fig. 11.10, it can be seen that the closed-loop system transfer matrix is

$$H = GC(I + GC)^{-1} \qquad (11.54)$$

Ideally, the decoupled system response transfer matrix should be in the form of

$$H = \begin{bmatrix} h_{11} & 0 & \cdots & 0 \\ 0 & h_{22} & 0 \cdots & 0 \\ 0 & \ddots & \ddots & 0 \\ 0 & \cdots & 0 & h_{mm} \end{bmatrix} \qquad (11.55)$$

where h_{ii} is a proper and stable transfer function, and $h_{ij} = 0$ for $i \neq j$, $i, j = 1, 2, \ldots, m$. That is to say, H should be a nonsingular diagonal transfer matrix, i.e., $H = \text{diag}[h_{ii}]_{m \times m}$ and $\det(H) \neq 0$.

Combining (11.54) and (11.55), a fundamental decoupling precondition can be ascertained as $\det[G(0)] \neq 0$, which means that the multivariable process required for decoupling regulation must be nonsingular in essence, or not ill-conditioned.

Moreover, it can be seen from (11.54) that the controller matrix C should be nonsingular and take the responsibility for maintaining $(I + GC)^{-1}$ stable. In addition, to facilitate system operation, there should be no cross interaction between tuning each column controllers in C because each column controllers have the same input signal and there is a postmultiplication relationship between G and C. Correspondingly, decoupling regulation for individual output variables can be implemented in a transparent manner.

11.2.2 Desired Closed-Loop Transfer Matrix

For a multivariable process, the individual output performance is constrained by the multiple time delays involved and RHP zeros in the process transfer matrix determinant (Morari and Zafiriou 1989; Skogestad and Postlethwaite 2005). It is therefore motivated to determine a desired closed-loop response transfer matrix corresponding to the achievable optimal output response performance. Consequently, the desired decoupling controller matrix can be inversely derived from the unity feedback control structure, as shown in Fig. 11.10.

11.2 Decoupling Control Design for Multiple-Input-Multiple-Output (MIMO) Processes

Note that the inverse of H in (11.55) is also a diagonal transfer matrix. Substituting (11.55) into (11.54), the controller matrix can be derived as

$$C = G^{-1}(H^{-1} - I)^{-1} = \frac{\text{adj}(G)}{\det(G)} \text{diag}\left\{\frac{h_{ii}}{1 - h_{ii}}\right\}_{m \times m} \tag{11.56}$$

where $\text{adj}(G) = [G^{ij}]_{m \times m}^T$ is the adjoint of G, G^{ij} denotes the complement minor corresponding to g_{ij} of G. Denote by $C = [c_{ij}]_{m \times m}$ the controller matrix. According to the postmultiplication relationship between a square matrix and a diagonal matrix, each column controllers in C (with the same subscript of i) can be obtained as

$$c_{ji} = \frac{G^{ij}}{\det(G)} \cdot \frac{h_{ii}}{1 - h_{ii}}, \quad i, j = 1, 2, \ldots, m. \tag{11.57}$$

Let

$$p_{ij} = \frac{G^{ij}}{\det(G)} = p_{0, ij} e^{L_{ij} s}, \quad i, j = 1, 2, \ldots, m. \tag{11.58}$$

where $p_{0,ij}$ denotes the "delay-free" part of p_{ij}, i.e., at least one term in either of the nominator and denominator polynomials of $p_{0,ij}$ does not include any time delay and thus is rational. It can be seen from (11.56) and (11.58) that $G^{-1} = [p_{ij}]_{m \times m}^T$.

Define the "inverse relative degree" of $p_{0,ij}$ as n_{ij} $(i, j = 1, 2, \ldots, m)$, which is the largest integer that satisfies

$$\lim_{s \to \infty} \frac{s^{n_{ij}-1}}{p_{0,ij}} = 0 \tag{11.59}$$

and let

$$N_i = \max\{n_{ij}; j = 1, 2, \ldots, m\}, \quad i = 1, 2, \ldots, m \tag{11.60}$$

$$\theta_i = \max\{L_{ij}; j = 1, 2, \ldots, m\}, \quad i = 1, 2, \ldots, m \tag{11.61}$$

It can be seen from (11.56) that each column controllers in C are related to the same diagonal element in H, which means that all c_{ji} $(j = 1, 2, \ldots, m)$ correspond to the same diagonal transfer function h_{ii} $(i = 1, 2, \ldots, m)$. Note that θ_i in (11.61) is positive, as can be verified from (11.58). If the desired diagonal transfer function, h_{ii}, for the ith output response does not include an equivalent time delay to balance θ_i, some or even all the ith column controllers c_{ji} $(j = 1, 2, \ldots, m)$ will not be physically realizable. Also, it can be seen from (11.57) that if the relative degree of the delay-free part in h_{ii} is lower than N_i, some or even all c_{ji} $(j = 1, 2, \ldots, m)$ will not be proper and thus not realizable. In addition, $\det(G)$ may contain RHP zeros and, therefore, if h_{ii} does not include these RHP zeros, all $c_{ji}(j=1,2,\ldots,m)$ would be bundled with them as unstable poles, which is definitely not allowed in practice.

Combining the H$_2$ optimal performance objective of the IMC theory (Morari and Zafiriou 1989) with the above implemental constraints, practically desired diagonal elements of the closed-loop system transfer matrix are proposed as

$$h_{ii} = \frac{e^{-\theta_i s}}{(\lambda_i s + 1)^{N_i}} \prod_{k=1}^{q_i} \frac{-s + z_k}{s + z_k^*}, \quad i = 1, 2, \ldots, m \quad (11.62)$$

where λ_i is an adjustable parameter for obtaining a desirable performance of the ith output variable and z_k ($k = 1, 2, \ldots, q_i$) are RHP zeros in det(G) excluding those canceled by the common RHP zeros of G^{ij} ($j=1,2,\ldots,m$), q_i is the number, and z_k^* is the complex conjugate of z_k.

With the above desired diagonal elements shown in (11.62), it can be ascertained from (11.57)–(11.61) that at least one of each column controllers in C can be practically implemented in a proper and rational form, while the other controllers of the corresponding column can be implemented using prescribed dead-time compensators. Thus, the desired diagonal closed-loop transfer matrix shown in (11.55) can be realized, leading to decoupling regulation of all output variables.

The number of RHP zeros in det(G) can be ascertained from the Nyquist curve of det(G) in the complex plane. For instance, if det(G) has no RHP pole, the number of times its Nyquist curve encircling the origin is equal to the number of RHP zeros in det(G), according to the Nyquist stability criterion. Alternatively, the RHP zeros in det(G) can be computed numerically using a mathematical software package, like the MATLAB toolbox.

For MIMO processes with multiple time delays, det(G) may have infinite many RHP zeros due to multiple time-delay terms involved. In the case where det(G) has infinite many RHP zeros but finite left-half-plane (LHP) zeros, the desired closed-loop system transfer matrix can be determined with the following diagonal elements:

$$h_{ii} = \frac{e^{-\theta_i s}}{(\lambda_i s + 1)^{N_i}} \cdot \frac{\phi(s) e^{(\theta_{\max} - \theta_{\min})s}}{\phi(-s)} \prod_{k=1}^{q_i} \frac{-s - z_k}{s - z_k^*}, \quad i = 1, 2, \ldots, m \quad (11.63)$$

where z_k ($k=1,2,\ldots,q_i$) denote the LHP zeros in det(G), excluding those equivalent to the complex conjugates of the common RHP zeros in G^{ij} ($j=1,2,\ldots,m$) and $\theta_{\min}$ is the minimum of all time-delay factors involved in det(G) and $\theta_{\max}$ is the corresponding maximum. The term, $\phi(s)$, is defined from the following formulation:

$$\det(G) = \frac{\phi(s) e^{-\theta_{\min} s}}{\psi(s)} \quad (11.64)$$

11.2 Decoupling Control Design for Multiple-Input-Multiple-Output (MIMO) Processes

where $\psi(s)$ is the least common denominator of all terms in $\det(G)$ and $\phi(s)$ is the corresponding numerator polynomial in which there is at least one term that does not contain any time delay and thus is rational. Apparently, $\det(G)$ has the same zeros with $\phi(s)$.

Note that $\phi(-s)$ in (11.63) is the complex conjugate of $\phi(s)$ and, correspondingly, all zeros in $\phi(-s)$ are located at the mirror zeros of $\phi(s)$ across the imaginary axis in the complex plane. Moreover, it can be seen that $\phi(-s)$ may include time prediction factors that are not allowed in a physical all-pass filter, of which $\theta_{max} - \theta_{min}$ is the maximal time prediction length. Hence, the second part of h_{ii} shown in (11.63)

$$\frac{\phi(s)e^{(\theta_{max}-\theta_{min})s}}{\phi(-s)} \prod_{k=1}^{q_i} \frac{-s-z_k}{s-z_k^*}$$

can be viewed as an all-pass filter, facilitating the realization of the H_2 optimal performance objective for individual output responses. Note that there inevitably exists RHP zero-pole cancelation in this filter, which, however, cannot be removed directly for implementation. A rational approximation, as introduced in (11.28), can be used for practical implementation.

It should be noted that although $\det[G(s)] / \det[G(-s)]$ can be directly used to configure the all-pass part of diagonal elements in the closed-loop system transfer matrix, an additional all-pass filter, $\psi(-s)/\psi(s)$, will be introduced unfavorably, which may degrade system performance and thus is not recommended.

In the case where $\det(G)$ has infinite many RHP and LHP zeros, it is suggested to use only dominant RHP zeros in $\det(G)$ to determine the desired closed-loop system transfer matrix. This will facilitate analytical design of the decoupling controller matrix in a transparent manner, though at the loss of certain output performance. Note that it has been well recognized in frequency-domain control theory (Morari and Zafiriou 1989; Ogunnaike and Ray 1994; Åström and Hägglund 1995; Shinskey 1996; Skogestad and Postlethwaite 2005) that off-dominant zeros in a control system characteristic equation have little impact on the achievable system performance in practice.

To sum up, by classifying four possible cases of the RHP zero distribution in $\det(G)$, desired diagonal system response transfer matrices are correspondingly listed in Table 11.1.

11.2.3 Decoupling Controller Matrix Design

According to the desired diagonal closed-loop system transfer matrix listed in Table 11.1, the ideally desired decoupling controller matrix C can be inversely derived from (11.56). For instance, in case 2 in Table 11.1 that $\det(G)$ has finite

Table 11.1 A summary of desired closed-loop transfer matrices and decoupling controller matrices

Case	$\det(G)$	$h_{ii}\,(i=1,2,\ldots,m)$	$c_{ji}\,(i,j=1,2,\ldots,m)$
1	No RHP zero	$\dfrac{e^{-\theta_i s}}{(\lambda_i s+1)^{N_i}}$	$\dfrac{D_{ij}e^{-(\theta_i-L_{ij})s}}{(\lambda_i s+1)^{N_i}}\cdot\dfrac{1}{1-\dfrac{e^{-\theta_i s}}{(\lambda_i s+1)^{N_i}}},\quad D_{ij}=p_{0,ij}$
2	Finite RHP zeros $[z_k\,(k=1,2,\ldots,q_i)=$ RHP zeros excluding those canceled by the common RHP zeros of $G^{ij}\,(j=1,2,\ldots,m)]$	$\dfrac{e^{-\theta_i s}}{(\lambda_i s+1)^{N_i}}\prod_{k=1}^{q_i}\dfrac{-s+z_k}{s+z_k^*}$	$\dfrac{D_{ij}e^{-(\theta_i-L_{ij})s}}{(\lambda_i s+1)^{N_i}}\prod_{k=1}^{q_i}(s+z_k^*)\cdot\dfrac{1}{1-\dfrac{e^{-\theta_i s}}{(\lambda_i s+1)^{N_i}}\prod_{k=1}^{q_i}\dfrac{-s+z_k}{s+z_k^*}}$, $D_{ij}=p_{0,ij}\prod_{k=1}^{q_i}(-s+z_k)$
3	Infinite RHP and LHP zeros $[z_k\,(k=1,2,\ldots,q_i)=$ dominant RHP zeros excluding those canceled by the common RHP zeros of $G^{ij}\,(j=1,2,\ldots,m)]$	$\dfrac{e^{-\theta_i s}}{(\lambda_i s+1)^{N_i}}\cdot\dfrac{\phi(s)e^{(\theta_{\max}-\theta_{\min})s}}{\phi(-s)}\prod_{k=1}^{q_i}\dfrac{-s-z_k}{s-z_k^*}$	$\dfrac{G^{ij}D_{ij}\psi(s)e^{(\theta_{\min}-\theta_i)s}}{(\lambda_i s+1)^{N_i}}\prod_{k=1}^{q_i}(s-z_k^*)\cdot\dfrac{1}{1-\dfrac{D_{ij}\phi(s)e^{-\theta_i s}}{(\lambda_i s+1)^{N_i}}\prod_{k=1}^{q_i}(s-z_k^*)}$, $D_{ij}=\dfrac{e^{(\theta_{\max}-\theta_{\min})s}}{\phi(-s)}\prod_{k=1}^{q_i}(-s-z_k)$
4	Infinite RHP but finite LHP zeros $[z_k\,(k=1,2,\ldots,q_i)=$ LHP zeros excluding those equal to the complex conjugates of the common RHP zeros of $G^{ij}\,(j=1,2,\ldots,m)]$		

11.2 Decoupling Control Design for Multiple-Input-Multiple-Output (MIMO) Processes

RHP zeros, each column controllers in the desired decoupling controller matrix can be derived accordingly as

$$C_{ideal.ji} = \frac{G^{ij}}{\det(G)} \cdot \frac{\frac{e^{-\theta_i s}}{(\lambda_i s+1)^{N_i}} \prod_{k=1}^{q_i} \frac{-s+z_k}{s+z_k^*}}{1 - \frac{e^{-\theta_i s}}{(\lambda_i s+1)^{N_i}} \prod_{k=1}^{q_i} \frac{-s+z_k}{s+z_k^*}}, \quad i,j = 1,2,\ldots,m. \quad (11.65)$$

For a MIMO process with multiple time delays, it can be seen from (11.58) that the first part in (11.65) is not a rational transfer function and thus difficult to be realized in practice. In addition, the RHP zeros in $\det(G)$ will cause RHP zero-pole cancelation in (11.65), entailing the decoupling controller matrix to behave in an unstable manner. A rational approximation is therefore needed for practical application.

Using (11.58)–(11.61), one can rearrange (11.65) in the form of

$$c_{ji} = \frac{D_{ij} e^{-(\theta_i - L_{ij})s}}{(\lambda_i s+1)^{N_i} \prod_{k=1}^{q_i}(s+z_k^*)} \cdot \frac{1}{1 - \frac{e^{-\theta_i s}}{(\lambda_i s+1)^{N_i}} \prod_{k=1}^{q_i} \frac{-s+z_k}{s+z_k^*}}, \quad i,j = 1,2,\ldots,m \quad (11.66)$$

where λ_i becomes the common adjustable parameter in each column controllers in C and

$$D_{ij} = p_{0,ij} \prod_{k=1}^{q_i}(-s+z_k) \quad (11.67)$$

which can be rationally approximated using (11.28) for practical implementation. Note that a physical constraint, $U - V \leq N_i + q_i$, is required to use (11.28) for approximation in order to maintain the properness of c_{ji} for implementation. Generally, V may be specified first, and then U can be taken as $U = V + N_i + q_i$ to obtain the best approximation. From a mathematical point of view, it is preferred to first formulate $p_{0,ij}$ in (11.58) as

$$p_{0,ij} = \frac{\alpha(s)\left[1 + \eta_1(s)e^{-\sigma_1 s} + \cdots + \eta_{m-\mu}(s)e^{-\sigma_{m-\mu} s}\right]}{\beta(s)\left[1 + \xi_1(s)e^{-\delta_1 s} + \cdots + \xi_{m-\nu}(s)e^{-\delta_{m-\nu} s}\right]}$$

where $\alpha(s)$ and $\beta(s)$ are rational polynomials, $\sigma_k > 0 (k = 1, 2, \ldots, m-\mu)$, $\delta_k > 0 (k = 1, 2, \ldots, m-\nu)$, $\mu < m$, and $\nu < m$. Then, U can be initially taken as the order of $\alpha(s)$ and V the order of $\beta(s)$, in view of that time-delay terms in both the nominator and the denominator decay much faster than $\alpha(s)$ and $\beta(s)$ when $s \to \infty$. It is obvious that increasing the orders of U and V will give a better approximation but at the cost of a higher computation effort and implementation complexity.

Fig. 11.11 Positive feedback control unit

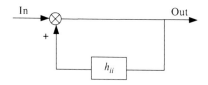

Note that the second multiplication term in c_{ji} has the following properties:

$$\lim_{s \to \infty} \frac{1}{1 - \frac{e^{-\theta_i s}}{(\lambda_i s+1)^{N_i}} \prod_{k=1}^{q_i} \frac{-s+z_k}{s+z_k^*}} = 1 \quad (11.68)$$

$$\lim_{s \to 0} \frac{1}{1 - \frac{e^{-\theta_i s}}{(\lambda_i s+1)^{N_i}} \prod_{k=1}^{q_i} \frac{-s+z_k}{s+z_k^*}} = \infty \quad (11.69)$$

Thus, it can be viewed as a special integrator with a relative degree of zero which can eliminate the output deviation from the corresponding set-point. In fact, this integrator can be realized using a positive feedback control unit, as shown in Fig. 11.11.

For the other cases of the RHP zero distribution in $\det(G)$, as listed in Table 11.1, the desired decoupling controller matrix can be correspondingly derived following a similar design procedure as above. The results are summarized in Table 11.1 for clarity, where D_{ij} in each case can be rationally approximated using (11.28) for practical implementation.

11.2.4 Robust Stability Analysis

As a rational approximation is utilized to achieve the ideally desired decoupling controller matrix in Table 11.1 for implementation, the stability of the resulting control system needs to be verified. Moreover, there usually exist unmodeled process dynamics in practice. Given specified bounds of the process uncertainties in practical applications, an evaluation of the control system robust stability is required so that admissible tuning ranges of the adjustable parameters in the decoupling controller matrix can be determined.

For the nominal control system (i.e., $G = \widehat{G}$, where $\widehat{G}$ denotes the process model identified), it can be derived from Fig. 11.10 that the transfer matrix relating the system input vectors, R, D_I, D_O, and N_O, to the output vectors, Y and U, is

$$\begin{bmatrix} Y \\ U \end{bmatrix} = \begin{bmatrix} GC(I+GC)^{-1} & (I+GC)^{-1}G & I-GC(I+GC)^{-1} & -GC(I+GC)^{-1} \\ C(I+GC)^{-1} & -C(I+GC)^{-1}G & C(I+GC)^{-1} & -C(I+GC)^{-1} \end{bmatrix} \begin{bmatrix} R \\ D_I \\ D_O \\ N_O \end{bmatrix}$$
$$(11.70)$$

11.2 Decoupling Control Design for Multiple-Input-Multiple-Output (MIMO) Processes

It can be seen that R, D_O, and N_O have similar impact on Y and U. Hence, stability analysis for the nominal system can be limited to the submatrix relating R and D_1 to Y and U.

Note that G has been assumed to be nonsingular and stable, and that there is an equivalent transformation,

$$GC(I+GC)^{-1} = I - (I+GC)^{-1} \qquad (11.71)$$

It can therefore be concluded that a sufficient and necessary condition for retaining the nominal system stability is that $(I+GC)^{-1}$ must be maintained stable. This condition can be checked graphically by using the Nyquist curve stability criterion or numerically by computing if $\det(I+GC)$ has any RHP zeros.

In the presence of process uncertainties, robust stability analysis is herein focused on the process additive, multiplicative input, and output uncertainties, as commonly encountered in practical applications. Generally, the process additive uncertainties, as shown in Fig. 11.3, may be viewed as parameter perturbations to the process transfer matrix identified. Correspondingly, the process family may be described by $\Pi_A = \{\widehat{G}_A(s) : \widehat{G}_A(s) = G(s) + \Delta_A\}$, where Δ_A is assumed to be stable. The process multiplicative input uncertainties, as shown in Fig. 10.3a, may be loosely interpreted as the process input actuator uncertainties, and correspondingly the process family may be described by $\Pi_I = \{\widehat{G}_I(s) : \widehat{G}_I(s) = G(s)(I + \Delta_I)\}$, where Δ_I is assumed to be stable. The process multiplicative output uncertainties, as shown in Fig. 10.3b, may be practically viewed as the process output measurement uncertainties, and correspondingly the process family may be described by $\Pi_O = \{\widehat{G}_O(s) : \widehat{G}_O(s) = (I + \Delta_O)G(s)\}$, where Δ_O is assumed to be stable.

By reorganizing the perturbed control system in the form of the standard $M - \Delta$ structure for robustness analysis, the transfer matrix from the outputs to inputs of Δ_A, Δ_I, and Δ_O can be derived, respectively, as

$$M_A = -C(I+GC)^{-1} \qquad (11.72)$$

$$M_I = -C(I+GC)^{-1}G \qquad (11.73)$$

$$M_O = -GC(I+GC)^{-1} \qquad (11.74)$$

Note that M_A, M_I, and M_O maintain stability provided that the nominal control system has been maintained stable, i.e., the transfer matrix shown in (11.70) has been maintained stable.

Then using the small-gain theorem, the robust stability constraints can be obtained, respectively, as

$$\left\| C(I+GC)^{-1} \right\|_\infty < \frac{1}{\|\Delta_A\|_\infty} \qquad (11.75)$$

$$\left\| C(I+GC)^{-1}G \right\|_\infty < \frac{1}{\|\Delta_I\|_\infty} \tag{11.76}$$

$$\left\| GC(I+GC)^{-1} \right\|_\infty < \frac{1}{\|\Delta_O\|_\infty} \tag{11.77}$$

The robust stability constraints shown in (11.75)–(11.77), however, are not analytical, and the computation effort for an H_∞ norm may be considerably large, especially for MIMO processes with multiple time delays. To mitigate the computation burden, the equivalent relationship between the small-gain theorem and the multivariable spectral radius stability criterion (Skogestad and Postlethwaite 2005) is therefore used, i.e.,

$$\|M\Delta\|_\infty < 1 \Leftrightarrow \rho(M\Delta) < 1, \forall \omega \in [0,\infty) \tag{11.78}$$

Thus, the above robust stability constraints can be reformulated, respectively, as

$$\rho\left(C(I+GC)^{-1}\Delta_A\right) < 1, \forall \omega \in [0,\infty) \tag{11.79}$$

$$\rho\left(C(I+GC)^{-1}G\Delta_I\right) < 1, \forall \omega \in [0,\infty) \tag{11.80}$$

$$\rho\left(GC(I+GC)^{-1}\Delta_O\right) < 1, \forall \omega \in [0,\infty) \tag{11.81}$$

Note that the spectral radius stability constraints shown in (11.79)–(11.81) can be checked graphically by observing if the magnitude plots of the left-hand sides of (11.79)–(11.81) fall below unity for $\omega \in [0,+\infty)$.

Given a specified bound of Δ_A, Δ_I, or Δ_O in practice, as will be illustrated in the examples later, the above spectral radius stability constraints can be used to intuitively evaluate the control system robust stability. In this way, admissible tuning ranges of the adjustable parameters (λ_i, $i=1,2,\ldots,m$) in the decoupling controller matrix (C) can be numerically determined for process operation.

Combined with (11.62), it can be seen that a smaller value of λ_i in the decoupling controller matrix will result in a faster response of the corresponding ith output variable, but the output energy of the ith column controllers in C and the corresponding actuators' signals will become larger, tending to surpass the output capacities in practice. Moreover, more aggressive dynamic behavior of the ith output response will occur in the presence of process uncertainties. On the contrary, increasing λ_i will slow down the ith output response, but the output energy of the ith column controllers in C and the corresponding actuators' actions will be required to be smaller. Consequently, a less aggressive dynamic behavior of the ith system output response will occur in the presence of process uncertainties. Therefore, tuning the adjustable parameters λ_i ($i=1,2,\ldots,m$) needs to meet a good trade-off between the achievable output performance and the output capacities of C and the corresponding actuators.

11.2 Decoupling Control Design for Multiple-Input-Multiple-Output (MIMO) Processes

Based on extensive simulations, it is generally suggested to initially take these adjustable parameters λ_i ($i = 1, 2, \ldots, m$) in the range of $(1.0 - 10)\theta_i$. Then, by monotonically varying them online, a desirable output performance can be gradually reached.

To cope with process uncertainties, it is suggested to increase monotonically these adjustable parameters λ_i ($i = 1, 2, \ldots, m$) in C so that the nominal system response will be slowed down in exchange for further enhanced system robust stability. Note that, if by doing so, the control system performance and robust stability are still not acceptable, process reidentification can be performed to obtain a more accurate process model for designing C so that the unmodeled process dynamics can be effectively reduced to obtain better system performance and robust stability.

11.2.5 Illustrative Examples

Two examples from the existing literature are used to demonstrate the effectiveness and merits of the presented MIMO decoupling control method, one with no RHP zero in $\det(G)$ and the other with infinite many RHP zeros in $\det(G)$.

Example 11.3. Consider the widely studied 3×3 industrial distillation column (Tyreus 1979),

$$G = \begin{bmatrix} \dfrac{1.986e^{-0.71s}}{66.7s + 1} & \dfrac{-5.24e^{-60s}}{400s + 1} & \dfrac{-5.984e^{-2.24s}}{14.29s + 1} \\ \dfrac{-0.0204e^{-0.59s}}{(7.14s + 1)^2} & \dfrac{0.33e^{-0.68s}}{(2.38s + 1)^2} & \dfrac{-2.38e^{-0.42s}}{(1.43s + 1)^2} \\ \dfrac{-0.374e^{-7.75s}}{22.22s + 1} & \dfrac{11.3e^{-3.79s}}{(21.74s + 1)^2} & \dfrac{9.811e^{-1.59s}}{11.36s + 1} \end{bmatrix}$$

The Nyquist curve of the process transfer matrix determinant is plotted in Fig. 11.12. It is seen that the Nyquist curve does not encircle the origin, indicating there is no RHP zero in $\det(G)$. According to the presented MIMO decoupling control method, it follows from (11.58) that $L_{11} = 0.71$, $L_{12} = 0.8$, and $L_{13} = -1.4$. Thus, $\theta_1 = 0.8$ can be determined in terms of the definition in (11.61). Then, using (11.59) obtains $n_{11} = 1$, $n_{12} = 1$, and $n_{13} = 0$. Hence, $N_1 = 1$ can be determined in terms of the definition in (11.60). Similarly, the use of (11.58)–(11.61) yields $\theta_2 = 0.68$ and $\theta_3 = 1.85$, and $N_2 = 2$ and $N_3 = 1$. According to the design formula for case 1 in Table 11.1, the diagonal elements of the desired system response transfer matrix can be determined as

$$h_{11} = \dfrac{e^{-0.8s}}{\lambda_1 s + 1}, \quad h_{22} = \dfrac{e^{-0.68s}}{(\lambda_2 s + 1)^2}, \quad h_{33} = \dfrac{e^{-1.85s}}{\lambda_3 s + 1}.$$

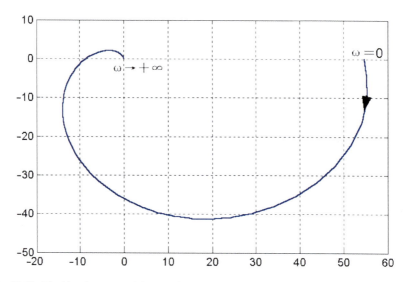

Fig. 11.12 The Nyquist curve of the transfer matrix determinant of Example 11.3

Hence, the decoupling controller matrix can be derived using the analytical design formulas given in Table 11.1, together with the approximation formulas in (11.28)–(11.32) for practical implementation. For illustration, to obtain similar controller orders with those of Wang et al. (2003) for comparison, the following executable controller forms in the decoupling controller matrix are derived:

$$c_{11} = f_1 \cdot \frac{14543s^2 + 256.3578s + 0.5502}{(\lambda_1 s + 1)(438.7353s + 1)} e^{-0.09s}$$

$$c_{21} = f_1 \cdot \frac{12391s^3 + 746.2116s^2 + 9.7508s + 0.0199}{(\lambda_1 s + 1)(3940.3s^2 + 447.8424s + 1)}$$

$$c_{31} = f_1 \cdot \frac{1736.5s^3 - 21.7287s^2 - 0.8474s - 0.002}{(\lambda_1 s + 1)(4815.4s^2 + 449.8302s + 1)} e^{-2.2s}$$

$$c_{12} = f_2 \cdot \frac{4773900s^6 - 6620600s^5 - 3286200s^4 - 532380s^3 - 41045s^2 - 526.1791s - 0.296}{(\lambda_2 s + 1)^2 (611700s^4 + 109510s^3 + 12128s^2 + 465.9313s + 1)} e^{-3.73s}$$

$$c_{22} = f_2 \cdot \frac{13471000s^6 + 3306200s^5 + 892990s^4 + 117120s^3 + 6709.9s^2 + 142.0148s + 0.3149}{(\lambda_2 s + 1)^2 (336570s^4 + 33465s^3 + 9959.2s^2 + 461.3811s + 1)}$$

$$c_{32} = f_2 \cdot \frac{-197040s^5 - 104730s^4 - 29099s^3 - 4024.9s^2 - 171.9233s - 0.374}{(\lambda_2 s + 1)^2 (257300s^4 + 55907s^3 + 10254s^2 + 461.9346s + 1)} e^{-2.2s}$$

11.2 Decoupling Control Design for Multiple-Input-Multiple-Output (MIMO) Processes

$$c_{13} = f_3 \cdot \frac{400930s^4 + 33536s^3 + 1342.3s^2 + 31.5279s + 0.2638}{(\lambda_3 s + 1)(33025s^3 + 3869.9s^2 + 447.5041s + 1)} e^{-1.79s}$$

$$c_{23} = f_3 \cdot \frac{16790s^3 + 1582.9s^2 + 39.2646s + 0.0885}{(\lambda_3 s + 1)(511.4853s^2 + 440.0233s + 1)}$$

$$c_{33} = f_3 \cdot \frac{2195s^3 + 212.3057s^2 + 5.2157s + 0.01}{(\lambda_3 s + 1)(1319.1s^2 + 441.8636s + 1)} e^{-0.26s}$$

where

$$f_1 = \frac{1}{1 - \frac{e^{-0.8s}}{\lambda_1 s + 1}}, \quad f_2 = \frac{1}{1 - \frac{e^{-0.68s}}{(\lambda_2 s + 1)^2}}, \quad f_3 = \frac{1}{1 - \frac{e^{-1.85s}}{\lambda_3 s + 1}}.$$

Note that f_1, f_2, and f_3 can be implemented using the feedback control unit shown in Fig. 11.11.

The adjustable parameters are taken as $\lambda_1 = 15$, $\lambda_2 = 12$, and $\lambda_3 = 18$ to obtain the similar rising speeds of set-point tracking with those of Wang et al. (2003). By adding a unit step change to the ternary set-point inputs at $t = 0$ (s), $t = 200$ (s), and $t = 400$ (s), respectively, and then adding a step change of load disturbance with a magnitude of 0.1 to all the three process inputs at $t = 600$ (s), the output responses are shown in Fig. 11.13. It is seen that there is no overshoot in the set-point responses using the presented MIMO decoupling control method, while the three process output responses are almost decoupled from each other. Moreover, obviously improved load disturbance rejection performance is obtained. Note that further improved system performance for both the set-point tracking and load disturbance rejection can be transparently obtained in the proposed method by gradually decreasing the adjustable parameters λ_1, λ_2, and λ_3, together with a higher-order controller approximation for implementation. Besides, it should be mentioned that the conventional PID controllers cannot be used to obtain acceptable output performance or even cannot stabilize the system output responses due to low approximation capacity for the ideally desired decoupling controller matrix, as shown in Table 11.1. The same conclusion was drawn by Wang et al. (2003) from a comparison of the Nyquist curve fitting.

To demonstrate the robustness of the presented MIMO decoupling control method, the perturbation tests in Wang et al. (2003) are performed here. That is, all static gains in the process transfer matrix are assumed to be actually 40% larger. In another case, all time constants in the process transfer matrix are assumed to be 40% larger to represent the unmodeled process dynamics. According to the robust stability analysis in Sect. 11.2.4, the magnitude plots of the spectral radius condition in (11.79) for checking the robust stability of the corresponding perturbed systems are shown in Fig. 11.14. It can be seen that both the peak values (dotted and dash dot lines) are much smaller than unity, indicating that the corresponding system

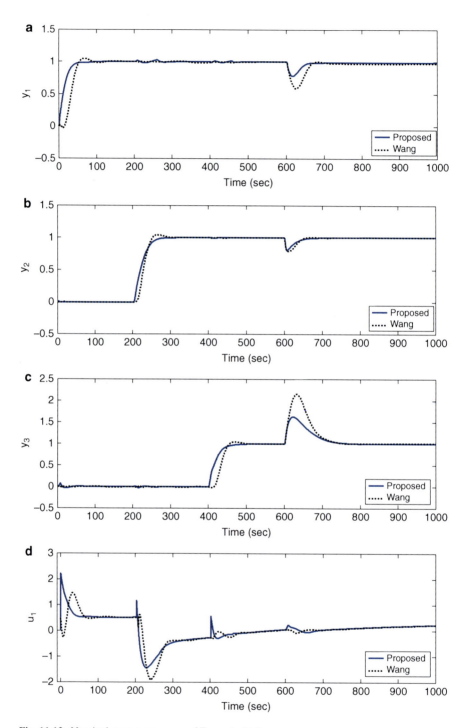

Fig. 11.13 Nominal output responses of Example 11.3

11.2 Decoupling Control Design for Multiple-Input-Multiple-Output (MIMO) Processes 403

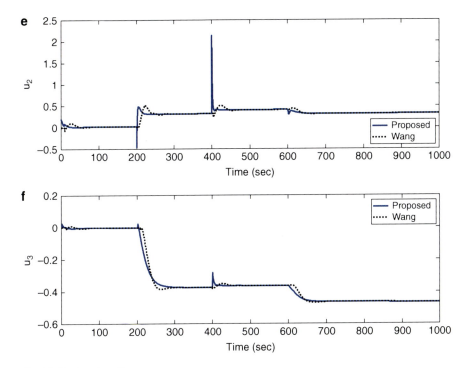

Fig. 11.13 (continued)

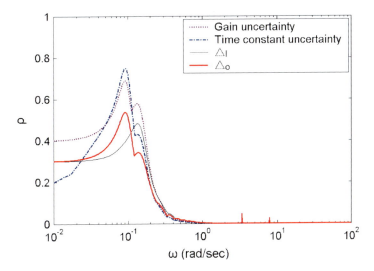

Fig. 11.14 The magnitude plots of spectral radius for Example 11.3

maintains good robust stability. Accordingly, the perturbed output responses are shown in Fig. 11.15a–c, d–f, respectively. Note that Fig. 11.15a–c also demonstrates that the perturbations of the process static gains do not affect decoupling regulation of the output responses, as also indicated by the analytical controller matrix design procedure presented in Sect. 11.2.3. The corresponding three control outputs indeed vary little compared to those shown in Fig. 11.13, and thus are omitted.

To further demonstrate the control system robust stability in terms of the presented MIMO decoupling control method, assume that there actually exist the process multiplicative input uncertainties, $\Delta_I = diag[(s+0.3)/(s+1),(s+0.2)/(s+1),(s+0.2)/(s+1)]_{3\times 3}$, which can be loosely interpreted as the first process input actuator has up to 100% uncertainty at high frequencies and almost 30% uncertainty in the low-frequency range, while the other two process input actuators have up to 100% uncertainty at high frequencies and almost 20% uncertainty in the low-frequency range. In another case, assume that there exist the process multiplicative output uncertainties, $\Delta_O = diag[-(s+0.2)/(2s+1), -(s+0.2)/(2s+1), -(s+0.3)/(2s+1)]_{3\times 3}$, which may be practically viewed as the first two process output measurements offered by the corresponding sensors decrease by up to 50% uncertainty at high frequencies and by almost 20% uncertainty in the low-frequency range, while the third process output measurement decreases by up to 50% uncertainty at high frequencies and by almost 30% uncertainty in the low-frequency range. Figure 11.14 shows the corresponding magnitude plots of the spectral radius conditions in (11.80) and (11.81) based on the assumed Δ_I (thin solid line) and Δ_O (thick solid line), which indicate that the proposed control system can maintain robust stability well. The corresponding perturbed output responses are shown in Fig. 11.16, well verifying the above robust stability analysis.

Example 11.4. Consider the binary process studied by Jerome and Ray (1992),

$$G = \begin{bmatrix} \dfrac{1.05e^{-4.58s}}{1.64s+1} & \dfrac{0.32}{(1.6s+1)(1.61s+1)} \\ \dfrac{1.18e^{-15.2s}}{3.6s+1} & \dfrac{0.9}{(4.5s+1)(4.51s+1)} \end{bmatrix}$$

It was ascertained by Jerome and Ray (1992) that there are infinite many RHP zeros and four LHP zeros in the process transfer matrix determinant, and the four LHP zeros are approximate roots of the following polynomial:

$$\chi(s) = (1.64s+1)(4.0542s+1)(40.459s^2+11.116s+1)$$

Thus, this process belongs to Case 4 in Table 11.1.

According to the presented MIMO decoupling control method, one can first write the process transfer matrix determinant in the form of

$$\det(G) = \frac{[0.945(1.6s+1)(1.61s+1)(3.6s+1) - 0.3776(1.64s+1)(4.5s+1)(4.51s+1)e^{-10.62s}]e^{-4.58s}}{(1.64s+1)(4.5s+1)(4.51s+1)(1.6s+1)(1.61s+1)(3.6s+1)}$$

11.2 Decoupling Control Design for Multiple-Input-Multiple-Output (MIMO) Processes 405

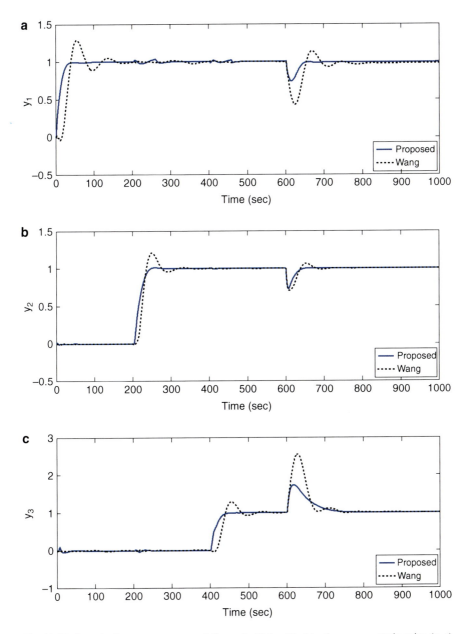

Fig. 11.15 Perturbed output responses of Example 11.3 subject to the process static gains (a–c) and time constants (d–f) variations

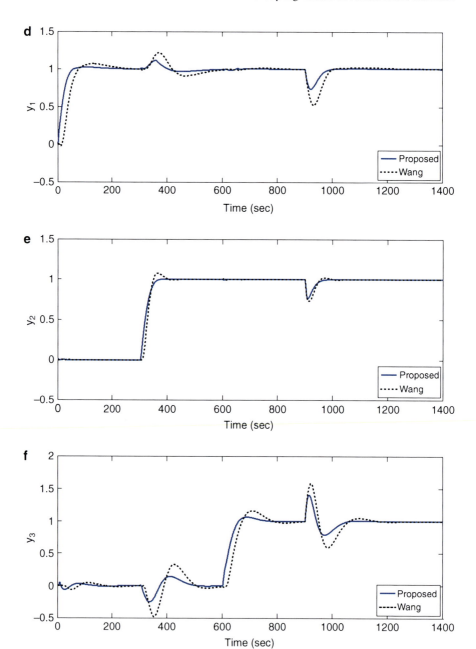

Fig. 11.15 (continued)

11.2 Decoupling Control Design for Multiple-Input-Multiple-Output (MIMO) Processes

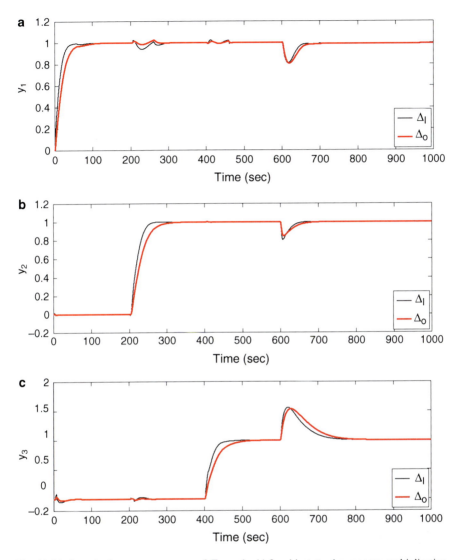

Fig. 11.16 Perturbed output responses of Example 11.3 subject to the process multiplicative uncertainties

It follows that $\theta_{\min} = 4.58$, $\theta_{\max} = 15.2$, and $\phi(s)$ is the polynomial in the square bracket of the numerator and $\psi(s)$ is the denominator polynomial.

Subsequently, using (11.58)–(11.61) yields $\theta_1 = \theta_2 = 4.58$ and $N_1 = N_2 = 2$. Thus, the diagonal elements of the desired system response transfer matrix can be determined as

$$h_{11} = \frac{D\phi(s)e^{-4.58s}}{\chi(s)(\lambda_1 s + 1)^2}, \quad h_{22} = \frac{D\phi(s)e^{-4.58s}}{\chi(s)(\lambda_2 s + 1)^2}.$$

where

$$D = \frac{(-1.64s+1)(-4.0542s+1)(40.459s^2-11.116s+1)}{0.945(-1.6s+1)(-1.61s+1)(-3.6s+1)e^{-10.62s}-0.3776(-1.64s+1)(-4.5s+1)(-4.51s+1)}$$

Note that D cannot be directly implemented due to the occurrence of RHP zero-pole cancelation. The analytical approximation formulas in (11.28)–(11.32) are therefore used to obtain a low-order approximation by taking $U = 2$ and $V = 1$,

$$D_{2/1} = \frac{20.1786s^2 + 10.0787s + 1.7624}{0.5868s + 1}$$

Then, the use of the design formula for case 4 in Table 11.1 gives the decoupling controller matrix,

$$C = D_c \cdot \begin{bmatrix} \frac{0.9F_1(1.64s+1)}{(\lambda_1 s+1)^2} & -\frac{0.32F_2(1.64s+1)(4.5s+1)(4.51s+1)}{(\lambda_2 s+1)^2(1.6s+1)(1.61s+1)} \\ -\frac{1.18F_1(1.64s+1)(4.5s+1)(4.51s+1)e^{-15.2s}}{(\lambda_1 s+1)^2(3.6s+1)} & \frac{1.05F_2(4.5s+1)(4.51s+1)e^{-4.58s}}{(\lambda_2 s+1)^2} \end{bmatrix}$$

where

$$D_c = \frac{D_{2/1}(1.6s+1)(1.61s+1)(3.6s+1)}{\chi(s)}, \quad F_1 = \frac{1}{1 - \frac{D_{2/1}\phi(s)e^{-4.58s}}{\chi(s)(\lambda_1 s+1)^2}},$$

$$F_2 = \frac{1}{1 - \frac{D_{2/1}\phi(s)e^{-4.58s}}{\chi(s)(\lambda_2 s+1)^2}}.$$

Note that both F_1 and F_2 can be practically implemented using the feedback control unit shown in Fig. 11.11.

Based on the standard IMC structure, Jerome and Ray (1992) suggested a controller matrix design that seemed to be able to move undesired dynamics of individual outputs to a single-output response to obtain apparently improved response performance in the other outputs. For comparison, the optimization of the second output at the cost of severely degraded response performance in the first output is performed here. In the presented MIMO decoupling control method, the adjustable parameters are taken as $\lambda_1 = 3.5$ and $\lambda_2 = 3.0$ to obtain similar rising speeds of set-point tracking with those of Jerome and Ray's method. By adding a unit step change to the binary set-point inputs at $t = 0$ (s) and 150 (s), respectively, and an inverse step change of load disturbance with a magnitude of 0.1 to both the process inputs at $t = 300$ (s), the output responses are shown in Fig. 11.17. It is seen that entirely decoupled output responses have been obtained using the presented MIMO decoupling control method (solid line), while the second output response is comparable with that of Jerome and Ray's method, barring a small time delay that is specifically adopted to obtain the decoupled output responses. Note that Jerome and Ray's method has resulted in severe oscillation in the first process

11.2 Decoupling Control Design for Multiple-Input-Multiple-Output (MIMO) Processes

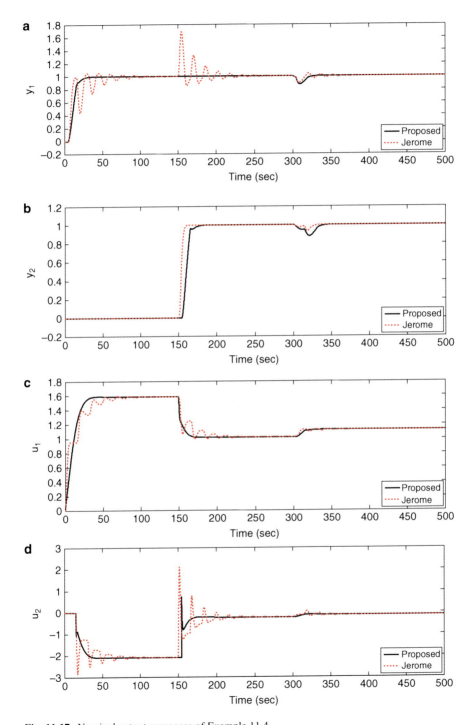

Fig. 11.17 Nominal output responses of Example 11.4

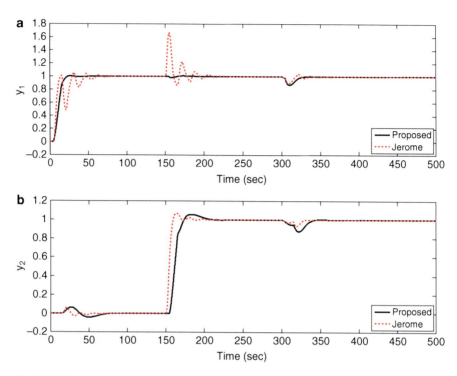

Fig. 11.18 Perturbed output responses of Example 11.4

output response and the oscillatory control outputs are far from acceptable from a practical viewpoint of application. It is therefore demonstrated that the idea of sacrificing dynamic response performance in one process output for improving the other process output responses is trivial for practical application when compared to the presented decoupling control method.

To demonstrate the control system robust stability, assume that all time constants in the process transfer matrix are actually 20% larger to introduce unmodeled dynamics. The perturbed system output responses are shown in Fig. 11.18. It is once again seen that the proposed decoupling control system maintains good robust stability in the presence of the severe process uncertainties. Note that the control outputs of the proposed control system have varied slightly and thus are omitted.

11.3 A 2DOF Decoupling Control Scheme for MIMO Processes

To facilitate separate optimization of the set-point tracking and load disturbance rejection for a MIMO process, a 2DOF control structure, which has been intensively studied for the control of SISO processes as presented in Chap. 8, can be explored

11.3 A 2DOF Decoupling Control Scheme for MIMO Processes

Fig. 11.19 Multivariable 2DOF decoupling control structure

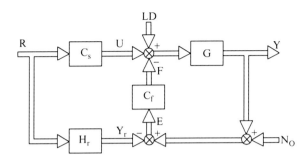

for application. Using the H_∞ norm optimization approach, a few 2DOF control methods have been reported in the literature (Limebeer et al. 1993; Prempain and Bergeon 1998; Lundström and Skogestad 1999), where an integrated controller matrix was eventually derived for both the set-point tracking and load disturbance rejection, and therefore, cannot implement independent regulation of the set-point tracking and load disturbance rejection online. Huang and Lin (2006) proposed an alternative 2DOF decoupling control structure to realize separate optimization of the load disturbance rejection, which showed evident improvement when compared to previous methods in the literature, but more than two controller matrices computed from numerical iteration algorithms were used, involving a considerable computation effort for practical application, especially for online tuning.

The 2DOF decoupling control method (Liu et al. 2007b) is presented here for practical application. The multivariable 2DOF control structure is shown in Fig. 11.19, where $g_{ij}(s) = g_{0,ij}(s)e^{-\theta_{ij}s}$, $i, j = 1, 2, \ldots, m$, and $g_{0,ij}(s)$ is a rational and stable transfer function. $C_s = [c_{ij}]_{m \times m}$ is a controller matrix for the set-point tracking; $H_r = \text{diag}[h_{r,i}]_{m \times m}$ is a diagonal transfer matrix offering the referential output trajectories, $Y_r = [y_{r,i}]_{m \times 1}$. The controller matrix, $C_f = [cf_{ij}]_{m \times m}$, installed in the feedback channels of the closed-loop structure set between the process inputs and outputs, is for rejecting load disturbance and eliminating output error. $LD = [d_i]_{m \times 1}$ denotes load disturbances entering into individual channels of the process, and $N_O = [n_i]_{m \times 1}$ is the output measurement noise vector, $R = [r_i]_{m \times 1}$ denotes the set-point vector, $Y = [y_i]_{m \times 1}$ is the output vector, and $U = [u_i]_{m \times 1}$ is the controller output vector.

For a general transfer matrix of a MIMO process with multiple time delays, as shown in (11.53), it can be seen from Fig. 11.19 that an "open-loop" control for the set-point tracking can be realized in the nominal case ($G = \hat{G}$) if

$$H_r = GC_s \qquad (11.82)$$

Note that without model mismatch, the output error signal, $E = [e_i]_{m \times 1}$, is zero if no load disturbance (LD) and measurement noise (N_O) occurs. In the presence of load disturbance and measurement noise, the error signal (E) will no longer be zero and thus trigger the closed-loop controller matrix, C_f, to adjust the process inputs for counteraction.

Hence, the 2DOF control structure shown in Fig. 11.19 allows separate regulation of the set-point tracking and load disturbance rejection. Note that the referential diagonal system transfer matrix, H_r, can be viewed as a frequency-domain model predictive control (MPC) strategy for the set-point tracking. For decoupling regulation, these two controller matrices, C_s and C_f, should be designed to realize diagonal transfer matrices for both the set-point response and load disturbance response.

Considering that the achievable system performance for a MIMO process with multiple time delays is practically constrained by the time delays in individual channels and possible NMP characteristics, as previously analyzed in Sect. 11.2, an analysis on the desired transfer matrices for the set-point tracking and load disturbance rejection is given as follows.

11.3.1 Desired Set-Point and Closed-Loop Transfer Matrices

From (11.82), the controller matrix for the set-point tracking can be inversely derived as

$$C_s = G^{-1} H_r = \frac{adj(G)}{det(G)} H_r \qquad (11.83)$$

where $adj(G) = \left[G^{ij}\right]^T_{m \times m}$ is the adjoint of G and G^{ij} denotes the complement minor corresponding to g_{ij} in G.

According to the postmultiplication relationship between a square matrix and a diagonal matrix, each column controllers in C_s (with the same subscript of i) can be obtained as

$$c_{ji} = \frac{G^{ij}}{det(G)} h_{r,i}, \quad i, j = 1, 2, \ldots, m. \qquad (11.84)$$

Let

$$p_{ij} = \frac{G^{ij}}{det(G)} = p_{0,ij} e^{L_{ij} s}, \quad i, j = 1, 2, \ldots, m. \qquad (11.85)$$

According to the "inverse relative degree" defined in (11.59), let

$$\theta_{r,i} = \max\{L_{ij}; j = 1, 2, \ldots, m\}, \quad i, j = 1, 2, \ldots, m. \qquad (11.86)$$

$$n_{r,i} = \max\{n_{ij}; j = 1, 2, \ldots, m\}, \quad i, j = 1, 2, \ldots, m. \qquad (11.87)$$

It can be seen from (11.84) that each column controllers in C_s are related to the same diagonal element in H_r, i.e., all $c_{ji} (j = 1, 2, \ldots, m)$ are corresponding to

11.3 A 2DOF Decoupling Control Scheme for MIMO Processes

the same $h_{r,i}$ for $i = 1, 2, \ldots, m$. Note that $\theta_{r,i}$ ($i = 1, 2, \ldots, m$) defined in (11.86) is positive, which can be verified from (11.85) using the algebra of linear matrix. Some or even all the ith column controllers c_{ji} ($j = 1, 2, \ldots, m$) derived in (11.84) will not be realizable if the corresponding diagonal element $h_{r,i}$ in H_r does not include an equivalent time delay to balance $\theta_{r,i}$. This constraint is due to the fact that each output can only respond to the corresponding set-point after certain time delay arising from the process response. It can also be seen from (11.84) that if the relative degree of the delay-free part in $h_{r,i}$ were lower than $n_{r,i}$, some or even all c_{ji} ($j = 1, 2, \ldots, m$) would not be proper and thus cannot be implemented in practice. In addition, $\det(G)$ possibly includes RHP zeros that are not canceled out by the common RHP zeros of G^{ij} ($j = 1, 2, \ldots, m$). If $h_{r,i}$ does not include these RHP zeros, each of c_{ji} ($j = 1, 2, \ldots, m$) would be bundled with them as unstable poles, which is definitely not allowed in practice.

Hence, similar to the determination of a desired system transfer matrix presented in Sect. 11.2.2, the desired diagonal elements in H_r are proposed as

$$h_{r,i} = \frac{e^{-\theta_{r,i} s}}{(\lambda_{c,i} s + 1)^{n_{r,i}}} \prod_{k=1}^{q_i} \frac{-s + z_k}{s + z_k^*}, \quad i = 1, 2, \ldots, m \tag{11.88}$$

where $\lambda_{c,i}$ is an adjustable parameter for tuning desirable set-point tracking performance for the ith process output and z_k ($k = 1, 2, \ldots, q_i$) are RHP zeros of $\det(G)$ excluding those canceled out by the common RHP zeros of G^{ij} ($j = 1, 2, \cdots, m$), q_i is the number, and z_k^* is the complex conjugate of z_k.

With the desired diagonal elements prescribed in (11.88), it can be ascertained from (11.84)–(11.87) that at least one of each column controllers in C_s can be implemented in a proper and rational form, while the other controllers of the corresponding column in C_s can be implemented using dead-time compensators prescribed. Thereby, decoupling regulation for the set-point tracking can be realized for all output variables.

Note that the closed-loop structure set between the process inputs and outputs is used for load disturbance rejection. The transfer matrix of load disturbance response (from LD to Y, as shown in Fig. 11.19) can be derived as

$$H_d = G(I + C_f G)^{-1} \tag{11.89}$$

Correspondingly, the closed-loop complementary sensitivity function matrix can be determined as

$$T_d = C_f G(I + C_f G)^{-1} \tag{11.90}$$

Note that it is in fact equivalent to the transfer matrix relating the load disturbances ($LD = [d_i]_{m \times 1}$) to the controller matrix output vector, $F = [f_i]_{m \times 1}$.

Ideally, it is expected that when a load disturbance, d_i, enters into the ith process input, the resulting output error, $E = [e_i]_{m \times 1}$, should be detected by the controller matrix C_f immediately after the process time delay. Then C_f turns out a control signal f_i (i.e., $F = [0, 0, \ldots, f_i, 0, \ldots, 0]^T_{1 \times m}$) to counteract the disturbance. To reject multiple load disturbances entering into the process inputs at different times, T_d is expected to be a diagonal transfer matrix, i.e., $T_d = diag\{t_{d,i}\}_{m \times m}$, such that the control signals, f_i ($i = 1, 2, \ldots, m$) can be separately tuned for disturbance rejection. In this way, all the load disturbance responses can be separately regulated in a transparent manner.

Note that the inverse of T_d, i.e., T_d^{-1}, is also a diagonal matrix if T_d itself is obtained as a diagonal transfer matrix. It follows from (11.90) that

$$C_f = (T_d^{-1} - I)^{-1} G^{-1} \qquad (11.91)$$

Correspondingly, each row controllers in C_f (with the same subscript of j) can be obtained as

$$cf_{ij} = \frac{t_{d,i}}{1 - t_{d,i}} \cdot \frac{G^{ji}}{\det(G)} \pm, \quad i, j = 1, 2, \ldots, m. \qquad (11.92)$$

It can be seen from (11.92) that each row controllers in C_f are related to the same diagonal element in T_d, i.e., all cf_{ij} ($j = 1, 2, \ldots, m$) correspond to the same $t_{d,i}$ for $i = 1, 2, \ldots, m$.

Using the definitions in (11.58) and (11.59), let

$$\theta_{d,i} = \max\{L_{ji}; j = 1, 2, \ldots, m\}, \quad i = 1, 2, \ldots, m. \qquad (11.93)$$

$$n_{d,i} = \max\{n_{ji}; j = 1, 2, \ldots, m\}, \quad i = 1, 2, \ldots, m. \qquad (11.94)$$

Hence, following a similar analysis as above, one can determine the desired forms of $t_{d,i}$ ($i = 1, 2, \ldots, m$) for implementation.

In fact, there are four possible cases of the RHP zero distribution in $\det(G)$, as categorized in Table 11.1. Accordingly, the desired forms of H_r and T_d are summarized in Table 11.2 for clarity.

11.3.2 Controller Matrix Design

With the desired H_r and T_d, as listed in Table 11.2, the ideally desired controller matrix C_s for set-point tracking and the desired closed-loop controller matrix C_f for load disturbance rejection can be inversely derived from (11.84) and (11.92), respectively. For instance, in case 2 that $\det(G)$ has finite RHP zeros, each column controllers in C_s and each row controllers in C_f can be derived, respectively, as

11.3 A 2DOF Decoupling Control Scheme for MIMO Processes

Table 11.2 A summary of desired transfer matrices and controller matrices

Case	$h_{r,i}^1$ ($i=1,2,\ldots,m$)	$t_{d,i}^1$ ($i=1,2,\ldots,m$)	c_{ji} ($i,j=1,2,\ldots,m$)	cf_{ij} ($i,j=1,2,\ldots,m$)
1	$h_{r,i}^1 = \dfrac{e^{-\theta_{r,i}s}}{(\lambda_{c,i}s+1)^{n_{r,i}}}$	$t_{d,i}^1 = \dfrac{e^{-\theta_{d,i}s}}{(\lambda_{f,i}s+1)^{n_{d,i}}}$	$\dfrac{D_{ij}e^{-(\theta_{r,i}-L_{ij})s}}{(\lambda_{c,i}s+1)^{n_{r,i}}}$, $D_{ij} = p_{0,ij}$	$\dfrac{D_{ji}e^{-(\theta_{d,i}-L_{ji})s}}{(\lambda_{f,i}s+1)^{n_{d,i}}} \cdot \dfrac{1}{1 - \dfrac{e^{-\theta_{d,i}s}}{(\lambda_{f,i}s+1)^{n_{d,i}}}}$, $D_{ji} = p_{0,ji}$
2 or 3	$h_{r,i}^1 \displaystyle\prod_{k=1}^{q_i}\dfrac{-s+z_k}{s+z_k^*}$	$t_{d,i}^1 \displaystyle\prod_{k=1}^{q_i}\dfrac{-s+z_k}{s+z_k^*}$	$\dfrac{D_{ij}e^{-(\theta_{r,i}-L_{ij})s}}{(\lambda_{c,i}s+1)^{n_{r,i}}}\displaystyle\prod_{k=1}^{q_i}(s+z_k^*)$, $D_{ij} = p_{0,ij}\displaystyle\prod_{k=1}^{q_i}(-s+z_k)$	$\dfrac{D_{ji}e^{-(\theta_{d,i}-L_{ji})s}}{(\lambda_{f,i}s+1)^{n_{d,i}}}\displaystyle\prod_{k=1}^{q_i}(s+z_k^*)$ $1 - \dfrac{e^{-\theta_{d,i}s}}{(\lambda_{f,i}s+1)^{n_{d,i}}}\displaystyle\prod_{k=1}^{q_i}\dfrac{-s+z_k}{s+z_k^*}$, $D_{ji} = p_{0,ji}\displaystyle\prod_{k=1}^{q_i}(-s+z_k)$
4	$\dfrac{h_{r,i}^1\phi(s)e^{(\theta_{\max}-\theta_{\min})s}}{\phi(-s)}\displaystyle\prod_{k=1}^{q_i}\dfrac{-s-z_k}{s-z_k^*}$	$\dfrac{t_{d,i}^1\phi(s)e^{(\theta_{\max}-\theta_{\min})s}}{\phi(-s)}\displaystyle\prod_{k=1}^{q_i}\dfrac{-s-z_k}{s-z_k^*}$	$\dfrac{G^{ij}D_{ij}\psi(s)e^{(\theta_{\min}-\theta_{r,i})s}}{(\lambda_{c,i}s+1)^{n_{r,i}}}\displaystyle\prod_{k=1}^{q_i}(s-z_k^*)$, $D_{ij} = \dfrac{e^{(\theta_{\max}-\theta_{\min})s}}{\phi(-s)}\displaystyle\prod_{k=1}^{q_i}(-s-z_k)$	$\dfrac{G^{ji}D_{ji}\psi(s)e^{(\theta_{\min}-\theta_{d,i})s}}{(\lambda_{f,i}s+1)^{n_{d,i}}}\displaystyle\prod_{k=1}^{q_i}(s-z_k^*)$ $1 - \dfrac{D_{ji}\phi(s)e^{-\theta_{d,i}s}}{(\lambda_{f,i}s+1)^{n_{d,i}}}\displaystyle\prod_{k=1}^{q_i}(s-z_k^*)$, $D_{ji} = \dfrac{e^{(\theta_{\max}-\theta_{\min})s}}{\phi(-s)}\displaystyle\prod_{k=1}^{q_i}(-s-z_k)$

$$c_{ji} = \frac{D_{ij} e^{-(\theta_{r,i} - L_{ij})s}}{(\lambda_{c,i} s + 1)^{n_{r,i}} \prod_{k=1}^{q_i} (s + z_k^*)}, \quad i, j = 1, 2, \ldots, m \qquad (11.95)$$

$$cf_{ij} = \frac{D_{ji} e^{-(\theta_{d,i} - L_{ji})s}}{(\lambda_{f,i} s + 1)^{n_{d,i}} \prod_{k=1}^{q_i} (s + z_k^*)} \cdot \frac{1}{1 - \frac{e^{-\theta_{d,i} s}}{(\lambda_{f,i} s + 1)^{n_{d,i}}} \prod_{k=1}^{q_i} \frac{-s + z_k}{s + z_k^*}}, \quad i, j = 1, 2, \ldots, m$$

$$(11.96)$$

where $\lambda_{c,i}$ is the common adjustable parameter in each column controllers in C_s and $\lambda_{f,i}$ is the common adjustable parameter in each row controllers in C_f, and

$$D_{ij} = p_{0,ij} \prod_{k=1}^{q_i} (-s + z_k) \qquad (11.97)$$

$$D_{ji} = p_{0,ji} \prod_{k=1}^{q_i} (-s + z_k) \qquad (11.98)$$

Obviously, it can be seen from (11.85) that both D_{ij} and D_{ji} are not of rational transfer function for a MIMO process with time delays, thus difficult to be implemented in practice. Moreover, the RHP zeros in det(G) will result in RHP zero-pole cancelation in D_{ij} and D_{ji}, entailing C_s and C_f to behave in an unstable manner. Analytical approximation formulas in (11.28)–(11.32) are therefore suggested for practical implementation of D_{ij} and D_{ji}.

Note that the second multiplication term in cf_{ij} shown in (11.96) satisfies the two conditions presented in (11.68) and (11.69) and, therefore, can be regarded as a special integrator with relative degree of zero which can eliminate output error. This integrator can be practically implemented using a positive feedback control unit shown in Fig. 11.11.

For the other cases of the RHP zero distribution in det(G), as categorized in Table 11.2, the corresponding executable forms of C_s and C_f can be derived analytically following a similar procedure as above. The results are summarized in Table 11.2.

Note that for the first three cases in Table 11.2, D_{ij} (or D_{ji}) can also be factorized into two parts as $D_{ij} = G^{ij} D_{0,ij}$ (or $D_{ji} = G^{ji} D_{0,ji}$) such that only the second part, $D_{0,ij}$ (or $D_{0,ji}$), needs to be approximated for implementation. Such exercise can give better approximation accuracy but at the cost of a higher complexity for implementation. Generally, a low-order approximation, e.g., second-order, is preferred to facilitate practical application.

11.3.3 Robust Stability Analysis

In the nominal case $(G = \hat{G})$, it can be seen from Fig. 11.19 that assessing the control system stability can be limited to the closed-loop structure set between the process inputs and outputs, provided that the controller matrix C_s for the set-point tracking has been configured stable. As far as the closed-loop structure is concerned, the inputs are U, LD, Y_r, and N_O, and the outputs are Y and F. Note that U and LD have similar impact on the closed-loop structure and so do Y_r and N_O. Hence, assessing the nominal system stability can be further limited to the transfer matrix relating LD and N_O to Y and F, i.e.,

$$\begin{bmatrix} Y \\ F \end{bmatrix} = \begin{bmatrix} G(I + C_f G)^{-1} & -GC_f(I + GC_f)^{-1} \\ C_f G(I + C_f G)^{-1} & C_f(I + GC_f)^{-1} \end{bmatrix} \begin{bmatrix} LD \\ N_O \end{bmatrix} \quad (11.99)$$

Obviously, if all elements in the transfer matrix shown in (11.99) are maintained stable, the closed-loop internal stability can be guaranteed so that the overall control system stability can be ensured. To relieve the computation effort for checking the stability of all elements in the above transfer matrix, a simplified stability condition is given below based on the stability theorem developed in the conventional unity feedback control structure (Zhou et al. 1996):

Corollary 11.1. *The nominal control system shown in Fig. 11.19 maintains internal stability if and only if* $(I + C_f G)^{-1}$ *is stable.*

Proof. Note that there exist the following equivalent relationships:

$$C_f G(I + C_f G)^{-1} = I - (I + C_f G)^{-1} \quad (11.100)$$

$$C_f G(I + C_f G)^{-1} = C_f(I + GC_f)^{-1} G \quad (11.101)$$

Substituting (11.100) and (11.101) into (11.99), it can be seen that the stability condition in Corollary 11.1 guarantees the internal stability of the nominal control system for a stable MIMO process, G. This completes the proof. □

Note that the stability of $(I + C_f G)^{-1}$ can be verified by checking if $\det(I + C_f G)$ has any RHP zeros, which can be performed using the Nyquist curve criterion or any numerical solving method like the MATLAB toolbox.

In the presence of process uncertainties, robust stability analysis can also be limited to the closed-loop structure, owing to the open-loop control manner for the set-point tracking. The process additive (Δ_A), multiplicative input (Δ_I), and output uncertainties (Δ_O), as shown in Fig. 11.20, are commonly encountered in practical applications. By rearranging the perturbed closed-loop structure shown in Fig. 11.20 in the form of the standard $M - \Delta$ structure for robustness analysis (Zhou et al.

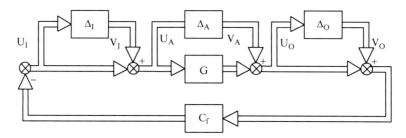

Fig. 11.20 The closed-loop structure with process additive, multiplicative input, and output uncertainties

1996), the transfer matrices relating the outputs to the inputs of Δ_I, Δ_A, and Δ_O can be derived as

$$\begin{bmatrix} U_I \\ U_A \\ U_O \end{bmatrix} = M \begin{bmatrix} V_I \\ V_A \\ V_O \end{bmatrix} \quad (11.102)$$

where

$$M = \begin{bmatrix} -(I+C_fG)^{-1}C_fG & -(I+C_fG)^{-1}C_f & -(I+C_fG)^{-1}C_f \\ (I+C_fG)^{-1} & -(I+C_fG)^{-1}C_f & -(I+C_fG)^{-1}C_f \\ (I+GC_f)^{-1}G & (I+GC_f)^{-1} & -(I+GC_f)^{-1}GC_f \end{bmatrix} \quad (11.103)$$

Note that there exist the following equivalent transformations:

$$(I+C_fG)^{-1}C_fG = I - (I+C_fG)^{-1} \quad (11.104)$$

$$(I+GC_f)^{-1} = I - (I+GC_f)^{-1}GC_f \quad (11.105)$$

Using the equivalent transformation,

$$C_f(I+GC_f) = (I+C_fG)C_f \quad (11.106)$$

yields

$$(I+C_fG)^{-1}C_f = C_f(I+GC_f)^{-1} \quad (11.107)$$

Besides, it follows from (11.101) that

$$(I+GC_f)^{-1}GC_f = G(I+C_fG)^{-1}C_f \quad (11.108)$$

11.3 A 2DOF Decoupling Control Scheme for MIMO Processes

Hence, it can be concluded by using (11.100), (11.101), and (11.104)–(11.108) that M retains stability if the nominal control system is maintained stable, i.e., $(I + C_f G)^{-1}$ is kept stable.

Then the multivariable spectral radius stability criterion (Skogestad and Postlethwaite 2005) can be used to obtain the robust stability constraint,

$$\rho(M\Delta) < 1, \forall \omega \in [0, \infty) \tag{11.109}$$

For instance, in the presence of the process additive uncertainties, the spectral radius stability constraint can be derived as

$$\rho\left((I + C_f G)^{-1} C_f \Delta_A\right) < 1, \forall \omega \in [0, \infty) \tag{11.110}$$

When assessing the control system robust stability with the process multiplicative input and output uncertainties, the spectral radius stability constraint can be derived as

$$\rho\left(\begin{bmatrix} -(I + C_f G)^{-1} C_f G & -(I + C_f G)^{-1} C_f \\ (I + G C_f)^{-1} G & -(I + G C_f)^{-1} G C_f \end{bmatrix} \begin{bmatrix} \Delta_I & 0 \\ 0 & \Delta_O \end{bmatrix}\right) < 1, \forall \omega \in [0, \infty) \tag{11.111}$$

For implementation, the spectral radius stability constraints shown in (11.110) and (11.111) can be checked graphically by observing if the magnitude plots of the spectral radius fall below unity for $\omega \in [0, +\infty)$. In this way, admissible tuning ranges of the adjustable parameters of C_f can be ascertained. Given a specified bound of Δ_A, Δ_I, or Δ_O in practice, (11.110) and (11.111) can be used to evaluate the control system robust stability, which will be illustrated in the examples later.

Combining the desired system transfer matrix shown in Table 11.2 with (11.82), one can see that when the adjustable parameter $\lambda_{c,i}$ ($i = 1, 2, \ldots, m$) in C_s is tuned to small, the corresponding ith output response becomes faster for the set-point tracking, but the outputs of the ith column controllers in C_s and the corresponding actuators will be required to be larger. More aggressive dynamic behavior of the ith output response will be turned out in the presence of process uncertainties. On the contrary, gradually increasing $\lambda_{c,i}$ will slow down the corresponding ith output response, but the outputs of the ith column controllers in C_s and the corresponding actuators can be reduced. Correspondingly, a less aggressive dynamic behavior of the ith system output response will occur in the presence of process uncertainties. Hence, tuning $\lambda_{c,i}$ ($i = 1, 2, \ldots, m$) aims at a good trade-off between the achievable set-point tracking performance and the output capacities of C_s and the corresponding actuators.

Following a similar analysis, one can conclude that decreasing the adjustable parameter $\lambda_{f,i}$ ($i = 1, 2, \ldots, m$) in C_f can enhance the closed-loop performance for rejecting load disturbance d_i entering into the ith process input, but the outputs of the ith row controllers in C_f and the corresponding actuators will be required to be larger, leading to lower robust stability of the closed-loop structure, and vice versa. Therefore, tuning $\lambda_{f,i}$ ($i = 1, 2, \ldots, m$) in C_f aims at a good trade-off between the nominal closed-loop performance for load disturbance rejection and its robust stability in the presence of process uncertainties.

Based on a large amount of simulation studies, it is recommended to initially take $\lambda_{c,i}$ and $\lambda_{f,i}$ in the range of $(1.0 - 10)\theta_{r,i}$ and $(1.0 - 10)\theta_{d,i}$ for $i = 1, 2, \ldots, m$, respectively. Then by varying them monotonically online, a desirable performance for the set-point tracking and load disturbance rejection can be reached in a transparent manner.

11.3.4 Illustrative Examples

Two examples from the existing literature are used to demonstrate the effectiveness and merits of the presented 2DOF decoupling control method, one with no RHP zero in $\det(G)$ and the other with a dual RHP zero in $\det(G)$.

Example 11.5. Consider again the 3×3 process of Example 11.3 for illustration. It has been introduced in Example 11.3 that this process belongs to case 1 in Table 11.1. For designing the controller matrix for the set-point tracking, it follows from the definitions in (11.85)–(11.87) that $\theta_{r,1} = 0.8$, $\theta_{r,2} = 0.68$, and $\theta_{r,2} = 1.85$ and $n_{r,1} = 1$, $n_{r,2} = 2$, and $n_{r,3} = 1$. According to the design formulas for case 1 in Table 11.2, the diagonal elements in the desired system response transfer matrix can be determined as

$$h_{r,1} = \frac{e^{-0.8s}}{\lambda_{c,1}s + 1}, \quad h_{r,2} = \frac{e^{-0.68s}}{(\lambda_{c,2}s + 1)^2}, \quad h_{r,3} = \frac{e^{-1.85s}}{\lambda_{c,3}s + 1}.$$

which are exactly the same as those of the presented MIMO decoupling control method for Example 11.3. Correspondingly, the controller matrix for the set-point tracking can be derived the same as those for Example 11.3, except for that the control unit, $f_i (i = 1, 2, 3)$, in all elements in the decoupling controller matrix should be removed.

For designing the closed-loop controller matrix for load disturbance rejection, it follows from the definitions in (11.93) and (11.94) that $\theta_{d,1} = 0.71$, $\theta_{d,2} = 1.85$, and $\theta_{d,3} = 1.59$ and $n_{d,1} = n_{d,3} = 1$ and $n_{d,2} = 2$. Thus, the closed-loop controller matrix can also be derived using the design formula for Case 1 in Table 11.2. For comparison with existing decoupling control methods (Wang et al. 2003; Huang and Lin 2006) in terms of similar controller orders, the controller elements in this controller matrix are derived as

$$cf_{11} = D_1 \cdot \frac{14543s^2 + 256.3578s + 0.5502}{(\lambda_{f,1}s + 1)(438.7353s + 1)}$$

11.3 A 2DOF Decoupling Control Scheme for MIMO Processes

$$cf_{12} = D_1 \cdot \frac{-5757900s^5 - 3439700s^4 - 562940s^3 - 41482s^2 - 526.426s - 0.296}{(\lambda_{f,1}s + 1)(615900s^4 + 116360s^3 + 12510s^2 + 466.7655s + 1)} e^{-3.76s}$$

$$cf_{13} = D_1 \cdot \frac{400930s^4 + 33536s^3 + 1342.3s^2 + 31.5279s + 0.2638}{(\lambda_{f,1}s + 1)(33025s^3 + 3869.9s^2 + 447.5041s + 1)} e^{-0.65s}$$

$$cf_{21} = D_2 \cdot \frac{12391s^3 + 746.2116s^2 + 9.7508s + 0.0199}{(\lambda_{f,2}s + 1)^2 (3940.3s^2 + 447.8424s + 1)} e^{-1.05s}$$

$$cf_{22} = D_2 \cdot \frac{13471000s^6 + 3306200s^5 + 892990s^4 + 117120s^3 + 6709.9s^2 + 142.0148s + 0.3149}{(\lambda_{f,2}s + 1)^2 (336570s^4 + 33465s^3 + 9959.2s^2 + 461.3811s + 1)} e^{-1.17s}$$

$$cf_{23} = D_2 \cdot \frac{16790s^3 + 1582.9s^2 + 39.2646s + 0.0885}{(\lambda_{f,2}s + 1)^2 (511.4853s^2 + 440.0233s + 1)}$$

$$cf_{31} = D_3 \cdot \frac{1736.5s^3 - 21.7287s^2 - 0.8474s - 0.002}{(\lambda_{f,3}s + 1)(4815.4s^2 + 449.8302s + 1)} e^{-2.99s}$$

$$cf_{32} = D_3 \cdot \frac{-197040s^5 - 104730s^4 - 29099s^3 - 4024.9s^2 - 171.9233s - 0.374}{(\lambda_{f,3}s + 1)(257300s^4 + 55907s^3 + 10254s^2 + 461.9346s + 1)} e^{-3.11s}$$

$$cf_{33} = D_3 \cdot \frac{2195s^3 + 212.3057s^2 + 5.2157s + 0.01}{(\lambda_{f,3}s + 1)(1319.1s^2 + 441.8636s + 1)}$$

$$D_1 = \frac{1}{1 - \frac{e^{-0.71s}}{\lambda_{f,1}s + 1}}, \quad D_2 = \frac{1}{1 - \frac{e^{-1.85s}}{(\lambda_{f,2}s + 1)^2}}, \quad D_3 = \frac{1}{1 - \frac{e^{-1.59s}}{\lambda_{f,3}s + 1}}$$

where D_1, D_2, and D_3 can be implemented using the control unit shown in Fig. 11.11.

For illustration, to obtain the similar rising speeds of set-point tracking with those of Wang et al. (2003) and Huang and Lin (2006), the adjustable parameters in C_s are taken as $\lambda_{c,1} = 8$, $\lambda_{c,2} = 10$, and $\lambda_{c,3} = 15$. To compare with Huang and Lin (2006) for load disturbance rejection in terms of a 2DOF control structure, the adjustable parameters in C_f are taken as $\lambda_{f,1} = 0.2$, $\lambda_{f,2} = 18$, and $\lambda_{f,3} = 15$.

By adding a unit step change to the three set-point inputs at $t = 0$ (s), 300 (s), and 600 (s), respectively, and then adding an inverse step change of load disturbance with a dynamics of $G_L = [1.986e^{-0.71s}/(66.7s + 1), -0.0204e^{-3.53s}/(11.49s + 1), -0.374e^{-7.75s}/(22.22s + 1)]^T$ to all the process outputs at $t = 900$ (s), as assumed by Huang and Lin (2006), the output responses are shown in Fig. 11.21. It is seen that the ternary set-point responses without overshoot have been obtained using the presented 2DOF decoupling control method (solid line), and the three process

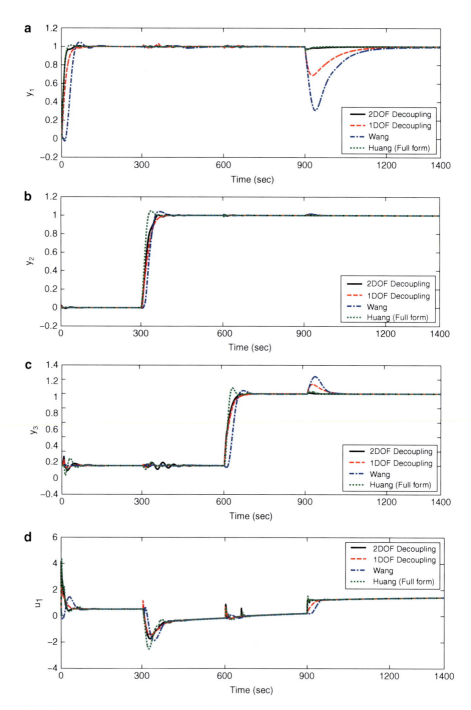

Fig. 11.21 Nominal output responses of Example 11.5

11.3 A 2DOF Decoupling Control Scheme for MIMO Processes

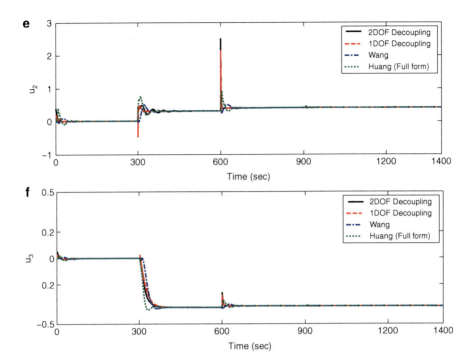

Fig. 11.21 (continued)

output responses are almost entirely decoupled from each other. Moreover, both the presented 2DOF control scheme and Huang and Lin (2006) have led to obviously enhanced load disturbance rejection owing to the use of a 2DOF control structure. The same set-point tracking performance is obtained by the presented 2DOF control method and the 1DOF decoupling method based on the unity feedback control structure, owing to the use of the same controller forms and adjustable parameters for the set-point tracking. Note that better system performance for both the set-point tracking and load disturbance rejection can be independently obtained in the presented 2DOF control scheme by monotonically decreasing the adjustable parameters of C_s and C_f.

To demonstrate the control system robust stability, one of the process parameter perturbation tests in Wang et al. (2003) is performed here. That is, all time constants in the process transfer matrix are assumed to be actually 40% larger to introduce the unmodeled process dynamics. The perturbed system output responses are shown in Fig. 11.22, which indicates that the presented 2DOF decoupling control system maintains robust stability well in the presence of the severe process parameter perturbations (solid line). The control signals have not varied much compared to the nominal case and thus are omitted. Note that better robust stability can be conveniently obtained in the presented 2DOF control system by monotonically increasing the adjustable parameters of C_f.

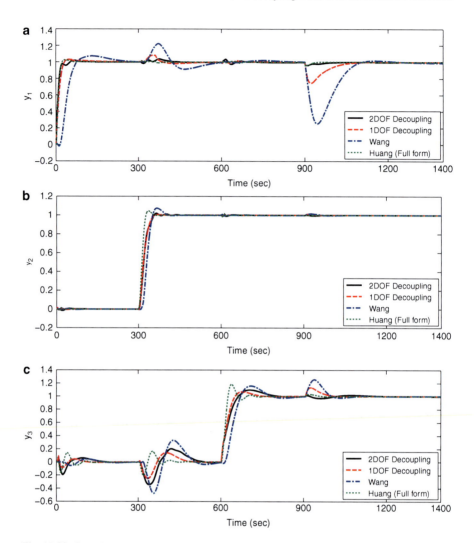

Fig. 11.22 Perturbed output responses of Example 11.5

Example 11.6. Consider the binary process studied in the literature (Jerome and Ray 1986; Wang et al. 2000b; Huang and Lin 2006),

$$G = \begin{bmatrix} \dfrac{(-s+1)e^{-2s}}{s^2+1.5s+1} & \dfrac{0.5(-s+1)e^{-4s}}{(2s+1)(3s+1)} \\ \dfrac{0.33(-s+1)e^{-6s}}{(4s+1)(5s+1)} & \dfrac{(-s+1)e^{-3s}}{4s^2+6s+1} \end{bmatrix}$$

11.3 A 2DOF Decoupling Control Scheme for MIMO Processes

Obviously there is a common RHP zero, $s = 1$, in each element in the process transfer matrix. Thus, it is a dual RHP zero of $\det(G)$. It can be verified that $\det(G)$ has no additional RHP zeros. It follows from the definitions in (11.85)–(11.87) that $\theta_{r,1} = 2$, $\theta_{r,2} = 3$, and $n_{r,1} = n_{r,2} = 1$. According to the design formula for case 2 in Table 11.2, the controller matrix for the set-point tracking can be derived as

$$C_s = D \cdot \begin{bmatrix} \dfrac{s^2 + 1.5s + 1}{(\lambda_{c,1}s + 1)(s + 1)} & -\dfrac{0.5(s^2 + 1.5s + 1)(4s^2 + 6s + 1)e^{-2s}}{(\lambda_{c,2}s + 1)(s + 1)(2s + 1)(3s + 1)} \\ -\dfrac{0.33(s^2 + 1.5s + 1)(4s^2 + 6s + 1)e^{-3s}}{(\lambda_{c,1}s + 1)(s + 1)(4s + 1)(5s + 1)} & \dfrac{4s^2 + 6s + 1}{(\lambda_{c,2}s + 1)(s + 1)} \end{bmatrix}$$

where

$$D = \dfrac{1}{1 - \dfrac{0.165(s^2+1.5s+1)(4s^2+6s+1)}{(2s+1)(3s+1)(4s+1)(5s+1)} e^{-5s}}$$

Note that D can be implemented using the control unit shown in Fig. 11.11.

For designing the closed-loop controller matrix for load disturbance rejection, it follows from the definitions in (11.93) and (11.94) that $\theta_{d,1} = 2$, $\theta_{d,2} = 3$, and $n_{d,1} = n_{d,2} = 1$. So the closed-loop controller matrix can be similarly derived as

$$C_f = D \cdot \begin{bmatrix} \dfrac{s^2 + 1.5s + 1}{(\lambda_{f,1}s + 1)(s + 1)} D_1 & -\dfrac{0.5(s^2 + 1.5s + 1)(4s^2 + 6s + 1)e^{-s}}{(\lambda_{f,1}s + 1)(s + 1)(2s + 1)(3s + 1)} D_1 \\ -\dfrac{0.33(s^2 + 1.5s + 1)(4s^2 + 6s + 1)e^{-4s}}{(\lambda_{f,2}s + 1)(s + 1)(4s + 1)(5s + 1)} D_2 & \dfrac{4s^2 + 6s + 1}{(\lambda_{f,2}s + 1)(s + 1)} D_2 \end{bmatrix}$$

where

$$D_1 = \dfrac{1}{1 - \dfrac{(-s+1)e^{-2s}}{(\lambda_{f,1}s+1)(s+1)}}, \quad D_2 = \dfrac{1}{1 - \dfrac{(-s+1)e^{-3s}}{(\lambda_{f,2}s+1)(s+1)}}.$$

Note that D_1 and D_2 can be implemented using the control unit shown in Fig. 11.11.

For comparison with Huang and Lin (2006), which had demonstrated superiority over Jerome and Ray (1986) and Wang et al. (2000b), the adjustable parameters of C_s are taken as $\lambda_{c,1} = 2$ and $\lambda_{c,2} = 2$ to obtain the similar rising speeds of set-point tracking, and the adjustable parameters of C_f are taken as $\lambda_{f,1} = 0.8$ and $\lambda_{f,2} = 1.5$ to obtain the similar load disturbance response peak.

By adding a unit step change to the binary set-point inputs at $t = 0$ (s) and 100 (s), respectively, and then adding an inverse step change of load disturbances with a dynamics of $G_L = [e^{-s}/(25s + 1), e^{-s}/(25s + 1)]^T$ to both the process outputs at $t = 200$ (s), as assumed by Huang and Lin (2006), the output responses are shown in Fig. 11.23. It is seen that the presented 2DOF decoupling control method results in entirely decoupled output responses with no overshoot for the set-point tracking (thick solid line). To allow a simple implementation of D included in

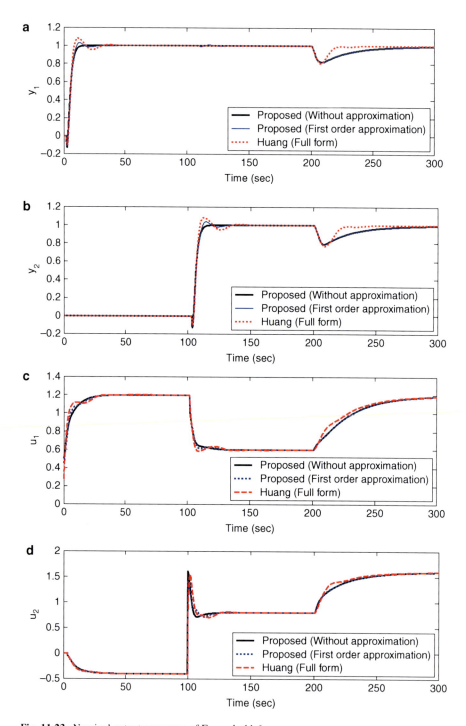

Fig. 11.23 Nominal output responses of Example 11.6

11.3 A 2DOF Decoupling Control Scheme for MIMO Processes

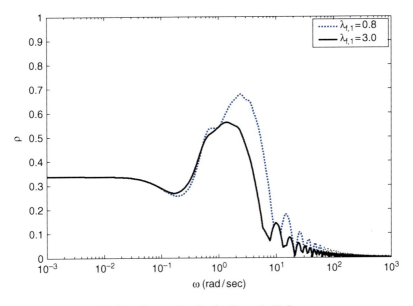

Fig. 11.24 The magnitude plots of spectral radius for Example 11.6

C_s and C_f, the first-order rational approximation formula in (11.33) is therefore used, obtaining $D_{1/1} = (8.227s + 1.1976)/(9.142s + 1)$. The corresponding output responses are also shown in Fig. 11.23 for comparison, which indicates that a negligible degradation in the output performance occurs.

To compare the control system robust stability, the perturbation test used in Wang et al. (2000b) is performed here. That is, the static gains and time delays in diagonal elements in the process transfer matrix are assumed to be actually increased by 20% and 30%, respectively. According to the robust stability constraint given in (11.110), the magnitude plot of the spectral radius for checking robust stability is shown in Fig. 11.24. It is seen that the peak value (dotted line) is evidently smaller than unity, indicating that the presented 2DOF decoupling control system can maintain good robust stability. The corresponding output responses are shown in Fig. 11.25. It is seen that obviously enhanced robust stability is therefore obtained, given the similar nominal output performance. Moreover, increasing the common adjustable parameter $\lambda_{c,1}$ in the first column controllers in C_s can gradually suppress the oscillation of the set-point response for the process output y_1, such as the case of $\lambda_{c,1} = 5$ shown in Fig. 11.25 (thick solid line). On the other hand, increasing the common adjustable parameter $\lambda_{f,1}$ in the first row controllers in C_f will gradually suppress the oscillation of load disturbance response for y_1, such as the case of $\lambda_{f,1} = 3$ shown in Fig. 11.25. Correspondingly, a smaller peak value of the spectral radius is seen, as shown also in Fig. 11.24 (solid line), which indicates further enhanced robust stability. Note that both the set-point response and load disturbance response of the process output y_2 have almost not been affected by the tuning of

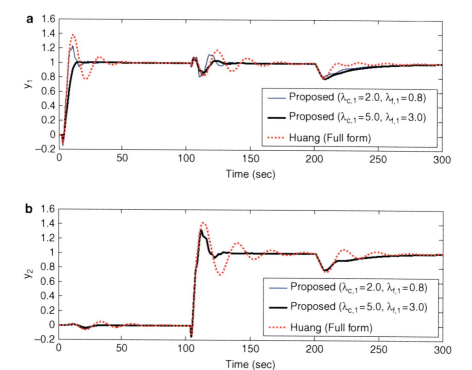

Fig. 11.25 Perturbed output responses of Example 11.6

$\lambda_{c,1}$ and $\lambda_{f,1}$. It is thus demonstrated that each of the adjustable parameters in C_s can be independently tuned online to optimize the set-point tracking performance of the corresponding output variable, and so does for tuning each of the adjustable parameters in C_f to optimize the load disturbance rejection performance of the corresponding output variable.

11.4 Summary

For the decoupling regulation of MIMO processes, in particular in the presence of multiple time delays involved with individual loops, three decoupling control methods have been presented, respectively, based on the IMC structure, the unity feedback control structure, and a 2DOF control structure.

For stable TITO processes, an analytical decoupling control design (Liu et al. 2006) based on the standard IMC structure has been presented, which can realize absolute decoupling regulation of the nominal binary output responses. Owing to the analytical design procedure, the computation effort is small in comparison with previous decoupling control methods based on numerical algorithms. Moreover,

11.4 Summary

there exists a quantitative tuning relationship between the adjustable control parameters and the nominal binary output responses, therefore facilitating the system operation in practice. Robust stability constraints developed in terms of the spectral radius stability criterion can be used to graphically determine an admissible tuning range of the single common adjustable parameter in each column controllers in the decoupling controller matrix, given specified upper bounds of the process additive or multiplicative uncertainties in practice.

For MIMO processes, an analytical decoupling controller matrix design (Liu et al. 2007a) has been presented in the framework of the unity feedback control structure that is widely used in engineering practice. It has been demonstrated that the proposed method can realize significant or even absolute decoupling regulation for the nominal system. The key lies with the formulation of a practically desired closed-loop system transfer matrix through an analysis of the NMP characteristics of the process inverse transfer matrix (i.e., G^{-1}). A new concept of "inverse relative degree" has been defined with respect to all elements in G^{-1}. Based on a classification of different cases of the RHP zero distribution in $\det(G)$, the decoupling controller matrix can be inversely derived from determining a desired closed-loop system transfer matrix. For practical implementation, an analytical approximation method has been given to achieve the ideally desired decoupling controller matrix, efficiently reducing the computation effort compared to existing decoupling control methods based on numerical iteration algorithms. Stability has been analyzed for both the nominal system and the perturbed system with process additive, multiplicative input, and output uncertainties. Tuning the decoupling controller matrix to meet a good trade-off between the nominal system performance and its robust stability can be conveniently executed owing to that each column controllers in the decoupling controller matrix are tuned in common by a single adjustable parameter in a monotonic manner.

To further enhance system performance for the set-point tracking and load disturbance rejection, a 2DOF decoupling control scheme (Liu et al. 2007b) has been presented for MIMO processes with multiple time delays. In the result, both the set-point tracking and load disturbance rejection can be separately regulated and optimized online, while significant or even absolute decoupling regulation performance can be obtained. In the controller matrix for the set-point tracking, each column controllers are tuned in common by a single adjustable parameter, while in the closed-loop controller matrix for load disturbance rejection, each row controllers are tuned in common by a different single adjustable parameter. Each of these adjustable parameters can be tuned online in a monotonic manner to reach the best trade-off between the nominal performance of the control system and its robust stability, therefore facilitating practical application. Robust tuning constraints have been derived in terms of the multivariable spectral radius stability criterion, which can be graphically checked through the corresponding magnitude plots.

Five examples from the existing literature have been used to illustrate these three decoupling control methods. The effectiveness and merits of these decoupling control methods have been well demonstrated in terms of different cases of the RHP zero distribution in the transfer function matrices of these examples.

References

Alevisakis G, Seborg DE (1973) An extension of the Smith predictor to multivariable linear systems containing time delays. Int J Control 3(17):541–557

Åström KJ, Hägglund T (1995) PID controller: theory, design, and tuning, 2nd edn. ISA Society of America, Research Triangle Park

Åström KJ, Johansson KH, Wang QG (2002) Design of decoupled PI controllers for two-by-two systems. IEE Process Control Theory Appl 149(1):74–81

Chen D, Seborg DE (2003) Design of decentralized PI control systems based on Nyquist stability analysis. J Process Control 13(1):27–39

Desbiens A, Pomerleau A, Hodouin D (1996) Frequency based tuning of SISO controllers for two-by-two processes. IEE Process Control Theory Appl 143(1):49–56

Gilbert AF, Yousef A, Natarajan K, Deighton S (2003) Tuning of PI controllers with one-way decoupling in 2×2 MIMO systems based on finite frequency response data. J Process Control 13(6):553–567

Holt BR, Morari M (1985) Design of resilient processing plants-V. The effect of deadtime on dynamic resilience. Chem Eng Sci 40(7):1229–1237

Huang HP, Lin FY (2006) Decoupling multivariable control with two degrees of freedom. Ind Eng Chem Res 45(9):3161–3173

Jerome NF, Ray WH (1986) High-performance multivariable control strategies for systems having time delays. AICHE J 32(6):914–931

Jerome NF, Ray WH (1992) Model-predictive control of linear multivariable systems having time delays and right-half-plane zeros. Chem Eng Sci 47(4):763–785

Lee J, Kim DH, Edgar TF (2005) Static decouplers for control of multivariable processes. AICHE J 51(10):2712–2720

Limebeer DJN, Kasenally EM, Perkins JD (1993) On the design of robust two degree of freedom controllers. Automatica 29(1):157–168

Liu T, Zhang W, Gu DY (2006) Analytical design of decoupling internal model control (IMC) scheme for two-input-two-output (TITO) processes with time delays. Ind Eng Chem Res 45(9):3149–3160

Liu T, Zhang W, Gao F (2007a) Analytical decoupling control strategy using a unity feedback control structure for MIMO processes with time delays. J Process Control 17(2):173–186

Liu T, Zhang W, Gao F (2007b) Analytical two-degrees-of-freedom (2-DOF) decoupling control scheme for multiple-input-multiple-output (MIMO) processes with time delays. Ind Eng Chem Res 46(20):6546–6557

Lundström P, Skogestad S (1999) Two-degree-of-freedom controller design for an ill-conditioned distillation process using μ-synthesis. IEEE Trans Autom Control 7(1):12–21

Luyben WL (1990) Process modeling, simulation, and control for chemical engineers. McGraw Hill, New York

Morari M, Zafiriou E (1989) Robust process control. Prentice Hall, Englewood Cliff

Ogunnaike BA, Ray WH (1979) Multivariable controller design for linear systems having multiple time delays. AICHE J 25(6):1043–1056

Ogunnaike BA, Ray WH (1994) Process dynamics, modeling, and control. Oxford University Press, New York

Perng MH, Ju JS (1994) Optimally decoupled robust control MIMO plants with multiple delays. IEE Process Control Theory Appl 141(1):49–56

Pomerleau D, Pomerleau A (2001) Guide lines for the tuning and the evaluation of decentralized and decoupling controllers for processes with recirculation. ISA Trans 40(4):341–351

Prempain E, Bergeon B (1998) A multivariable two-degree-of-freedom control methodology. Automatica 34(12):1601–1606

Seborg DE, Edgar TF, Mellichamp DA (2004) Process dynamics and control, 2nd edn. Wiley, Hoboken

Shinskey FG (1996) Process control system, 4th edn. McGraw Hill, New York

References

Shiu SJ, Hwang SH (1998) Sequential design method for multivariable decoupling and multiloop PID controllers. Ind Eng Chem Res 37(1):107–119

Skogestad S, Postlethwaite I (2005) Multivariable feedback control: analysis and design, 2nd edn. Wiley, Chichester

Toh WH, Rangaiah GP (2002) A methodology for autotuning of multivariable systems. Ind Eng Chem Res 41(18):4605–4615

Tyreus BD (1979) Multivariable control system design for an industrial distillation column. Ind Eng Chem Process Des Dev 18(2):177–182

Waller M, Waller JB, Waller KV (2003) Decoupling revisited. Ind Eng Chem Res 42(20):4575–4577

Wang QG, Huang B, Guo X (2000a) Auto-tuning of TITO decoupling controllers from step tests. ISA Trans 39(4):407–418

Wang QG, Zou B, Zhang Y (2000b) Decoupling Smith predictor design for multivariable systems with multiple time delays. Chem Eng Res Des Trans Inst Chem Eng Part A 78(4):565–572

Wang QG, Zhang Y, Chiu MS (2003) Non-interacting control design for multivariable industrial processes. J Process Control 13(3):253–265

Watanabe K, Ishiyama Y, Ito M (1983) Modified Smith predictor control for multivariable systems with delays and unmeasurable step disturbances. Int J Control 37(5):959–973

Wood RK, Berry MW (1973) Terminal composition control of binary distillation column. Chem Eng Sci 28(10):1707–1717

Zhou KM, Doyle JC, Glover K (1996) Robust and optimal control. Prentice Hall, Englewood Cliff

Chapter 12
Batch Process Control

12.1 The Implementation Requirements

Batch processes have been widely applied in modern industries to manufacture a large quantity of products with good consistency and high efficiency. Typical batch processes include robotic manipulators, semiconductor product lines, injection molding, pharmaceutical crystallization, etc. Generally, a batch process is defined as "a process that leads to the production of finite quantities of material by subjecting quantities of input materials to an ordered set of processing activities over a finite period of time using one or more pieces of equipment" (Instrument Society of America 1995).

For batch process operation, there are fundamental requirements as follows:

1. A sequential execution of the processing activities to turn out expected products or output performance
2. A finite operating time in each batch, specifically called cycle
3. Resetting initial process conditions to zero or fixed nonzero values for running each cycle

Accordingly, the control tasks for batch process operation are

1. Realize perfect tracking of the desired set-point trajectory in each cycle for time-invariant batch processes
2. Eliminate the influence of repetitive load disturbance occurring from cycle to cycle
3. Robustly track the desired set-point trajectory as close as possible, in the presence of time-varying uncertainties in each cycle or cycle-to-cycle uncertainties
4. Maintain control system robust stability against process uncertainties in both the time direction in each cycle and the batchwise direction from cycle to cycle
5. Comply with the process input and output constraints for implementation

To meet the above requirements, different control methodologies have been developed in the past three decades (Bonvin et al. 2006; Wang et al. 2009). Among these methodologies, iterative learning control (ILC) has been widely recognized and practiced in recent years for various industrial and chemical batch processes (Moore 1993; Ahn et al. 2007; Wang et al. 2009). This methodology is in principle based on using repetitive operating information of a batch process from historical cycles to progressively improve tracking performance from cycle to cycle. As surveyed by Bonvin et al. (2006), Ahn et al. (2007), and Wang et al. (2009), quite a number of ILC methods have been developed in both continuous- and discrete-time domains which can realize perfect tracking for time-invariant linear or nonlinear batch processes. Presently, the challenges for practical application of ILC are primarily related to robust convergence and stability against process uncertainties.

To deal with process uncertainties, a number of robust ILC methods have been developed in recent years for delay-free or fixed-delay batch processes (Lee et al. 2000; Xiong and Zhang 2003; Shi et al. 2005, 2006; Harte et al. 2005; Nagy and Braatz 2003; Nagy et al. 2007; Wijdeven et al. 2009). As time delay is usually associated with process operation, which may be uncertain or even time-varying from cycle to cycle, Xu et al. (2001) extended the Smith predictor control structure that is well known for superior control of a time delay SISO process to improve tracking performance of ILC for a batch process with time delay. A state-space ILC method (Li et al. 2005) was developed to allow for fixed state and control delays, based on a two-dimensional (2D) linear continuous-discrete Roesser's model of the process. By comparison, Tan et al. (2009) developed a phase lag compensation method to perform ILC in the presence of the input delay.

In view of that time-delay mismatch is usually involved with other process uncertainties in operating a time-delay batch process in practical applications, a robust ILC method based on the IMC structure is therefore presented for practical application. Sufficient conditions for the convergence of ILC are explored for time-delay batch processes with or without model uncertainties. To facilitate the controller design, a unified controller structure using the standard IMC controller form is proposed for implementation.

12.2 An IMC-Based Iterative Learning Control (ILC) Scheme

It is well known that perfect tracking can be obtained by the standard IMC structure (Morari and Zafiriou 1989) only when there exists no process uncertainty and load disturbance. Given a constant or step-type set-point as commonly practiced, the IMC structure can be used to guarantee no steady-state offset for a batch process if the cycle time (T_P) is long enough. Moreover, the IMC structure can be used to maintain the closed-loop system robust stability, that is to say, accommodating process uncertainties. These merits of IMC motivate the development of an IMC-based ILC control scheme here. Note that given any bounded signals entering into an IMC system subject to process uncertainties, including the feedforward control signal of ILC that may be independently designed for perfect tracking, only bounded output will be in the result if the IMC system maintains robust stability.

12.2 An IMC-Based Iterative Learning Control (ILC) Scheme

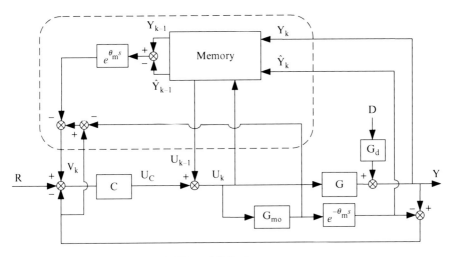

Fig. 12.1 A block diagram of the IMC-based ILC scheme

12.2.1 The IMC-Based ILC Structure

The IMC-based ILC control scheme is shown in Fig. 12.1, where the transfer function block diagram encircled by a dashed line is the ILC part, and the rest is the standard IMC structure; R (i.e., r in time domain) denotes the set-point, Y (i.e., y in time domain) is the process output, and D is the load disturbance. C is the IMC controller and also used for implementation of ILC, $G_m = G_{mo}e^{-\theta s}$ is the process model, and G_d is the load disturbance transfer function. "Memory" is a storage used for recording the current cycle information of process output (Y_k), model output ($\hat{Y}_k = G_m U_k$), and control input (U_k) to the process, while providing the last cycle information of process output (y_{k-1}), model output ($\hat{y}_{k-1}$), and control input (U_{k-1}). V_k is the ILC updating information used to compute the control increment (U_C) for adjusting U_k.

For implementation, the initial run ($k = 1$) of this control scheme is exactly an IMC strategy with a zero initialization of the ILC control law (i.e., $V_1 = U_0 = 0$). Starting from the second run ($k \geq 2$), the whole control structure is implemented as an ILC scheme, based on the IMC control law (U_1) stored in the "Memory." The key idea behind such an implementation is that, based on an IMC control law prescribed to allow for the process uncertainties with certain stability margin, the ILC control law, which is essentially of feedforward control, is subsequently implemented to progressively realize perfect tracking, according to the linear superposition principle. Hence, relative independence can be obtained for designing the IMC control law to maintain the control system robust stability and the ILC control law to realize perfect tracking, respectively.

A time-delay process denoted as G in Fig. 12.1 is generally modeled in the transfer function form of

$$G_m(s) = G_{mo}e^{-\theta_m s}, \qquad (12.1)$$

where G_{mo} is a rational proper transfer function and θ_m indicates the process response delay. For operation of open-loop stable batch processes, as considered here for ILC design, one can further express G_{mo} as

$$G_{mo}(s) = k_p \frac{B_+(s)B_-(s)}{A(s)}, \qquad (12.2)$$

where k_p denotes the process static gain, $A(0) = B_+(0) = B_-(0) = 1$, all zeros of $A(s)$ and $B_-(s)$ are located in the complex LHP, and all zeros of $B_+(s)$ are located in the complex RHP. Denote $\deg\{A(s)\} = m$, $\deg\{B_-(s)\} = n_1$, and $\deg\{B_+(s)\} = n_2$, respectively. Generally, there exists $n_1 + n_2 < m$ in practice, which indicates that G_{mo} is strictly proper.

Denote Y_d (i.e., y_d in time domain) as the desired output trajectory. It is assumed that the initial resetting condition is satisfied, i.e., $r(0) = y_d(0) = y_k(0)$, where k denotes the cycle number. Without loss of generality, zero initial values of $y_d(0) = y_k(0) = 0$ are considered here for the convenience of analysis, since nonzero initial constant values can be easily converted to zero for study.

12.2.2 The IMC Design

With the process description in (12.1) and (12.2), in order to achieve the H_2 optimal set-point tracking performance, it follows from the IMC design presented in Sect. 7.3 that the controller should be configured as

$$C_{IMC}(s) = \frac{A(s)}{k_p B_-(s) B_+^*(s)(\lambda_c s + 1)^{m-n_1}}, \qquad (12.3)$$

where λ_c is an adjustable parameter for controller tuning and $B_+^*(s)$ denotes the complex conjugate of $B_+(s)$, corresponding to all zeros in LHP. It can be easily verified that the above controller is bi-proper and therefore executable in practice.

Assume that the process is located in a family of $\Pi = \{G : |G(j\omega) - G_m(j\omega)| / |G_m(j\omega)| \leq |\Delta_m(j\omega)|\}$, where $\Delta_m(j\omega)$ denotes the process multiplicative uncertainty. According to the small-gain theory (Zhou et al. 1996), the IMC system maintains robust stability if and only if

$$|\Delta_m(j\omega)T(j\omega)| < 1, \quad \forall \omega \in [0, \infty), \qquad (12.4)$$

where $T = G_m C_{IMC}$ is the closed-loop system transfer function in the nominal case ($G = G_m$).

12.2 An IMC-Based Iterative Learning Control (ILC) Scheme

Substituting (12.3) and the process model of (12.1) into (12.4) yields the robust tuning constraint,

$$(\lambda_c^2 \omega^2 + 1)^{\frac{m-n_1}{2}} > |\Delta_m(j\omega)|, \quad \forall \omega \in [0, \infty). \tag{12.5}$$

Hence, given a specified upper bound of $\Delta_m(j\omega)$ in practice, the closed-loop system stability can be assessed graphically by observing if the magnitude plot of the left-hand side of (12.5) is above that of the right-hand side for $\omega \in [0, \infty)$.

To facilitate the subsequent implementation of ILC for perfect tracking, the above IMC design for an initial run of the proposed control scheme should of primary aim at holding the control system stability with good robustness margin.

12.2.3 Robust ILC Design

The objective of an ILC control law is to progressively achieve perfect tracking along the cycles, which can be mathematically expressed as

$$\lim_{k \to \infty} \|y_d - y_k\| = 0, \tag{12.6}$$

where y_d denotes the desired time-domain output trajectory which is usually identical with the set-point (r) unless the set-point is not a continuous signal. For instance, if r is a step change as often adopted in practice, a low-pass filter can be used to turn out a smooth trajectory, i.e., $Y_d = FR$, where F can be practically taken as a first-order stable transfer function with a small time constant.

Due to the existence of the process response delay, the above objective should be modified as

$$\lim_{k \to \infty} \|y_d e^{-\theta s} - y_k\| = 0, \tag{12.7}$$

where θ denotes the process response delay.

If θ cannot be known exactly in practice, the objective function for design of an ILC updating law is proposed accordingly as

$$\lim_{k \to \infty} \|y_d e^{-\theta_m s} - y_k\| = 0. \tag{12.8}$$

Since the ILC strategy shown in Fig. 12.1 is essentially of feedforward control based on using the last cycle information of process output (Y_{k-1}) and control input (U_{k-1}), the control input in the current cycle can be derived as

$$U_k = U_{k-1} + C\left[R - \left(Y_k - \hat{Y}_k\right) + V_k\right]. \tag{12.9}$$

Considering that the IMC control law has been stored as U_1 after an initial run of the IMC strategy as shown in Fig. 12.1, the output error information, $R - \left(Y_k - \hat{Y}_k\right)$, should be removed from the above ILC updating information in the current cycle so that relative independence can be obtained for designing the control system stability (through IMC) and the convergence rate (through ILC) for perfect tracking.

It follows from (12.8) that the tracking error in the current cycle can be evaluated as

$$E_k = Y_d e^{-\theta_m s} - Y_k \tag{12.10}$$

which can also be assessed in a predictive manner by multiplying $e^{\theta_m s}$ to both sides of (12.10),

$$E_k e^{\theta_m s} = Y_d - Y_k e^{\theta_m s}. \tag{12.11}$$

Due to the fact that $Y_k e^{\theta_m s}$ cannot be known in advance for computation, the above error prediction can be estimated using the model response as

$$\hat{E}_k e^{\theta_m s} = Y_d - \hat{Y}_k e^{\theta_m s} \tag{12.12}$$

which can therefore be used to compute the control increment (U_C) for adjusting U_k so that perfect tracking can be realized against the influence from the process time delay.

Besides, the influence of a repetitive-type load disturbance, which can be estimated from the last cycle as

$$G_d D = Y_{k-1} - \hat{Y}_{k-1}, \tag{12.13}$$

where $\hat{Y}_{k-1} = G_m U_{k-1}$, should be included in the ILC updating information (V_k) in the current cycle for load disturbance rejection.

Based on the above analysis, the ILC updating information is proposed as

$$V_k = Y_d - R + \left(Y_k - \hat{Y}_k\right) - e^{\theta_m s}\left(\hat{Y}_k + Y_{k-1} - \hat{Y}_{k-1}\right). \tag{12.14}$$

By substituting (12.14) into (12.9), the ILC control law can be derived as

$$U_k = U_{k-1} + C\left[Y_d - e^{\theta_m s}\left(\hat{Y}_k + Y_{k-1} - \hat{Y}_{k-1}\right)\right]. \tag{12.15}$$

In the case where $Y_d = R$, as shown in Fig. 12.1, (12.14) and (12.15) can be simplified, respectively, as

$$V_k = Y_k - \hat{Y}_k - e^{\theta_m s}\left(\hat{Y}_k + Y_{k-1} - \hat{Y}_{k-1}\right), \tag{12.16}$$

12.2 An IMC-Based Iterative Learning Control (ILC) Scheme

$$U_k = U_{k-1} + C\left[R - e^{\theta_m s}\left(\hat{Y}_k + Y_{k-1} - \hat{Y}_{k-1}\right)\right]. \qquad (12.17)$$

Multiplying both sides of (12.15) by G and using the following equalities,

$$Y_k = GU_k + G_d D, \qquad (12.18)$$

$$G_m(Y_k - Y_{k-1}) = G\left(\hat{Y}_k - \hat{Y}_{k-1}\right), \qquad (12.19)$$

together with the feature of a repetitive-type load disturbance,

$$G_d D = Y_k - \hat{Y}_k = Y_{k-1} - \hat{Y}_{k-1}, \qquad (12.20)$$

one can obtain

$$Y_k = \frac{1 + C\left(G_{mo} - Ge^{\theta_m s}\right)}{1 + G_{mo}C} Y_{k-1} + \frac{GC}{1 + G_{mo}C} Y_d. \qquad (12.21)$$

It follows from (12.11) that

$$Y_k = Y_d e^{-\theta_m s} - E_k, \qquad (12.22)$$

$$Y_{k-1} = Y_d e^{-\theta_m s} - E_{k-1}. \qquad (12.23)$$

Substituting (12.22) and (12.23) into (12.21) yields

$$E_k = \frac{1 + C\left(G_{mo} - Ge^{\theta_m s}\right)}{1 + G_{mo}C} E_{k-1}. \qquad (12.24)$$

It can be easily verified that (12.24) holds as well in the case where $Y_d = R$. Denote the transfer function in (12.24) as

$$Q(s) = 1 - \frac{GCe^{\theta_m s}}{1 + G_{mo}C}. \qquad (12.25)$$

It can be seen that the tracking error will not be enlarged from cycle to cycle, if C is designed to keep $Q(s)$ stable.

Note that $Q(s)$ in (12.25) is free of delay in the denominator, thus inheriting the important merit of a Smith predictor scheme for obtaining a linear characteristic equation for a time-delay process. Hence, the closed-loop pole assignment can be easily configured in terms of the dynamic response requirements of system operation.

A sufficient condition to the convergence of the proposed ILC method is given in the following theorem:

Theorem 12.1. *If* $\|Q(s)\|_\infty = \sup_{\omega \in [0,\infty)} |Q(j\omega)| < 1$, *perfect tracking, i.e.,* $\lim_{k \to \infty} E_k = 0$, *can be realized by the ILC scheme shown in Fig. 12.1.*

Proof. Taking the 2-norm for both sides of (12.24) gives

$$\|E_k\|_2 = \left\| \frac{1 + C\left(G_{mo} - Ge^{\theta_m s}\right)}{1 + G_{mo}C} \right\|_2 \|E_{k-1}\|_2. \qquad (12.26)$$

According to the well-known Parseval's theorem, there exists

$$\|E_k(j\omega)\|_2 = \|e_k(t)\|_2, \qquad (12.27)$$

where $e_k(t) = y_d(t) - y_k(t)$ is the time domain tracking error in the kth cycle. Note that the following norm relationship holds for a SISO system,

$$\left\| \frac{1 + C\left(G_{mo} - Ge^{\theta_m s}\right)}{1 + G_{mo}C} \right\|_2 \leq \left\| \frac{1 + C\left(G_{mo} - Ge^{\theta_m s}\right)}{1 + G_{mo}C} \right\|_\infty. \qquad (12.28)$$

Substituting (12.25), (12.27), and (12.28) into (12.26) yields

$$\|e_k(t)\|_2 \leq \|Q(s)\|_\infty \|e_{k-1}(t)\|_2. \qquad (12.29)$$

Using the induction method to (12.29), it follows that

$$\|e_k(t)\|_2 \leq \|Q(s)\|_\infty^k \|e_0(t)\|_2. \qquad (12.30)$$

Hence, using the sufficient condition given in Theorem 12.1, the conclusion follows accordingly. This completes the proof. □

In the nominal case, $G = G_m$, a sufficient condition to the convergence of this ILC method can be similarly obtained from (12.25) as

Corollary 12.1. *If* $\sup_{\omega \in [0,\infty)} |1/(1 + G_{mo}C)| < 1$, *perfect tracking, i.e.,* $\lim_{k \to \infty} E_k = 0$, *can be realized by the ILC scheme shown in Fig. 12.1 for batch processes without model uncertainty* $(G = G_m)$.

If a batch process is in a family described by a multiplicative uncertainty, $\Pi = \{G : |G(j\omega) - G_m(j\omega)| / |G_m(j\omega)| \leq |\Delta_m(j\omega)|\}$, a sufficient condition to the convergence of the proposed ILC method can be obtained from (12.25) and (12.30) as

12.2 An IMC-Based Iterative Learning Control (ILC) Scheme

Corollary 12.2. *If* $|1 - G_{mo}(j\omega)C(j\omega)\Delta_m(j\omega)| < |1 + G_{mo}(j\omega)C(j\omega)|$, $\forall \omega \in [0, \infty)$, *perfect tracking, i.e.,* $\lim_{k \to \infty} E_k = 0$, *can be realized by the ILC scheme shown in Fig. 12.1 for a batch process in the family of* $\Pi = \{G : |G(j\omega) - G_m(j\omega)| / |G_m(j\omega)| \le |\Delta_m(j\omega)|\}$.

With the sufficient conditions given in the above theorem and corollaries for convergence analysis, one can go on to analyze whether the above IMC controller can be used for implementation of ILC or not.

In the nominal case, $G = G_m$, denote

$$S_2(s) = \frac{1}{1 + G_{mo}C}. \tag{12.31}$$

Substituting (12.3) into (12.31) yields

$$S_2(s) = \frac{1}{1 + \frac{B_+(s)}{B_+^*(s)(\lambda_c s + 1)^{m-n_1}}}. \tag{12.32}$$

It can be easily verified that

$$S_2(0) = \frac{1}{2}, \tag{12.33}$$

$$S_2(\infty) = 1. \tag{12.34}$$

Therefore, the sufficient condition given in Corollary 12.1 can only be satisfied in the low-frequency range. In fact, the tracking error from cycle to cycle will not contain high-frequency components if the initial run of IMC does not generate an oscillatory output response containing high-frequency components and $Q(s)$ in (12.25) is maintained stable with a proper design of C. Moreover, the desired output trajectory in practical applications is usually prescribed in the low-frequency range, that is to say, accurate tracking is usually required within the closed-loop bandwidth (ω_b) of interest for system operation.

Note that the convergence rate is fixed as shown in (12.33) for the direct-current (DC) component of the tracking error if the above IMC controller is used, which can heavily affect the overall convergence rate of ILC because the DC component usually takes a large percentage in the desired output trajectory such as a step-type profile. To overcome this deficiency, the above IMC controller is slightly modified for implementation of ILC, i.e.,

$$C_{ILC}(s) = \frac{k_c A(s)}{k_p B_-(s) B_+^*(s)(\lambda_c s + 1)^{m-n_1}}. \tag{12.35}$$

Correspondingly, it follows that

$$S_2(0) = \frac{1}{1+k_c}. \tag{12.36}$$

It is seen that $S_2(0)$ is inversely proportional to k_c in C_{ILC}.

Compared to the IMC controller shown in (12.3), there exists an adjustable gain, k_c, in the above ILC controller. By letting $k_c = 1$, the above ILC controller is reduced to the IMC controller in (12.3). Therefore, the ILC controller shown in (12.36) can be taken as a unified form for implementation of the IMC-based ILC scheme.

In the case where the process is in the family of $\Pi = \{G : |G(j\omega) - G_m(j\omega)| / |G_m(j\omega)| \le |\Delta_m(j\omega)|\}$, one can denote

$$S_3(s) = \frac{1 - G_{mo} C \Delta_m}{1 + G_{mo} C}. \tag{12.37}$$

It follows from substituting (12.3) into (12.37) that

$$S_3(0) = \frac{1 - \Delta_m(0)}{2}. \tag{12.38}$$

which implies that $-1 < \Delta_m(0) < 3$ is required to ensure $|S_3(0)| < 1$ for the convergence of ILC.

If the above ILC controller is used, it follows that

$$S_3(0) = \frac{1 - k_c \Delta_m(0)}{1 + k_c}. \tag{12.39}$$

The first derivative with respect to k_c can be obtained as

$$\frac{d S_3(0)}{d k_c} = \frac{-1 - \Delta_m(0)}{(1 + k_c)^2}. \tag{12.40}$$

It can be verified that

$$\min |S_3(0)| = \begin{cases} 1, & \Delta_m(0) \le -1; \\ |\Delta_m(0)|, & -1 < \Delta_m(0) < 0; \\ 0, & \Delta_m(0) > 0. \end{cases} \tag{12.41}$$

Hence, it can be concluded from (12.41) that $\Delta_m(0)$ is not allowed to be smaller than negative unity for the convergence of ILC. Note that for $-1 < \Delta_m(0) < 0$, $S_3(0)$ monotonically decreases with respect to k_c and $\min |S_3(0)|$ can be reached only when $k_c \to \infty$. In the case where $\Delta_m(0) > 0$, $\min |S_3(0)| = 0$ can be reached when $k_c = 1/\Delta_m(0)$.

12.2 An IMC-Based Iterative Learning Control (ILC) Scheme

To quantify the tuning constraint of the adjustable time constant, λ_c, in the above ILC controller, a necessary condition for the convergence of ILC can be used,

$$|Q(j\omega)| < 1, \quad \forall \omega \leq \omega_b, \tag{12.42}$$

where ω_b denotes the closed-loop bandwidth of the IMC system, which can be estimated from the nominal closed-loop sensitivity function, i.e.,

$$|1 - G_m(j\omega_b)C_{IMC}(j\omega_b)| \leq \frac{1}{\sqrt{2}}. \tag{12.43}$$

Substituting (12.3) into (12.43) yields

$$\left|1 - \frac{B_+(j\omega_b)e^{-j\theta_m \omega_b}}{B_+^*(j\omega_b)(j\lambda_c\omega_b + 1)^{m-n_1}}\right| \leq \frac{1}{\sqrt{2}}. \tag{12.44}$$

Based on the above estimation of the closed-loop bandwidth, a tuning constraint of λ_c for the above ILC controller can be quantitatively established.

In the nominal case, $G = G_m$, by substituting (12.31) and (12.35) into (12.42), one can obtain

$$\left|1 + \frac{k_c B_+(j\omega)}{B_+^*(j\omega)(j\lambda_c\omega + 1)^{m-n_1}}\right| > 1, \quad \forall \omega \leq \omega_b \tag{12.45}$$

In the case where the process is in the family of $\Pi = \{G : |G(j\omega) - G_m(j\omega)|/|G_m(j\omega)| \leq |\Delta_m(j\omega)|\}$, by substituting (12.35) and (12.37) into (12.42), one can obtain

$$\left|1 + \frac{k_c B_+(j\omega)}{B_+^*(j\omega)(j\lambda_c\omega + 1)^{m-n_1}}\right| > \left|1 - \frac{k_c B_+(j\omega)\Delta_m(j\omega)}{B_+^*(j\omega)(j\lambda_c\omega + 1)^{m-n_1}}\right|, \quad \forall \omega \leq \omega_b \tag{12.46}$$

Note that the above tuning constraints can be graphically checked similar to the robust tuning constraint in (12.5). Therefore, the convergence performance can be assessed intuitively.

Owing to the unified controller form for both IMC and ILC in the presented control scheme, the adjustable parameters, λ_c and k_c, can be conveniently tuned to satisfy the robust stability constraint of IMC and the convergence conditions of ILC, respectively. For the convenience of implementation, λ_c may be first tuned in terms of the robust stability constraint of IMC and then used for ILC. If the convergence rate is preferred to be faster, k_c can be subsequently increased for this purpose. If the corresponding convergence stability cannot be guaranteed, λ_c may be retuned in terms of the convergence conditions shown in (12.42)–(12.46).

12.2.4 Implementation Against Measurement Noise

Measurement noise may bring false output error into the updating information of ILC, causing confusion in the control action or even jeopardize the control system stability. It is therefore necessary to evaluate the ILC robustness against measurement noise for practical application.

It can be seen from (12.14) that the ILC updating information is likely affected by measurement noise, which may hinder the convergence as can be verified from the ILC control law shown in (12.15). Nevertheless, the proposed controller formula in (12.35) has a low-pass property, which facilitates reducing the influence of measurement noise.

To enhance the convergence robustness against measurement noise, it is suggested to use an online noise-spike filtering strategy (Seborg et al. 2004) for filtering Y_k in the current cycle. In view of that all the process output data and ILC control law in the last cycle are available when computing the ILC updating law in the current cycle, an off-line denoising strategy can be used for the computation through a low-pass Butterworth filter,

$$\text{Butter}(n_f, f_c) = \frac{b_0 + b_1 z^{-1} + b_2 z^{-2} + \cdots + b_{n_f} z^{-n_f}}{1 + a_1 z^{-1} + a_2 z^{-2} + \cdots + a_{n_f} z^{-n_f}}, \quad (12.47)$$

where n_f is a user-specified filter order and f_c is the cut-off frequency. That is, the uncorrupted output data and ILC control law in the last cycle can be recovered by filtering the corresponding data with the same low-pass Butterworth filter in both the forward and reverse directions so that no phase lag or amplitude distortion will be created.

Owing to the fact that measurement noise is mainly of high frequency, the guideline for choosing the cut-off frequency is suggested as

$$f_c \geq \frac{(10 \sim 20)\omega_b}{\pi}, \quad (12.48)$$

where ω_b can be estimated from (12.43).

12.3 Illustrative Examples

Two examples from the existing literature are used to demonstrate the effectiveness and merits of the presented IMC-based ILC method. Example 12.1 is given to demonstrate the tracking performance for a desired output trajectory, together with perturbation tests including time-delay mismatch for illustrating the convergence robustness. Example 12.2 is given to show the effectiveness of the presented ILC method for batch process operation against load disturbance, including a

12.3 Illustrative Examples

measurement noise test. The simulation solver option is chosen as ode5 (Dormand–Prince), and the step size is fixed as $T_s = 0.01$ (s) throughout all tests. For assessing the tracking performance, the mean-square-error (MSE) fitting criterion is used, namely,

$$\text{MSE} = \frac{1}{N_p} \sum_{i=1}^{N_p} [y_d(iT_s) - y(iT_s)]^2,$$

where $y_d(iT_s)$ and $y(iT_s)$ denote the desired output trajectory and the process output in a cycle time of T_p, respectively, and $N_s = T_p/T_s$ is the number of sampled data points in the cycle time.

Example 12.1 Consider the time-delay batch process studied by Xu et al. (2001),

$$G_1(s) = \frac{1}{s+1} e^{-s}.$$

Using a process model, $G_m(s) = e^{-2s}/(s+1)$, Xu et al. (2001) gave an ILC control algorithm based on a Smith predictor control structure with a PD controller, $C = 0.5(s + 1)$, for tracking a desired output trajectory,

$$y_d(t) = \begin{cases} 0, & t \leq 1; \\ 1.5(t-1), & 1 < t \leq 7; \\ 9, & 7 < t \leq 8. \end{cases}$$

This method can realize almost perfect tracking after 11 cycles, based on an initial run of the Smith predictor control scheme using the PD controller.

For comparison, using the above process model, a unified controller for both IMC and ILC can be determined from the controller formula in (12.35) as

$$C(s) = \frac{k_c(s+1)}{\lambda_c s + 1}.$$

By taking $\lambda_c = 1$ and $k_c = 1$ for an initial run of IMC and also for the subsequent ILC, i.e., $C(s) = 1$, the tracking results are shown in Fig. 12.2. It is seen that the presented IMC-based ILC method results in almost perfect tracking after 10 cycles. Figure 12.2c demonstrates that apparently faster convergence is obtained by the proposed method. As shown in Fig. 12.2, the initial run of IMC gives enhanced tracking performance compared to the PD controller of Xu et al. (2001). Figure 12.3 shows the magnitude plot of Q with respect to the frequency, which demonstrates that $Q(\omega_b = 0.25) = 0.5116$ can guarantee the convergence though $\max_\omega |Q| = 1.218 > 1$ occurs beyond the low-frequency range.

To demonstrate the convergence robustness in the presence of the time-delay mismatch, assume that the process time delay randomly fluctuates in a range of

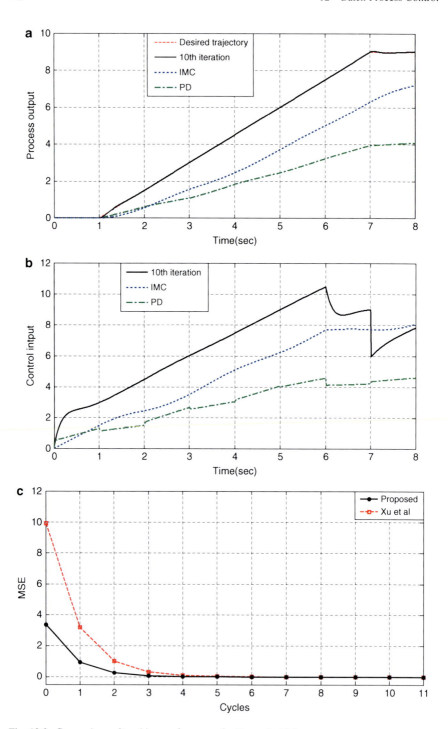

Fig. 12.2 Comparison of tracking performance for Example 12.1

12.3 Illustrative Examples

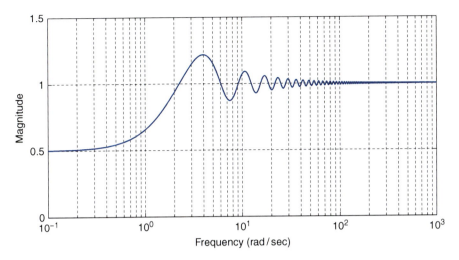

Fig. 12.3 The magnitude plot of Q for Example 12.1

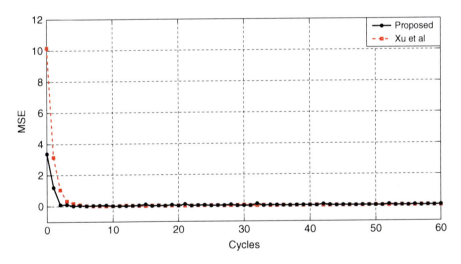

Fig. 12.4 The MSE plot for Example 12.1 with uncertain time delay

[0.8, 1.2](s) from cycle to cycle. Figure 12.4 shows the MSE plot with respect to the cycle number, which demonstrates good robustness of the proposed ILC method.

Note that the above ILC design is based on exact modeling of the rational part in the process transfer function, as assumed in Xu et al. (2001) for test. To demonstrate the convergence robustness in the presence of an entire model mismatch as likely encountered in practice, assume that the rational part of the process is actually perturbed to $G_o(s) = 2/(1.5s + 1)$, together with the above time-delay variation. The corresponding MSE results are plotted in Fig. 12.5. It is seen that the presented

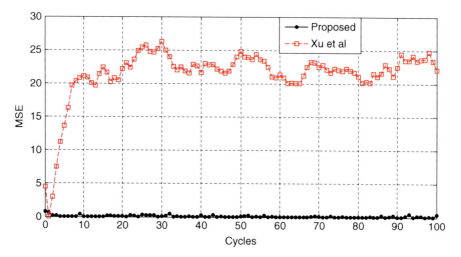

Fig. 12.5 The MSE plot for Example 12.1 with model mismatch and time delay variation

ILC method still maintains good convergence robustness with respect to the cycle number.

Example 12.2 Consider the batch process studied by Chin et al. (2004),

$$G_2(s) = \frac{2.5}{300s^2 + 35s + 1} e^{-\theta s}.$$

Based on a process model, $G_m(s) = 1.5 e^{-\theta_m s}/(270s^2 + 33s + 1)$, Chin et al. (2004) gave an ILC scheme for the delay-free case, $\theta = \theta_m = 0$, to overcome a repetitive load disturbance with a slow dynamics of $G_d(s) = 1/(10s + 1)$. For illustration, $\theta = 3$ and $\theta_m = 10$ are first assumed for test, together with a control input limit in a range of $[-10, 10]$. According to the controller formulae in (12.35), the unified controller can be determined as

$$C(s) = \frac{k_c \left(270s^2 + 33s + 1\right)}{1.5(\lambda_c s + 1)^2}.$$

The cycle time is assumed to be $T_p = 400$ (s). The set-point is a unit step change, and a low-pass filter, $F(s) = 1/(3s + 1)$, is used to shape a desired output trajectory. A repetitive load disturbance with the above dynamics and a magnitude of -0.5 is added at $t = 200$ (s). By taking $\lambda_c = 5$ and $k_c = 1$ for an initial run of IMC and taking $\lambda_c = 8$ and $k_c = 1.5$ for the subsequent ILC, the tracking results are shown in Fig. 12.6. It is seen that perfect tracking can be obtained almost after 20–30 cycles. The corresponding MSE plot is shown in Fig. 12.7 (solid line). Note that the MSE value converges to a very small constant rather than zero is due to the time-delay mismatch between the desired output trajectory and the process output response.

12.3 Illustrative Examples

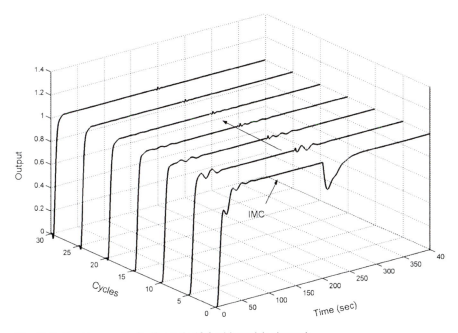

Fig. 12.6 Tracking results for Example 12.2 with model mismatch

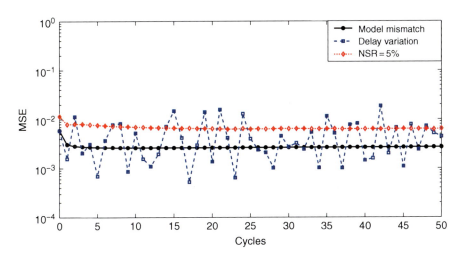

Fig. 12.7 The MSE plots of three tests for Example 12.2

To demonstrate the convergence robustness in the presence of the time-delay variation, assume that the process time delay randomly fluctuates in a range of [0, 5](s) from cycle to cycle. Figure 12.7 shows the MSE plot with respect to the cycle number (dashed line), which once again demonstrates good robustness of the presented ILC method against the time-delay variation.

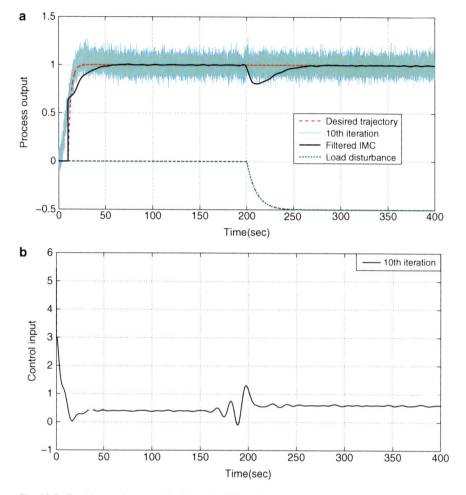

Fig. 12.8 Tracking performance for Example 12.2 against measurement noise

To demonstrate the convergence robustness in the presence of measurement noise, assume that a random noise $N\left(0, \sigma_N^2 = 0.38\%\right)$, causing NSR = 5%, is added to the process output measurement. The worst case of the time-delay mismatch, i.e., $\theta = 0$ and $\theta_m = 10$, is considered for test. According to the denoising strategy presented in Sect. 12.2.4, the ILC updating law is computed based on filtering the noisy output and ILC control law in the last cycle with a low-pass Butterworth filter using a cutoff frequency, $f_c = 0.3$ (Hz), in both forward and reverse directions. The tracking result is shown in Fig. 12.8, which demonstrates that good tracking can thus be obtained after 10 cycles. The filtered output response in terms of the initial run of IMC is also shown in Fig. 12.8, indicating good denoising effect. The corresponding MSE plot is also shown in Fig. 12.7 (dotted line) for comparison. It can be seen that the MSE value converges almost to a small constant which is indeed close to the variance of the measurement noise, thus demonstrating good convergence robustness.

12.4 Summary

For the control of industrial batch processes, especially in the presence of time-delay and time-varying uncertainties from cycle to cycle, an IMC-based ILC method (Liu et al. 2010) has been presented for robust tracking of the desired output trajectory. Relative independence is therefore obtained for designing the IMC control law to maintain the control system robust stability and the ILC control law to realize perfect tracking, respectively. Sufficient conditions for the convergence of ILC have been derived. It is a remarkable merit that the developed characteristic equation of the ILC transfer function connecting the tracking errors from cycle to cycle is free of delay, which can facilitate the pole assignment of the closed-loop system and the corresponding controller design.

For the convenience of implementation, a unified controller form has been proposed for implementing IMC in the initial run and the subsequent ILC for perfect tracking. To cope with process uncertainties, robust tuning constraints of the unified controller have been derived, respectively, for maintaining the control system stability and the convergence for tracking a desired output trajectory. For practical application, the unified controller can be monotonically tuned through a single adjustable parameter to meet a good trade-off between the tracking performance and robust stability of the closed-loop system.

To deal with measurement noise that may hinder the convergence of ILC in practice, a denoising strategy has been presented for filtering each cycle data to compute the ILC updating law.

Two examples from existing literature have been used to illustrate the presented ILC method. It has been demonstrated that perfect tracking can be efficiently realized by the presented ILC method for time-invariant batch processes, no matter whether there exists model mismatch or not. In the presence of time-varying uncertainties from cycle to cycle, including measurement noise, the presented ILC method has been shown to maintain good system stability and convergence robustness.

References

Ahn H-S, Chen YQ, Moore KL (2007) Iterative learning control: brief survey and categorization. IEEE Trans Syst Man Cybern C Appl Rev 37:1099–1121

Bonvin D, Srinivasan B, Hunkeler D (2006) Control and optimization of batch processes. IEEE Trans Control Syst Mag 26(1):34–45

Chin I, Qin SJ, Lee KS, Cho M (2004) A two-stage iterative learning control technique combined with real-time feedback for independent disturbance rejection. Automatica 40:1913–1922

Harte TJ, Hätönen J, Owens DH (2005) Discrete-time inverse model-based iterative learning control stability, monotonicity and robustness. Int J Control 78(8):577–586

Instrument Society of America (1995) Batch control. Part 1, models and terminology. ISA, Research Triangle Park

Lee JH, Lee KS, Kim WC (2000) Model-based iterative learning control with a quadratic criterion for time-varying linear systems. Automatica 36(5):641–657

Li X-D, Chow TWS, Ho JKL (2005) 2-D system theory based iterative learning control for linear continuous systems with time delays. IEEE Trans Circuits Syst I Regul Pap 52(7):1421–1430

Liu T, Gao F, Wang YQ (2010) IMC-based iterative learning control for batch processes with uncertain time delay. J Process Control 20(2):173–180

Moore KL (1993) Iterative learning control for deterministic systems. Springer, London

Morari M, Zafiriou E (1989) Robust process control. Prentice Hall, Englewood Cliff

Nagy ZK, Braatz RD (2003) Robust nonlinear model predictive control of batch processes. AICHE J 49(7):1776–1786

Nagy ZK, Mahn B, Franke R, Allgöwer F (2007) Efficient output feedback nonlinear model predictive control for temperature control of industrial batch reactors. Control Eng Pract 15(7):839–850

Seborg DE, Edgar TF, Mellichamp DA (2004) Process dynamics and control, 2nd edn. Wiley, Hoboken

Shi J, Gao F, Wu T-J (2005) Robust design of integrated feedback and iterative learning control of a batch process based on a 2D Roesser system. J Process Control 15(8):907–924

Shi J, Gao F, Wu T-J (2006) Robust iterative learning control design for batch processes with uncertain perturbations and initialization. AICHE J 52(6):2171–2187

Tan KK, Zhao S, Huang S, Lee TH, Tay A (2009) A new repetitive control for LTI systems with input delay. J Process Control 19(4):711–716

Wang YQ, Gao F, Doyle FJ III (2009) Survey on iterative learning control, repetitive control, and run-to-run control. J Process Control 19(10):1589–1600

Wijdeven J, Donkers T, Bosgra O (2009) Iterative learning control for uncertain systems: robust monotonic convergence analysis. Automatica 45(10):2383–2391

Xiong Z, Zhang J (2003) Product quality trajectory tracking in batch processes using iterative learning control based on time-varying perturbation models. Ind Eng Chem Res 42:6802–6814

Xu JX, Hu Q, Lee TH, Yamamoto S (2001) Iterative learning control with Smith time delay compensator for batch processes. J Process Control 11:321–328

Zhou KM, Doyle JC, Glover K (1996) Robust and optimal control. Prentice Hall, Englewood Cliff

Chapter 13
Concluding Remarks

In this monograph, we have presented two series of our research results on industrial process identification and control system design. The contents are divided into two parts. Part I (Process Identification) presents a series of continuous-time model identification methods for describing the dynamic response characteristics of open-loop stable, integrating, and unstable industrial processes. The widely practiced step response test and relay feedback test have been considered for developing the proposed identification methods, together with robust identification methods that allow for unsteady or unknown initial process conditions and load disturbance in practical applications. Part II (Control System Design), corresponding to the control-oriented model identification methods in Part I, presents a series of model-based control methods developed by the authors for single-input-single-output (SISO) processes, cascade control processes, multiple-input-multiple-output (MIMO) processes, and batch processes.

The main contributions of Part I can be summarized as follows:

1. Process frequency response estimation under an open-loop or closed-loop step test, or a relay feedback test.
2. For the use of an open-loop step test to identify a stable or integrating process, identification methods have been detailed for obtaining the low-order process models of first-order-plus-time-delay (FOPDT) and second-order-plus-time-delay (SOPDT), which have been most widely used for control system design and controller tuning in industrial engineering practices. A few higher-order model identification algorithms have also been given for obtaining a very special model with more parameters to facilitate advanced control design for industrial processes with special requirements.
3. For the use of an open-loop step test subject to unsteady or unknown initial process conditions and unexpected load disturbance, robust identification algorithms have been presented for obtaining FOPDT and SOPDT models based on a modified implementation of the step test. Moreover, a piecewise model identification method has been presented for simultaneously identifying the

process model and the disturbance model in the presence of a deterministic (inherent) type of load disturbance.

4. For the use of a closed-loop step test, the guidelines for implementing the identification test with a simple proportional (P), proportional-integral (PI), or proportional-integral-derivative (PID) controller for closed-loop stabilization have been provided. The corresponding identification algorithms have been detailed for obtaining FOPDT and SOPDT models for stable, integrating, and unstable processes, respectively.

5. For the use of a relay feedback test, the guidelines for model structure selection have been provided together with a list of various relay response shapes for reference. By deriving analytical relay response expressions, it has been clarified that the steady oscillation can definitely be formed for stable and integrating processes. A limiting condition to forming steady oscillation for unstable processes has been revealed. Given a biased or unbiased relay test, identification algorithms have been detailed for obtaining FOPDT and SOPDT models for stable, integrating, and unstable processes, respectively. Moreover, a generalized relay identification method has been developed for identifying any order models for stable processes, from which the process static gain can be separately identified, independent of whether a biased or unbiased relay test is used.

The main contributions of Part II can be summarized as follows:

1. Based on a review of the internal model control (IMC) theory, an enhanced IMC design for load disturbance rejection has been presented together with a set of IMC-based PID tuning formulae.

2. For SISO processes, advanced two-degrees-of-freedom (2DOF) control schemes have been presented for separate optimization of set-point tracking and load disturbance rejection. Based on a classification of different cases that a load disturbance may enter into a process of stable, integrating, or unstable type, enhanced IMC design formulae have been correspondingly developed for improving the load disturbance rejection performance.

3. Advanced cascade control strategies have been presented for time delay or slow processes to improve the load disturbance rejection performance based on the available measurement of a secondary (intermediate) output of such a process for feedback control. An important advantage of these control schemes is that both the set-point tracking and load disturbance rejection can be separately tuned and optimized, compared to a conventional cascade control structure.

4. For multiloop control of MIMO processes, an analytical multiloop PI/PID controller design method has been presented based on a two-input-two-output (TITO) process description that is mostly established for the implementation convenience of multiloop control. The multiloop structure controllability has been discussed, along with the conclusion that it is generally impractical for entire decoupling regulation to be realized by a multiloop structure.

5. For decoupling control of MIMO processes with time delays, an analytical decoupling control design based on the standard IMC structure has been

presented for stable TITO processes with time delays. The control design can realize absolute decoupling regulation of the nominal binary output responses. Moreover, an analytical MIMO decoupling controller matrix design has been presented in the frame of the unity feedback control structure that is most widely used in engineering practice, which can realize significant or even absolute decoupling regulation for the nominal system. In addition, a 2DOF MIMO decoupling control scheme has been developed to realize separate regulation and optimization of the set-point tracking and load disturbance rejection for individual channels.
6. For industrial batch processes, an IMC-based iterative learning control (ILC) method has been presented for robust tracking of the desired output trajectory, especially in the presence of time delay and time-varying uncertainties. Relative independence is therefore obtained for designing the IMC control law and the ILC control law, the former for maintaining the robust stability of the control system and the latter for realizing the perfect tracking.

Although we have tried to make this monograph as self-contained as possible, there indeed exist a number of underdeveloped or untouched research topics related to our discussed topics. We therefore conclude by offering some suggestions and open issues for future research exploration:

1. Robust frequency response estimation in the middle- to high-frequency range required for advanced control of some industrial processes in the presence of measurement noise. Modified step or relay tests that are simple enough for practical application are desired for developing such frequency response estimation methods.
2. Consistent parameter estimation for model identification using multiple step or relay tests subject to nonstatic or time-varying load disturbance, including the transient type of disturbance that exists for only a short duration.
3. Design of open-loop or closed-loop step test(s) for effective excitation of the dynamic response of a MIMO process with no diagonal dominance to facilitate control-oriented model identification.
4. Design of relay feedback test(s) for the sustainable oscillation of a MIMO process to form the limit cycle, together with the stability conditions for relay feedback identification of MIMO processes in terms of a specified multiloop or decoupling control structure.
5. Step or relay identification of non-square MIMO processes that are associated with input–output pairing and no diagonal dominance.
6. Model identification or parameter estimation of nonlinear processes like the Hammerstein- or Wiener-type from simple or modified step or relay test(s).
7. Anti-windup IMC designs for SISO processes subject to the input or output constraints, together with the robust stability conditions.
8. Advanced cascade control methods to deal with different cases of load disturbance entering into the process, together with the control input constraints.
9. Multiloop control design for non-square MIMO processes, in particular for the PID tuning methods that facilitate economic operation.

10. Decoupling control design for non-square or unstable MIMO processes with time delays, including the robust stability conditions.
11. Two-dimensional (2D) ILC design for batch processes to realize robust tracking in both the time direction during a cycle and the batchwise direction from cycle to cycle, especially in the presence of time delay (including variation) in the control implementation or the output measurement.
12. High-performance discretization of continuous-time control systems for practical applications in industrial computers and distributed digital controllers (DDC), especially in the presence of an input or output time delay that is not an integer multiple of the sampling period.

References

Ahmed S, Huang B, Shah SL (2006) Parameter and delay estimation of continuous-time models using a linear filter. J Process Control 16:323–331

Ahmed S, Huang B, Shah SL (2007) Novel identification method from step response. Control Eng Pract 15:545–556

Ahmed S, Huang B, Shah SL (2008) Identification from step responses with transient initial conditions. J Process Control 18:121–130

Ahn H-S, Chen YQ, Moore KL (2007) Iterative learning control: brief survey and categorization. IEEE Trans Systems, Man, and Cybernetics-Part C: Applications and Reviews 37:1099–1121

Alevisakis G, Seborg DE (1973) An extension of the Smith predictor to multivariable linear systems containing time delays. Int J Control 3(17):541–557

Ananth I, Chidambaram M (1999) Closed loop identification to transfer function model for unstable systems. J Frankl Inst 336(7):1055–1061

Åström KJ, Hägglund T (1995) PID controller: theory, design, and tuning, 2nd edn. ISA Society of America, Research Triangle Park

Åström KJ, Hägglund T (2005) Advanced PID control. ISA Society of America, Research Triangle Park

Åström KJ, Wittenmark B (1997) Computer-controlled system: theory and design. Prentice Hall, Upper Saddle River

Atherton DP (1982) Oscillations in relay systems. Trans Inst Meas Control (London) 3:171–184

Atherton DP (2006) Relay autotuning: an overview and alternative approach. Ind Eng Chem Res 45:4075–4080

Bi Q, Cai WJ, Lee EL, Wang QG, Hang CC, Zhang Y (1999) Robust identification of first-order plus dead-time model from step response. Control Eng Pract 7:71–77

Bonvin D, Srinivasan B, Hunkeler D (2006) Control and optimization of batch processes. IEEE Trans Control Syst Mag 26(1):34–45

Braatz RD (1995) Internal model control: the control handbook. CRC Press, Boca Raton

Bristol EH (1966) On a new measure of interaction for multivariable process control. IEEE Trans Autom Control 11(1):133–134

Camacho EF, Bordons C (2004) Model predictive control. Springer, London

Campo PJ, Morari M (1994) Achievable closed-loop properties of systems under decentralized control: involving the steady-state gain. IEEE Trans Autom Control 39(3):932–943

Cha S, Chun D, Lee J (2002) Two-step IMC-PID method for multiloop control system design. Ind Eng Chem Res 41(12):3037–3041

Chen T, Bruce F (1995) Optimal sampled-data control systems. Springer, Tokyo

Chen J, Gu G (2000) Control-oriented system identification: an $H\infty$ approach. Wiley, New York

Chen D, Seborg DE (2002) Multiloop PI/PID controller design based on Gershgorin bands. IEE Process Control Theory Appl 149(1):68–73

Chen D, Seborg DE (2003) Design of decentralized PI control systems based on Nyquist stability analysis. J Process Control 13(1):27–39

Cheres E (2006) Parameter estimation of an unstable system with a PID controller in a closed loop configuration. J Frankl Inst 343(2):204–209

Chien IL, Huang HP, Yang JC (1999) A simple multiloop tuning method for PID controllers with no proportional kick. Ind Eng Chem Res 38(4):1456–1468

Chien I-L, Peng SC, Liu JH (2002) Simple control method for integrating processes with long deadtime. J Process Control 12(3):391–404

Chin I, Qin SJ, Lee KS, Cho M (2004) A two-stage iterative learning control technique combined with real-time feedback for independent disturbance rejection. Automatica 40:1913–1922

Chiu MS, Arkun Y (1992) A methodology for sequential design of robust decentralized control systems. Automatica 28(5):997–1002

Co T (2010) Relay stabilization and bifurcations of unstable SISO processes with time delay. IEEE Trans Autom Control 55(5):1131–1141

Cui H, Jacobsen EW (2002) Performance limitations in decentralized control. J Process Control 12:485–494

Desbiens A, Pomerleau A, Hodouin D (1996) Frequency based tuning of SISO controllers for two-by-two processes. IEE Process Control Theory Appl 143(1):49–56

Deshpande PB (1980) Process identification of open-loop unstable systems. AIChE J 26(2):305–308

Forssell U, Ljung L (2000) Identification of unstable systems using output error and Box–Jenkins model structures. IEEE Trans Autom Control 45(1):137–141

Garcia P, Albertos P (2008) A new dead-time compensator to control stable and integrating processes with long dead-time. Automatica 44:1062–1071

Garcia P, Albertos P, Hägglund T (2006) Control of unstable non-minimum-phase delayed systems. J Process Control 16:1099–1111

Garnier H, Wang L (eds) (2008) Identification of continuous-time models from sampled data. Springer, London

Gilbert AF, Yousef A, Natarajan K, Deighton S (2003) Tuning of PI controllers with one-way decoupling in 2×2 MIMO systems based on finite frequency response data. J Process Control 13(6):553–567

Goodwin GC, Graebe L, Salgado ME (2001) Control system design. Prentice Hall, Upper Saddle River

Gündes AN, Özgüler AB (2002) Two-channel decentralized integral-action controller design. IEEE Trans Autom Control 47(12):2084–2088

Halevi Y, Palmor ZJ, Efrati T (1997) Automatic tuning of decentralized PID controllers for MIMO processes. J Process Control 7(2):119–128

Hang CC, Loh AP, Vasnani VU (1994) Relay feedback auto-tuning of cascade controllers. IEEE Trans Control Syst Technol 2(1):42–45

Hang CC, Åström KJ, Wang QG (2002) Relay feedback auto-tuning of process controllers—A tutorial review. J Process Control 12:143–163

Hang CC, Wang QG, Yang XP (2003) A modified Smith predictor for a process with an integrator and long dead time. Ind Eng Chem Res 42:484–489

Harte TJ, Hätönen J, Owens DH (2005) Discrete-time inverse model-based iterative learning control stability, monotonicity and robustness. Int J Control 78(8):577–586

He MJ, Cai WJ, Ni W, Xie L-H (2009) RNGA based control system configuration for multivariable processes. J Process Control 19:1036–1042

Ho WK, Lee TH, Gan OP (1997) Tuning of multiloop proportional-integral-derivative controllers based on gain and phase margin specification. Ind Eng Chem Res 36:2231–2238

Holt BR, Morari M (1985) Design of resilient processing plants-V. The effect of deadtime on dynamic resilience. Chem Eng Sci 40(7):1229–1237

Horn IG, Arulandu JR, Braatz RD (1996) Improved filter design in internal model control. Ind Eng Chem Res 35(10):3437–3441

Hovd M, Skogestad S (1993) Improved independent design of robust decentralized controllers. J Process Control 3:43–51

Huang H-P, Chen C-C (1997) Control-system synthesis for open-loop unstable process with time delay. IEE Process Control Theory Appl 144(4):334–346

Huang CT, Huang MF (1993) Estimation of the second-order parameters from the process transients by simple calculation. Ind Eng Chem Res 32:228–230

Huang HP, Lin FY (2006) Decoupling multivariable control with two degrees of freedom. Ind Eng Chem Res 45(9):3161–3173

Huang HP, Ohshima M, Hashimoto L (1994) Dynamic interaction and multiloop control system design. J Process Control 4(1):15–22

Huang HP, Chien I-L, Lee YC (1998) Simple method for tuning cascade control systems. Chem Eng Commun 165:89–121

Huang HP, Lee MW, Chen CL (2001) A system of procedures for identification of simple models using transient step response. Ind Eng Chem Res 40:1903–1915

Huang HP, Jeng JC, Chiang CH, Pan W (2003) A direct method for multi-loop PI/PID controller design. J Process Control 13(8):769–786

Huang HP, Jeng JC, Luo KY (2005) Auto-tune system using single-run relay feedback test and model-based controller design. J Process Control 15:713–727

Hwang C, Cheng YC (2005) A note on the use of the Lambert W function in the stability analysis of time-delay systems. Automatica 41(11):1979–1985

Hwang SH, Lai ST (2004) Use of two-stage least-squares algorithms for identification of continuous systems with time delay based on pulse responses. Automatica 40:1561–1568

Ingimundarson A, Hägglund T (2001) Robust tuning procedures of dead-time compensating controllers. Control Eng Pract 9:1195–1208

Instrument Society of America (1995) Batch control. Part 1, models and terminology. ISA, Research Triangle Park

Jensen N, Fisher DG, Shah SL (1986) Interaction analysis in multivariable control systems. AIChE J 32(6):959–970

Jerome NF, Ray WH (1986) High-performance multivariable control strategies for systems having time delays. AIChE J 32(6):914–931

Jerome NF, Ray WH (1992) Model-predictive control of linear multivariable systems having time delays and right-half-plane zeros. Chem Eng Sci 47(4):763–785

Jin HP, Heung IP, Lee I-B (1998) Closed-loop on-line process identification using a proportional controller. Chem Eng Sci 53(9):1713–1724

Johnson MA, Moradi MH (2005) PID control: new identification and design. Springer, London

Jung J, Choi JY, Lee J (1999) One-parameter method for a multiloop control system design. Ind Eng Chem Res 38:1580–1588

Kaya I (2001) Improving performance using cascade control and a Smith predictor. ISA Trans 40:223–234

Kaya I (2004) Two-degree-of-freedom IMC structure and controller design for integrating processes based on gain and phase-margin specifications. IEE Process Control Theory Appl 151(4):401–407

Kaya I (2006) Parameter estimation for integrating processes using relay feedback control under static load disturbances. Ind Eng Chem Res 45:4726–4731

Kaya I, Atherton DP (2001) Parameter estimation from relay autotuning with asymmetric limit cycle data. J Process Control 11:429–439

Kwak HJ, Sung SW, Lee I-B (1997) On-line process identification and autotuning for integrating processes. Ind Eng Chem Res 36(12):5329–5338

Kwak HJ, Whan S, Lee IB (2001) Modified Smith predictor for integrating processes: comparisons and proposition. Ind Eng Chem Res 40:1500–1506

Lee J, Edgar TF (2000) Phase conditions for stability of multi-loop control systems. Comput Chem Eng 23:1623–1630

Lee J, Edgar TF (2004) Dynamic interaction measures for decentralized control of multivariable processes. Ind Eng Chem Res 43(2):283–287

Lee J, Cho W, Edgar TF (1998a) Multiloop PI controller tuning for interacting multivariable processes. Comput Chem Eng 22(11):1711–1723

Lee Y, Park S, Lee M, Brosilow C (1998b) PID controller tuning for desired closed-loop responses for SI/SO systems. AIChE J 44(1):106–115

Lee YH, Park SW, Lee MY (1998c) PID controller tuning to obtain desired closed loop responses for cascade control systems. Ind Eng Chem Res 37(5):1859–1865

Lee JH, Lee KS, Kim WC (2000a) Model-based iterative learning control with a quadratic criterion for time-varying linear systems. Automatica 36(5):641–657

Lee Y, Lee J, Park S (2000b) PID controllers tuning for integrating and unstable processes with time delay. Chem Eng Sci 55:3481–3493

Lee YH, Oh SG, Park SW (2002) Enhanced control with a general cascade control structure. Ind Eng Chem Res 41(11):2679–2688

Lee J, Kim DH, Edgar TF (2005) Static decouplers for control of multivariable processes. AIChE J 51(10):2712–2720

Li W, Eskinat E, Luyben WL (1991) An improved autotune identification method. Ind Eng Chem Res 30:1530–1541

Li SY, Cai WJ, Mei H, Xiong Q (2005a) Robust decentralized parameter identification for two-input two-output process from closed-loop step responses. Control Eng Pract 13:519–531

Li X-D, Chow TWS, Ho JKL (2005b) 2-D system theory based iterative learning control for linear continuous systems with time delays. IEEE Trans Circuits Syst I: Regul Pap 52(7):1421–1430

Limebeer DJN, Kasenally EM, Perkins JD (1993) On the design of robust two degree of freedom controllers. Automatica 29(1):157–168

Lin C, Wang QG, Lee TH (2004) Relay feedback: a complete analysis for first-order systems. Ind Eng Chem Res 43:8400–8402

Liu T, Gao F (2008a) Alternative identification algorithms for obtaining a first-order stable/unstable process model from a single relay feedback test. Ind Eng Chem Res 47(4):1140–1149

Liu T, Gao F (2008b) Identification of integrating and unstable processes from relay feedback. Comput Chem Eng 32(12):3038–3056

Liu T, Gao F (2008c) Robust step-like identification of low-order process model under nonzero initial conditions and disturbance. IEEE Trans Autom Control 53:2690–2695

Liu T, Gao F (2009) A generalized approach for relay identification of time delay and nonminimum phase processes. Automatica 45(4):1072–1079

Liu T, Gao F (2010a) A frequency domain step response identification method for continuous-time processes with time delay. J Process Control 20(7):800–809

Liu T, Gao F (2010b) Closed-loop step response identification of integrating and unstable processes. Chem Eng Sci 65(10):2884–2895

Liu T, Gao F (2010c) New insight into internal model control filter design for load disturbance rejection. IET Control Theory Appl 4(3):448–460

Liu T, Gao F (2011) Enhanced IMC design of load disturbance rejection for integrating and unstable processes with slow dynamics. ISA Trans 50(2):239–248

Liu T, Cai YZ, Gu DY, Zhang WD (2005a) New modified Smith predictor scheme for controlling integrating and unstable processes. IEE Process Control Theory Appl 152(2):238–246

Liu T, Gu DY, Zhang WD (2005b) Decoupled two-degree-of-freedom control strategy for cascade control systems. J Process Control 15(2):159–167

Liu T, Zhang W, Gu DY (2005c) Analytical multiloop PI/PID controller design for two-by-two processes with time delays. Ind Eng Chem Res 44(6):1832–1841

Liu T, Zhang WD, Gu DY (2005d) Analytical two-degree-of-freedom tuning design for open-loop unstable processes with time delay. J Process Control 15:559–572

Liu T, Zhang WD, Gu DY (2005e) New IMC-based control strategy for open-loop unstable cascade processes. Ind Eng Chem Res 44(4):900–909

References

Liu T, Zhang W, Gu DY (2006) Analytical design of decoupling internal model control (IMC) scheme for two-input-two-output (TITO) processes with time delays. Ind Eng Chem Res 45(9):3149–3160

Liu M, Wang QG, Huang B, Hang CC (2007a) Improved identification of continuous-time delay processes from piecewise step tests. J Process Control 17:51–57

Liu T, Zhang W, Gao F (2007b) Analytical decoupling control strategy using a unity feedback control structure for MIMO processes with time delays. J Process Control 17(2):173–186

Liu T, Zhang W, Gao F (2007c) Analytical two-degrees-of-freedom (2-DOF) decoupling control scheme for multiple-input-multiple-output (MIMO) processes with time delays. Ind Eng Chem Res 46(20):6546–6557

Liu T, Gao F, Wang YQ (2008) A systematic approach for on-line identification of second-order process model from relay feedback test. AIChE J 54(6):1560–1578

Liu T, Yao K, Gao F (2009) Identification and autotuning of temperature control system with application to injection molding. IEEE Trans Control Syst Technol 17(6):1282–1294

Liu T, Gao F, Wang YQ (2010a) IMC-based iterative learning control for batch processes with uncertain time delay. J Process Control 20(2):173–180

Liu T, Zhou F, Yang Y, Gao F (2010b) Step response identification under inherent-type load disturbance with application to injection molding. Ind Eng Chem Res 49(22):11572–11581

Ljung L (1999) System identification: theory for the user, 2nd edn. Prentice Hall, Englewood Cliff

Loh AP, Hang CC, Quek CK, Vasnani VU (1993) Autotuning of multiloop proportional-integral controllers using relay feedback. Ind Eng Chem Res 32(6):1102–1107

Lu X, Yang YS, Wang QG, Zheng WX (2005) A double two-degree-of-freedom control scheme for improved control of unstable delay processes. J Process Control 15:605–614

Lundström P, Skogestad S (1999) Two-degree-of-freedom controller design for an ill-conditioned distillation process using μ-synthesis. IEEE Trans Autom Control 7(1):12–21

Luyben WL (1986) Simple method for tuning SISO controllers in multivariable systems. Ind Eng Chem Process Des Dev 25:654–660

Luyben WL (1987) Derivation of transfer functions for highly nonlinear distillation columns. Ind Eng Chem Res 26:2490–2495

Luyben WL (1990) Process modeling, simulation, and control for chemical engineers. McGraw Hill, New York

Luyben WL (2001) Getting more information from relay feedback tests. Ind Eng Chem Res 40:4391–4402

Ma M, Zhu X (2006) A simple auto-tuner in frequency domain. Comput Chem Eng 30:581–586

Majhi S (2007a) Relay-based identification of a class of non-minimum phase SISO processes. IEEE Trans Autom Control 52:134–139

Majhi S (2007b) Relay based identification of processes with time delay. J Process Control 17(2):93–101

Majhi S, Atherton DP (2000a) Obtaining controller parameters for a new Smith predictor using autotuning. Automatica 36:1651–1658

Majhi S, Atherton DP (2000b) Online tuning of controllers for an unstable FOPDT process. IEE Process Control Theory Appl 147(4):421–427

Marchetti G, Scali C, Lewin DR (2001) Identification and control of open-loop unstable processes by relay methods. Automatica 37(12):2049–2055

Mataušek MR, Micic AD (1999) On the modified Smith predictor for controlling a process with an integrator and long dead-time. IEEE Trans Autom Control 44(8):1603–1606

McAvoy TJ (1983) Interaction analysis. ISA Society of America, Research Triangle Park

Mikleš J, Fikar M (2007) Process modelling, identification, and control. Springer, Berlin

Moore KL (1993) Iterative learning control for deterministic systems. Springer, London

Morari M, Zafiriou E (1989) Robust process control. Prentice Hall, Englewood Cliff

Nagrath D, Prasad V, Bequette BW (2002) A model predictive formulation for control of open-loop unstable cascade systems. Chem Eng Sci 57:365–378

Nagy ZK, Braatz RD (2003) Robust nonlinear model predictive control of batch processes. AIChE J 49(7):1776–1786

Nagy ZK, Mahn B, Franke R, Allgöwer F (2007) Efficient output feedback nonlinear model predictive control for temperature control of industrial batch reactors. Control Eng Pract 15(7):839–850

Normey-Rico JE, Camacho EF (2002) A unified approach to design dead-time compensators for stable and integrative processes with dead-time. IEEE Trans Autom Control 47(2):299–305

Normey-Rico JE, Camacho EF (2008) Simple robust dead-time compensator for first-order plus dead-time unstable processes. Ind Eng Chem Res 47:4784–4790

Normey-Rico JE, Camacho EF (2009) Unified approach for robust dead-time compensator design. J Process Control 19:38–47

Ogunnaike BA, Ray WH (1979) Multivariable controller design for linear systems having multiple time delays. AIChE J 25(6):1043–1056

Ogunnaike BA, Ray WH (1994) Process dynamics, modeling, and control. Oxford University Press, New York

Padhy PK, Majhi S (2006) Relay based PI-PD design for stable and unstable FOPDT processes. Comput Chem Eng 30:790–796

Palmor ZJ, Halevi Y, Krasney N (1995) Automatic tuning of decentralized PID controllers for TITO processes. Automatica 31(7):1001–1010

Panda RC (2006) Estimation of parameters of underdamped second order plus dead time processes using relay feedback. Comput Chem Eng 30:832–837

Panda RC, Yu CC (2003) Analytical expressions for relay feed back responses. J Process Control 13:489–501

Panda RC, Yu CC (2005) Shape factor of relay response curves and its use in autotuning. J Process Control 15:893–906

Paraskevopoulos PN, Pasgianos GD, Arvanitis KG (2004) New tuning and identification methods for unstable first order plus dead-time. IEEE Trans Control Syst Technol 12(3):455–464

Park JH, Sung SW, Lee I-B (1998) An enhanced PID control strategy for unstable processes. Automatica 34(6):751–756

Perng MH, Ju JS (1994) Optimally decoupled robust control MIMO plants with multiple delays. IEE Process Control Theory Appl 141(1):49–56

Pintelon R, Schoukens J (2001) System identification: a frequency domain approach. IEEE Press, New York

Pomerleau D, Pomerleau A (2001) Guide lines for the tuning and the evaluation of decentralized and decoupling controllers for processes with recirculation. ISA Trans 40(4):341–351

Prempain E, Bergeon B (1998) A multivariable two-degree-of-freedom control methodology. Automatica 34(12):1601–1606

Rake H (1980) Step response and frequency response methods. Automatica 16:519–526

Ramakrishnan V, Chidambaram M (2003) Estimation of a SOPTD transfer function model using a single asymmetrical relay feedback test. Comput Chem Eng 27:1779–1784

Rangaiah GP, Krishnaswamy PR (1996) Estimating second-order dead time parameters from underdamped process transients. Chem Eng Sci 51:1149–1155

Rao AS, Chidambaram M (2007) Simple analytical design of modified Smith predictor with improved performance for unstable first-order plus time delay (FOPTD) processes. Ind Eng Chem Res 46:4561–4571

Richard YC, Michael GS (1998) Robust control toolbox user's guide, 3rd edn. The MathWorks Inc., Natick, USA

Salgado ME, Conley A (2004) MIMO interaction measure and controller structure selection. Int J Control 77(4):367–383

Saraf V, Zhao FT, Bequette BW (2003) Relay autotuning of cascade-controlled open-loop unstable reactors. Ind Eng Chem Res 42(20):4488–4494

Seborg DE, Edgar TF, Mellichamp DA (2004) Process dynamics and control, 2nd edn. Wiley, Hoboken

Shamsuzzoha M, Lee M (2007) IMC-PID controller design for improved disturbance rejection of time-delayed processes. Ind Eng Chem Res 46(7):2077–2091

References

Shen SH, Yu CC (1994) Use of relay-feedback test for automatic tuning of multivariable systems. AIChE J 40(4):627–646

Shen SH, Wu JS, Yu CC (1996) Use of biased-relay feedback for system identification. AIChE J 42:1174–1180

Shi J, Gao F, Wu T-J (2005) Robust design of integrated feedback and iterative learning control of a batch process based on a 2D Roesser system. J Process Control 15(8):907–924

Shi J, Gao F, Wu T-J (2006) Robust iterative learning control design for batch processes with uncertain perturbations and initialization. AIChE J 52(6):2171–2187

Shinskey FG (1996) Process control system, 4th edn. McGraw Hill, New York

Shiu SJ, Hwang SH (1998) Sequential design method for multivariable decoupling and multiloop PID controllers. Ind Eng Chem Res 37(1):107–119

Shiu SJ, Hwang SH, Li ML (1998) Automatic tuning of systems with one or two unstable poles. Chem Eng Commun 167:51–72

Skogestad S (2003) Simple analytical rules for model reduction and PID controller tuning. J Process Control 13(4):291–309

Skogestad S, Postlethwaite I (2005) Multivariable feedback control: analysis and design, 2nd edn. Wiley, Chichester

Söderström T, Stoica P (1989) System identification. Prentice Hall, New York

Song SH, Cai WJ, Wang YG (2003) Auto-tuning of cascade control systems. ISA Trans 42(1):63–72

Sree RP, Chidambaram M (2006) Improved closed loop identification of transfer function model for unstable systems. J Frankl Inst 343(2):152–160

Srinivasan K, Chidambaram M (2003) Modified relay feedback method for improved system identification. Comput Chem Eng 27:727–732

Sung SW, Lee I-B (1996) Limitations and countermeasures of PID controllers. Ind Eng Chem Res 35(8):2596–2610

Sung SW, Lee J, Lee I-B (2009) Process identification and PID control. Wiley, Singapore

Tan KK, Wang QG, Lee TH (1998) Finite spectrum assignment control of unstable time delay processes with relay tuning. Ind Eng Chem Res 37:1351–1357

Tan KK, Lee TH, Ferdous R (2000) Simultaneous online automatic tuning of cascade control for open loop stable processes. ISA Trans 39:233–242

Tan KK, Huang SN, Jiang X (2001) Adaptive control of ram velocity for the injection moulding machine. IEEE Trans Control Syst Technol 9:663–671

Tan W, Marquez HJ, Chen T (2003) IMC design for unstable processes with time delays. J Process Control 13:203–213

Tan KK, Zhao S, Huang S, Lee TH, Tay A (2009) A new repetitive control for LTI systems with input delay. J Process Control 19(4):711–716

Thyagarajan T, Yu CC (2003) Improved autotuning using the shape factor from relay feedback. Ind Eng Chem Res 42:4425–4440

Tian YC, Gao F (1998) Double-controller scheme for control of processes with dominant delay. IEE Process Control Theory Appl 145(5):479–484

Tian YC, Gao F (1999a) Control of integrator processes with dominant time delay. Ind Eng Chem Res 39:2979–2983

Tian YC, Gao F (1999b) Injection velocity control of thermoplastic injection molding via a double controller scheme. Ind Eng Chem Res 38:3396–3406

Toh WH, Rangaiah GP (2002) A methodology for autotuning of multivariable systems. Ind Eng Chem Res 41(18):4605–4615

Tsypkin YZ (1984) Relay control system. Cambridge University Press, Oxford

Tyreus BD (1979) Multivariable control system design for an industrial distillation column. Ind Eng Chem Process Des Dev 18(2):177–182

Visioli A (2006) Practical PID Control. Springer, London

Vivek S, Chidambaram M (2005) An improved relay auto tuning of PID controllers for unstable FOPTD systems. Comput Chem Eng 29:2060–2068

Waller M, Waller JB, Waller KV (2003) Decoupling revisited. Ind Eng Chem Res 42(20):4575–4577

Wang QG, Zhang Y (2001) Robust identification of continuous systems with dead-time from step responses. Automatica 37:377–390

Wang QG, Hang CC, Zou B (1997a) Low-order modelling from relay feedback. Ind Eng Chem Res 36:375–381

Wang QG, Zou B, Lee TH (1997b) Auto-tuning of multivariable PID controllers from decentralized relay feedback. Automatica 33(3):319–330

Wang QG, Lee TH, Zhang Y (1998) Mutiloop version of the modified Ziegler-Nichols method for two input two output processes. Ind Eng Chem Res 37:4725–4733

Wang QG, Huang B, Guo X (2000a) Auto-tuning of TITO decoupling controllers from step tests. ISA Trans 39(4):407–418

Wang QG, Zou B, Zhang Y (2000b) Decoupling Smith predictor design for multivariable systems with multiple time delays. Chem Eng Res Des Trans Inst Chem Eng Part A 78(4):565–572

Wang QG, Guo X, Zhang Y (2001) Direct identification of continuous time delay systems from step responses. J Process Control 11:531–542

Wang QG, Lee TH, Lin C (2003a) Relay feedback: analysis, identification and control. Springer, London

Wang QG, Zhang Y, Chiu MS (2003b) Non-interacting control design for multivariable industrial processes. J Process Control 13(3):253–265

Wang QG, Liu M, Hang CC, Tang W (2006) Robust process identification from relay tests in the presence of nonzero initial conditions and disturbance. Ind Eng Chem Res 47:4063–4070

Wang QG, Liu M, Hang CC, Zhang Y, Zheng WX (2008) Integral identification of continuous-time delay systems in the presence of unknown initial conditions and disturbances from step tests. Ind Eng Chem Res 47:4929–4936

Wang YQ, Gao F, Doyle FJ III (2009) Survey on iterative learning control, repetitive control, and run-to-run control. J Process Control 19(10):1589–1600

Watanabe K, Ishiyama Y, Ito M (1983) Modified Smith predictor control for multivariable systems with delays and unmeasurable step disturbances. Int J Control 37(5):959–973

Wijdeven J, Donkers T, Bosgra O (2009) Iterative learning control for uncertain systems: robust monotonic convergence analysis. Automatica 45(10):2383–2391

Wood RK, Berry MW (1973) Terminal composition control of binary distillation column. Chem Eng Sci 28(10):1707–1717

Xiong Z, Zhang J (2003) Product quality trajectory tracking in batch processes using iterative learning control based on time-varying perturbation models. Ind Eng Chem Res 42:6802–6814

Xu JX, Hu Q, Lee TH, Yamamoto S (2001) Iterative learning control with Smith time delay compensator for batch processes. J Process Control 11:321–328

Yang XP, Wang QG, Hang CC, Lin C (2002) IMC-based control system design for unstable processes. Ind Eng Chem Res 41(17):4288–4294

Yao K, Gao F (2007) Optimal start-up control of injection molding barrel temperature. Polym Eng Sci 47(3):254–261

Yu CC (2006) Autotuning of PID controllers: a relay feedback approach, 2nd edn. Springer, London

Zhang WD, Xi YG, Yang GK, Xu XM (2002a) Design PID controllers for desired time-domain or frequency-domain response. ISA Trans 41(4):511–520

Zhang Y, Wang QG, Åström KJ (2002b) Dominant pole placement for multi-loop control systems. Automatica 38(7):1213–1220

Zhang WD, Gu DY, Wang W, Xu X (2004) Quantitative performance design of a modified Smith predictor for unstable processes with time delay. Ind Eng Chem Res 43:56–62

Zhang WD, Gu DY, Rieber JM (2008) Optimal dead-time compensator design for stable and integrating processes with time delay. J Process Control 18:449–457

Zheng WX (1996) Identification of closed-loop systems with low-order controllers. Automatica 32(12):1753–1757

References

Zhong Q-C, Mirkin L (2002) Control of integral processes with dead time. Part II: quantitative analysis. IEE Process Control Theory Appl 149(4):291–296

Zhong Q-C, Normey-Rico JE (2002) Control of integral processes with dead time. Part I: disturbance observer-based 2DOF control scheme. IEE Process Control Theory Appl 149(4):285–290

Zhou KM, Doyle JC, Glover K (1996) Robust and optimal control. Prentice Hall, Englewood Cliff

Zhu Y (2001) Multivariable system identification for process control. Elsevier Science, London

Index

A

Acceptable control (performance), 363, 401, 403, 412
Actuator, 332, 359, 360, 364, 375, 382, 397, 398, 404, 419
Additive uncertainty, 250, 355–358, 361, 381, 382
Adjoint (matrix), 373, 393, 414
Adjustable parameter, 111, 189, 214, 253, 254, 257, 262, 265, 272, 273, 285, 287, 293, 294, 296, 309, 327, 328, 332, 333, 340, 343, 345, 348, 358, 360–363, 365, 368, 374, 375, 377, 383, 384, 388, 394, 397, 398, 400, 401, 403, 410, 415, 418, 421, 423, 425, 427, 429–431, 438, 445, 453
All-pass, 253, 326, 327, 395
Analytical approximation, 271, 295, 298, 317, 319, 333, 342, 346, 360, 364, 378, 381, 384, 410, 418, 431
Analytical tuning, 14, 79, 82, 116, 194, 197, 216, 265, 271, 272, 278, 297, 317, 319, 333, 357, 368, 383, 384, 400, 430, 431, 456, 457
Angular frequency, 20, 24, 33, 36, 95, 122, 129, 160, 162, 166, 175, 180, 182
Anti-windup, 457
asymptotic canceling, 276
Asymptotic constraint, 255, 257, 264, 267, 286, 287, 289, 294, 297, 299
ATV, 164, 168
Augmented process (system), 3, 8, 371

B

Backward discretization operator, 69, 112, 189, 214, 337
Bandwidth, 249, 352, 443, 445
Batch process, 6, 52, 245, 435–453, 455, 457, 458
Biased (asymmetrical) relay, 7, 119, 194
Bi-proper, 253, 288, 376, 377, 438
Blending system, 3, 5
Butterworth filter, 68, 109, 121, 122, 162, 166, 178, 182, 185, 188, 205, 237, 446, 452

C

Cascade control, 323–349, 455–457
Causal, 374
Characteristic equation, 257, 395, 453
Chemical processes, 6, 8, 85, 323, 384
Closed-loop control, 14, 69, 70, 79, 85, 87, 88, 93, 106, 111, 247, 271, 351, 371, 373, 413, 416, 422, 427, 431
Closed-loop identification, 8, 85–87, 116, 119, 217
Closed-loop step test, 14, 70–80, 82, 85, 87, 101, 102, 104, 106, 116, 276, 456, 457
Closed-loop transfer function, 65, 73, 86, 93, 110, 111, 189, 213, 251, 253, 254, 256, 278, 287, 289, 299
Complementary sensitivity function, 248, 249, 257, 264, 267, 286, 287, 290, 328, 329, 332, 333, 415
Complement minor, 373, 393, 414
Complex conjugate, 327, 374, 394–396, 415, 438
Complex plane, 248, 251, 394, 395
Composition controller, 4
Composition transmitter, 4
Condition number, 354
Consistent estimation, 26–31, 33, 46, 48, 57, 58, 82, 90, 102

468   Index

Continuous stirred tank reactor (CSTR), 85, 330, 340
Continuous-time, 45, 455, 458
Controllability, 354–356, 368, 456
Controlled output, 7, 119
Controller design, 246, 253, 256, 267, 269, 271, 272, 283, 285, 289, 293–301, 315, 319, 325–329, 332–333, 342, 346, 348, 349, 356, 357, 360, 364, 368, 436, 453, 456
Controller parameterization, 255
Controller tuning, 9, 24, 52, 53, 79, 82, 88, 104, 116, 172, 180, 186, 216, 240, 246, 278, 284, 293, 311, 319, 367, 438, 455
Control signal, 65, 112, 113, 313, 416, 425
Control structure, 66, 77, 85, 87, 109, 111, 247, 253, 255, 256, 271, 281, 283–285, 293, 300, 318, 323–326, 329–340, 342, 345–348, 354, 355, 361, 368, 372, 390, 392, 412–414, 419, 423, 425, 430, 431, 436, 437, 447, 456, 457
Control system design, 6, 14, 15, 24, 29, 52, 54, 112, 197, 216, 217, 240, 245, 246, 252, 278, 284, 293, 331, 371, 455
Convergence, 81, 82, 101, 149, 180, 436, 440, 442–447, 449–453
Critically damped, 41, 123, 124, 134–136, 140, 141, 165, 166, 168, 169, 189, 194
Current cycle, 437, 439, 440, 446
Cutoff frequency, 68, 185, 188, 452

D

Damping factor, 14, 16, 72, 81, 96, 177
Dead-time compensator, 374, 378, 415
Decentralized control, 354
Decoupler, 356, 363, 367, 368, 371, 372
Decoupling control, 355, 371–431, 456–458
Degrees-of-freedom (DOF), 63, 281–319, 456
Derivative time (constant), 13, 94, 267, 445
Describing function, 130, 131, 133, 197, 218, 230
Determinant, 353, 373, 383, 388, 392, 401, 402, 406
Detuning factor, 359
Diagonal Dominance, 352, 355, 356, 363, 367, 368, 371, 457
Differential system equation, 81
Distillation column, 323, 330, 363, 383, 385, 401
Disturbance model, 52, 53, 58–61, 63, 64, 66, 69, 82, 283, 456
Disturbance rejection, 3, 6, 63, 65, 66, 75, 246, 256–273, 275, 276, 278, 281–283, 286, 292, 294, 295, 297, 302–305, 307–310, 312, 315–319, 323, 324, 328, 330, 331, 335, 338, 345, 347, 348, 366, 367, 403, 412–416, 422, 423, 425, 427, 430, 431, 440, 456, 457
Disturbance response, 52–54, 58–60, 62, 67–69, 191, 256–259, 264, 266–268, 273, 274, 276, 277, 281, 283, 289, 292, 299, 302, 305, 306, 309, 310, 312, 314, 317, 319, 324, 328, 330, 332, 333, 335, 336, 338, 340, 343, 347, 348, 384, 414–416, 427, 429
Disturbance response peak (DP), 258, 259, 265–268, 273, 274, 276, 291, 305, 312, 427
Dynamic detuning matrix, 357–359, 386
Dynamic response, 3, 5–7, 15, 20, 29, 39, 52, 53, 68, 82, 85, 88, 95, 97, 119–121, 123, 194, 218, 250, 284, 292, 331, 352, 412, 441, 455, 457

E

Error band, 9, 189, 214, 246, 259

F

Fast Fourier transform (FFT), 16, 130, 176, 177, 185
Feedback control, 8, 63, 69, 77, 85, 87, 102, 112, 116, 122, 162, 178, 182, 188, 237, 247, 253, 255, 256, 271, 281, 283, 286, 287, 289, 294, 296, 299, 300, 317–319, 323, 330, 331, 343, 347, 348, 376, 390, 392, 398, 403, 410, 418, 419, 425, 430, 431, 456, 457
Feedforward control, 63, 65, 69, 281, 283, 285, 293, 294, 331, 436, 437, 439
First-order-plus-dead-time (FOPDT), 14, 18–20, 23, 29–31, 33, 36–38, 41, 48, 51, 54, 58–60, 62, 63, 66, 69, 75–77, 79–82, 87, 88, 90, 92–96, 99–102, 104, 105, 107, 110, 112, 116, 122–134, 160–164, 168–170, 182, 186–188, 194, 195, 197–200, 203–206, 210–212, 214, 216–219, 228–232, 235, 237–240, 257–263, 272, 278, 283, 285, 293, 455, 456
First-principle, 3, 5, 10
Fitting accuracy, 10, 15, 21, 23–25, 37, 46, 51, 57, 62, 63, 74, 75, 79, 82, 90, 96, 98–102, 107, 125, 133, 134, 160, 180, 186, 195, 206, 210, 228, 230, 231, 234

Index

FOPDT. *See* First-order-plus-dead-time (FOPDT)
Fourier transform, 15, 16, 130, 175, 186
Frequency domain, 7, 9, 10, 14, 25, 29, 76, 116, 245, 247–249, 253, 278, 281, 395, 414
Frequency response, 6, 9, 14–18, 20, 23–26, 28–33, 35–37, 71–82, 93, 94, 96, 99, 102, 105, 107, 116, 126, 130, 139, 146, 148, 155, 158, 160, 175–178, 180, 182, 185–187, 195, 205, 216, 229, 232, 240, 276, 356, 357, 371, 455, 457

G

Gain crossover frequency, 248
Gain margin (GM), 248

H

Higher-order model, 14, 15, 18–20, 22–23, 30, 81, 82, 188
Historical cycle, 436
H_2 optimal performance, 254, 256, 287, 326, 358, 374, 394
Hypothesis testing, 29, 30
Hysteresis, 87

I

Identification effectiveness, 52, 59, 68, 109, 189, 217
Identity matrix, 352, 359
Ill-conditioned, 352, 354, 392
IMC. *See* Internal model control (IMC)
Inherent-type load disturbance, 52–54, 58, 60, 62, 63, 66, 67, 82
Injection molding, 67–69, 82, 108–115, 188–194, 213, 214, 216
Injection velocity, 52, 67–71
Instrumental variables (IV), 46
Integral-of-absolute-error (IAE), 246
Integral-of-squared-error (ISE), 246
Integral-of-time-weighted-absolute-error (ITAE), 246
Integral-of-time-weighted-squared-error (ITSE), 246
Integral property, 295, 333, 360
Integrating processes, 87–95, 116, 197–216, 283–292, 456
Interaction, 189, 351, 352, 354, 356–358, 371, 392
Internal model control (IMC), 63, 65, 66, 69, 82, 110–112, 116, 169, 170, 191, 195, 252–276, 278, 281–283, 285, 287, 293, 302, 305, 310, 312, 317–319, 329, 330, 334, 356, 357, 372, 381, 382, 387, 390, 394, 410, 430, 436–448, 456, 457
  filter, 257, 258, 264, 265, 267, 273–275, 283
Internal stability, 86, 255, 256, 281, 282, 286, 319, 376, 419
Inverse relative degree, 393, 414, 431
Inverse response, 6, 7, 45, 171, 182, 186, 197
Iterative learning control (ILC), 436–453, 457, 458

L

Laplace transform, 4, 9, 14–17, 26, 27, 63, 72, 81, 95, 177, 195, 217, 247, 258, 285, 328, 375
Lead-lag controller, 296, 376, 378
Least-squares (LS), 14, 21–25, 31, 42–46, 48, 53, 56, 59, 68, 75, 82, 89–91, 98–105, 116, 180, 181, 186, 195
Left-half-plane (LHP), 7, 394, 395, 406, 438
Limit cycle, 9, 121, 122, 124–127, 129, 131–134, 136, 138, 140–145, 147–151, 155, 157, 159–165, 167, 172, 175, 178, 182–185, 189, 197–206, 209–214, 217–219, 222, 226, 229–231, 233–238, 457
Linear fractional approximation, 379, 380
Linear fractional transformation, 356
Linear interpolation, 265
Linearly independent, 21, 25, 45, 56, 75, 89, 98, 181
Linear regression, 81, 180
Linear superposition principle, 43, 52, 54, 127, 135, 173, 199
Linear time invariant (LTI), 126, 436
Load disturbance, 3, 13, 85, 170, 197, 245, 281, 323, 364, 388, 435, 455
Lower bound, 17, 29
Low frequency range, 15, 23, 35, 79, 80, 95, 99, 107, 126, 163, 180, 186, 195, 235, 240, 334, 366, 406, 443, 447
Low-order model, 5, 6, 14, 15, 38, 82, 88–92, 116, 122, 123, 125–170, 172, 194, 197, 210, 218, 278, 293, 319, 331
Low-pass filter, 24, 63, 122, 253, 327, 439, 450
LTI. *See* Linear time invariant (LTI)

M

Maclaurin series, 43, 44, 81, 271, 360, 379
Manipulated variable, 3, 4, 351, 352

Mean-square-error (MSE), 447–453
Measurement noise, 6, 7, 13, 18, 24, 26–34, 36, 39, 40, 46, 48–51, 57–61, 65, 73, 77, 78, 81, 82, 89–92, 101, 102, 104, 109, 112, 116, 119, 121, 122, 160, 162, 164, 166, 167, 175, 178, 181, 184, 194, 205, 234, 238, 249, 344, 349, 390, 413, 446, 447, 452, 453, 457
MIMO. *See* Multiple-input-multiple-output (MIMO)
Minimum-phase (MP), 45, 253
Model fitting, 9–10, 14, 15, 20, 21, 23, 25, 26, 29, 35, 46, 50, 53, 57, 59, 63, 74, 79, 95, 96, 99, 102, 125–170, 178–181
Model identification, 3, 6–10, 13–49, 52–80, 85, 87–92, 99, 109, 116, 119, 121, 125, 134, 160, 163, 171, 182, 189, 211, 214, 235, 237, 238, 455, 457
Model predictive control (MPC), 414
Model uncertainty, 263, 442
Monte Carlo tests, 33, 34, 60, 77, 102, 104, 107
MP. *See* Minimum-phase (MP)
MPC. *See* Model predictive control (MPC)
MSE. *See* Mean-square-error (MSE)
Multiloop control, 351–368, 371, 456, 457
Multiple-input-multiple-output (MIMO), 252, 390–430, 455–458
Multiplicative uncertainty, 66, 250, 251, 260, 290, 291, 300, 329, 333, 334, 442

N

Newton-Raphson method, 132, 140, 147, 180, 230
Noise-to-signal ratio (NSR), 30, 31, 50, 61, 103, 121, 162, 167, 178, 184, 205, 238
Nominal case, 65, 111, 252, 253, 254, 279, 284, 322, 330, 370, 371, 411, 417, 423, 436, 440, 441, 443
Nominal performance, 345, 431
Nominal stability, 361
Nominal system, 170, 336, 338, 340, 342, 343, 347, 362, 364, 365, 368, 373, 381–383, 399, 401, 419, 431, 457
Non-minimum-phase (NMP), 45, 171, 181, 194, 339, 414, 431
Nyquist curve, 79, 80, 248, 377, 378, 380, 383, 387, 388, 394, 399, 401–403
Nyquist stability theorem, 251, 252, 360

O

One-dimensional search, 101, 133, 134, 141, 147, 149, 150, 157, 159, 210, 211
Online tuning, 6, 14, 119, 194, 197, 216, 217, 413
Open-loop control, 65, 324, 332, 333, 372, 413, 419
Optimal control, 327
Orthogonal property, 327
Output error (deviation), 9
Output trajectory, 438, 439, 443, 446, 447, 450, 453, 457
Overdamped, 41, 123, 124, 126, 134, 142, 143, 147, 149, 150, 164, 165, 168, 194
Overshoot, 67–69, 112, 247, 254, 255, 285, 328, 336, 338, 340, 342, 375, 384, 388, 403, 423, 427

P

Padé expansion, 257, 272, 295, 319, 333, 360, 378, 379
Pairing rule, 352
Parameter estimation, 13, 14, 21, 24–26, 28, 30, 34, 35, 45, 46, 48, 56, 57, 59, 75, 77, 81, 89, 99, 101, 102, 104, 116, 126, 180, 457
Parseval's theorem, 442
Perfect tracking, 435–437, 439, 440, 442, 443, 447, 450, 453, 457
Performance optimization, 282, 302, 348
Performance specification, 189, 214, 285, 295, 326
Perturbation, 361, 403, 425, 429, 446
Perturbed system, 171, 302, 304, 305, 308, 309, 311, 316–318, 336, 337, 339, 341, 343–348, 366, 382, 387, 390, 412, 425, 431
Phase crossover frequency, 248
Phase lag, 122, 130, 436, 446
Phase margin (PM), 249
Piecewise model identification, 6, 52–70, 82
PRBS. *See* Pseudo-random binary signal (PRBS)
Process modeling, 3–6, 273, 324, 373
Process uncertainties, 6, 66, 88, 189, 246, 250, 263, 273, 290, 292, 310, 313, 319, 326, 330, 333, 335, 340, 344, 349, 362, 363, 367, 381, 383, 398–401, 412, 419, 421, 422, 435–437, 453
Proportional (P) controller, 14, 77, 456

Proportional-integral (PI) controller, 14, 456
Proportional-integral-derivative (PID) controller, 9, 14, 271–272, 456
Pseudo-random binary signal (PRBS), 7

Q

Quadratic discriminant, 268
Quantitative tuning, 259, 265, 278, 291, 431

R

Ramp signal, 85
Rational approximation, 295, 333, 360, 378, 395, 397, 398, 429
Recovery time, 259, 260, 273, 274, 276, 291, 292, 305, 340
Recursive least-squares (RLS), 181, 186, 195
Regression variables, 46
Relative degree, 287, 393, 398, 414, 415, 418, 431
Relative gain array (RGA), 351–353, 368
Relay chattering, 185, 194
Relay feedback, 7–10, 77, 80, 87, 104, 106, 119–195, 197–240, 276, 357, 371, 455–457
Relay function, 7, 119, 120
Relay module, 119
Repetitive control, 245
Right-half-plane (RHP), 7, 35, 41, 50, 171, 256, 286, 287, 294–299, 301, 312, 315, 317, 319, 326, 327, 333, 335, 357, 358, 360, 374–378, 380, 383, 387, 388, 390, 392–399, 401, 406, 410, 415, 416, 418, 419, 422, 427, 431, 438
Rise time, 246, 249, 254, 255, 286, 294, 328, 332, 342, 384
Robustness, 10, 14, 17, 24, 30, 31, 33, 35, 36, 49–51, 57, 59, 60, 63, 73, 77, 81, 82, 92, 102, 104, 106, 116, 160, 162, 166, 168, 170, 178, 182, 194, 195, 205, 234, 237, 302, 315, 317, 343, 345, 349, 362, 399, 403, 419, 439, 446, 447, 449–453
Robust performance, 292
Robust stability, 66, 245, 249–252, 254, 260–262, 268–272, 278, 290–292, 300–302, 305, 309, 315, 319, 329–330, 333–336, 339, 340, 343, 345, 347, 348, 360–363, 368, 381–384, 387, 390, 398–401, 403, 406, 412, 419–422, 425, 429, 431, 435–438, 445, 453, 457, 458
Routh–Hurwitz stability, 297, 380

S

Sampling period, 9, 16, 17, 23, 26, 31, 33, 34, 59, 67, 69, 76, 91, 92, 102, 109, 112, 133, 177, 181, 210, 231, 234, 337, 458
Secondary (intermediate) process, 323, 324, 347, 456
Second-order-plus-dead-time (SOPDT), 14, 88, 122, 197, 217, 257, 283, 455
Sensitivity function, 66, 86, 248, 249, 257, 264, 267, 286, 287, 290, 292, 328–330, 332, 333, 335, 415, 445
Sequential tuning, 371
Set-point response, 170, 254, 281, 283, 285, 286, 293, 294, 302, 308, 310, 316, 317, 319, 324, 327, 328, 330, 332, 333, 336, 338, 340, 342, 345, 348, 357, 364, 367, 384, 403, 414, 423, 429
Set-point tracking, 3, 6, 63, 65, 69, 88, 105, 246, 281–285, 293, 303, 305, 310, 312, 315, 318, 325, 326, 331–333, 345, 348, 367, 374, 388, 403, 410, 412–416, 419, 421–423, 425, 427, 430, 431, 438, 456, 457
Settling time, 16, 23, 36, 37, 60, 76, 101, 113, 246
Signal-to-noise ratio (SNR), 7, 24, 30, 35, 182, 185
Single-input-single-output (SISO), 246–279, 461
Singular value, 351, 353–354
Singular value decomposition (SVD), 351, 353–354
SISO. *See* Single-input-single-output
Small gain theorem, 66, 252, 260, 269, 271, 272, 300, 319, 333, 355, 382, 399, 400
Smith predictor, 301, 323, 357, 371, 436, 441, 447
SNR. *See* Signal-to-noise ratio (SNR)
Spectrum radius, 361, 364, 366, 368, 382, 383, 387, 400, 403, 405, 406, 421, 429, 431
Stability margin, 437
Stable processes, 13–82, 119–195, 257–264, 335, 348, 456
State space, 106, 125, 197, 217, 238, 436
Static gain, 12, 18, 20, 22, 41, 54, 60, 122–124, 130–134, 139–141, 146, 147, 149, 155, 157, 159, 171, 177, 179, 182, 186, 194, 352, 353, 356, 368, 371, 384, 390, 403, 406
Statistical averaging principle, 13, 57, 77
Steady-state deviation (offset), 9
Step-like, 38–40

Step test, 7, 13–39, 41, 44, 45, 47, 49, 50, 52, 54, 55, 57, 58, 62, 66–70, 72–77, 79–82, 85, 87–89, 92, 101, 102, 104–106, 116, 214, 216, 276, 455–457
Strict properness, 22
Suboptimal controller, 327
SVD. *See* Singular value decomposition (SVD)
System operation, 3, 10, 85, 194, 282, 354, 368, 392, 431, 443

T

Taylor expansion, 149
Time compensator, 374, 378, 415
Time delay, 4, 13, 87, 122, 198, 217, 254, 284, 323, 357, 371, 436, 455
Time-delay process, 171, 438, 441
Time domain, 9, 10, 14, 23, 25, 26, 28, 30, 31, 40, 42, 43, 49, 50, 54, 55, 58, 76, 77, 89, 92, 102, 133, 180, 197, 233, 245–249, 254, 258, 265, 285, 332, 375, 384, 394, 436–439, 442
Time integral, 17, 73, 116
Time prediction, 395
Time scaling, 17, 28
Time shift, 16, 41, 54, 55, 72, 127–129, 135–137, 142, 152, 173, 175, 199, 221
Transfer function, 3, 22, 86, 176, 213, 247, 281, 326, 352, 372, 437
Transfer matrix, 255, 256, 352, 354–359, 361–363, 367, 371–374, 381–384, 387, 388, 390, 392, 394, 395, 398, 399, 401–403, 406, 409, 412–416, 419, 421, 422, 427, 429, 431

Transient response, 5, 9, 13, 16, 38, 39, 44, 49, 53, 59, 60, 62–64, 68, 80, 82, 87, 90, 95, 102, 104, 105, 107, 109, 116, 384
Trapezoidal rule, 130, 176
Tuning formula, 286, 302, 307, 308, 315, 317, 328, 456
Tuning procedure, 324
Two-degrees-of-freedom (2DOF), 63, 66, 69, 82, 281–319, 324–330, 332, 335, 337, 339, 340, 345, 348, 349, 412–431, 456, 457
Two-input-two-output (TITO), 353, 371–390, 456

U

Ultimate frequency, 125
Unbiased (symmetrical) relay, 7, 119, 194
Underdamped, 41, 123, 125, 134, 150–152, 157, 159, 165, 166, 169, 194
Unstable processes, 8, 85–116, 217–240, 256, 282, 292–301, 317, 319, 335, 348, 349, 456
Upper bound, 6, 66, 249, 250, 262, 272, 439

W

Water-bed effect, 319, 330, 348

Z

Zero-pole cancellation, 257, 295, 298, 299, 312, 319, 333, 360, 378, 388, 395, 410, 418
Ziegler–Nichols (ZN) tuning, 112

Other titles published in this series (continued):

Soft Sensors for Monitoring and Control of Industrial Processes
Luigi Fortuna, Salvatore Graziani, Alessandro Rizzo and Maria G. Xibilia

Adaptive Voltage Control in Power Systems
Giuseppe Fusco and Mario Russo

Advanced Control of Industrial Processes
Piotr Tatjewski

Process Control Performance Assessment
Andrzej W. Ordys, Damien Uduehi and Michael A. Johnson (Eds.)

Modelling and Analysis of Hybrid Supervisory Systems
Emilia Villani, Paulo E. Miyagi and Robert Valette

Process Control
Jie Bao and Peter L. Lee

Distributed Embedded Control Systems
Matjaž Colnarič, Domen Verber and Wolfgang A. Halang

Precision Motion Control (2nd Ed.)
Tan Kok Kiong, Lee Tong Heng and Huang Sunan

Optimal Control of Wind Energy Systems
Julian Munteanu, Antoneta Iuliana Bratcu, Nicolaos-Antonio Cutululis and Emil Ceangă

Identification of Continuous-time Models from Sampled Data
Hugues Garnier and Liuping Wang (Eds.)

Model-based Process Supervision
Arun K. Samantaray and Belkacem Bouamama

Diagnosis of Process Nonlinearities and Valve Stiction
M.A.A. Shoukat Choudhury, Sirish L. Shah and Nina F. Thornhill

Magnetic Control of Tokamak Plasmas
Marco Ariola and Alfredo Pironti

Real-time Iterative Learning Control
Jian-Xin Xu, Sanjib K. Panda and Tong H. Lee

Deadlock Resolution in Automated Manufacturing Systems
ZhiWu Li and MengChu Zhou

Model Predictive Control Design and Implementation Using MATLAB®
Liuping Wang

Predictive Functional Control
Jacques Richalet and Donal O'Donovan

Fault-tolerant Flight Control and Guidance Systems
Guillaume Ducard

Fault-tolerant Control Systems
Hassan Noura, Didier Theilliol, Jean-Christophe Ponsart and Abbas Chamseddine

Detection and Diagnosis of Stiction in Control Loops
Mohieddine Jelali and Biao Huans (Eds.)

Stochastic Distribution Control System Design
Lei Guo and Hong Wang

Dry Clutch Control for Automotive Applications
Pietro J. Dolcini, Carlos Canudas-de-Wit and Hubert Béchart

Advanced Control and Supervision of Mineral Processing Plants
Daniel Sbárbaro and René del Villar (Eds.)

Active Braking Control Design for Road Vehicles
Sergio M. Savaresi and Mara Tanelli

Active Control of Flexible Structures
Alberto Cavallo, Giuseppe de Maria, Ciro Natale and Salvatore Pirozzi

Induction Motor Control Design
Riccardo Marino, Patrizio Tomei and Cristiano M. Verrelli

Fractional-order Systems and Controls
Concepcion A. Monje, YangQuan Chen,
Blas M. Vinagre, Dingyu Xue and Vincente
Feliu

*Model Predictive Control of Wastewater
Systems*
Carlos Ocampo-Martinez

Wastewater Systems
Carlos Ocampo-Martinez

Tandem Cold Metal Rolling Mill Control
John Pitter and Marwan A. Simaan